Lecture Notes in Artificial Intelligence   8543

Subseries of Lecture Notes in Computer Science

LNAI Series Editors

Randy Goebel
University of Alberta, Edmonton, Canada
Yuzuru Tanaka
Hokkaido University, Sapporo, Japan
Wolfgang Wahlster
DFKI and Saarland University, Saarbrücken, Germany

LNAI Founding Series Editor

Joerg Siekmann
DFKI and Saarland University, Saarbrücken, Germany

Stephen M. Watt   James H. Davenport
Alan P. Sexton   Petr Sojka
Josef Urban (Eds.)

Intelligent
Computer Mathematics

International Conference, CICM 2014
Coimbra, Portugal, July 7-11, 2014
Proceedings

 Springer

Volume Editors

Stephen M. Watt
The University of Western Ontario
London, ON, Canada
E-mail: stephen.watt@uwo.ca

James H. Davenport
University of Bath, UK
E-mail: j.h.davenport@bath.ac.uk

Alan P. Sexton
University of Birmingham
Edgbaston, Birmingham, UK
E-mail: a.p.sexton@cs.bham.ac.uk

Petr Sojka
Masaryk University
Brno, Czech Republic
E-mail: sojka@fi.muni.cz

Josef Urban
Radboud University Nijmegen
GL Nijmegen, The Netherlands
E-mail: josef.urban@gmail.com

ISSN 0302-9743   e-ISSN 1611-3349
ISBN 978-3-319-08433-6   e-ISBN 978-3-319-08434-3
DOI 10.1007/978-3-319-08434-3
Springer Cham Heidelberg New York Dordrecht London

Library of Congress Control Number: 2014941699

LNCS Sublibrary: SL 7 – Artificial Intelligence

Typesetting: Camera-ready by author, data conversion by Scientific Publishing Services, Chennai, India

Printed on acid-free paper

Springer is part of Springer Science+Business Media (www.springer.com)

Preface

As computers and communications technology advance, greater opportunities arise for intelligent mathematical computation. While computer algebra, automated deduction, mathematical publishing and mathematical user interfaces individually have long and successful histories, we are now seeing increasingly fruitful interaction among these areas. For the past several years, the Conferences on Intelligent Computer Mathematics (CICM) event has been a primary venue for discussing these areas and their interplay.

CICM was first held as a joint meeting in 2008, co-locating related conferences and workshops, and has been held annually since, evolving to a multi-track conference. Previous CICM meetings have been held in Birmingham (UK 2008), Grand Bend (Canada 2009), Paris (France 2010), Bertinoro (Italy 2011), Bremen (Germany 2012) and Bath (UK 2013). This volume contains the papers presented at CICM 2014, held 7–11 July 2014 at the University of Coimbra, Portugal.

The CICM 2014 meeting was organized with five invited presentations, four main tracks, a number of workshops, a doctoral mentoring program and an informal track to share work in progress. The program of the meeting, as well as additional materials, have been made available at http://cicm-conference.org/2014/cicm.php.

Invited Speakers

The meeting was pleased to have five distinguished invited speakers to making presentations on a set of subjects, each touching on several CICM topics. Each of the invited speakers has been kind enough to provide a record in this volume:

- Yves Bertot, Inria, "Links between homotopy theory and type theory"
- Jaime Carvalho e Silva, University of Coimbra, "What international studies say about the importance and limitations of using computers to teach mathematics in secondary schools"
- António Leal Duarte, University of Coimbra, "Teaching tiles"
- Herbert Van de Sompel, Los Alamos National Laboratory, "Towards robust hyperlinks for web-based scholarly communication"
- Eric Weisstein, Wolfram|Alpha, "Computable data, mathematics, and digital libraries in Mathematica and Wolfram|Alpha"

The presentations of Yves Bertot and António Leal Duarte are represented by abstracts following this preface, while those of Jaime Carvalho e Silva, Herbert Van de Sompel and Eric Weisstein are recorded in these proceedings as full invited papers. The invited presentation of António Leal Duarte was held jointly with the 10th International Workshop on Automated Deduction in Geometry.

Main tracks

The main tracks of the CICM meeting this year were Calculemus, Digital Mathematics Libraries (DML), Mathematical Knowledge Management (MKM), and Systems and Projects (S&P).

The *Calculemus* track of CICM examines the integration of symbolic computation and mechanized reasoning. This year, the track continued the tradition of eliciting and publishing papers on the boundary of theorem-proving and computer algebra, using a variety of tools, often within individual contributions as well as across the range. A novelty this year was the papers applying symbolic reasoning, viewed broadly, to new application areas.

The *Digital Mathematics Libraries* track has evolved from the DML workshop series as a forum for the development of math-aware technologies, standards, algorithms and processes towards the fulfillment of the dream of a global DML. As in previous years, a blended mix of papers by computer scientists (D), mathematicians (M) and librarians of the digital age (L) was accepted to complement the DML related reports and papers in the S&P and MKM tracks, and an invited paper by Herbert Van de Sompel on the topic of robust scholarly communication in the digital era.

The *Mathematical Knowledge Management* track of CICM is concerned with all aspects of managing mathematical knowledge in the informal, semi-formal, and formal settings. This year a relatively large number of the MKM papers were devoted to novel methods for managing formal mathematical libraries developed with proof assistants such as HOL Light, Mizar, Isabelle and Coq. This was complemented by papers devoted to search over informal mathematical corpora, querying repositories of geometric knowledge, and integration of MKM services into user applications. A number of MKM-related system and project descriptions were also submitted to the Systems and Projects track of CICM. The track featured an invited talk by Yves Bertot on the relation between homotopy theory and type theory.

The *Systems and Projects* track provides a medium to present short descriptions of existing systems or on-going projects in the areas of the other tracks of the conference. All accepted papers in this track were presented in the conference via short talks, posters and, most importantly, demonstrations, thus providing a hands-on view of the developments and applications in this field. This year the three other tracks were well represented with very high quality submissions, testifying to the vibrancy of practical and applied work in the areas of mathematical knowledge management, theorem proving and computer algebra, and digital mathematics libraries.

Prior to the creation of the CICM, two of the present tracks already had a significant history: there had been 15 previous Calculemus meetings and 6 MKM conferences. In 2007, when Calculemus and MKM were held together in Hagenberg, Austria, as part of the RISC Summer, it was decided to continue to hold these meetings together. This led to the first CICM in 2008. The DML track has been present since that first CICM, at first as a workshop, and the Systems and Projects track was added in 2011.

This year there were 55 articles submitted for consideration in response to the call for papers. Of these, 41 were full papers submitted to the Calculemus, DML and MKM tracks, and 14 were short descriptions submitted to the Systems and Projects track. A small number of papers were moved between tracks when it was felt there would be a more natural fit. Each submission received at least three reviews, and in several cases one or two further opinions were obtained. The review cycle included a response period, in which authors could clarify points raised by the referees. This made for a highly productive round of deliberations before the final decisions were taken. In the end, the track Program Committees decided to accept 26 full papers and 9 Systems and Projects descriptions.

The Program Committee work for the main tracks was managed using the EasyChair system. This allowed Committee members to declare actual or potential conflicts of interest, and thereby be excluded from any deliberations on those papers. Submissions on which track chairs had conflicts were handled by the general program chair. In this way, Committee members could (and did!) submit papers for consideration without compromising the peer review process.

Workshops and Related Programs

As in previous years, several workshops and informal programs were organized in conjunction with the CICM. This year these were:

- *CICM Doctoral Program*, providing a dedicated forum for PhD students to present their on-going or planned research and receive feedback, advice and suggestions from a dedicated research advisory board.
- *CICM Work-in-Progress Session*, a forum for the presentation of original work not yet in a suitable form for communication as a formal paper.
- *Workshop on Mathematical User Interfaces* to share ideas and experiences on how users interact with mathematics represented on a computer.
- *OpenMath Workshop*, centered on the refinement of the OpenMath data language for the exchange of mathematical expressions, its deployment in various contexts, and developing mathematical "content dictionaries".
- *Workshop on The Notion of Proof*, devoted to discussions on the state-of-the-art of automated theorem proving from the computer science, mathematical and philosophical points of view, with a special emphasis on proof checking, proof representation and the intuitive notion of mathematical proof.
- *ThEdu 2014: Theorem Proving Components for Educational Software*, with the goal to combine and focus systems from theorem proving, computer algebra and dynamic geometry to enhance existing educational software and the design of the next generation of mechanised mathematics assistants.

Appreciation

We thank all those who have contributed to this meeting. We are grateful for the support of our generous Portuguese sponsors — the Center for Informatics and Systems of the University of Coimbra (CISUC), the Center for Research and Development in Mathematics and Applications (CIDMA) of the University of Aveiro, the Fundação para a Ciência e a Tecnologia (FCT) of the Ministério da Educação e Ciência — and of Maplesoft. We thank Andrei Voronkov for the EasyChair system which we found indispensable, and for his amazing responsiveness in dealing with it. Most fundamentally, we thank the invited speakers, the contributing authors, the referees, the members of the Program Committee and the local organizers, all of whose efforts contributed to the practical and scientific success of the meeting.

May 2014   Stephen M. Watt
                                                          James H. Davenport
                                                              Alan P. Sexton
                                                                Petr Sojka
                                                               Josef Urban

Organization

CICM Steering Committee

Serge Autexier (Publicity/Workshop Officer)
Thierry Bouche (DML Delegate)
Jacques Carette (CICM PC Chair 2013)
Bill Farmer (Treasurer)
Michael Kohlhase (Secretary)
Florian Rabe (MKM Delegate)
Renaud Rioboo (Calculemus Delegate)
Stephen Watt (CICM PC Chair 2014)

CICM 2014 Organizing Committee

General Program Chair

Stephen M. Watt

Local Arrangements Chair

Pedro Quaresma

Calculemus Track Chair

James H. Davenport

DML Track Chair

Petr Sojka

MKM Track Chair

Josef Urban

S&P Track Chair

Alan P. Sexton

Doctoral Program Chair

David Wilson

Publicity and Workshops Chair

Serge Auxtexier

CICM 2014 Local Arrangements Committee

Reinhard Kahle	New University of Lisbon, Portugal
Pedro Quaresma	University of Coimbra, Portugal
Eugénio Rocha	University of Aveiro, Portugal
Vanda Santos	CISUC Coimbra, Portugal
Carlota Simões	University of Coimbra, Portugal

Calculemus Track Program Committee

James H. Davenport	University of Bath, UK
Matthew England	University of Bath, UK
Dejan Jovanović	New York University, USA
Laura Kovacs	Chalmers University of Technology, Sweden
Assia Mahboubi	Inria, France
Adam Naumowicz	University of Białystok, Poland
Grant Passmore	University of Cambridge and University of Edinburgh, UK
Florian Rabe	Jacobs University Bremen, Germany
Claudio Sacerdoti Coen	University of Bologna, Italy
Freek Wiedijk	Radboud University Nijmegen, The Netherlands

DML Track Program Committee

Akiko Aizawa	National Institute of Informatics, Japan
Łukasz Bolikowski	ICM, University of Warsaw, Poland
Thierry Bouche	Université Joseph Fourier, Grenoble, France
Yannis Haralambous	Inst. Mines-Télécom/Télécom Bretagne & CNRS, France
Janka Chlebíková	University of Portsmouth, UK
Michael Kohlhase	Jacobs University Bremen, Germany
Jiří Rákosník	Institute of Mathematics AS CR, Czech Republic
David Ruddy	Cornell University, USA
Petr Sojka	Masaryk University Brno, Czech Republic
Volker Sorge	University of Birmingham, UK
Frank Tompa	University of Waterloo, Canada
Richard Zanibbi	Rochester Institute of Technology, USA

MKM Track Program Committee

Rob Arthan	Queen Mary University of London, UK
David Aspinall	University of Edinburgh, UK
Michael Beeson	San Jose State University, USA

Thomas Hales	University of Pittsburgh, USA
Johan Jeuring	Open Universiteit Nederland/Universiteit Utrecht, The Netherlands
Peter Jipsen	Chapman University, USA
Cezary Kaliszyk	University of Innsbruck, Austria
Michael Kohlhase	Jacobs University Bremen, Germany
Christoph Lange	University of Bonn/Fraunhofer IAIS, Germany
Paul Libbrecht	Weingarten University of Education, Germany
Ursula Martin	University of Oxford, UK
Bruce Miller	National Institute of Standards and Technology, USA
Adam Naumowicz	University of Bialystok, Poland
Florian Rabe	Jacobs University Bremen, Germany
Claudio Sacerdoti Coen	University of Bologna, Italy
Alan P. Sexton	University of Birmingham, UK
Enrico Tassi	Inria, France
Josef Urban	Radboud University Nijmegen, The Netherlands
Stephen M. Watt	University of Western Ontario, Canada
Makarius Wenzel	Université Paris-Sud 11, France
Freek Wiedijk	Radboud University Nijmegen, The Netherlands

Systems and Projects Track Program Committee

Jesse Alama	Technical University of Vienna, Austria
Rob Arthan	Queen Mary University of London, UK
Deyan Ginev	Jacobs University Bremen, Germany
Jónathan Heras	University of Dundee, UK
Mateja Jamnik	University of Cambridge, UK
Predrag Janičić	University of Belgrade, Serbia
Christoph Lange	University of Bonn and Fraunhofer IAI, Germany
Christoph Lüth	DFKI and University of Bremen, Germany
Bruce Miller	National Institute of Standards and Technology, USA
Alan P. Sexton	University of Birmingham, UK
Hendrik Tews	TU Dresden, Germany

Additional Referees

Accattoli B.
Adams M.
Brain M.J.
Caprotti O.
Chyzak F.
Eder C.
Ferreira M.
Graham-Lengrand S.
Guidi F.
Heijltjes W.B.
Iancu M.
Ion P.
Jucovschi C.
Kahl W.
Khan-Afshar S.
Kohlhase A.
Kühlwein D.

Krebbers R.
Lee M.
Lelay C.
Líška M.
Mörtberg A.
O'Connor R.
van Oostrom V.
Peltier N.
Prodescu C.-C.
Recio T.
Růžička M.
Schürmann C.
So C.
Subramaniam M.
Verbeek F.
Whiteside I.
Zengler C.

Links between Homotopy Theory and Type Theory

Yves Bertot

Inria
2004 route des lucioles
06902 Sophia Antipolis
France

In the recent history of computer verified proof, we observe the conjunction of two domains that are seemingly extraordinarily distant: homotopy theory and type theory.

Type theory [6] is the realm of the motto "proof as programs, propositions as types", also often referred to as the *Curry-Howard isomorphism*. In this approach, proofs are objects of the logical discourse like any other object (number, program, statement) and their type is the proposition that they prove. For instance, a bounded integer can be encoded as a pair of an integer and a proof that this integer is within the prescribed bounds. This novelty also brings new problems: while we understand perfectly that there are several numbers in the same type of natural numbers, it is not immediately clear what we should do with several (distinct) proofs of the same statement. For instance, what should we do with two instances $(3, p_1)$ and $(3, p_2)$ describing the value 3 bounded by 4, where p_1 and p_2 are distinct proofs that 3 is smaller than 4.

The question of proof unicity (also called *proof irrelevance*) is especially important for proofs of equality between two values. This was studied at the end of the 1990s by Hoffmann and Streicher [5], and the conclusion was twofold. First, the unicity of proofs of equality is not a consequence of type theory's inherent foundations so that this property should be added as an axiom (which they named axiom K) if it is desired for all types; second, the type of equalities between elements of a given type can be endowed with a structure already known in mathematics as a *groupoid* structure. At about the same time, Hedberg [4] showed that proofs of equalities are unique in any type where equality is decidable. This result is put to efficient use in *Mathematical Components* [2].

Meanwhile, mathematics is also traversing crises of its own, first because some mathematical facts are now established with the help of intensive computations performed mechanically [3], second because some areas of mathematics reach a level of internal complexity where too few experts are competent to exert a critical eye. Bringing in computer verified proofs is a means to improve the confidence one can have in proofs that are otherwise refereed by too few people.

Homotopy theory is one of the topics where computer verified proof is being tested. Homotopy theory is about paths between objects, and it is well know that collections of paths also respect a groupoid structure. It thus feels natural to attempt to use directly the equality types of type theory to represent the paths used in homotopy theory. It turns out that this experiment works surprisingly well [1, 8]. For example, homotopy theory naturally considers higher-dimension paths (paths between paths) but similarly type theory is very well suited to consider equalities between equality proofs. Most of the notions around homotopies can be described around equality types and the whole system seems to provide a nice context to perform *synthetic homotopy theory* in the same spirit that Euclid's, Hilbert's, or Tarski's axiom systems can be used to perform *synthetic geometry*.

Type theory was already trying to pose as a new foundation for mathematics, but the question of giving a status to several proofs of the same statement was an unresolved issue. Insights from homotopy theory help clarifying this question: multiplicity of proofs should be encouraged, even though the collection of types where unicity of equality proofs is guaranteed should be given a specific status, as it simply corresponds to the sets of traditional mathematics. For these sets where unicity of equality proofs are guaranteed, one should be able to benefit from most of the machinery already developed in Mathematical Components for types with decidable equality.

New experiments are made possible in this homotopy type theory and new questions arise. In turn, these questions lead to the proposal of new extensions [7]. One such extension is the univalence axiom, which simply states that two homotopically equivalent types should be considered equal; another extension is the concept of higher inductive type, where inductive types can be equiped with paths between constructors. These extensions bring questions of consistency and computability that are still being studied.

References

1. Awodey, S., Warren, M.A.: Homotopy theoretic models of identity types. Math. Proc. Cambridge Philos. Soc. 146(1), 45–55 (2009)
2. Gonthier, G., et al.: A machine-checked proof of the odd order theorem. In: Blazy, S., Paulin-Mohring, C., Pichardie, D. (eds.) ITP 2013. LNCS, vol. 7998, pp. 163–179. Springer, Heidelberg (2013)
3. Hales, T.C.: Cannonballs and honeycombs. Notices of the AMS 47(4), 440–449 (2000)
4. Hedberg, M.: A Coherence Theorem for Martin-Löf's Type Theory. Journal of Functional Programming 8(4), 413–436 (1998)
5. Hofmann, M., Streicher, T.: The groupoid interpretation of type theory. In: Twenty-five Years of Constructive Type Theory (Venice, 1995). Oxford Logic Guides, vol. 36, pp. 83–111. Oxford Univ. Press, New York (1998)
6. Martin-Löf, P.: Intuitionistic type theories. Bibliopolis (1984)
7. The Univalent Foundations Program. Homotopy Type Theory: Univalent Foundations of Mathematics. Institute for Advanced Study (2013), http://homotopytypetheory.org/book
8. Voevodsky, V.: Univalent foundations project (2010), http://www.math.ias.edu/~vladimir/Site3/Univalent_Foundations_files/univalent_foundations_project.pdf

Teaching Tiles (*Azulejos que Ensinam*)

António Leal Duarte

CMUC/Departament of Mathematics, University of Coimbra
P-3001-454 Coimbra, Portugal

Abstract. The mathematical glazed tiles (*azulejos*) of the Jesuit College in the city of Coimbra, in Portugal, are remarkable and unique artifacts. They seem to be the only known example of glazed tiles for classroom use displaying geometrical diagrams of true mathematical (Euclidian) demonstrations. Scientific motifs as decorative elements in buildings were widely used in Europe and, in particular, in spaces built by the Society of Jesus. Panels of *azulejos* using ornamental mathematical motifs are well known in Portugal and elsewhere. But the mathematical *azulejos* of Coimbra are unique in that they are genuine didactical aids to the teaching of mathematics and not merely decorative artifacts.

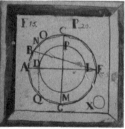

The majority of the Coimbra mathematical *azulejos* display strictly geometrical (Euclidian) matters, while a few concern other scientific matters. The Euclidian diagrams are drawn from one of André Tacquet's famous and very popular editions of Euclid's Elements, which were extensively used in Jesuit schools. The first edition, with the title *Elementa geometriæ planæ ac solidæ quibus accedunt selecta ex Archimede theoremata* was published in 1654; many other editions and translations were published in the next decades.

A Portuguese translation of Tacquet's Elements appeared in 1735, thus roughly at the same time as when the tiles were created, and it is tempting to relate the two events and assume that they were drawn from this edition. However, a closer examination reveals that this is not the case. It is clear today that the diagrams in the *azulejos* were copied from one (or more) of Tacquet's Latin editions.

Table of Contents

Systems and Projects

What International Studies Say about the Importance and Limitations of Using Computers to Teach Mathematics in Secondary Schools

Jaime Carvalho e Silva

Departamento de Matemática, Universidade de Coimbra,
Apartado 3008, EC Santa Cruz, 3001-501 Coimbra, Portugal
jaimecs@mat.uc.pt
http://www.mat.uc.pt/~jaimecs/pessoal/

Abstract. The use of technology in schools has been one of the most debated topics around mathematics education. In some countries there is a huge investment, in others there is a downscaling. Malaysia decided in 2013 to put its 10 million students to use Google laptops and Google apps, while Australia in the same year decided it would not continue funding their own high school laptop program. Who is right from the educational point of view? The last major curriculum document written in the world to date, the Common Core State Standards-CCSS in the United States, whose mathematics part is coordinated by the well known mathematician William McCallum, sets as one of its standards for mathematical practice: "Mathematically proficient students consider the available tools when solving a mathematical problem. These tools might include pencil and paper, concrete models, a ruler, a protractor, a calculator, a spreadsheet, a computer algebra system, a statistical package, or dynamic geometry software." Strong moves need substantiation from research, including the analysis of the existing situation in different countries. What does research say about the use of computers in schools in present time and the use of different pieces of software from spreadsheets to computer algebra systems?

Keywords: mathematics education.

1 Introduction

It is extremely frequent to see heated debates on the pros and cons of using computers in the classroom, be it the primary school classroom, the middle school classroom, the secondary school classroom or even the higher education classroom.

For example [12] in the *The New York Times*, on June 10, 2010, Randall Stross, a professor of business at San Jose State University, quotes some studies about the impact of computers and technology on student's test scores to conclude that

S.M. Watt et al. (Eds.): CICM 2014, LNAI 8543, pp. 1–11, 2014.
© Springer International Publishing Switzerland 2014

Middle school students are champion time-wasters. And the personal computer may be the ultimate time-wasting appliance. Put the two together at home, without hovering supervision, and logic suggests that you wont witness a miraculous educational transformation.

One of the studies he quotes was published in the *Quarterly Journal of Economics* and carried out in Romania, comparing families that in 2009 received vouchers to buy computers against families that did not receive that voucher. Results showed that students in the first group showed *significantly lower school grades in math, English and Romanian*. This is a good example of bad use of statistics. You carry out a large scale collection of data, isolating a few variables and ignoring most of them and then imply there is a cause-effect relation between the variables you isolated. Anybody that works closely within education knows you cannot draw simple conclusions from complex data. Another study of a similar kind arrives at similar conclusions: a working paper published by the *National Bureau of Economic Research* correlates data relative to the introduction of broadband services in the american state of North Carolina between 2000 and 2005 to the results of middle school test scores during that period. This time they conclude that the "negative effects" were limited to low income households.

In a recent analysis [11] published also in *The New York Times*, Carlo Rotella, director of American studies at Boston College, discusses in detail the position of Joel Klein, chief executive of an IT company and former chancellor of New York Citys public schools from 2002 to 2011. In short, Joel Klein believes that teachers and students need new and interesting tools that help them teach and learn. When asked about evidence for his claims he just says that tablets "will help teachers do" what educational research shows is important, that "an individual student will learn more if you can tailor the curriculum to match her learning style, pace and interests." Of course, he did not prove that tablets will indeed accomplish this goal.

Lots of countries are investing hard in introducing some kind of high tech tools in the classroom. John Koetsier reports [5], on the digital news site *VentureBeat*, that the Malaysian government is investing massively in introducing computers, internet access software and "Google Apps" in all 10 thousand schools in the country totaling 10 million students, teachers, and parents. Why was this choice made? Because "Google Apps" are, for educational use, completely free. Is this the best software for educational use? Will students learn better mathematics with this environment? This is not stated.

It is clear that we have lots of options on the use of IT technology in schools (including not using it at all), but we need clear ideas before accepting or rejecting hardware, software and communications in the classroom.

2   International Studies

We live in a society where technology is in a rapid evolution, with new tools arriving at the consumer market every year. It is more than natural that these tools are also offered at the school level. Two main reasons can be stated in favor of this: first of all the school has never been as efficient as the society desires and so new approaches are normally welcome (at least by most people); secondly, if the school is to prepare students for "real life" and for some professional activity, then teaching should somehow incorporate the technological tools that students will find someday in their adult life.

In education, things are never simple. Several authors, like Luc Trouche [13], already pointed out we should make a distinction between an artifact and an instrument. This means that you may have some artifact in the classroom (like a technological tool, hardware or software) but it may have no effect at all unless you are able to integrate it in your activity, and then it becomes an instrument for you. This is not a useless distinction because we have important examples of this difference and the PISA OECD studies give us one of these.

2.1   Digital Reading

PISA is a program conducted by the OECD to study the extent to which 15-year-old students (normally near the end of compulsory education) have acquired the knowledge and skills that are essential for full participation in modern society, focusing mainly in mathematics, reading and science. This program began in the year 2000 and is applied in numerous countries every three years. A lot of data is collected about the students, the teachers, the schools and the student's environment. Also some other studies are conducted in parallel, at least in some of the countries participating in the main PISA study. In 2012 a total of 65 countries and economies participated in the PISA data collecting but only 44 countries and economies participated in a computer-based assessment of problem solving; 32 of them also participated in a computer-based assessment of reading and mathematics.

For the first time, the PISA 2009 survey also assessed 15-year-old students ability to read, understand and apply digital texts. These texts are very different from printed ones, namely at the level of their organisation. In 19 countries and economies students were given questions via computer to assess this ability. The PISA 2009 results [9] about digital reading show something striking. They show that even when guidance on navigation is explicit, significant numbers of students still cannot locate crucial pages.

Digital reading poses new problems to users: indexing and retrieval techniques are new because of the virtual nature of page contents and formats; also hyperlinks are introduced and new multipage documents are used in a networked structure that may confuse the reader. So, a completely new environment comes up and PISA digital reading assessment offers powerful evidence that todays 15-year-olds, the "digital natives", do not automatically know how to operate effectively in the digital environment, contrarily to what we could have thought.

2.2 Computer-Based Assessment of Mathematics

For the first time in 2012, PISA included an optional computer-based assessment of mathematics. 32 of the 65 countries and economies participated in this. Specially designed PISA questions were presented on a computer, and students responded on the computer, although they could also use pencil and paper as they worked out through the test questions. The PISA 2012 report [10] justifies this part of the PISA program:

> (...) computer-based items can be more interactive, authentic and engaging than paper-based items. They can be presented in new formats (e.g. drag-and-drop), include real-world data (such as a large, sortable dataset), and use colour, graphics and movement to aid comprehension. Students may be presented with a moving stimulus or representations of three-dimensional objects that can be rotated, or have more flexible access to relevant information. New item formats can expand response types beyond verbal and written, giving a more rounded picture of mathematical literacy. (...) computers have become essential tools for representing, visualising, exploring, and experimenting with all kinds of mathematical objects, phenomena and processes, not to mention for realising all types of computations at home, at school, and at work.[10]

Fourty one specially designed computer-based items were developed for this assessment. These items were designed so that mathematical reasoning and processes would take precedence over the ability of using the computer as a tool. The report details the approach used:

> Each computer-based item involves three aspects:
> - the mathematical demand (as for paper-based items);
> - the general knowledge and skills related to information and communication technologies (ICT) that are required (e.g. using keyboard and mouse, and knowing common conventions, such as arrows to move forward). These are intentionally kept to a minimum;
> - competencies related to the interaction of mathematics and ICT, such as making a pie chart from data using a simple "wizard", or planning and implementing a sorting strategy to locate and collect desired data in a spreadsheet.[10]

The conclusion of this part of the study is that "there is a high degree of consistency in student performance on items delivered on paper and by computer" but with some important exceptions:

> In the field of mathematics, one participant (Shanghai-China) saw a large difference, of around 50 score points, in favour of the paper based format. Three other countries and economies showed substantial differences in the same direction - Poland (28-point difference), Chinese Taipei (22-point difference) and Israel (20-point difference). Conversely, there are

also countries for which computer delivery of the assessment appears to have been advantageous. The largest difference, of about 30 score points, was seen in Brazil. Colombia also saw a difference of about 20 points in the same direction. The United States, the Slovak Republic and Italy also saw marked, albeit smaller, differences in favour of the computer delivery of the assessment. Across OECD countries, the performance advantage of the computer-based assessment is slightly higher for boys than for girls. ([10], p. 491)

This is a quite recent report and these differences are not yet discussed in terms of the nature of the tasks, of the mode of delivery, or of the student familiarity with computers. In the PISA 2015 program, the computer-based assessment will be the primary mode of delivery for mathematics literacy and all the other domains, but the use of paper-based assessment instruments is an option for countries choosing to do so. In some years we will have then more data for our discussion.

2.3   Improvements in Performance

Another important question we need to answer, and is raised by a lot of people, is wether students perform better or worse in a computer environment (at school and at home). Some PISA studies also address this question. The PISA 2003 study discusses the relation between the frequency of use of computers at home and student performance in mathematics. And the conclusion is very clear:

(...) in every country, students reporting rare or no use of computers at home (on average 18% of students) score much lower than their counterparts reporting moderate use or frequent use. [7]

The PISA 2006 study compares the PISA scores and the use of ICT. In [8] students are grouped according to frequency of ICT use and then the average performances of each group are compared. Of course this does not tell the whole story because some factors that affect computer use also affect student performance. In order to give a clear picture the PISA 2006 study includes questions about the location and frequency of student computer use. Trying to include a number of relevant variables the PISA 2006 study concludes that:

A higher frequency of computer use is associated with higher average science scores in all countries considered. Among OECD countries, the largest effect of using a computer almost every day was found in Iceland, Japan, The Netherlands, Norway, Poland and Spain. Among partner countries, the largest effect of using computer almost every day was found in Bulgaria; Macao, China; and Slovenia. ([8], p. 150) (...) in a large majority of countries, the benefits from higher computer use tend to be greater at home than at school. Therefore, despite the better environment and support that schools are expected to provide, computer use tends to have less impact at school than at home. ([8], p. 156)

Having this in mind the PISA study recommends concrete actions regrading ICT use in schools:

> (...) the analysis has shown that computer use increases student performance but that this increase is not the same for all students. (...) as the benefits from computer use depend on the characteristics of each student, policies to increase ICT use need to be tailored to students. (...) the positive effects of computer use on student performance are greatest when they are supported by a sufficient level of capital. Skills, interests and attitudes affect students engagement with ICT, the activities they carry out on the computer and how well. An increase in ICT use that is not supported by an increase in capital would have a lower impact on student performance. ([8], p. 156)

Of course, it is clear from these studies that the simple use of ICT does not guarantee an improvement in performance:

> (...) the apparently negative association between performance and some kinds of computer usage, shown by PISA 2003 and now PISA 2006, carries a warning not to assume that more is better for students performance. ([8], p. 158)

It is clear form these studies that the use of ICT has generally very positive effects on student performance at the mathematics and science level. Only incomplete studies will conclude that the use of ICT has a negative influence in student performance. These OECD big scale studies examine the educational situation in great detail and include very different political and social realities in the big number of countries involved so that their conclusions are very reliable. What we loose in these huge statistical studies is the detail. We need now to know what works and what does not work in each situation.

3   ICMI Studies

ICMI, the International Commission on Mathematical Instruction, founded in 1908 to foster efforts to improve the quality of mathematics teaching and learning worldwide, has produced two large studies that discuss in detail the impact and use of ICT in mathematics education. These were:

- ICMI Study 1. The Influence of Computers and Informatics on Mathematics and its Teaching
  Study Conference held in Strasbourg, France, March 1985.
  Study Volume published by Cambridge University Press, 1986, eds: R.F. Churchhouse et al. (ICMI Study Series)
  Second edition published by UNESCO, 1992, eds: Bernard Cornu and Anthony Ralston. (Science and Technology Education No. 44)

- ICMI Study 17. Digital Technologies and Mathematics Teaching and Learning: Rethinking the Terrain
  Study Conference held in Hanoi, Vietnam, December 2006.
  Study Volume published by Springer, 2010: Mathematics Education and Technology-Rethinking the Terrain. The 17th ICMI Study Series: New ICMI Study Series, Vol. 13. Hoyles, Celia; Lagrange, Jean-Baptiste (Eds.) (New ICMI Study Series 13)

These studies point out some directions for the integration of ICT in mathematics education, but it is also clear that much more research needs to be done:

> The way digital technologies can support and foster today collaborative work, at the distance or not, between students or between teachers, and also between teachers and researchers, and the consequences that this can have on students learning processes, on the evolution of teachers practices is certainly one essential technological evolution that educational research has to systematically explore in the future. ([4], p. 473)

Numerous examples are described and quoted in this 500-page volume but we need to have in mind what I consider to be the main conclusion:

> Making technology legitimate and mathematically useful requires modes of integration (...) requires tasks and situations that are not simple adaptation of paper and pencil tasks, often tasks without equivalent in the paper and pencil environment, thus tasks not so easy to design when you enter in the technological world with your paper and pencil culture. ([4], p. 468)

The range of hardware and software considered in this Study is huge, from Dynamic Geometry Environments to Computer Algebra Systems, including Animation Microworlds, Games and Spreadsheets, showing that the use of ICT in the mathematics classroom is not limited to any particular kind of software and offers thus many possibilities for mathematics teaching and learning.

Another important point visible in this Study is the need to find an answer to the "wrong-doing" of certain technologies. How to deal with the pitfalls of numerical analysis, namely dealing with rounding errors? How to correctly identify a tangent to a circle in a Dynamic Geometry Environment that has difficulties with the continuity? The Study calls for a reasonable

> (...) basic understanding of the inner representation of mathematics (e.g., numbers, equations, stochastics, graphical representations, and geometric figures) within a computer and a global awareness of problems related to the difference between conceptual and computational mathematics. ([4], p. 153)

4 First Conclusion

What we discussed from these international studies allows us to conclude that the CCSS are right in investing decidedly in the use of ICT in the classroom:

Standards for Mathematical Practice (...) 5 Use appropriate tools strategically. Mathematically proficient students consider the available tools when solving a mathematical problem. These tools might include pencil and paper, concrete models, a ruler, a protractor, a calculator, a spreadsheet, a computer algebra system, a statistical package, or dynamic geometry software. Proficient students are sufficiently familiar with tools appropriate for their grade or course to make sound decisions about when each of these tools might be helpful, recognizing both the insight to be gained and their limitations. For example, mathematically proficient high school students analyze graphs of functions and solutions generated using a graphing calculator. They detect possible errors by strategically using estimation and other mathematical knowledge. When making mathematical models, they know that technology can enable them to visualize the results of varying assumptions, explore consequences, and compare predictions with data. Mathematically proficient students at various grade levels are able to identify relevant external mathematical resources, such as digital content located on a website, and use them to pose or solve problems. They are able to use technological tools to explore and deepen their understanding of concepts.[6]

Most of the countries in the world have a clear vision of what needs to be done. For example the official curriculum for Singapore reads:

AIMS OF MATHEMATICS EDUCATION IN SCHOOLS: (...) (6) Make effective use of a variety of mathematical tools (including information and communication technology tools) in the learning and application of mathematics. (...)
(...) The use of manipulatives (concrete materials), practical work, and use of technological aids should be part of the learning experiences of the students.
SKILLS: (...) Skill proficiencies include the ability to use technology confidently, where appropriate, for exploration and problem solving.[3]

What happens in the real classroom is not so simple.

5   A Difficult Task

In a national examination in Portugal for the 12th grade, a mathematical modeling problem involved the study of the function:

$$d(x) = 149.6(1 - 0.0167\cos x) \qquad (1)$$

Graphing Calculators (GC) are allowed in national examinations in Portugal and so the students can use them to study this function. The biggest challenge here for the student, and it has been proven to be a big obstacle, is to find a viewing window to obtain the graph for this function. Of course you can get the

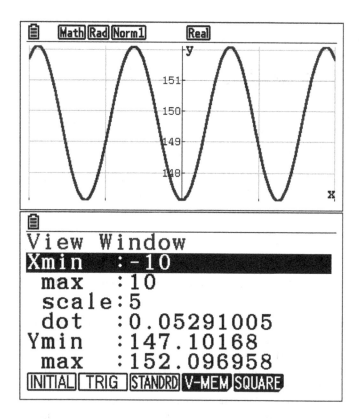

Fig. 1. Using the AUTO feature to draw a simple graph of $d(x)$ in a GC

help of the AUTO function of the calculator but then you are not sure you get all the details you need in graph that shows up.

The second more difficult group of questions in the national 12th grade examinations in Portugal, were the ones requiring the use of graphing calculators, some of them also involving modeling problems. In a previous study [1] we concluded that these problems all involve the need to choose a viewing window. There is no algorithm that can guarantee you get the best viewing window. You can produce a table of values to help you but you will need always to experiment or know some properties of the function in order to be sure you get a "complete" graph. In more difficult situations you may need to use more than one graph to capture the details of the graph of the function you want to study.

Another similar difficulty is discussed by Luc Trouche in his paper [13] in the journal *Educational Studies in Mathematics*. If a student tries to use a GC to study the limit of a function when the independent variable goes to $+\infty$ he will try to graph the function "as far as possible". But if he is faced with a function like

$$f(x) = \ln x + 10 \sin x \tag{2}$$

he will think it will not have a limit, when the limit is really $+\infty$. The graph will give him a dangerous message:

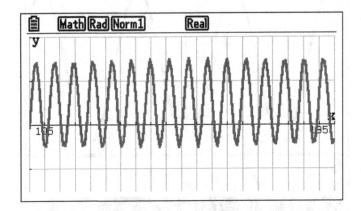

Fig. 2. Graphing function $f(x)$ in the viewing window $[100, 200] \times [-15, 20]$

Luc Trouche concludes that the complexity of the transformation of this new artifact into a useful instrument for the work of the student is related with the sophistication of the artifact, namely when it comes to a GC with CAS-Computer Algebra System. This is a big educational challenge and should be kept in mind when some hardware or software is selected to be used in the classroom.

In the same paper Luc Trouche warns against the lack of investment in the use of ICT in the classroom, observing that the use in the classroom is too limited, in France and other countries, and consequently the "learning of the use of instruments is made most of the time alone or between friends" ([13], p. 190) with all its dangers. It is clear from the international studies that the use of ICT can produce an improvement in student performance, but to arrive at that point of improvement, a lot of research, experimentation and planning must be made.

We conclude with a recommendation made by Seymour Papert, the inventor of the educational programming language LOGO, in the plenary talk he gave at the Study Conference included in ICMI Study 17. Seymour Papert ended his talk asking us to spend reasonable part of our time and energy thinking about possible futures, freeing our minds from the current constraints.

References

1. Balsa, J., Carvalho e Silva, J.: Uma análise do papel da calculadora gráfica nos exames nacionais. In: Actas do XIII Seminário de Investigação em Educação Matemática, pp. 237–242. APM, Lisboa (2002)
2. Carvalho e Silva, J.: A formação de professores em novas tecnologias da informação e comunicação no contexto dos novos programas de Matemática do Ensino Secundário. Comunicação apresentada no 2. Simpósio "Investigação e Desenvolvimento de Software Educativo", D.E.I., Coimbra (1997), http://lsm.dei.uc.pt/simposio/pdfs/c11.pdf

3. Curriculum Planning and Development Division, Secondary Mathematics Syllabuses, Ministry of Education, Singapore (2006)
4. Hoyles, C., Lagrange, J.-B. (eds.): Mathematics Education and Technology - Rethinking the Terrain: The 17th ICMI Study (New ICMI Study Series). Springer, Berlin (2010)
5. Koetsier, J.: Google: 10 million Malaysian students, teachers, and parents will now use Google Apps for Education. VentureBeat, April 10 (2013), http://venturebeat.com/2013/04/10/google-10-million-malaysian-students-teachers-and-parents-will-now-use-google-apps-for-education/
6. McCallum, W., et al.: Common Core State Standards for Mathematics. National Governors Association Center for Best Practices & Council of Chief State School Officers, Washington, DC (2010)
7. OECD, Are Students Ready for a Technology-Rich World? What PISA Studies Tell Us. OECD Publishing, Paris (2006)
8. OECD, Are the New Millennium Learners Making the Grade?, Technology Use and Educational Performance in PISA 2006. OECD Publishing, Paris (2010)
9. OECD, PISA 2009 Results: Students on Line: Digital Technologies and Performance (vol. VI). OECD Publishing, Paris (2011)
10. OECD, PISA 2012 Results: What Students Know and Can Do - Student Performance in Mathematics, Reading and Science (vol. I, Revised edition, February 2014). OECD Publishing, Paris (2014)
11. Rotella, C.: No Child Left Untabled. The New York Times, September 12 (2013), http://www.nytimes.com/2013/09/15/magazine/no-child-left-untabled.html
12. Stross, R.: Computers at Home: Educational Hope vs. Teenage Reality. The New York Times, July 10 (2010), http://www.nytimes.com/2010/07/11/business/11digi.html
13. Trouche, L.: Environnements Informatisés et Mathématiques: quels usages pour quels apprentissages? Educational Studies in Mathematics 55, 181–197 (2004)

Towards Robust Hyperlinks for Web-Based Scholarly Communication

Herbert Van de Sompel*, Martin Klein, and Harihar Shankar

Los Alamos National Laboratory, Los Alamos NM 87545, USA
{herbertv,mklein,harihar}@lanl.gov

Abstract. As the scholarly communication system evolves to become natively web-based, hyperlinks are increasingly used to refer to web resources that are created or used in the course of the research process. These hyperlinks are subject to reference rot: a link may break or the linked content may drift and eventually no longer be representative of the content intended by the link. The Hiberlink project quantifies the problem and investigates approaches aimed at alleviating it. The presentation will provide an insight in the project's findings that result from mining a massive body of scholarly literature spanning the period from 1997 to 2012. It will also provide an overview of components of a possible solution: pro-active web archiving, links with added attributes, and the Memento "Time Travel for the Web" protocol.

Keywords: scholarly communication, web archiving, reference rot, digital preservation.

1 Introduction: The Brittleness of Web-Based Scholarly Communication

Traditionally, references in scholarly communication point to published articles or books. But, as the scholarly communication system evolves to become natively web-based, hyperlinks are increasingly used to refer to web resources that are created or used in the course of the research process. This includes scholarly artifacts such as software, datasets, websites, presentations, blogs, videos, scientific workflows, and ontologies. In addition, as exemplified by the work on Research Objects [1], scholarly communication objects become increasingly compound, consisting of multiple inter-related constituent resources.

The dynamic, compound, and inter-related nature of scholarly communication objects yields significant challenges for the fulfilment of the archival function that is essential for any system of scholarly communication [2]. In order to illustrate this, consider a comparison between the print era of the journal system and the natively web-based scholarly communication system that is emerging.

Figure 1 depicts the journal system era. A published journal article references other articles, published in the same or other journals. The referencing as well

---

* http://public.lanl.gov/herbertv/

S.M. Watt et al. (Eds.): CICM 2014, LNAI 8543, pp. 12–25, 2014.

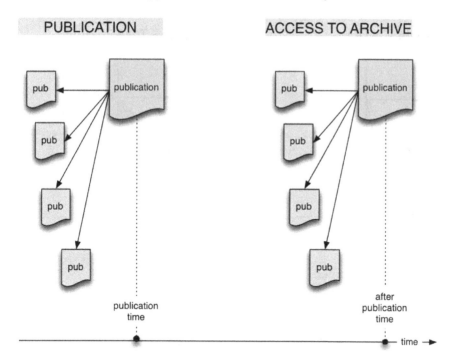

Fig. 1. Publications link to publications

as the referenced articles are preserved as part of journal collections managed by libraries, worldwide. In order to revisit the article and its context of referenced articles some time after publication, it suffices to pay a visit to the appropriate library stacks and pull the relevant journal issues. Since the content was printed on paper and is fixed, the combination of the article and its surrounding context of referenced articles remains the same as it was on the day of the article's publication. Gathering the entire context surrounding the referencing article may require some travel, but the original information bundle can accurately be recreated.

Figure 2 reconsiders this scenario for web-based scholarly communication. The scenario is conservative in that it still takes a journal article, rather than another scholarly object such as software, as its starting point. Nevertheless, it still serves as a sufficient illustration. The web-based article not only references other articles but also links to a variety of web resources including software, data, scientific blogs, project web sites. Recreating the information bundle made up of the article and its surrounding context some time after publication is now far less trivial. Challenges are introduced by the dynamic nature of the web, the malleability of digital content, and the dynamic nature of scholarly objects, especially the ones created in the course of the research process. The links in the article are subject to reference rot, a term coined in the Hiberlink[1] project to refer to the combination of link rot and content drift. Link rot, also known as

---

[1] http://hiberlink.org

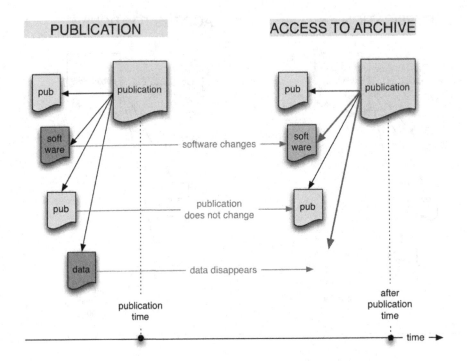

Fig. 2. Publications link to a variety of dynamic resources

"404 Page Not Found", describes the scenario where a link is no longer accessible and the HTTP request results in an error, typically a 404. Content drift describes the evolution of a web resource's content away from what it was at the moment it was linked, possibly up to a point that it becomes unrepresentative of the content intended by the link.

2   Hiberlink: Investigating Reference Rot

The Mellon-funded Hiberlink project, a collaboration between the Los Alamos National Laboratory and the University of Edinburgh, explores the aforementioned reference rot along a research track and a solutions track.

2.1   Hiberlink Research Track

A research track aims at characterizing and quantifying reference rot in web based scholarly communication at unprecedented scale. This work is inspired by a 2011 pilot study [5] that investigated 160,000 URIs extracted from two open access scholarly collections: arXiv.org and the electronic thesis repository of the University of Northern Texas. It was by far the most extensive link rot study ever conducted. Hiberlink goes beyond that and mines vast open access

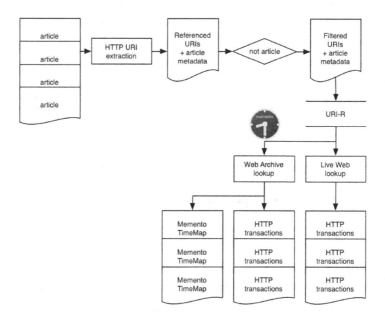

Fig. 3. The workflow used for assessing the current and archival status of URI references

and commercial scholarly corpora spanning publication dates between 1997 and 2012 for millions of HTTP URIs. In a process as depicted in Figure 3 the current and archival status of these URI references are investigated:

- Based on the assumption that the core scholarly literature is adequately archived by platforms such as Portico[2] and CLoCKSS[3], URI references to journal articles are discarded and only URIs of web at large resources are maintained.
- For each remaining reference, the combination of a URI and the referencing datetime, the current status of the URI is checked: is the URI still operational or not?
- For each remaining reference, web archives are also consulted using the Memento "Time Travel for the Web" protocol [4], to determine whether archived versions of the referenced URI exist with an archival datetime that is within 1 day, 7 days, 14 days, 30 days, and 6 months of the referencing date, respectively. Understanding the dynamic nature of web resources, the representativeness of these archival resources diminishes as the time between referencing and archiving increases.

The findings are dramatic and indicate that millions of scholarly papers suffer from reference rot.

---

[2] http://portico.org
[3] http://www.clockss.org/

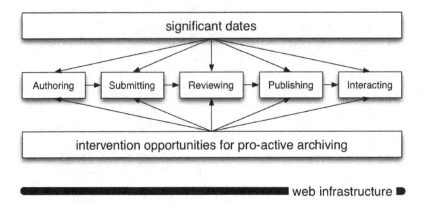

Fig. 4. Potential points of intervention for pro-active archiving in an article's lifecycle

2.2 Hiberlink Solutions Track

A solutions track aims at identifying and prototyping approaches that can ameliorate the reference rot problem. This work has two strands.

A first strand investigates possible approaches to pro-actively archive web resources that are referenced in scholarly articles. Various significant moments in an article's lifecycle are considered as possible intervention points (Figure 4) and the approach in which archiving will be conducted at each may differ. The time at which pro-active archiving occurs has an obvious impact on the accuracy of the archived snapshot vis-à-vis the version of the referenced resource that the author observed: the closer the snapshot time is to the observation time, the better. These are examples of archival approaches for different intervention points:

- During the note-taking or authoring phase of a paper, an extension for a bibliographic reference manager used by an author can not only record information about a web resource but also deposit a snapshot in web archive. This approach was prototyped[4] for the Zotero[5] browser extension.
- When a paper draft is submitted to a manuscript management system, a special-purpose module can extract the URIs of referenced web resources and archive them. Experiments are ongoing using the Open Journal Systems[6] platform.
- When a new paper becomes available in a repository, a third party archival service can be notified. The service can then collect the paper, extract referenced URIs and archive them. This approach is followed by the HiberActive system (Figure 5) developed at LANL [3].

---

[4] Wincewicz, R. https://www.youtube.com/v/ZYmi_Ydr65M%26vq
[5] https://www.zotero.org/
[6] http://pkp.sfu.ca/ojs/

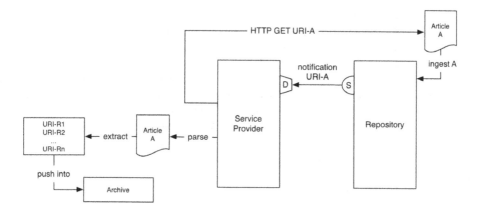

Fig. 5. HiberActive architecture for archiving referenced web resources

– Resources can also be archived as the paper that references them is being interacted with. For example, a tweet about a paper can trigger a process that is similar to the aforementioned HiberActive.

A second strand investigates what links to referenced resources should look like if both the actual URI and the URI of an archived version exist. Further details about the challenge involved and possible approaches to address it are provided in the next section.

3   The Missing Link Proposal

This section details Hiberlink's Missing Link Proposal aimed at extending HTML's anchor element with an attribute that contains information aimed at increasing link robustness over time.

3.1   Motivation

A common approach to address reference rot consists of the combination of:

1. *Creating a snapshot* of the linked resource using an on-demand web archiving service as provided by, among others, WebCite[7], archive.today[8], perma.cc[9]. The result of creating a snapshot, is a copy of the linked resource with the content that was intended to be linked to and, obviously, a URI for the snapshot in the web archive. For example, `http://perma.cc/0Hg62eLdZ3T` is an archived snapshot of `http://blogs.law.harvard.edu/futureoftheinternet/2013/09/22/perma/` taken on October 2nd 2013 and hosted by the perma.cc web archive.

---

[7] `http://www.webcitation.org/`
[8] `http://archive.today`
[9] `http://perma.cc`

2. *Linking to the snapshot* instead of to the original resource. This approach is illustrated in the New York Times article on link rot: In Supreme Court Opinions, Web Links to Nowhere[10]. Towards the end of this article, the link on the text **a new, permanent link** points to the archived snapshot `http://perma.cc/0Hg62eLdZ3T`, but not to the original resource `http://blogs.law.harvard.edu/futureoftheinternet/2013/09/22/perma/`. This common replacement approach is illustrated by the difference between **Original Link** and **Link Replaced by Link to Archived Snapshot** in Figure 6.

There is good news and bad news regarding this common approach to address reference rot:

- The good news is that a snapshot of the linked resource is taken and stored in a web archive. The more snapshots of a resource, the better our record that documents its evolution over time.
- The bad news is that in order for the link to the archived snapshot to work, the archive that hosts the snapshot (perma.cc in the ongoing example) needs to remain permanently operational, a rather unrealistic expectation. If that archive goes temporarily or permanently off-line, the link to the archived snapshot stops working: one link rot problem was replaced by another.

Not surprisingly, archived snapshots of the original resource also exist in archives other than perma.cc. They can be found by searching web archives for the original URI `http://blogs.law.harvard.edu/futureoftheinternet/2013/09/22/perma/`. For example:

- *Internet Archive* has result for that search at `https://web.archive.org/web/*/http://blogs.law.harvard.edu/futureoftheinternet/2013/09/22/perma/`
- *archive.today* has result for that search at `http://archive.today/blogs.law.harvard.edu/futureoftheinternet/2013/09/22/perma/`

More generally, every existing web archive supports searching for archived snapshots by means of the URI of the original resource of which the archive took (a) snapshot(s). Also, all major web archives support the Memento protocol [4] that allows accessing archived snapshots by using the URI of the original resource. In addition, some web archives have bespoke APIs that provide such functionality. Using the ongoing example, this means that, in case perma.cc would temporarily or permanently go off-line, snapshots can still exist in other web archives.

However, in the common approach to address reference rot, the original URI is replaced by the URI of the archived snapshot. As a result, this approach prohibits finding snapshots in other web archives, making its functioning totally dependent on the continued existence of the web archive that assigned the URI

---

[10] `http://www.nytimes.com/2013/09/24/us/politics/in-supreme-court-opinions-clicks-that-lead-nowhere.html`

for the archived snapshot. From the perspective of web persistence this can hardly be regarded as a satisfactory solution. In order to maximize chances of future retrieval of snapshots of a linked resource, that resource's original URI must be maintained when linking to it.

3.2   Citing Web Resources

When citing web resources in scholarly literature or in Wikipedia articles, a practice exists to not only include the URI of the cited resource in the citation but also one or more of the following information elements:

- The date the cited resource was visited.
- The URI of an archival version of the cited resource, typically available from a web archive.
- The date the cited resource was archived.

Indeed, in order to cite a web resource, Wikipedia's Simple Citation Template[11] and most scholarly citation styles including the Chicago citation style[12], the American Psychological Association style[13], and the American Chemical Society style[14] list the URI and the access date of the cited resource as necessary information. In addition, Wikipedia's Full Citation Template[15] includes the URI of an archival version of the resource. The latter also includes the date the archival version of the cited resource was created.

3.3   Reference Rot and Temporal Context

Inclusion of this temporal context information - access date, archive url, and archive date - is motivated by the understanding that the web is ephemeral, that the URI of the cited resource - from now on referred to as **linkedurl** - is subject to reference rot.

The temporal context information provided in these web citations serves the following purposes:

- The date the cited resource was visited, from now on referred to as **versiondate**, serves as a reminder that one should not assume that the content at the cited resource will be the same when visiting it some time after that date.
- The URI of an archival version of the cited resource, from now on referred to as **versionurl**, allows revisiting a version of the page as it was when it was cited.
- The date the cited resources was archived is somehow informative from a documentation perspective but seems less crucial as this information is typically also available when visiting the archived version of the resource.

---

[11] http://en.wikipedia.org/wiki/Template:Citation#Simple_citation
[12] http://www.bibme.org/citation-guide/Chicago/website
[13] http://www.studygs.net/citation.htm
[14] http://library.williams.edu/citing/styles/acs.php
[15] http://en.wikipedia.org/wiki/Template:Citation#Full_citation_parameters

3.4   The Case for Structured Temporal Context on Links

This temporal context information has, so far, been included in a way that is helpful for human consumption only. Despite the many variations in expressing the information that is relevant for a web citation, a user can interpret it and connect the dots. Also, temporal context information has so far only been included in formal web citations. However, since all links are subject to reference rot, addition of such information should not be limited to formal citations of web resources, but should rather be applicable to all links to web resources.

There are compelling reasons to express temporal context information in a structured manner on links to support use by applications such as browsers, crawlers, search engines:

- The many variations in expressing web citation information makes machine interpretation challenging.
- In the current representation of information, the linkedurl and the versionurl look like two independent URIs despite the tight - temporal - relationship between them.
- The approach used for formal web citations can not be used for links in general because it would e.g. require adding two links to the same anchor text.
- The versionurl, if provided in a structured manner, can be used by applications such as browsers, to indicate and provide the option to retrieve the archived snapshot of the linked resource.
- The combination of the linkedurl and the versiondate, if provided in a structured manner, can be used by applications such as browsers, to indicate and provide the option to obtain an archived snapshot of the linked resource that is temporally near to the versiondate, even if no versionurl is provided. The Memento protocol [4] that specifies content negotiation in the datetime dimension provides this functionality in an interoperable manner, but it could also be provided by leveraging bespoke APIs of web archives.

The question then arises how to best convey the temporal context information so that applications can use it. And how to do so in a uniform manner, i.e. a manner that is independent on the venue conveying the information. With this regard, it is interesting to observe that in 1995, the definition of the anchor element[16] included an optional URN attribute, possibly/likely provided to address concerns regarding web persistence. The attribute was deprecated and it is probably a fair guess that this happened because no infrastructure existed to act upon URNs. The meanwhile disappeared HTML5 development page for the anchor element[17] included a reminder that the URN attribute is obsolete.

---

[16] http://www.w3.org/MarkUp/1995-archive/Elements/A.html

[17] Original URI http://dev.w3.org/html5/markup/a.html, Snapshot URI dated January 1st 2014 https://web.archive.org/web/20140101022041/http://dev.w3.org/html5/markup/a.html

There are several reasons to revisit the inclusion of attributes related to web persistence in select HTML elements, most importantly the anchor element.

First, there is a growing concern regarding persistence at least in some pockets of the web:

- Wikipedia has an active Link rot thread[18] looking into the problem domain.
- The Hiberlink project and CrossRef's OpCit explore the problem for scholarly communication. The pilot study [5] that led to Hiberlink found disconcerting percentages of link rot and lack of archival versions for web resources referenced in the arXiv.org preprint collection and the thesis repository of the University of North Texas.
- Reference rot has become a significant concern in legal cases that depend on web resources [7].
- The Modern Language Association style[19] for citing web resources no longer mandates the inclusion of the cited URI because **Web addresses are not static.**

Second, infrastructure has emerged that can play a role in achieving an increased degree of web persistence including:

- Web archives such as the Internet Archive[20], the UK Web Archive[21], the Icelandic web archive[22].
- Versioning systems such as MediaWiki[23] and GitHub[24].
- Approaches for pro-active archiving of resources as they are being cited as intended by Wikipedia's WebCiteBOT effort[25], and Hiberlink's Solutions Track.
- Services that support on-demand archiving for web authors that link to resources as provided by WebCite[26], archive.today[27], perma.cc[28].
- The Memento protocol [4] that uses content negotiation in the datetime dimension to provide access to prior versions of resources by means of their original URI and the date of the desired version.

---

[18] http://en.wikipedia.org/wiki/Wikipedia_talk:Link_rot
[19] https://owl.english.purdue.edu/owl/resource/747/08/
[20] http://archive.org/web/
[21] http://www.webarchive.org.uk/ukwa/info/about#what_uk_archive
[22] http://vefsafn.is/index.php?page=english
[23] http://www.mediawiki.org/wiki/MediaWiki
[24] https://github.com/
[25] https://en.wikipedia.org/wiki/User:WebCiteBOT
[26] http://www.webcitation.org/
[27] http://archive.today
[28] http://perma.cc

3.5   Structured Expression of Temporal Context on Links

When attempting to express temporal context information for links in HTML, a major design characteristic should be that the linkedurl be considered the central information element, to be conveyed as the value of the href attribute of the anchor element. The major motivation for this approach is that the linkedurl is the URI by which the linked resource is known throughout the web, including in web archives. Displacing the linkedurl as central information element is akin to throwing away the resource's key.

Another significant consideration in the design of a technical solution is the extent to which it stands a credible chance for standardization and hence support in browsers. Although reference rot is considered a significant problem in some pockets of the web, many consider it only remotely relevant. Hence, a generic technical approach that allows addressing issues related to web persistence but also enables other novel use cases may be more acceptable for the web community at large:

- The early thinking in the Hiberlink project was along the lines of a specific solution, dedicated at addressing reference rot only. For example, one proposal[29], illustrated as **Link Augmented with Attributes from Missing Link Proposal** in Figure 6, consisted of adding both a **versiondate** and **versionurl** attribute with aforementioned semantics to HTML's anchor element.
- However, independent from the above Hiberlink perspective, the Internet Robustness project at Harvard University[30] developed a similar reasoning regarding the need for an additional attribute for HTML's anchor element to be able to point at a cached version of a resource in order to allow a user to still access required information even if the hosting server is under a distributed denial of service attack. While both pointing to an archived and a cached resource resort aim at increased web persistence, they are significantly different cases, featuring augmented links with different semantics.
- This insight led to the consideration that the most generic solution might be one that allows expressing **links about links** (typed links that pertain to the URI expressed in the anchor's href attribute). Indeed, inspired by the way that Web Links [6] allow a resource to express typed links that pertain to itself in HTTP headers, HTML could allow expressing typed links pertaining to resources that are linked from a page. Such an approach would, for example, also allow expressing links to alternate representations of the resource provided in href, including links to mobile versions. This approach is illustrated as **Link Augmented with Typed Link Proposal to an Archived Snapshot** in Figure 6, whereby the **memento** relation type originates in the Memento protocol [4] and indicates an archived snapshot.

---

[29] http://mementoweb.org/missing-link/#option1
[30] http://cyber.law.harvard.edu/research/internetrobustness

– A precedent for the extension of HTML for the purpose of referencing alternates exists in the recently proposed srcset[31] extension that allows conveying alternatives for embedded images.

At the time of writing, a broader collaboration has been launched that includes Hiberlink partners as well as Harvards perma.cc web archive, Harvards Berkman Center, and Old Dominion University and that aims at devising and standardizing a solution that allows augmenting links in HTML. The primary focus is on adding information in support of web persistence, but the eventual technical solution might cover a broader range of use cases.

3.6 Applications

Temporal context information can be put to use in various applications. Search engines could use it, for example, to highlight frequently referenced snapshots of a resource. Also, when a web author adds versiondate information to a link, this could be taken as a hint for a pro-active archiving application that the linked resource should be archived. Such an application could be integrated with a desktop authoring tool, or a content management system like a MediaWiki, or it could be a crawler-based third party service, for example operated by a web archive that is looking to extend its archival collection.

A major use case for temporal context information is browsers. A challenge in using the information is the way in which to make it actionable by the user. One way to do so is by means of a right click on a linked resource. As illustrated in a demonstration video[32], the Memento extension for Chrome[33] uses this approach to present the following options in the context menu:

– Retrieve a snapshot of the resource as it existed around the date set in a calendar user interface element.
– Retrieve the most recent snapshot of the resource, a feature that is especially helpful in case of 404 responses.

Using the nomenclature of the Missing Link proposal, the following options could be added to the context menu to leverage temporal context information:

– Retrieve the archived snapshot with the provided versionurl.
– Retrieve the archived snapshot that is temporally closest to the provided versiondate.

---

[31] http://www.w3.org/TR/html-srcset/
[32] Van de Sompel, H. Memento extension for Chrome: A preview
   https://www.youtube.com/watch?v=WtZHKeFwjzk
[33] Shankar, H. Memento Time Travel for Chrome
   http://bit.ly/memento-for-chrome

Original Link

It allows writers and editors to capture and fix transient information on the Web with

`<a href="http://blogs.law.harvard.edu/futureoftheinternet/2013/09/22/perma/">`

a new , permanent link`</a>`

Link Replaced by Link to Archived Snapshot

It allows writers and editors to capture and fix transient information on the Web with

`<a href="http://perma.cc/0Hg62eLdZ3T">`

a new , permanent link`</a>`

Link Augmented with Attributes from Missing Link Proposal

It allows writers and editors to capture and fix transient information on the Web with

`<a href="http://blogs.law.harvard.edu/futureoftheinternet/2013/09/22/perma/"`
`   versionurl="http://perma.cc/0Hg62eLdZ3T"`
`   versiondate="2013-10-02T03:10:00Z">`

a new , permanent link`</a>`

Link Augmented with Typed Link to an Archived Snapshot

It allows writers and editors to capture and fix transient information on the Web with

`<a href="http://blogs.law.harvard.edu/futureoftheinternet/2013/09/22/perma/"`
`   lset="http://perma.cc/0Hg62eLdZ3T memento 2013-10-02T03:10:00Z">`

a new , permanent link`</a>`

Fig. 6. Existing and proposed approaches to link to archived snapshots of resources

4   Conclusion

The transition of the scholarly communication system to a natively web-based endeavor is ongoing and brings along challenges that did not exist in its paper-based predecessor but are inherited from the very nature of the web. Reference rot is one such challenge. It is a result of the dynamic and ephemeral nature of the web. While reference rot may be acceptable to certain web communities, it is of significant concern from the perspective of the long-term integrity of the web-based scholarly record. The Hiberlink project quantifies reference rot at a vast and unprecedented scale but also explores solutions aimed at ameliorating the problem. This paper, which accompanies an invited conference presentation, focused on the latter and paid special attention to challenges related to linking to resources for which archived snapshots are pro-actively created. The status quo with this regard is unsatisfactory as it is itself subject to reference rot. Efforts are underway aimed at augmenting links with information that can improve web persistence and hence the persistence of the web-based scholarly record.

Acknowledgments. Hiberlink is funded by the Andrew W. Mellon Foundation. The authors would like to acknowledge the Hiberlink partners at the University of Edinburgh: the Language Technology Group at Informatics for their contributions to the Research Track and Edina for their developments under the Solutions Track. The authors would like to acknowledge their colleagues at Harvard University - the perma.cc effort at the Law Library and the Internet Robustness effort at the Berkman Center - for the collaboration on enhancing HTML with robustness information. The authors would also like to thank Robert Sanderson, formerly of the Los Alamos National Laboratory and now at Stanford University, for the Pilot Study that inspired Hiberlink and for contributions to the Missing Link Proposal.

References

1. Bechhofer, S., De Roure, D., Gamble, M., Goble, C., Buchan, I.: Research Objects: Towards Exchange and Reuse of Digital Knowledge. Nature Precedings (2010), http://dx.doi.org/10.1038/npre.2010.4626.1
2. Roosendaal, H., Geurts, P.: Forces and functions in scientific communication: an analysis of their interplay. In: Proceedings of Cooperative Research Information Systems in Physics 1997 (CRISP 1997) (1997), http://www.physik.uni-oldenburg.de/conferences/crisp97/roosendaal.html
3. Klein, M., Shankar, H., Wincewicz, R., Van de Sompel, H.: HiberActive: Pro-Active Archiving of Web References from Scholarly Articles. Open Repositories (to appear, 2014)
4. Van de Sompel, H., Nelson, M.L., Sanderson, R.: RFC7089: HTTP Framework for Time-Based Access to Resource States – Memento (2013), http://tools.ietf.org/html/rfc7089
5. Sanderson, R., Phillips, M., Van de Sompel, H.: Analyzing the Persistence of Referenced Web Resources with Memento (2011), http://arxiv.org/abs/1105.3459
6. Nottingham, M.: RFC5988: Web Linking (2010), http://tools.ietf.org/html/rfc5988
7. Zittrain, J., Albert, K., Lessig, L.: Perma: Scoping and Addressing the Problem of Link and Reference Rot in Legal Citations. Harvard Law Review (2014), http://harvardlawreview.org/2014/03/perma-scoping-and-addressing-the-problem-of-link-and-reference-rot-in-legal-citations/

Computable Data, Mathematics, and Digital Libraries in *Mathematica* and Wolfram|Alpha

Eric Weisstein

Wolfram|Alpha
Champaign, IL, 61820, USA
`eww@wolfram.com`

Abstract. This talk will focus on the infrastructure developed for representing and accessing data (especially mathematical data) in Wolfram|Alpha, as well as on the technologies and language extensions developed in the most recent version of *Mathematica* for making this data even more computationally accessible. Based on experiences using these technologies to create a prototype semantic digital library for a subset of mathematics, we believe the ambitious dream of creating of a semantic digital library for all of mathematics is now within reach.

1 Introduction

Wolfram|Alpha (`http://www.wolframalpha.com`) is a freely available website that contains hand-curated data sets taken from hundreds of technological, scientific, sociological, and other domains, including a core set of mathematical ones. This data has hitherto been accessible either directly via the website, through its API, or through a number of other specialized sources (such as various apps and SIRI). More recently, a large portion of this information has been exposed through the Wolfram Language itself via a set of built-in functions centered around an entity-property approach to information representation. The technology developed for Wolfram|Alpha has also recently been used and extended with the help of funding from the Sloan and Wolfram Foundations to create a prototype digital mathematics library covering known results and identities in the specific area of continued fractions. A recent US National Research Council report has identified approaches it believes could enable the creation of a substantial digital mathematics library, and we are currently investigating partnerships and technologies that could help turn this ambitious dream into a reality.

2 Computable Data in Wolfram|Alpha

Wolfram|Alpha was unveiled in 2009. In the ensuing years, it has become known for its ability to perform an extensive variety of computations in mathematics as well as many other fields. It currently answers millions of users queries per day. Whereas the traditional use of mathematical software is to carry out computations and the traditional use of encyclopedias is to give static information about

S.M. Watt et al. (Eds.): CICM 2014, LNAI 8543, pp. 26–29, 2014.

a certain entity or property, the goal of this website is to bring these two modes (purely dynamic and computational versus purely static and informational) together to dynamically generate knowledge about known structures.

Wolfram|Alpha's knowledge comes from a combination of *Mathematica* computations, roughly 1000 curated data sets, and links to a number of real-time data sources. Mathematical domains known to Wolfram|Alpha include graphs, groups, polyhedra, knots, curves, surfaces, and others. For querying computational knowledge, Wolfram|Alpha implements natural language encoding and processing. Finally, while results are by default displayed into a web browser, they are also available in a number of different formats including text, MathML, LaTeX, XML, images, together with *Mathematica* and its several data formats.

3   Computable Data in *Mathematica*

Starting in Version 6 (released in 2007), *Mathematica* itself has included a set of approximately 20 curated data collections covering mathematical, scientific, geographic, and a number of other domains. Five years after the release of Wolfram|Alpha and seven years after the released of *Mathematica* 6, the original concept of *Mathematica* data collections and the extensive additional functionality, coverage, and development work done for Wolfram|Alpha have been reunited in the recently released *Mathematica* 10.

There are several components to this integration, the first being the extensive augmentation of the set of available data collections. However, rather than bundling all the additional data into *Mathematica* itself, the integration has been accomplished using the Wolfram|Alpha API to expose a selected set of its data sets to *Mathematica* over an internet connection. One benefit of this approach is that data is updated, extended, and improved on the Wolfram|Alpha site much more frequently (usually weekly) than *Mathematica* itself is released.

An even greater step forward is the introduction of entity, entity class, property, and related built-in symbols as a means to represent and manipulate computable data in *Mathematica*. Each curated object in an available data set is assigned a domain (say "PlaneCurve") and a canonical name (say "Ellipse"). Using this framework, objects can be easily referenced and acted open using functions such as EntityValue[Entity["PlaneCurve", "Ellipse"], EntityProperty["PlaneCurve", "Area"], EntityValue[Entity ["PlaneCurve", "Ellipse"], "Classes"], and so on. Similarly, a command like EntityList[EntityClass["PlaneCurve", "Conic"]] can be used to list entities in the plane curve domain belonging to entity class conic.

There are also a number of convenient ways to construct or discover canonical entity, entity class, and property names from within *Mathematica*. The first is a revamped implementation of *Mathematica's* "free-form input" functionality. To wit, by preceding an input with a special character or keystroke (= for a simple Wolfram|Alpha result, == for a full result including all output pods, and CTRL-= for an in-line result), it becomes a natural language query to a Wolfram|Alpha server whose result is returned directly into the current notebook. For example, simply typing `CTRL-= ellipse` into a *Mathematica* front-end returns the expression for the ellipse plane curve entity, while typing `CTRL-= ellipse area` gives the corresponding entity-property expression. As a trivial example, in the latter case, the resulting expression can be directly evaluated to give the expected formula `Function[{a, b}, π a b]`.

Not only does free-form input provide a simple interface for users to access data, it also provides a disambiguation mechanism in the event that multiple interpretations are available. For example, `CTRL-= mercury` defaults to a chemical element but presents the user with a set of assumptions for the plant, periodical, word, city, and given name. In a more computational setting, a programmatic approach is available using either `SemanticInterpretation` (which returns a single best semantic interpretation of the specified free-form string as a Wolfram Language expression) or `Interpreter` (which tries to interpret the natural language input as an object of the specified form).

The resulting synthesis of data representation, exposure, and access provides a powerful, flexible, and extensible framework which is practically applicable to virtually any domain of interest.

4   Prototype Semantic Digital Math Library: The eCF Project

Given the existence of the Wolfram|Alpha framework, it is natural to ask how difficult it would be to create from scratch a semantic digital library covering some specific domain of interest.

Precisely this question was addressed in the recently completed eCF ("e-Continued Fraction") project, undertaken from March 2012 to September 2013. The project resulted in the collection, semantic encoding, and exposure on the internet of significant results from the mathematical corpus concerning continued fractions. This work was supported by the Sloan Foundation with the goal of creating a new type of free digital archive for mathematical data that both ensures preservation and promotes dissemination of a targeted segment of mathematical knowledge for the public good.

Continued fractions presented an ideal subject for this proof-of-concept as they constitute a subset of mathematics that is historically rich, well-defined, and nontrivial, yet at the same time manageable in scope. Work completed includes a nearly exhaustive collection of continued fraction identities, a normalized representative bibliographic database of relevant books and articles, and an extensive collection of hand-curated theorems and results. All of these entities can be queried using a natural language syntax and provide additional linking and cross-entity entraining. In addition, many offer both visualizations and traditionally typeset versions, thus combining familiar traditional mathematical markup with modern tools for computational exploration.

This work was implemented using extensions of the framework developed for the Wolfram|Alpha computational knowledge engine and website. As such, it is generalizable to any area of knowledge where information is encodable and computable. It differed from previous efforts by treating individual results (not papers) as entities of interest. Our methodology consisted of the following: 1) mine papers from archives of $\sim$ 800 historical articles, together with results from books and the newer literature, 2) extract theorems and other results, encode them in semantic form, and store them in computer-readable (and if possible, computational) form, 3) tag author, publication, reference, and subject information, 4) link to the original literature, 5) present in a coherent and unified form, 6) verify by human and computer, and 7) encode and access all data using extensions of the framework developed for Wolfram|Alpha.

At the completion of this work last year, approximately 400 theorems, conjectures, and other results were encoded and exposed. Results also include the first ever comprehensive table of continued fraction identities, containing $\sim$ 1,300 core and $\sim$ 11,000 derived continued fractions. All results are searchable using a natural language interface and are easily and freely accessible via Wolfram|Alpha.

5 Future Work

Our experience both with eCF and in other domains for which we have previously curated computable data in Wolfram|Alpha suggests it is feasible to develop tools and processes that allow a significant portion of mathematical knowledge to be mined, encoded, and exposed semi-automatically via crowdsourcing.

The Future World Heritage Digital Mathematics Library symposium took place in Washington, DC on June 1–3, 2012. After nearly two years of consideration, the National Research Council has now published their final report, which is available the arXiv e-print service (http://arxiv.org/abs/1404.1905).

While the NRC report is very detailed, touches on many aspects of relevance to the realization of a WMDHL, and in particular identifies approaches it believes could enable the creation of a substantial digital mathematics library, concrete steps that could be undertaken in the near-term to turn this dream into a reality remain elusive. A a result, the Sloan Foundation and Wolfram Foundation are currently investigating partnerships, technologies, and constituent components that could help turn the ambitious dream of creating a successful, comprehensive, and authoritative digital library for mathematics into a reality.

Acknowledgments. I thank the CICM organizers for the opportunity to share this work. I also express appreciation to Daniel Goroff, the Alfred P. Sloan Foundation, Stephen Wolfram, and the Wolfram Foundation for their support of the eCF project. I thank my eCF co-investigators Michael Trott, Oleg Marichev, Todd Rowland, and intern Christopher Stover. Finally, I thank Michael Trott and André Kuzniarek for helping spearhead the nascent effort to make the giant leap from "continued fractions" to "all of mathematics."

Towards the Formal Reliability Analysis
of Oil and Gas Pipelines

Waqar Ahmed[1], Osman Hasan[1],
Sofiène Tahar[2], and Mohammad Salah Hamdi[3]

[1] School of Electrical Engineering and Computer Science (SEECS)
National University of Sciences and Technology (NUST)
Islamabad, Pakistan
{12phdwahmad,osman.hasan}@seecs.nust.edu.pk
[2] Electrical and Computer Engineering Department
Concordia University, Montreal, Canada
tahar@ece.concordia.ca
[3] Information Systems Department
Ahmed Bin Mohammed Military College, Doha, Qatar
mshamdi@abmmc.edu.qa

Abstract. It is customary to assess the reliability of underground oil
and gas pipelines in the presence of excessive loading and corrosion effects to ensure a leak-free transport of hazardous materials. The main
idea behind this reliability analysis is to model the given pipeline system as a Reliability Block Diagram (RBD) of segments such that the
reliability of an individual pipeline segment can be represented by a random variable. Traditionally, computer simulation is used to perform this
reliability analysis but it provides approximate results and requires an
enormous amount of CPU time for attaining reasonable estimates. Due
to its approximate nature, simulation is not very suitable for analyzing
safety-critical systems like oil and gas pipelines, where even minor analysis flaws may result in catastrophic consequences. As an accurate alternative, we propose to use a higher-order-logic theorem prover (HOL) for
the reliability analysis of pipelines. As a first step towards this idea, this
paper provides a higher-order-logic formalization of reliability and the
series RBD using the HOL theorem prover. For illustration, we present
the formal analysis of a simple pipeline that can be modeled as a series
RBD of segments with exponentially distributed failure times.

Keywords: Reliability Block Diagrams, Formal Methods, Theorem
Proving, Oil and Gas pipeline.

1 Introduction

On April 20, 2010, methane gas leakage on the Deepwater Horizon oil rig operated by Transocean, a subcontractor of British Petroleum (BP), caused a big
explosion [1]. This leakage not only killed 11 workers instantly but destroyed
and sank the rig, and caused millions of gallons of oil to pour into the Gulf of

S.M. Watt et al. (Eds.): CICM 2014, LNAI 8543, pp. 30–44, 2014.
© Springer International Publishing Switzerland 2014

Mexico. The gushing well, about a mile under the sea, was finally brought under control after more than three months of frenetic attempts. The spill, which is considered to be the largest accidental marine oil spill in the history of the petroleum industry, caused extensive damage to marine and wildlife habitats as well as the Gulf's fishing and tourism industries and its impact still continues. Just like the BP pipeline, there are tens of thousands of miles long oil and gas pipelines around the world. All of these pipelines are aging and are becoming more and more susceptible to failures, which may lead to disasters like the BP one. Hence, it is very important to do rigorous reliability analysis of oil and gas pipelines to detect and rectify potential problems.

The reliability analysis of a pipeline system involves a three-step process: (i) partitioning the given pipeline into segments and constructing its equivalent reliability block diagram (RBD), (ii) assessing the reliability of the individual segments and (iii) evaluating the reliability of the complete pipeline system based on the RBD and the reliability of its individual segments. The reliability of an individual segment is usually expressed in terms of its failure rate λ and a random variable, like exponential [2] or Weibull random variable [3], which models the failure time. A single oil or gas pipeline can be simply modeled as a series RBD [2]. However, in many cases, these pipeline systems have either reserved components or subsystems and such pipeline systems exhibit a combination of series and parallel RBDs [4].

The reliability analysis of oil and gas pipelines has predominantly been accomplished by first gathering data from in-line inspection tools to detect cracks, corrosion or damage [5, 6]. This information is then manipulated using the paper-and-pencil based analytical analysis and computer simulations to deliver diagnostics and insightful pipeline integrity reports (e.g. [2, 4, 7]). However, due to the complex nature of large pipeline system analysis, paper-and-pencil proof methods are error prone and the exhaustive testing of all possible system behaviors using simulation is almost impossible. Thus, these traditional analysis techniques cannot guarantee accurate results, which is a severe limitation in the case of oil and gas pipelines as an uncaught system bug may endanger human and animal life or lead to a significant financial loss.

The inaccuracy limitations of traditional analysis techniques can be overcome by using formal methods [8], which use computerized mathematical reasoning to precisely model the system's intended behavior and to provide irrefutable proof that a system satisfies its requirements. Both model checking and theorem proving have been successfully used for the precise probabilistic analysis of a broad range of systems (e.g. [9–13]). However, to the best of our knowledge, no formal analysis approach has been used for the reliability analysis of oil and gas pipelines so far. The foremost requirement for conducting the formal reliability analysis of underground oil and gas pipelines is the ability to formalize RBDs recursively and continuous random variables. Model checking is a state-based formal method technique. The inherent limitations of model checking is the state-space explosion problem and the inability to model complex datatypes such as trees, lists and recursive definitions [14]. On the other hand, higher-order logic [15] is a system of deduction with a precise semantics and can be used to formally model any system that can be described mathematically including recursive definitions, random variables, RBDs, and continuous components. Similarly, interactive theorem provers are computer based formal reasoning tools that allow us to verify higher-order-logic properties under user guidance. Higher-order-logic theorem provers can be used to reason about recursive definitions using induction methods [16]. Thus, higher-order-logic theorem proving can be used to conduct the formal analysis of oil and gas pipelines.

A number of higher-order-logic formalizations of probability theory are available in higher-order logic (e.g. [17–19]). Hurd's formalization of probability theory [17] has been utilized to verify sampling algorithms of a number of commonly used discrete [17] and continuous random variables [20] based on their probabilistic and statistical properties [21, 22]. Moreover, this formalization has been used to conduct the reliability analysis of a number of applications, such as memory arrays [23], soft errors [24] and electronic components [25]. However, Hurd's formalization of probability theory only supports having the whole universe as the probability space. This feature limits its scope and thus this probability theory cannot be used to formalize more than a single continuous random variable. Whereas, in the case of reliability analysis of pipelines, multiple continuous random variables are required. The recent formalizations of probability theory by Mhamdi [18] and Hölzl [19] are based on extended real numbers (including $\pm\infty$) and provide the formalization of Lebesgue integral for reasoning about advanced statistical properties. These theories also allow using any arbitrary probability space that is a subset of the universe and thus are more flexible than Hurd's formalization. However, to the best of our knowledge, these foundational theories have not been used to formalize neither reliability and RBDs nor continuous random variables so far.

In this paper, we use Mhamdi's formalization of probability theory [18], which is available in the HOL theorem prover [26], to formalize reliability and the commonly used series RBD, where its individual segments are modeled as random variables. Our formalization includes various formally verified properties of reliability and series RBD that facilitate formal reasoning about the reliability of some simple pipelines using a theorem prover. To analyze more realistic models of pipelines, it is required to formalize other RBDs, such as parallel, series-parallel and parallel-series [27]. In order to illustrate the utilization and effectiveness of the proposed idea, we utilize the above mentioned formalization to analyze a simple pipeline that can be modeled as a series RBD with an exponential failure time for individual segments.

2   Preliminaries

In this section, we give a brief introduction to theorem proving in general and the HOL theorem prover in particular. The intent is to introduce the main ideas behind this technique to facilitate the understanding of the paper for the reliability analysis community. We also summarize Mhamdi's formalization of probability theory [18] in this section.

2.1   Theorem Proving

Theorem proving [28] is a widely used formal verification technique. The system that needs to be analysed is mathematically modelled in an appropriate logic and the properties of interest are verified using computer based formal tools. The use of formal logics as a modelling medium makes theorem proving a very flexible verification technique as it is possible to formally verify any system that can be described mathematically. The core of theorem provers usually consists of some well-known axioms and primitive inference rules. Soundness is assured as every new theorem must be created from these basic or already proved axioms and primitive inference rules.

The verification effort of a theorem in a theorem prover varies from trivial to complex depending on the underlying logic [29]. For instance, first-order logic [30] utilizes the propositional calculus and terms (constants, function names and free variables) and is semi-decidable. A number of sound and complete first-order logic automated reasoners are available that enable completely automated proofs. More expressive logics, such as higher-order logic [15], can be used to model a wider range of problems than first-order logic, but theorem proving for these logics cannot be fully automated and thus involves user interaction to guide the proof tools. For reliability analysis of pipelines, we need to formalize (mathematically model) random variables as functions and their distribution properties are verified by quantifying over random variable functions. Henceforth, first-order logic does not support such formalization and we need to use higher-order logic to formalize the foundations of reliability analysis of pipelines.

2.2   HOL Theorem Prover

HOL is an interactive theorem prover developed at the University of Cambridge, UK, for conducting proofs in higher-order logic. It utilizes the simple type theory of Church [31] along with Hindley-Milner polymorphism [32] to implement higher-order logic. HOL has been successfully used as a verification framework for both software and hardware as well as a platform for the formalization of pure mathematics.

The HOL core consists of only 5 basic axioms and 8 primitive inference rules, which are implemented as ML functions. Soundness is assured as every new theorem must be verified by applying these basic axioms and primitive inference rules or any other previously verified theorems/inference rules.

We utilized the HOL theories of Booleans, lists, sets, positive integers, *real* numbers, measure and probability in our work. In fact, one of the primary motivations of selecting the HOL theorem prover for our work was to benefit from these built-in mathematical theories. Table 1 provides the mathematical interpretations of some frequently used HOL symbols and functions, which are inherited from existing HOL theories, in this paper.

Table 1. HOL Symbols and Functions

HOL Symbol	Standard Symbol	Meaning
$\wedge$	and	Logical and
$\vee$	or	Logical or
$\neg$	not	Logical $negation$
::	$cons$	Adds a new element to a list
++	$append$	Joins two lists together
HD L	$head$	Head element of list L
TL L	$tail$	Tail of list L
EL n L	$element$	n^{th} element of list L
MEM a L	$member$	True if a is a member of list L
λx.t	$\lambda x.t$	Function that maps x to $t(x)$
SUC n	$n+1$	Successor of a num
lim(λn.f(n))	$\lim_{n \to \infty} f(n)$	Limit of a $real$ sequence f

2.3 Probability Theory and Random Variables in HOL

Mathematically, a measure space is defined as a triple (Ω, Σ, μ), where Ω is a set, called the sample space, Σ represents a σ-algebra of subsets of Ω, where the subsets are usually referred to as measurable sets, and μ is a measure with domain Σ. A probability space is a measure space (Ω, Σ, Pr), such that the measure, referred to as the probability and denoted by Pr, of the sample space is 1. In Mhamdi's formalization of probability theory [18], given a probability space p, the functions space and subsets return the corresponding Ω and Σ, respectively. This formalization also includes the formal verification of some of the most widely used probability axioms, which play a pivotal role in formal reasoning about reliability properties.

Mathematically, a random variable is a measurable function between a probability space and a measurable space. A measurable space refers to a pair $(S, \mathcal{A})$, where S denotes a set and $\mathcal{A}$ represents a nonempty collection of sub-sets of S. Now, if S is a set with finite elements, then the corresponding random variable is termed as a discrete random variable and else it is called a continuous one. The probability that a random variable X is less than or equal to some value x, $Pr(X \leq x)$ is called the cumulative distribution function (CDF) and it characterizes the distribution of both discrete and continuous random variables. Mhamdi's formalization of probability theory [18] also includes the formalization of random variables and the formal verification of some of their classical properties using the HOL theorem prover.

3 Reliability

In reliability theory [27], reliability $R(t)$ of a system or component is defined as the probability that it performs its intended function until some time t.

$$R(t) = Pr(X > t) = 1 - Pr(X \leq t) = 1 - F_X(t) \qquad (1)$$

where $F_X(t)$ is the CDF. The random variable X, in the above definition, models the time to failure of the system. Usually, this time to failure is modeled by the exponential random variable with parameter λ that represents the failure rate of the system. Now, the CDF can be modeled in HOL as follows:

Definition 1: *Cumulative Distributive Function*
⊢ ∀ p X x. CDF p X x = distribution p X {y | y ≤ Normal x}

where p represents the probability space, X is the random variable and x represents a *real* number. The function Normal converts a *real* number to its corresponding value in the *extended−real* data-type, i.e, the *real* data-type including the positive and negative infinity. The function distribution accepts a probability space p, a random variable X and a set and returns the probability of X acquiring all the values of the given set in the probability space p. Now, Definition 1 can be used to formalize the reliability definition, given in Equation 1, as follows:

Definition 2: *Reliability*
⊢ ∀ p X x. Reliability p X x = 1 - CDF p X x

We used the above mentioned formal definition of reliability to formal verify some of the classical properties of reliability in HOL. The first property in this regard relates to the fact that the reliability of a good component is 1, i.e., maximum, prior to its operation, i.e., at time 0. This property has been verified in HOL as the following theorem.

Theorem 1: *Maximum Reliability*
⊢ ∀ p X. prob_space p ∧ (events p = POW (p_space p)) ∧
        (∀ y. X y ≠ NegInf ∧ X y ≠ PosInf) ∧
        (∀ z. 0 ≤ z ⇒ (λx. CDF p X x) contl z) ∧
        (∀ x. Normal 0 ≤ X x) ⇒
        (Reliability p X 0 = 1)

The first two assumptions of the above theorem ensure that the variable p represents a valid probability space based on the formalization of Mhamdi's probability theory [18]. The third assumption constraints the random variable to be well-defined, i.e., it cannot acquire negative or positive infinity values. The fourth assumption states that the CDF of the random variable X is a continuous function, which means that X is a continuous random variable. This assumption utilizes the HOL function contl, which accepts a lambda abstraction function and a real value and ensures that the function is continuous at the given value. The last assumption ensures that the random variable X can acquire positive values only since in the case of reliability this random variable always models time, which cannot be negative. The conclusion of the theorem represents our desired property that reliability at *time=0* is *1*.

The proof of the Theorem 1 exploits some basic probability theory axioms and the following property according to which the probability of a continous random variable at a point is zero.

The second main characteristic of the reliability function is its decreasing monotonicity, which is verified as the following theorem in HOL:

Theorem 2: *Reliability is a Monotone Function*
⊢ ∀ p X a b. prob_space p ∧ (events p = POW (p_space p)) ∧
      (∀ y. X y ≠ NegInf ∧ X y ≠ PosInf) ∧
      (∀ x. Normal 0 ≤ X x) ∧ a ≤ b ⇒
      (Reliability p X (b)) ≤ (Reliability p X (a))

The assumptions of this theorem are the same as the ones used for Theorem 1 except the last assumption, which describes the relationship between variables a and b. The above property clearly indicates that the reliability cannot increase with the passage of time.

The formal reasoning about the proof of Theorem 2 involves some basic axioms of probability theory and a property that the CDF is a monotonically increasing function.

Finally, we verified that the reliability tends to 0 as the time approaches infinity. This property is verified under the same assumptions that are used for Theorem 1.

Theorem 3: *Reliability Tends to Zero As Time Approaches Infinity*
⊢ ∀ p X. prob_space p ∧ (events p = POW (p_space p)) ∧
    (∀ y. X y ≠ NegInf ∧ X y ≠ PosInf) ∧ (∀ x. Normal 0 ≤ X x) ⇒
    (lim (λn. Reliability p X (&n)) = 0)

The HOL function lim models the limit of a real sequence. The proof of Theorem 3 primarily uses the fact that the CDF approches to 1 as its argument approaches infinity.

These three theorems completely characterize the behavior of the reliability function on the positive real axis as the argument of the reliability is time and thus cannot be negative. The formal verification of these properties based on our definition ensure its correctness. Moreover, these formally verified properties also facilitate formal reasoning about reliability of systems, as will be demonstrated in Section 5 of this paper. The proof details about these properties can be obtained from our proof script [33].

4 Formalization of Series Reliability Block Diagram

In a serially connected system [27], depicted in Figure 1, the reliability of the complete system mainly depends upon the failure of a single component that has the minimum reliability among all the components of the system. In other words, the system stops functioning if any one of its component fails. Thus, the operation of such a system is termed as reliable at any time t, if all of its components are functioning reliably at this time t. If the event $A_i(t)$ represents the reliable functioning of the ith component of a serially connected system with N components at time t then the overall reliability of the system can be mathematically expressed as [27]:

$$R_{series}(t) = Pr(A_1(t) \cap A_2(t) \cap A_3(t) \cdots \cap A_N(t)) \tag{2}$$

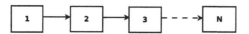

Fig. 1. System with a Series Connection of Components

Using the assumption of mutual independence of individual reliability events of a series system [27], the above equation can be simplified as:

$$R_{series}(t) = \prod_{i=1}^{N} R_i(t) \tag{3}$$

Moreover, an intrinsic property of a series system is that its overall reliability is always less than or equal to the reliability of the sub-component with the least reliability.

$$R_{series}(t) \leq min(R_i(t)) \tag{4}$$

We proceed with the formalization of the series RBD by first formalizing the notion of mutual independence of more than two random variables, which is one of the most essential prerequisites for reasoning about the simplified expressions for RBD. Two events A and B are termed as mutually independent iff $Pr(A \cap B) = Pr(A)Pr(B)$. All the events involved in reliability modeling are generally assumed to be mutually independent. Since we often tackle the reliability assessment of systems with more than two components, we formalize the mutual independence of a list of random variables in this paper as follows:

Definition 3: *Mutual Independence of Events*
⊢ ∀ p L. mutual_indep p L =
  ∀ L1 n. PERM L L1 ∧ 2 ≤ n ∧ n ≤ LENGTH L ⇒
    prob p (inter_set p (TAKE n L1)) =
    list_prod (list_prob p (TAKE n L1))

The function **mutual_indep** takes a list of events or sets L along with the probability space p as input and returns True if the given list of events are mutually independent in p. The formal definitions for the HOL functions used in the above definition are given in Table 1. The predicate **PERM** ensures that its two list arguments form a permutation of one another, the function **LENGTH** returns the length of a list, the function **TAKE** returns a list that contains the first n elements of its argument list, the function **inter_set** performs the intersection of all the sets in a list of sets and returns the probability space in case of an empty list argument, the function `list_prob` returns a list of probabilities associated with the given list of events in the given probability space and the function `list_prod` recursively multiplies all the elements of its argument list of real numbers. Thus, using these functions the function `mutual_indep` ensures that for any 2 or more elements n, taken in any order, of the given list of events L, the property $Pr(\bigcap_{i=0}^{n} L_i) = \prod_{i=0}^{n} Pr(L_i)$ holds.

Table 2. HOL Functions used in Definition 3

Function Name	HOL Definition
PERM	⊢ ∀ L1 L2. PERM L1 L2 = ∀ x. FILTER ($= x) L1 = FILTER ($= x)L2
LENGTH	⊢ (LENGTH [] = 0) ∧ ∀ h t. LENGTH (h::t) = SUC (LENGTH t)
TAKE	⊢ (∀ n. TAKE n [] = []) ∧ ∀ n x xs. TAKE n (x::xs) = if n = 0 then [] else x::TAKE (n - 1) xs
inter_set	⊢ (∀ p. inter_set p [] = p_space p) ∧ ∀ p h t. inter_set p (h::t) = h ∩ inter_set p t
list_prod	⊢ (∀ list_prod [] = 1) ∧ ∀ h t. list_prod (h::t) = h * list_prod t
list_prob	⊢ (∀ p. list_prob p [] = []) ∧ ∀ p h t. list_prob p (h::t) = prob p (h ∩ p_space p) * list_prob p t
min	⊢ ∀ x y. min x y = if x ≤ y then x else y
min_rel	⊢ (∀ f. min_rel f [] = 1) ∧ ∀ f h t. min_rel f (h::t) = min (f h) (min_rel f t)

Next, we propose to formalize the RBDs in this paper by using a list of events, where each event models the proper functioning of a single component at a given time based on the corresponding random variable. This list of events can be modeled as follows:

Definition 4: *Reliability Event List*
⊢ ∀ p x. rel_event_list p [] x = [] ∧
  ∀ p x h t. rel_event_list p (h::t) x =
PREIMAGE h {y | Normal x < y} ∩ p_space p :: rel_event_list p t x

The function `rel_event_list` accepts a list of random variables, representing the time to failure of individual components of the system, and a *real* number x, which represents the time index where the reliability is desired, and returns a list of sets corresponding to the events that the individual components are functioning properly at the given time x. This list of events can be manipulated, based on the structure of the complete system, to formalize various RBDs.

Similarly, the individual reliabilities of a list of random variables can be modeled as the following recursive function:

Definition 5: *Reliability of a List of Random Variables*
⊢ ∀ p x . rel_list p [] x = [] ∧
  ∀ p h t x. rel_list p (h::t) x =
  Reliability p h x :: rel_list p t x

The function rel_list takes a list of random variables and a *real* number x, which represents the time index where the reliability is desired, and returns a list of the corresponding reliabilities at the given time x. It is important to note that all the above mentioned definitions are generic enough to represent the behavior of any RBD, like series, parallel, series-parallel and parallel-series.

Now, using Equation (2), the reliability of a serially connected structure can be defined as:

Definition 6: *System with a Series Connection of Components*
⊢ ∀ p L. rel_series p L = prob p (inter_set p L)

The function rel_series takes a list of random variables L, representing the failure times of the individual components of the system, and a probability space p as input and returns the intersection of all the events corresponding to the reliable functioning of these components using the function inter_set, given in Table 2. Based on this definition, we formally verified the result of Equation (2) as follows:

Theorem 4: *Reliability of a System with Series Connections*
⊢ ∀ p L x. prob_space p ∧ (events p = POW (p_space p)) ∧
    0 ≤ x ∧ 2 ≤ LENGTH (rel_event_list p L x) ∧
    mutual_indep p (rel_event_list p L x) ⇒
  (rel_series p (rel_event_list p L x) = list_prod (rel_list p L x))

The first two assumptions ensure that p is a valid probability space based on Mhamdi's probability theory formalization [18]. The next one ensures that the variable x, which models time, is always greater than or equal to 0. The next two assumptions of the above theorem guarantee that we have a list of at least two mutually exclusive random variables (or a system with two or more components). The conclusion of the theorem represents Equation (2) using Definitions 4 and 6. The proof of Theorem 4 involves various probability theory axioms, the mutual independence of events and the fact that the probability of any event that is in the returned list from the function rel_event_list is equivalent to its reliability. More proof details can be obtained from our proof script [33].

Similarly, we verified Equation (4) as the following theorem in HOL:

Theorem 5: *Reliability of a System depends upon the minimum reliability of the connected components*
⊢ ∀ p L x. prob_space p ∧ (events p = POW (p_space p)) ∧
    0 ≤ x ∧ 2 ≤ LENGTH (rel_event_list p L x) ∧
    mutual_indep p (rel_event_list p L x) ⇒
  (rel_series p (rel_event_list p L x) ≤
    min_rel (λ L. Reliability p L x) L)

The proof of the Theorem 5 uses several probability theory axioms and the fact that any subset of a mutually independent set is also mutually independent.

The definitions, presented in this section, can be used to model parallel RBD [27] and formally verify the corresponding simplified reliability relationships as well. The major difference would be the replacement of the function inter_set in Definition 6 by a function that returns the union of a given list of events.

5   Reliability Analysis of a Pipeline System

A typical oil and gas pipeline can be partitioned into a series connection of N segments, where these segments may be classified based on their individual failure times. For example, a 60 segment pipeline is analyzed in [2] under the assumption that the segments, which exhibit exponentially distributed failure rates, can be sub-divided into 3 categories according to their failure rates (λ), i.e., 30 segments with $\lambda = 0.0025$, 20 segments with $\lambda = 0.0023$ and 10 segments with $\lambda = 0.015$. The proposed approach for reliability analysis of pipelines allows us to formally verify generic expressions involving any number of segments and arbitrary failure rates. In this section, we formally verify the reliability of a simple pipeline, depicted in Figure 2, with N segments having arbitrary exponentially distributed failure times.

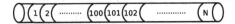

Fig. 2. A Simple Pipeline

We proceed with the formal reliability analysis of the pipeline, shown in Figure 2, by formalizing the exponential random variable in HOL.

Definition 7: *Exponential Distribution Function*
⊢ ∀ p X l. exp_dist p X l =
    ∀ x. (CDF p X x = if 0 ≤ x then 1 - exp (-1 * x) else 0)

The predicate exp_dist ensures that the random variable X exhibits the CDF of an exponential random variable in probability space p with failure rate l. We classify a list of exponentially distributed random variables based on this definition as follows:

Definition 8: *List of Exponential Distribution Functions*
⊢ ∀ p L. list_exp p [] L = T ∧
    ∀ p h t L. list_exp p (h::t) L =
    exp_dist p (HD L) h ∧ list_exp p t (TL L)

The list_exp function accepts a list of failure rates, a list of random variables L and a probability space p. It guarantees that all elements of the list L are exponentially distributed with corresponding failure rates given in the other list within the probability space p. For this purpose, it utilizes the list functions HD and TL, which return the *head* and *tail* of a list, respectively.

Next, we model the pipeline, shown in Figure 2, as a series RBD as follows:

Definition 9: *Reliability of Series Pipeline System*
⊢ ∀ p L . pipeline p L = rel_series p L

Now, we can use Definition 8 to guarantee that the random variable list argument of the function `pipeline` contains exponential random variables only and thus verify the following simplified expression for the pipeline reliability.

Theorem 6: *Series Pipeline System*
⊢ ∀ p L x C. prob_space p ∧ (events p = POW (p_space p)) ∧
   0 ≤ x ∧ 2 ≤ LENGTH (rel_event_list p L x) ∧
   mutual_indep p (rel_event_list p L x) ∧
    list_exp p C L ∧ (LENGTH C = LENGTH L) ⇒
    (pipeline p (rel_event_list p L x) = exp (-list_sum C * x))

The first five assumptions are the same as the ones used in Theorem 5. The sixth assumption list_exp p C L ensures that the list of random variable L contains all exponential random variables with corresponding failure rates given in list C. The next assumptions guarantees that the lengths of the two lists L and C are the same. While the conclusion of Theorem 6 represents desired reliability relationship for the given pipeline model. Here the function list_sum recursively adds the elements of its list argument and is used to add the failure rates of all exponentially distributed random variables, which are in turn used to model the individual segments of the series RBD of the pipeline. The proof of Theorem 6 is based on Theorem 4 and some properties of the exponential function exp. The reasoning was very straightforward (about 100 lines of HOL code) compared to the reasoning for the verification of Theorem 4 [33], which involved probability-theoretic guidance. This fact illustrates the usefulness of our core formalization for conducting the reliability analysis of pipelines.

The distinguishing features of this formally verified result include its generic nature, i.e., all the variables are universally quantified and thus can be specialized to obtain the reliability of the given pipeline for any given parameters, and its guaranteed correctness due to the involvement of a sound theorem prover in its verification, which ensures that all the required assumptions for the validity of the result are accompanying the theorem. Another point worth mentioning is that the individual failure rates of the pipeline segments can be easily provided to the above theorem in the form of a list, i.e., C. The above mentioned benefits are not shared by any other computer based reliability analysis approach for oil and gas pipelines and thus clearly indicate the usefulness of the proposed approach.

6 Conclusions

Probabilistic analysis techniques have been widely utilized during the last two decades to assess the reliability of oil and gas pipelines. However, all of these probability theoretic approaches have been utilized using informal system analysis methods, like simulation or paper-and-pencil based analytical methods, and thus do not ensure accurate results. The precision of results is very important in the area of oil and gas pipeline condition assessment since even minor flaws in the analysis could result in the loss of human lives or heavy damages to the environment. In order to achieve this goal and overcome the inaccuracy limitation of the traditional probabilistic analysis techniques, we propose to build upon our proposed formalization of RBDs to formally reason about the reliability of oil and gas pipelines using higher-order-logic theorem proving.

Building upon the results presented in this paper, the formalization of other commonly used RBDs, including parallel, series-parallel and parallel-series, and the Weibull random variable is underway. These advanced concepts are widely used in the reliability analysis of pipelines. However, their formalization requires some advanced properties of probability theory. For example, for formalizing the reliability block diagrams of the series-parallel and parallel-series structures, we need to first formally verify the principle of inclusion exclusion [34]. We also plan to formalize the underlying theories to reason about more realistic series pipeline systems, such as multi-state variable piping systems, where each subcomponent of the pipeline system consists of many irreversible states from good to worst. We also plan to investigate artificial neural networks in conjunction with theorem proving to develop a hybrid semi-automatic pipeline reliability analysis framework. Besides the pipeline reliability analysis, the formalized reliability theory foundation presented in this paper, may be used for the reliability analysis of a number of other applications, including hardware and software systems.

Acknowledgments. This publication was made possible by NPRP grant # [5 - 813 - 1 134] from the Qatar National Research Fund (a member of Qatar Foundation). The statements made herein are solely the responsibility of the author[s].

References

1. BP Leak the World's Worst Accidental Oil Spill, London Telegraph (August 03, 2010), http://www.telegraph.co.uk/finance/newsbysector/energy/oilandgas /7924009/bp-leak-the-worlds-worst-accidental-oil-spill.html (2014)
2. Zhang, Z., Shao, B.: Reliability Evaluation of Different Pipe Section in Different Period. In: Service Operations and Logistics, and Informatics, pp. 1779–1782. IEEE (2008)
3. Kolowrocki, K.: Reliability and Risk Analysis of Multi-State Systems With Degrading Components. Electronic Journal of International Group on Reliability 2(1), 86–104 (2009)

4. Soszynska, J.: Reliability and Risk Evaluation of a Port Oil Pipeline Transportation System in Variable Operation conditions. International Journal of Pressure Vessels and Piping 87(2-3), 81–87 (2010)
5. Pipeline Integrity Solution GE-Energy (2014), http://www.ge-energy.com/products_and_services/services/pipeline_integrity_services/
6. Pipecheck - Pipeline Integrity Assessment Software (2014), http://www.creaform3d.com/en/ndt-solutions/pipecheck-damage-assessment-software
7. Pandey, D.: Probabilistic Models for Condition Assessment of Oil and Gas Pipelines. Independent Nondestructive Testing and Evaluation International 31(3), 349–358 (1998)
8. Boca, P., Bowen, J., Siddiqi, J.: Formal Methods: State of the Art and New Directions. Springer (2009)
9. Hasan, O., Tahar, S.: Performance Analysis of ARQ Protocols using a Theorem Prover. In: International Symposium on Performance Analysis of Systems and Software, pp. 85–94. IEEE Computer Society (2008)
10. Kwiatkowska, M., Norman, G., Parker, D.: Probabilistic Model Checking for Systems Biology. In: Symbolic Systems Biology, pp. 31–59. Jones and Bartlett (2010)
11. Elleuch, M., Hasan, O., Tahar, S., Abid, M.: Formal Analysis of a Scheduling Algorithm for Wireless Sensor Networks. In: Qin, S., Qiu, Z. (eds.) ICFEM 2011. LNCS, vol. 6991, pp. 388–403. Springer, Heidelberg (2011)
12. Hasan, O., Patel, J., Tahar, S.: Formal Reliability Analysis of Combinational Circuits using Theorem Proving. J. Applied Logic 9(1), 41–60 (2011)
13. Fruth, M.: Formal Methods for the Analysis of Wireless Network Protocols. PhD thesis, Oxford University, UK (2011)
14. Kaufman, M.: Some Key Research Problems in Automated Theorem Proving for Hardware and Software Verification. Revista de la Real Academia de Ciencias Exactas, Físicas y Naturales. Serie A: Matemáticas 98(1), 181 (2004)
15. Brown, C.: Automated Reasoning in Higher-order Logic. College Publications (2007)
16. Kapur, D., Subramaniam, M.: Lemma Discovery in Automating Induction. In: McRobbie, M.A., Slaney, J.K. (eds.) CADE 1996. LNCS, vol. 1104, pp. 538–552. Springer, Heidelberg (1996)
17. Hurd, J.: Formal Verification of Probabilistic Algorithms. PhD Thesis, University of Cambridge, UK (2002)
18. Mhamdi, T., Hasan, O., Tahar, S.: On the Formalization of the Lebesgue Integration Theory in HOL. In: Kaufmann, M., Paulson, L.C. (eds.) ITP 2010. LNCS, vol. 6172, pp. 387–402. Springer, Heidelberg (2010)
19. Hölzl, J., Heller, A.: Three Chapters of Measure Theory in Isabelle/HOL. In: van Eekelen, M., Geuvers, H., Schmaltz, J., Wiedijk, F. (eds.) ITP 2011. LNCS, vol. 6898, pp. 135–151. Springer, Heidelberg (2011)
20. Hasan, O., Tahar, S.: Formalization of Continuous Probability Distributions. In: Pfenning, F. (ed.) CADE 2007. LNCS (LNAI), vol. 4603, pp. 3–18. Springer, Heidelberg (2007)
21. Hasan, O., Tahar, S.: Verification of Tail Distribution Bounds in a Theorem Prover. In: Numerical Analysis and Applied Mathematics, vol. 936, pp. 259–262. American Institute of Physics (2007)
22. Hasan, O., Abbasi, N., Akbarpour, B., Tahar, S., Akbarpour, R.: Formal Reasoning about Expectation Properties for Continuous Random Variables. In: Cavalcanti, A., Dams, D.R. (eds.) FM 2009. LNCS, vol. 5850, pp. 435–450. Springer, Heidelberg (2009)

23. Hasan, O., Tahar, S., Abbasi, N.: Formal Reliability Analysis using Theorem Proving. IEEE Transactions on Computers 59(5), 579–592 (2010)
24. Abbasi, N., Hasan, O., Tahar, S.: Formal Analysis of Soft Errors using Theorem Proving. In: Symbolic Computation in Software Science. EPTCS, vol. 122, pp. 75–84 (2013)
25. Abbasi, N., Hasan, O., Tahar, S.: An Approach for Lifetime Reliability Analysis using Theorem Proving. Journal of Computer and System Sciences 80(2), 323–345 (2014)
26. Slind, K., Norrish, M.: A Brief Overview of HOL4. In: Mohamed, O.A., Muñoz, C., Tahar, S. (eds.) TPHOLs 2008. LNCS, vol. 5170, pp. 28–32. Springer, Heidelberg (2008)
27. Bilintion, R., Allan, R.: Reliability Evaluation of Engineering System. Springer (1992)
28. Gordon, M.: Mechanizing Programming Logics in Higher-Order Logic. In: Current Trends in Hardware Verification and Automated Theorem Proving, pp. 387–439. Springer (1989)
29. Harrison, J.: Formalized Mathematics. Technical Report 36, Turku Centre for Computer Science (1996)
30. Fitting, M.: First-Order Logic and Automated Theorem Proving. Springer (1996)
31. Church, A.: A Formulation of the Simple Theory of Types. Journal of Symbolic Logic 5, 56–68 (1940)
32. Milner, R.: A Theory of Type Polymorphism in Programming. Journal of Computer and System Sciences 17, 348–375 (1977)
33. Ahmad, W.: Formalization of Reliability Block Diagram for Analyzing Oil and Gas Pipelines (2014), http://save.seecs.nust.edu.pk/wahmad/frsaogp.html
34. Trivedi, K.S.: Probability and Statistics with Reliability, Queuing and Computer Science Applications, 2nd edn. John Wiley and Sons Ltd., Chichester (2002)

Problem Formulation for Truth-Table Invariant Cylindrical Algebraic Decomposition by Incremental Triangular Decomposition

Matthew England[1], Russell Bradford[1], Changbo Chen[2], James H. Davenport[1], Marc Moreno Maza[3], and David Wilson[1]

[1] University of Bath, Bath, BA2 7AY, U.K.
[2] Chongqing Key Laboratory of Automated Reasoning and Cognition, Chongqing Institute of Green and Intelligent Technology, CAS, Chongqing, 400714, China
[3] University of Western Ontario, London, N6A 5B7, Canada
{R.J.Bradford,J.H.Davenport,M.England,D.J.Wilson}@bath.ac.uk,
changbo.chen@hotmail.com, moreno@csd.uwo.ca

Abstract. Cylindrical algebraic decompositions (CADs) are a key tool for solving problems in real algebraic geometry and beyond. We recently presented a new CAD algorithm combining two advances: truth-table invariance, making the CAD invariant with respect to the truth of logical formulae rather than the signs of polynomials; and CAD construction by regular chains technology, where first a complex decomposition is constructed by refining a tree incrementally by constraint. We here consider how best to formulate problems for input to this algorithm. We focus on a choice (not relevant for other CAD algorithms) about the order in which constraints are presented. We develop new heuristics to help make this choice and thus allow the best use of the algorithm in practice. We also consider other choices of problem formulation for CAD, as discussed in CICM 2013, revisiting these in the context of the new algorithm.

Keywords: cylindrical algebraic decomposition, truth table invariance, regular chains, triangular decomposition, problem formulation.

1 Introduction

A *cylindrical algebraic decomposition* (CAD) is: a *decomposition* of $\mathbb{R}^n$, meaning a collection of cells which do not intersect and whose union is $\mathbb{R}^n$; *cylindrical*, meaning the projections of any pair of cells with respect to a given variable ordering are either equal or disjoint; and, *(semi)-algebraic*, meaning each cell can be described using a finite sequence of polynomial relations.

CAD was introduced by Collins in [11], such that a given set of polynomials had constant sign on each cell. This meant that a single sample point for each cell was sufficient to conclude behaviour on the whole cell and thus it offered a constructible solution to the problem of quantifier elimination. Since then a range of other applications have been found for CAD including robot motion planning

S.M. Watt et al. (Eds.): CICM 2014, LNAI 8543, pp. 45–60, 2014.

[23], epidemic modelling [8], parametric optimisation [18], theorem proving [22] and reasoning with multi-valued functions and their branch cuts [14].

In [3] the present authors presented a new CAD algorithm combining two recent advances in CAD theory: construction by first building a cylindrical decomposition of complex space, incrementally refining a tree by constraint [9]; and the idea of producing CADs such that given formulae has invariant truth on each cell [4]. Experimental results in [3] showed this new algorithm to be superior to its individual components and competitive with the state of the art. We now investigate the choices that need to be made when using the new algorithm.

We conclude the introduction with the necessary background theory and then in Section 2 we demonstrate how constraint ordering affects the behaviour of the algorithm. No existing heuristics discriminate between these orderings and so we develop new ones, which we evaluate in Section 3. In Section 4 we consider other issues of problem formulation, revisiting [6] in the context of the new algorithm.

1.1 Background on CAD

The first CAD algorithm, introduced by Collins [11] with a full description in [1], works in two phases. First in the *projection* phase a projection operator is repeatedly applied to the set of polynomials (starting with those in the input), each time producing another set in one fewer variables. Then in the *lifting* phase CADs are built incrementally by dimension. First $\mathbb{R}^1$ is decomposed according to the real roots of the univariate polynomials. Then $\mathbb{R}^2$ is decomposed by repeating the process over each cell in $\mathbb{R}^1$ using the bivariate polynomials evaluated at a sample point, and so on. Collins' original projection operator was chosen so that the CADs produced could be concluded *sign-invariant* with respect to the input polynomials, meaning the sign of each polynomial on each cell is constant.

Such decompositions can contain far more information than required for most applications, which motivated CAD algorithms which consider not just polynomials but their origin. For example, when using CAD for quantifier elimination partial CAD [13] will avoid lifting over a cell if the solution there is already apparent. Another key adaptation is to make use of an *equational constraint* (EC): an equation logically implied by an input formula. The algorithm in [21] ensures sign-invariance for the polynomial defining an EC, any any other polynomials only when that constraint is satisfied. A discussion of the first 20 years of CAD research is given in [12]. Some of the subsequent developments are discussed next, with others including the use of certified numerics when lifting [19, 24].

1.2 TTICAD by Regular Chains

In [3] we presented a new CAD algorithm, referred to from now on as RC-TTICAD. It combined the following two recent advances.

Truth-table invariant CAD: A TTICAD is a CAD produced relative to a list of formulae such that each has constant truth value on every cell.

The first TTICAD algorithm was given in [4], where a new projection operator was introduced which acted on a set of formulae, each with an EC.

TTICADs are useful for applications involving multiple formulae like branch cut analysis (see for example Section 4 of [16]), but also for building truth-invariant CADs for a single formula if it can be broken into sub-formulae with ECs. The algorithm was extended in [5] so that not all formulae needed ECs, with savings still achieved if at least one did. These algorithms were implemented in the freely available MAPLE package PROJECTIONCAD [17].

CAD by regular chains technology: A CAD may be built by first forming a *complex cylindrical decomposition* (CCD) of $\mathbb{C}^n$ using triangular decomposition by regular chains, which is refined to a CAD of $\mathbb{R}^n$.

This idea to break from projection and lifting was first proposed in [10]. In [9] the approach was improved by building the CCD incrementally by constraint, allowing for competition with the best projection and lifting implementations. Both algorithms are implemented in the MAPLE REGULARCHAINS Library, with the algorithm from [10] currently the default CAD distributed with MAPLE.

RC-TTICAD combined these advances by adapting the regular chains computational approach to produce truth-table invariant CCDs and hence CADs. This new algorithm is specified in [3] where experimental results showed a MAPLE implementation in the REGULARCHAINS Library as superior to the two advances independently, and competitive with the state of the art. The CCD is built using a tree structure which is incrementally refined by constraint. ECs are dealt with first, with branches refined for other constraints in a formula only is the ECs are satisfied. Further, when there are multiple ECs in a formula branches can be removed when the constraints are not both satisfied. See [3, 9] for full details.

The incremental building of the CCD offers an important choice on problem formulation: in what order to present the constraints? Throughout we use $A \rightarrow B$ to mean that A is processed before B, where A and B are polynomials or constraints defined by them. Existing CAD algorithms and heuristics do not discriminate between constraint orderings [6, 15] and so a new heuristic is required to help make an intelligent choice.

2   Constraint Ordering

The theory behind RC-TTICAD allows for the constraints to be processed in any order. However, the algorithm as specified in [3] states that **equational constraints should be processed first**. This is logical as we need only consider the behaviour of non-ECs when corresponding ECs are satisfied, allowing for savings in computation.

We also advise **processing all equational constraints from a formula in turn**, i.e. not processing one, then moving to a different formula before returning to another in the first. Although not formally part of the algorithm specification, this should avoid unnecessary computation by identifying when ECs have a mutual solution before more branches have been created.

There remain two questions to answer with regards to constraint ordering:
Q1) In what order to process the formulae?
Q2) In what order to process the equational constraints within each formula?

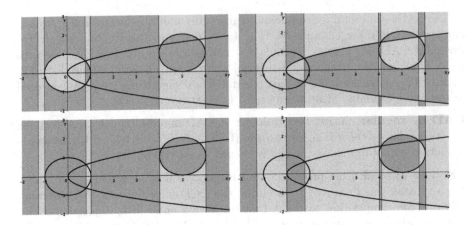

Fig. 1. Visualisations of the four TTICADs which can be built using `RC-TTICAD` for Example 1. The figures on the top have $\phi_1 \rightarrow \phi_2$ and those on the bottom $\phi_2 \rightarrow \phi_1$. The figures on the left have $f_1 \rightarrow f_2$ and those on the right $f_2 \rightarrow f_1$.

2.1 Illustrative Example

The following example illustrates why these questions matter.

Example 1. We assume the ordering $x \prec y$ and consider

$$f_1 := x^2 + y^2 - 1, \qquad f_2 := 2y^2 - x, \qquad f_3 := (x-5)^2 + (y-1)^2 - 1,$$
$$\phi_1 := f_1 = 0 \wedge f_2 = 0, \qquad \phi_2 := f_3 = 0.$$

The polynomials are graphed within the plots of Figure 1 (the circle on the left is f_1, the one on the right f_3 and the parabola f_2). If we want to study the truth of ϕ_1 and ϕ_2 (or a parent formula $\phi_1 \vee \phi_2$) we need a TTICAD to take advantage of the ECs. There are two possible answers to each of the questions above and so four possible inputs to `RC-TTICAD`. The corresponding outputs are[1]:

$\phi_1 \rightarrow \phi_2$ and $f_1 \rightarrow f_2$: 37 cells in 0.095 seconds.

$\phi_1 \rightarrow \phi_2$ and $f_2 \rightarrow f_1$: 81 cells in 0.118 seconds.

$\phi_2 \rightarrow \phi_1$ and $f_1 \rightarrow f_2$: 25 cells in 0.087 seconds.

$\phi_2 \rightarrow \phi_1$ and $f_2 \rightarrow f_1$: 43 cells in 0.089 seconds.

The plots in Figure 1 show the two-dimensional cells in each of these TTICADs.

First compare the induced CADs of $\mathbb{R}^1$ (how the real line is dissected). Observe the following similarities in all four images:

- The points $\frac{1}{4}(-1 \mp \sqrt{17})$ (approximately -1.28 and 0.78) are always identified.

The latter is at the intersection of f_1 and f_2 and so is essential for the output to be correct as the truth of ϕ_1 changes here. The former is the other root of the resultant of f_1 and f_2 and so marks an intersection with complex y-value.

---

[1] All timings in this paper were obtained on a Linux desktop (3.1GHz Intel processor, 8.0Gb total memory) using MAPLE 18.

– The points 4 and 6 are always identified. These mark the endpoints of f_3, required for cylindricity and obtained as roots of the discriminant of f_3.

Now we observe the differences in the induced CADs of the real line:

– If $f_1 \to f_2$ we identify ± 1, marking the ends of the circle f_1. Similarly if $f_2 \to f_1$ then we identify 0, marking the end of the parabola f_2.

These are identified as the roots of a discriminant and are included to ensure cylindricity. If f_1 is processed first then $f_2 = 0$ is only considered when $f_1 = 0$. Since their intersection is only a single value of x the discriminant of f_2 is not required (or more accurately it is trivial). Similarly, if we process f_2 first then the discriminant of f_1 is only calculated for two values of x, where it is constant.

– If $f_2 \to f_1$ then we identify the two real roots of the resultant of f_2 and f_3 (approximately 4.10 and 5.72) marking the real intersection of those curves.

If we process f_2 first and then f_1 the algorithm ensures their intersections are identified to maintain truth-table invariance. For this example it is not necessary since when we process f_1 we find there are no intersections of the set where ϕ_1 is true with f_3, but this was not known beforehand. If instead $f_1 \to f_2$ since there is no intersection the extra cells are avoided. However, note that the resultant of f_1 and f_3 is still calculated and the complex tree is split accordingly. This may explain why the timings for the orderings with $\phi_2 \to \phi_1$ are so similar.

Finally compare the CADs of $\mathbb{R}^2$. We see that in all four TTICADs the output is sign invariant for f_3, while if $\phi_1 \to \phi_2$ then the output is also sign invariant for whichever of f_1 and f_2 was processed first. The first constraint to be processed will always be sign-invariant for the output. The tree is initially refined into cases where its polynomial is zero or not and although these branches are split further that invariance is maintained. Similarly, the first constraint from a formula to be processed will usually be sign-invariant in the output, but this may be avoided if a formula has more than one EC. In this case the tree may be refined to the cases where either both are satisfied or not (as with ϕ_1 in this example).

2.2   Developing a Heuristic for Equational Constraint Ordering

The following propositions are illustrated by Example 1 and can be verified from the algorithm specification in [3].

Proposition 1. *The output of RC-TTICAD is always sign-invariant with respect to the discriminant of the first EC in each formula.*

Other discriminants will be calculated, but their impact is lesser. E.g. the discriminant of the second EC in a formula will be considered modulo the first.

Proposition 2. *The output of RC-TTICAD is always sign-invariant with respect to the cross-resultants of the set of first ECs in each formula.*

Other resultants will be calculated. Some of them have lesser impact, such as the resultant of the first EC in formula A with the second in formula B which is considered modulo the first EC in formula B. Other will be considered for all constraint orderings, such as the resultant of a pair of ECs in a formula.

Considering (1) and (2) leads us to suggest minimising the following sets under some measure when making a choice about constraint ordering.

Definition 1. *For a given constraint ordering o let P be the set of ECs which are ordered first in each formula. Then define the* **constraint ordering set C_o** *as the discriminants and cross resultants in the main variable of the ordering:*

$$C_o := \left(\bigcup_{p \in P} \{\mathrm{disc}(p)\} \right) \cup \left(\bigcup_{p,q \in P, p \neq q} \{\mathrm{res}(p,q)\} \right).$$

For Example 1 the constraint ordering sets are

$$\begin{aligned}
C_{f_1 \to f_2} &= \{\mathrm{disc}_y(f_1), \mathrm{disc}_y(f_3), \mathrm{res}_y(f_1, f_3)\} \\
&= \{-4x^2 + 4, -4x^2 + 40x - 96, 104x^2 - 520x + 672\}, \\
C_{f_2 \to f_1} &= \{\mathrm{disc}_y(f_2), \mathrm{disc}_y(f_3), \mathrm{res}_y(h, f_2)\} \\
&= \{8x, -4x^2 + 40x - 96, 4x^4 - 76x^3 + 561x^2 - 1908x + 2500\}.
\end{aligned}$$

A natural way to measure these would be to compare sotds, the *sum of total degrees of each monomial in each polynomial*, since this was shown to help with other CAD choices [6,15]. For Example 1 the sets above have sotd 8 and 14 respectively and thus the ordering $f_1 \to f_2$ is suggested. Regardless of which formula is processed first, this is the better choice. However, the following example demonstrates that sotd may not be so suitable in general.

Example 2. **[2014 x-axis ellipse problem]** A well studied test problem for CAD is the *x-axis ellipse problem* defined in [2] and specialising a problem in [20]. They concern an ellipse and seek to determine for what values of its parameters (the principal semi-axes and centre) it lies within the unit circle.

We propose a new problem, inspired by the original but requiring multiple formulae and hence a TTICAD. Suppose the ellipse is centred at $(c, 0)$ with principal semi-axes $a \in (0, 2)$ and $b = 1$. The problem is to determine for what values of (c, a) the ellipse intersects either of a pair of unit circles, centred on the x-axis at ± 2. Define the polynomials

$$f_1 := (x - 2)^2 + y^2 - 1, \quad f_2 := (x + 2)^2 + y^2 - 1, \quad h := (x - c)^2 + a^2 y^2 - a^2.$$

Then we seek to eliminate the quantifiers from $\Phi := (\exists y)(\exists x)\phi_1 \vee \phi_2$ where

$$\phi_1 := (f_1 = 0 \wedge h = 0 \wedge a > 0 \wedge a < 2), \quad \phi_2 := (f_2 = 0 \wedge h = 0 \wedge a > 0 \wedge a < 2).$$

The problem can be solved using a TTICAD for ϕ_1 and ϕ_2. We assume variable ordering $y \succ x \succ a \succ c$. There are eight possible constraint orderings for RC-TTICAD as listed in Table 1. The best choice is to process h first in each formula (then the formula ordering makes no difference) which is logical since h has

Table 1. Details on the TTICADs that can be built using `RC-TTICAD` for Example 2

Constraint Ordering o			TTICAD		C_o	
Formula order	ϕ_1 order	ϕ_2 order	Cells	Time (sec)	sotd	deg
$\phi_1 \to \phi_2$	$h \to f_1$	$h \to f_2$	24545	86.082	16	2
$\phi_1 \to \phi_2$	$h \to f_1$	$f_2 \to h$	73849	499.595	114	8
$\phi_1 \to \phi_2$	$f_1 \to h$	$h \to f_2$	67365	414.314	114	8
$\phi_1 \to \phi_2$	$f_1 \to h$	$f_2 \to h$	105045	1091.918	8	6
$\phi_2 \to \phi_1$	$h \to f_1$	$h \to f_2$	24545	87.378	16	2
$\phi_2 \to \phi_1$	$h \to f_1$	$f_2 \to h$	67365	401.598	114	8
$\phi_2 \to \phi_1$	$f_1 \to h$	$h \to f_2$	73849	494.888	114	8
$\phi_2 \to \phi_1$	$f_1 \to h$	$f_2 \to h$	105045	1075.568	8	6

no intersections with itself to identify. However, using sotd as a measure on the constraint ordering set will lead us to select the very worst ordering. Consider the constraint ordering sets for these two cases:

$$C_{f_1 \to h, f_2 \to h} = \{\mathrm{disc}_y(f_1), \mathrm{disc}_y(f_2), \mathrm{res}_y(f_1, f_2)\}$$
$$= \{-4x^2 + 16x - 12, -4x^2 - 16x - 12, 64x^2\},$$
$$C_{h \to f_1, h \to f_2} = \{\mathrm{disc}_y(h), \mathrm{res}_y(h, h)\} = \{4a^2(a^2 - c^2 + 2cx - x^2), 0\}.$$

Although the first has three polynomials and the second only one, this one has a higher sotd than the first three summed. This is because only h contained the parameters (a, c) while f_1 and f_2 did not, but their presence was not as significant as the complexity in x alone. A more suitable measure would be the sum of degrees in x alone (shown in the final column in Table 1) in which the first has 6 and the second only 2.

Remark 1. It is not actually surprising that sotd is inappropriate here while working well in [6,15]. In those studies sotd was measuring projection sets (either the whole set or at one stage in the projection) while here we are measuring only the subset which changes the most with the ordering. Sotd is principally a measure of sparseness. Sparseness of the entire projection set indicates less complexity while sparseness of one level is likely to lead to sparseness at the next. However, the constraint ordering set being sparse does not indicate that the other polynomials involved at that stage or subsequent ones will be.

Heuristic Definition 1. *Define the **EC ordering heuristic** as selecting the first EC to be processed in each formula such that the corresponding constraint ordering set has lowest sum of degrees of the polynomials within (all taken in the second variable of the ordering).*

Heuristic 1 follows from the analysis above and we evaluate it in Section 3. We can already see three apparent shortcomings:

(i) How to break ties if the sum of degrees are the same?
(ii) What to do if the complex geometry is different to the real geometry?
(iii) How to order remaining equational constraints?

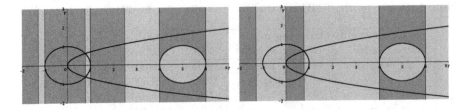

Fig. 2. Visualisations of two TTICADs built using RC-TTICAD for Example 3. They both have $\phi_2 \to \phi_1$, with the first having $f_1 \to f_2$ and the second $f_2 \to f_1$.

One answer to (i) is to break ties with sotd. A tie with Heuristic 1 is a good indication that the complex geometry in the highest dimension is of the same complexity and so further discrimination will require lower dimensional components. In fact, these are also needed to address (iii). Suppose a formula contained three ECs and we had determined which to process first. Then the choice of which is second means comparing the resultant of the first with each of the others modulo the first. In our experience such formulae tend to give similar output for the different orderings due to the simplifications in the tree so many ECs offer.

Heuristic 1 can be misled as suggested by (ii) and demonstrated next.

Example 3. Consider the polynomials and formulae from Example 1 but with f_2 and g_2 shifted under $y \mapsto y + 1$. The possible outputs from RC-TTICAD are:

$\phi_1 \to \phi_2$ and $f_1 \to f_2$: 39 cells in 0.094 seconds.
$\phi_1 \to \phi_2$ and $f_2 \to f_1$: 49 cells in 0.081 seconds.
$\phi_2 \to \phi_1$ and $f_1 \to f_2$: 27 cells in 0.077 seconds.
$\phi_2 \to \phi_1$ and $f_2 \to f_1$: 23 cells in 0.073 seconds.

Since f_2 no longer intersects h the best choice is the fourth instead of the third. Figure 2 compares these two TTICADs. The only difference now is whether the endpoints of the left circle or the parabola are identified. Since the parabola has only one endpoint it becomes the better choice. However, the constraint ordering set has the same degree in x or sotd and so still suggests $f_1 \to f_2$.

Heuristic 1 is misled here because the degree is a measure only of the behaviour in complex space, which did not change significantly between the examples. In [6] we demonstrated similar issues for CAD (and TTICAD) by projection and lifting. There we devised an alternative heuristic: the *number of distinct real roots of the univariate polynomials* (ndrr) which meant essentially comparing the induced CADs of the real line. However, RC-TTICAD does not start by directly computing all polynomials involved in the computation (the projection phase). Example 3 is in only two dimensions and so the ndrr could easily be applied to the univariate constraint ordering sets to suggest the best ordering. But for higher dimensional examples it is not so clear what or how to measure. Further, the complex geometry does have a direct effect on RC-TTICAD not present in the projection and lifting algorithms since we first build a CCD.

2.3   Developing a Heuristic for Formulae Ordering

Heuristic 1 helps with ordering ECs in formulae but not how to order the formulae themselves. In Example 1 the main difference between formulae orderings was which polynomial is ensured sign-invariant in the output. In Example 1 there was a clear choice to process ϕ_2 first since its sole EC would be sign-invariant regardless. In general we advise **placing a formula with only one EC first.**

Remark 2. In fact, the analysis so far suggests that the best choice would be to process a non-EC from a formula with no ECs first. This is because all the non-ECs in such a formula will always be sign-invariant in the output and so dealing with them first would occur no cost but possibly allow savings from another formulae with multiple ECs. The algorithm as specified in [3] does not allow this but we intend to investigate this possibility in future work.

We now seek a heuristic to help with formulae ordering when no obvious choice is available. Ideally, we require an (efficient) measure of how large the (real) projection is of a polynomial out of its main variable, but such a measure is not clear to us. Instead we explore an alternative approach. As discussed, the CAD algorithms based on regular chains technology first build a CCD before refining to a CAD. It has been observed that the refinement to real space usually takes the most time (involving real root isolation), but that the timings of the two stages are correlated. Hence, we consider building the CCD first for multiple orderings and then choosing the smallest one.

Heuristic Definition 2. *Define the **CCD size heuristic** as selecting a constraint ordering by constructing the CCD for each, extracting the set of polynomials used in each tree, and choosing the one to refine to a CAD whose set has the lowest sum of degree of the polynomials within (each taken in the main variable of that polynomial).*

We evaluate this heuristic in the next section. It clearly requires far more computation than Heuristic 1 and so the relative costs will have to be considered. This leads us to suggest a third heuristic combining the approaches.

Heuristic Definition 3. *Define the **constraint ordering heuristic** as using Heuristic 1 to suggest the best subset of constraint orderings and then having Heuristic 2 pick from these, splitting any further ties by picking lexicographically.*

3   Evaluating the Heuristics

3.1   Experiments and Data

We tested the effectiveness of the heuristic using 100 random systems of the form

$$\phi_1 := (f_1 = 0 \wedge f_2 = 0 \wedge g_1 > 0), \qquad \phi_2 := (f_3 = 0 \wedge f_4 = 0 \wedge g_2 > 0).$$

Table 2. How the CADs selected by the heuristics compare on cell count

Heuristic	Cell Count	Saving	% Saving
Heuristic 1	1589.67	428.61	26.73
Heuristic 2	1209.10	809.18	47.70
Heuristic 3	1307.63	710.65	40.97

Table 3. How the CADs selected by the heuristics compare on timings (in seconds)

Heuristic	Timing	Saving	% Saving	Net Saving	% Net Saving
Heuristic 1	14.48	22.02	37.17	22.01	37.12
Heuristic 2	9.02	27.47	49.45	-150.59	-215.31
Heuristic 3	9.42	27.08	43.84	-20.02	0.77

The polynomials were randomly generated using MAPLE's `randpoly` command as sparse polynomials in the ordered variables $x \prec y \prec z$ with maximum degree 3 and integer coefficients . Each problem has three questions of constraint ordering:
- Process ϕ_1 first or ϕ_2?   - Process f_1 first or f_2?   - Process f_3 first or f_4?

Hence each problem has eight possible orderings. We build TTICADs using `RC-TTICAD` for each ordering and compare the number of cells and computation time. We set a time limit of 40 minutes per problem (so an average of 5 minutes per CAD) in which 92 of the problems could be studied. The average CAD had 2018.3 cells and was computed in 36.5 seconds, but there were several outliers bringing the average up. The median values were 1554.1 cells and 6.1 seconds.

For each problem we considered how Heuristics 1, 2 and 3 performed. We start by comparing cell counts in Table 2. For each problem we calculated:

(a) The average cell count of the 8 TTICADs computed.
(b) The average cell count of the TTICADs selected by each heuristic (but note that Heuristic 3 always selects only one).
(c) The average saving from using each heuristic, computed as (a)−(b).
(d) The average percentage saving to the cell count, calculated as 100(c)/(a).

The figures in Table 2 show the values of (b)−(d) for each heuristic, averaged over the 92 problems. To compare timings we calculated (a′)−(d′) as the equivalent of (a)−(d) for timings. Then for each problem we also calculated:

(e′) The time taken to run each heuristic.
(f′) The net saving calculated as (a′)−(b′)−(e′).
(g′) The net percentage saving calculated as 100(f′)/(a′).

Table 3 shows the values of (b′)−(d′),(f′),(g′) averaged over the 92 problems.

Tables 4 and 5 shows where the selections made by each heuristic lie on the spread of possible outputs (where 1 is the CAD with the smallest and 8 the one with the biggest). In the event of two CADs having the same value a selection is recorded with the higher ranking. Since Heuristics 1 and 2 can pick more than one ordering we also display the figures as percentages. So for example, a selection by the first heuristic was the very best ordering 24% of the time.

Table 4. How the heuristics selections rank out of the possible CADs for cell counts

Heuristic		1	2	3	4	5	6	7	8	Total
Heuristic 1	#	60	46	44	26	21	17	17	15	246
	%	24.39	18.70	17.89	10.57	8.54	6.91	6.91	6.10	100.01
Heuristic 2	#	55	19	12	5	5	2	0	0	98
	%	56.12	19.39	12.24	5.10	5.10	2.04	0	0	99.99
Heuristic 3	#	44	22	7	6	4	4	3	2	92
	%	47.83	23.91	7.61	6.52	4.35	4.35	3.26	2.17	100.00

Table 5. How the heuristics selections rank out of the possible CADs for timings

Heuristic		1	2	3	4	5	6	7	8	Total
Heuristic 1	#	64	51	33	24	23	25	13	13	246
	%	26.02	20.73	13.41	9.76	9.35	10.16	5.29	5.29	100.01
Heuristic 2	#	44	29	12	4	1	4	1	3	98
	%	44.90	29.59	12.24	4.08	1.02	4.08	1.02	3.06	99.99
Heuristic 3	#	37	26	9	4	1	6	7	2	92
	%	40.22	28.26	9.78	4.35	1.09	6.52	7.61	2.17	100.00

3.2   Interpreting the Results

First we observe that all three heuristics will on average make selections on constraint ordering with substantially lower cell counts and timings than the problem average. As expected, the selections by Heuristic 2 are on average better than those by Heuristic 1. In fact, the measure used by Heuristic 2 seems to be correlated to both the cell counts and timings in the final TTICAD.

To consider the correlation we recorded the value of the measures used by Heuristics 1 and 2 and paired these with the corresponding cell counts and timings. This was done for each CAD computed (not just those the heuristics selected). The values were scaled by the maximum for each problem. (Note that Heuristic 3 did not have its own measure, it was a combination of the two.) Figure 3 shows the plots of these data. The correlation coefficients for the first measure were 0.43 with cell count and 0.40 with timing, while for the second measure 0.78 and 0.68. Since the second measure essentially completes the first part of the algorithm the correlation may not seem surprising. However, it suggests that on average the geometry of the real and complex decomposition are more closely linked than previously thought. This will be investigated in future work.

Although Heuristic 2 makes good selections, its cost is usually larger than any potential time savings (roughly 6 times larger on average). Further, this cost will rise with the number of orderings far quicker than the cost of the others. We note that the magnitude of this cost is inflated by the outliers, the average cost being 178.06 seconds while the median only 13.43. Heuristic 1 is far cheaper, essentially zero. Although the savings were not as high they were still significant, with most selections being among the best. We recommend Heuristic 1 as a cheap test to use before running the algorithm and it will likely become part of the default implementation.

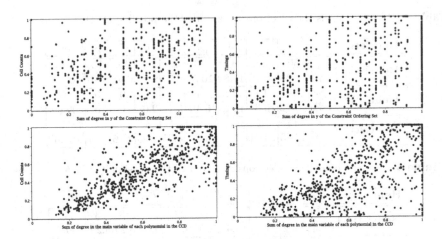

Fig. 3. These plots compare the measures used by the heuristics with the CADs computed in Section 3. The plots on the left have cell count on the vertical axis, and those on the right timings. The horizontal axes have the sum of degrees of polynomials in a set. On the top this is the constraint ordering set and on the bottom the polynomials in the CCD. All values are scaled to the problem they originate from.

The results for Heuristic 3 which used a mixture of the approaches are particularly interesting. It offers substantially more savings than Heuristic 1, almost achieving those those Heuristic 2 but its cost is on average 47.10 seconds (with a median value of 7.55), far less than those of Heuristic 2. On average Heuristic 3 took more time in total to compute than its time savings, but when we consider the percentage saving the average is (just) positive. This is not a mistake: the results are as stated because a number of outliers had a very high cost while for most examples the cost was significantly less than the savings.

We can see situations where all three heuristics could be of use:

Use Heuristic 1 if lowest computation time is prioritised, for example if many CADs must be computed or this is just a small step in a larger calculation.

Use Heuristic 2 if lowest cell count is prioritised, for example if only one CAD must be computed but then much work will be performed with its cells.

Use Heuristic 3 for a mixed approach, for example if a low cell count is required but the problem size makes Heuristic 2 infeasible.

4   Other Issues of Problem Formulation

For the original TTICAD algorithm (by projection and lifting) [4] the ordering of the constraints is not important, but other issues are, as investigated in [6]. We revisit two of those issues to see if further consideration is needed for RC-TTICAD.

4.1 Equational Constraint Designation

The TTICAD algorithm by projection and lifting [4] made use of a single *designated* EC per formula (any others were treated the same as non-ECs). Indeed, this projection operator generalised the one in [21] for a formula with one EC and in either case the user needs to make this designation before running the algorithm. RC-TTICAD [3] (and the algorithm in [9]) can take advantage of more than one EC per formula and so the user only needs to choose the order they are used in. We observe that the choice of which EC to process first is analogous to choosing which to designate. For example, consider two formulae of the form

$$\phi_i := f_1 = 0 \wedge f_2 = 0 \wedge g_1 < 0, \quad \phi_2 := f_3 = 0 \wedge g_2 = 0.$$

Then the resultants and discriminants that must be calculated for the first projection phase using the operator in [21] are

$$\{\mathrm{res}(f_i, f_j), \mathrm{res}(f_i, g_1), \mathrm{res}(f_i, f_3), \mathrm{disc}(f_i), \mathrm{disc}(f_3)\}$$

if f_i is designated and f_j not. All polynomials from the constraint ordering set are contained here, as can be shown for the general case. A good choice of designation for the projection and lifting algorithm is hence likely to correspond to a good choice of which EC from a formula to process first in the regular chains algorithm. We hope to investigate this further in the future.

4.2 Composing Sub-formulae

Consider $\Phi := (f_1 = 0 \wedge \psi_1) \vee (f_2 = 0 \wedge \psi_2)$. where ψ_1, ψ_2 are conjunctions. We seek a truth-invariant CAD for Φ but neither of the equations are ECs (at least not explicitly without knowledge of ψ_1 and ψ_2). One option would be to use $f_1 f_2 = 0$ as an EC (this is logically implied by Φ). Another option is to define

$$\phi_1 := f_1 = 0 \wedge \psi_1, \qquad \phi_2 := f_2 = 0 \wedge \psi_2$$

and construct a TTICAD for them (any TTICAD for ϕ_1, ϕ_2 is truth-invariant for Φ). For the projection and lifting algorithms the second approach is preferable as the projection set for the latter is contained in the former. RC-TTICAD requires as input semi-algebraic systems each representing a single conjunctive formula. Hence here there is not even an analogue of the former approach.

However, there was a similar question posed in [6, Section 4] which we now investigate in reference to RC-TTICAD. Consider the single conjunctive formulae, $\hat{\Phi} := f_1 = 0 \wedge \psi_1 \wedge f_2 = 0 \wedge \psi_2$, where ψ_1, ψ_2 are again conjunctions. We could build a CAD for $\hat{\Phi}$ or a TTICAD for ϕ_1, ϕ_2 as above. While the projection set for the latter is in general smaller, the following example gives an exception.

Example 4 (Example 6 in [6]). Let $x \prec y$ and consider the formula $\hat{\Phi}$ above with

$$f_1 := (y - 1) - x^3 + x^2 + x, \qquad \psi_1 := g_1 < 0, \qquad g_1 := y - \tfrac{x}{4} + \tfrac{1}{2},$$
$$f_2 := (-y - 1) - x^3 + x^2 + x, \qquad \psi_2 := g_2 < 0, \qquad g_2 := -y - \tfrac{x}{4} + \tfrac{1}{2}.$$

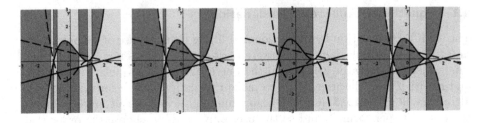

Fig. 4. Visualisations of CADs that can be built for Example 4

The polynomials are plotted in the images of Figure 4 where the solid curve is f_1, the solid line g_1, the dashed curve f_2 and the dashed line g_2. Various CADs may be computed for this problem:

- A CAD for $\hat{\Phi}$ using projection and lifting with the operator from [21] designating f_1: 39 cells as visualised in the first image.
- As above but designating f_2: 39 cells. Similar to first image but 2 dimensional cell divisions over f_2 instead of f_1.
- A TTICAD for ϕ_1, ϕ_2 using projection and lifting with the operator from [4]: 31 cells as visualised in the second image.
- A CAD for $\hat{\Phi}$ using RC-TTICAD (equivalent here to the algorithm in [9]): 9 cells (under any constraint ordering) as visualised in the third image.
- A TTICAD for ϕ_1, ϕ_2 using RC-TTICAD [3]: 29 cells (under any constraint ordering) as visualised in the fourth image.

The important observation is that f_1 has many intersections with g_2 and f_2 many intersections with g_1. The projection and lifting algorithms can avoid considering those pairs together by splitting into sub-formulae. In the first image only one EC is sign-invariant while in the second both are, but this was a price worth paying to avoid the intersections. It is not necessary for RC-TTICAD as this can take advantage of multiple ECs in a formula. It first identifies the intersection of f_1 and f_2 and the non-ECs are only considered modulo this set. Hence, even though they are in the same formula, those intersections are never computed.

In general, RC-TTICAD requires a parent formula to be broken into conjunctive sub-formulae before use, but would never benefit from further decomposition.

5   Final Thoughts

We developed new heuristics to choose constraint orderings for RC-TTICAD [3], finding that the choice of which to use may depend on the priorities of the user. A dynamic heuristic (such as one like Heuristic 2 but making choices based on the CCT after each increment) may offer further improvements and will be a topic of future work.

We also revisited other questions of problem formulation questions from [6], finding that they did not require further consideration for RC-TTICAD. However, there was one important issue we did not address, the variable ordering, for which there may be a free or constrained choice. For example, when using CAD for quantifier elimination we must order the variables as they are quantified but we may change the ordering within quantifier blocks. It has long been noted that problems which are easy in one variable ordering can be infeasible in another, with [7] giving problems where one variable ordering leads to a CAD with a cell count constant in the number of variables and another a cell count doubly exponential. The analysis was valid for any CAD regardless of the algorithm that produced it and so affects RC-TTICAD. A key area of our future work will be to analyse how best to choose a variable ordering, and to investigate whether an existing heuristic, or the new ones developed here can help.

Acknowledgements. This work was supported by the EPSRC (EP/J003247/1), the NSFC (11301524), and the CSTC (cstc2013jjys0002).

References

1. Arnon, D., Collins, G.E., McCallum, S.: Cylindrical algebraic decomposition I: The basic algorithm. SIAM J. Comput. 13, 865–877 (1984)
2. Arnon, D.S., Mignotte, M.: On mechanical quantifier elimination for elementary algebra and geometry. J. Symb. Comp. 5(1-2), 237–259 (1988)
3. Bradford, R., Chen, C., Davenport, J.H., England, M., Moreno Maza, M., Wilson, D.: Truth table invariant cylindrical algebraic decomposition by regular chains (submitted, 2014), Preprint: http://opus.bath.ac.uk/38344/
4. Bradford, R., Davenport, J.H., England, M., McCallum, S., Wilson, D.: Cylindrical algebraic decompositions for boolean combinations. In: Proc. ISSAC 2013, pp. 125–132. ACM (2013)
5. Bradford, R., Davenport, J.H., England, M., McCallum, S., Wilson, D.: Truth table invariant cylindrical algebraic decomposition (submitted, 2014), Preprint: http://opus.bath.ac.uk/38146/
6. Bradford, R., Davenport, J.H., England, M., Wilson, D.: Optimising problem formulation for cylindrical algebraic decomposition. In: Carette, J., Aspinall, D., Lange, C., Sojka, P., Windsteiger, W. (eds.) CICM 2013. LNCS (LNAI), vol. 7961, pp. 19–34. Springer, Heidelberg (2013)
7. Brown, C.W., Davenport, J.H.: The complexity of quantifier elimination and cylindrical algebraic decomposition. In: Proc. ISSAC 2007, pp. 54–60. ACM (2007)
8. Brown, C.W., El Kahoui, M., Novotni, D., Weber, A.: Algorithmic methods for investigating equilibria in epidemic modelling. J. Symbolic Computation 41, 1157–1173 (2006)
9. Chen, C., Moreno Maza, M.: An incremental algorithm for computing cylindrical algebraic decompositions. In: Proc. ASCM 2012. Springer (2012) (to appear), Preprint: arXiv:1210.5543v1
10. Chen, C., Moreno Maza, M., Xia, B., Yang, L.: Computing cylindrical algebraic decomposition via triangular decomposition. In: Proc. ISSAC 2009, pp. 95–102. ACM (2009)
11. Collins, G.E.: Quantifier elimination for real closed fields by cylindrical algebraic decomposition. In: Brakhage, H. (ed.) GI-Fachtagung 1975. LNCS, vol. 33, pp. 134–183. Springer, Heidelberg (1975)

12. Collins, G.E.: Quantifier elimination by cylindrical algebraic decomposition – 20 years of progress. In: Quantifier Elimination and Cylindrical Algebraic Decomposition. Texts & Monographs in Symbolic Computation, pp. 8–23. Springer (1998)
13. Collins, G.E., Hong, H.: Partial cylindrical algebraic decomposition for quantifier elimination. J. Symb. Comp. 12, 299–328 (1991)
14. Davenport, J.H., Bradford, R., England, M., Wilson, D.: Program verification in the presence of complex numbers, functions with branch cuts etc. In: Proc. SYNASC 2012, pp. 83–88. IEEE (2012)
15. Dolzmann, A., Seidl, A., Sturm, T.: Efficient projection orders for CAD. In: Proc. ISSAC 2004, pp. 111–118. ACM (2004)
16. England, M., Bradford, R., Davenport, J.H., Wilson, D.: Understanding Branch Cuts of Expressions. In: Carette, J., Aspinall, D., Lange, C., Sojka, P., Windsteiger, W. (eds.) CICM 2013. LNCS (LNAI), vol. 7961, pp. 136–151. Springer, Heidelberg (2013)
17. England, M.: An implementation of CAD in Maple utilising problem formulation, equational constraints and truth-table invariance. Uni. Bath, Dept. Comp. Sci. Tech. Report Series, 2013-04 (2013), http://opus.bath.ac.uk/35636/
18. Fotiou, I.A., Parrilo, P.A., Morari, M.: Nonlinear parametric optimization using cylindrical algebraic decomposition. In: Proc. CDC-ECC 2005, pp. 3735–3740 (2005)
19. Iwane, H., Yanami, H., Anai, H., Yokoyama, K.: An effective implementation of a symbolic-numeric cylindrical algebraic decomposition for quantifier elimination. In: Proc. SNC 2009, pp. 55–64 (2009)
20. Kahan, W.: Problem #9: an ellipse problem. SIGSAM Bull. 9(3), 11–12 (1975)
21. McCallum, S.: On projection in CAD-based quantifier elimination with equational constraint. In: Proc. ISSAC 1999, pp. 145–149. ACM (1999)
22. Paulson, L.C.: MetiTarski: Past and future. In: Beringer, L., Felty, A. (eds.) ITP 2012. LNCS, vol. 7406, pp. 1–10. Springer, Heidelberg (2012)
23. Schwartz, J.T., Sharir, M.: On the "Piano-Movers" Problem: II. General techniques for computing topological properties of real algebraic manifolds. Adv. Appl. Math. 4, 298–351 (1983)
24. Strzeboński, A.: Cylindrical algebraic decomposition using validated numerics. J. Symb. Comp. 41(9), 1021–1038 (2006)

A Tableaux-Based Decision Procedure
for Multi-parameter Propositional Schemata

David Cerna

Technical University of Vienna*, Austria
cernadavid1@logic.at
http://www.logic.at/people/cernadavid1/

Abstract. The class of *regular propositional schemata*, discovered by Aravantinos et al. [4], is a major advancement towards more expressive classes of inductive theorems with a decidable satisfiability problem. Though more expressive than previously known decidable classes outlined by Kapur & Giesl[17], it still requires the burdensome restriction of induction with only one free parameter. In general, unrestricted usage of multiple free parameters in schematic formulae is undecidable for satisfiability [2]. In later work, Aravantinos et al. [6] introduced *normalized clause sets* which have a decision procedure for satisfiability and allow for restricted usage of multiple parameters. In our work, we investigate classes of propositional schemata which allow for multiple free parameters and are more expressive than regular schemata. Specifically, the classes we investigate have a decision procedure for satisfiability testing without requiring the additional theoretical machinery of normalized clause sets. Thus, allowing one to avoid conversion to CNF formulae. Both of the classes we introduce, *linked schemata* and *pure overlap schemata* use the machinery introduced in the earlier works of Aravantinos et al.[4] with only a slight change to the decision procedure.

1 Introduction

The concept of schema has been pervasive throughout the history of logic [14]. First-order Peano arithmetic's usage of an induction schema is a well known example of schema in mathematical logic [22]. There are many other less known examples where schemata were used in both propositional and first-order logic to attain proof theoretic results. For example, results pertaining to proof length, unification, construction of 'proof skeletons', and first-order schematic Hilbert-type systems [7,9,21,19,20]. Also, in the analysis of the Fürstenberg's proof of the infinitude of primes [8], cut elimination resulted in a schema of proofs where the free parameter indexed the number of prime numbers. Very recently, work has been done on schematizing cut-elimination so that an arbitrary number of cuts can be eliminated without instantiating the free parameter of the proof [15].

The usage of schemata that we will focus on for the majority of this paper is schemata as an object level construction iterating propositional formulae. This

---

* This work was funded by the Vienna PhD School of Informatics.

S.M. Watt et al. (Eds.): CICM 2014, LNAI 8543, pp. 61–75, 2014.

work was pioneered by Aravantinos et al. [4] with application to the field of repeated circuit verification. This construction has also resulted in discoveries in the field of inductive theorem proving, namely, construction of classes of inductively defined formulae which have a decidable satisfiability problem and are more expressive than those currently known in the field [4,17,18]. Namely, theses classes are *bounded-linear schemata, regular schemata, nested regular schemata,* and a multiple free parameter generalization of regular schemata through normalized clause set representation, [6]. Decidable classes of inductive theorems discovered by Kapur et al. [17,18], the most prominent work in this area, were mainly universally quantified. Propositional schemata can express both bounded existential and universal quantification through $\wedge$ and $\vee$ iterations.

Our main results are formalizations of the class of linked schemata and pure overlap schemata, both being multiple free parameter extension of regular schema, together with a decision procedure for both classes. This decision procedure is a simple extension of the ST procedure for STAB[4]. Our decision procedure allows one to avoid the conversion of propositional schemata to normalized clause sets [6]. Our work is of a similar vein as Gentzen's work [16] in which he provided a method to fuse multiple inductions together in Peano arithmetic.

Though both classes of schemata we introduce are subclasses of the schemata representable by normalized clause sets and benefit from the satisfiability procedure of normalized clauses sets [6], the existence of a tableaux-based decision procedure for satisfiability testing for these classes remained an open problem. The benefit of a tableaux-based decision procedure is that one does not need to convert the propositional schemata into CNF form to test satisfiability. Note that, if one wants to keep logical equivalence between the original formula and the CNF form, the conversion can result in an exponential increase in formula size.

In this paper, we consider multiple regular schemata (each with its own parameter) such that the propositional symbols of one schema are not found in the other regular schemata. When this property holds, we can use the parts (i.e. the iterations and propositional variables not found in the iterations) to construct a formula with multiple free parameters–we refer to this class as the class of linked schemata. Essentially, we build formulae using the pieces of several regular schemata. Although, this idea is quite simple, it provides a class of schemata extending regular schemata which still has a tableaux-based decision procedure for satisfiability.

Next we investigate when it is possible for the propositional symbols to occur in two or more *linked* regular schemata, i.e. the same propositional symbol has occurrences indexed by two different parameters. To answer this question, we develop the concept of relative pure literals, literals which are pure when considering occurrences indexed by another parameter. This concept is used to construct the class of pure overlap schemata.

Both linked and pure overlap schemata are extensions of regular schemata, but after applying several tableaux extension rules to the constructed tableau, It is possible to reduce the branches of the constructed tableau to tableaux branches which are decidable using the decision procedure for regular schemata. Essentially, they are both propositional extensions of the class. It is not completely clear if these classes of schemata are the most expressive classes such that their satisfiability problem can be reduced to the satisfiability problem for regular schemata. An open problem regarding this point is whether the purity constraint can be relaxed and retain the reduction– results of Aravantinos et al. [4] (Thm. 6.2) suggests that this is not going to be the case.

Overall, our paper provides a simpler and more natural alternative to normalized clause set representation when deciding satisfiability for certain classes of multiple-parameter schemata.

The rest of this paper is structured as follows, Sec. 2 will be necessary background material from Aravantinos et al. [4], in Sec. 3 we formalize the construction of linked schemata, in Sec. 4 we formalize the construction of pure overlap schemata , in Sec. 5 we provide a decision procedure for the satisfiability problem of pure overlap schemata. Finally, in Sec. 6 we conclude the paper and shortly discuss the open problems.

2   Background

2.1   Propositional Schemata

The indexing language for standard schematic propositional logic as considered in Aravantinos et al. [4] is the set of *linear arithmetic terms* (denoted by $\mathcal{Z}$) built using the language $\{0, s(\cdot), +, -\}$ and a countably infinite set of variables $\mathcal{V}$. Multiplication is considered as a shorthand for terms of the form $x+x+x+x = 4 \cdot x$ and is not a real operator in the language, nor is it a necessary one. To stick to the framework of Aravantinos et al. [4] $\mathbb{Z}$ is considered as the standard model of the terms in $\mathcal{Z}$.

Definition 1 (Indexed Proposition[4]). *Let $\mathcal{P}$ be a fixed and countably infinite set of propositional symbols. An* indexed proposition *is an expression of the form $p_{\mathbf{a}}$ where $p \in \mathcal{P}$ and $\mathbf{a} \in \mathcal{Z}$. An indexed proposition $p_{\mathbf{a}}$ s.t. $\mathbf{a} \in \mathbb{Z}$ is called a propositional variable.*

Definition 2 (Formula Schemata[4]). *The set of* formula schemata *is the smallest set satisfying the following properties.*

- $\bot, \top$ *are formula schemata.*
- *If $\mathbf{a}, \mathbf{b} \in \mathcal{Z}$ then $\mathbf{a} < \mathbf{b}$ is a formula schema.*
- *Each indexed proposition is a formula schema.*
- *If ϕ_1, ϕ_2 are formula schemata then $\phi_1 \wedge \phi_2$, $\phi_1 \vee \phi_2$, $\neg\phi_1$ are formula schemata.*
- *If ϕ is a formula schema not containing $<$, and if $\mathbf{a}, \mathbf{b} \in \mathcal{Z}$, where i is an arithmetic variable, then $\bigwedge_{i=\mathbf{a}}^{\mathbf{b}} \phi$, $\bigvee_{i=\mathbf{a}}^{\mathbf{b}} \phi$ are formula schemata.*

Example 1. Consider the formula:

$$\varphi = q_1 \wedge \bigwedge_{i=0}^{n} \left(p_{i+2n} \wedge \left(\bigvee_{j=n}^{2n+1} \neg q_{n-j} \vee q_{j+1} \right) \right) \wedge 0 \leq n$$

φ is a formula schema.

Formula schemata are inherently finite. We will label the indexed propositions, $\top$, $\bot$ and statements of the form $a < b$, as *atoms*. Formula schemata of the form $\bigwedge_{i=\mathbf{a}}^{\mathbf{b}} \phi$ and $\bigvee_{i=\mathbf{a}}^{\mathbf{b}} \phi$ will be called *iterations*. A formula schema whose constituents are any of the following: $\top$, $\bot$, and $a < b$, is an *arithmetic formula*. Also, it is taken as a standard that arithmetic formulae of the form $a < b$ can only occur outside of iterations. This constraint is necessary being that $a < b$ is interpreted as an iteration, i.e.

$$a < b \equiv \bigvee_{i=a+1}^{b} \top \tag{1}$$

Also, we use $a = b$ as an abbreviation for $\neg(b < a) \wedge \neg(a < b)$ and $a \leq b$ as an abbreviation for $\neg(b < a)$. Iterations have both *free* and *bound* variables, where free variable and *parameter* are synonymous. A bound variable i is a variable in the scope of an iteration $\Pi_{i=a}^{b}\phi_i$ where $\Pi = \{\bigvee, \bigwedge\}$. A *substitution* is a function mapping all the free variables to linear expressions. If a substitution σ is applied to a schema φ, i.e $\varphi\sigma$ such that the domain of σ is every free variable in φ, then the linear expressions of φ are integer terms, i.e. all indices in φ are variable free.

Definition 3 (Interpretation [4]). *An* interpretation of the schematic language *is a function mapping every parameter to an integer and every propositional variable to a truth value* $\boldsymbol{T}$ *or* $\boldsymbol{F}$*. The substitution and interpretation will be denoted as* σ *and* $\mathcal{I}$*, respectively.*

Example 2. An Interpretation $\mathcal{I}$ such that φ from Ex. 1 is modelled by $\mathcal{I}$ would be $\sigma \equiv \{n \leftarrow 0\}$ and $q_1 = \mathbf{T}, p_0 = \mathbf{T}, q_0 = \mathbf{T}, q_{-1} = \mathbf{T}, q_1 = \mathbf{F}, q_2 = \mathbf{T}$

Definition 4 (Semantics of Schematic Formulae [4]). *The semantics of a schematic formula* φ *in a given interpretation* I*, denoted by* $[\![\varphi]\!]_{\mathcal{I}}$*, is defined as follows:*

- $[\![\top]\!]_{\mathcal{I}} = \mathbf{T}$ *and* $[\![\bot]\!]_{\mathcal{I}} = \mathbf{F}$
- $[\![\mathbf{a} < \mathbf{b}]\!]_{\mathcal{I}} = \mathbf{T} \Leftrightarrow [\![\mathbf{a}]\!]_{\mathcal{I}} <_{\mathbb{Z}} [\![\mathbf{b}]\!]_{\mathcal{I}}$
- $[\![P_{\mathbf{a}}]\!]_{\mathcal{I}} = \mathcal{I}(P_{[\![\mathbf{a}]\!]_{\mathcal{I}}})$ *for* $P \in \mathcal{P}$
- $[\![\neg\varphi]\!]_{\mathcal{I}} = \mathbf{T} \Leftrightarrow [\![\varphi]\!]_{\mathcal{I}} = \mathbf{F}$
- $[\![\varphi \vee \psi]\!]_{\mathcal{I}} = \mathbf{T} \Leftrightarrow [\![\varphi]\!]_{\mathcal{I}} = \mathbf{T}$ *or* $[\![\psi]\!]_{\mathcal{I}} = \mathbf{T}$
- $[\![\varphi \wedge \psi]\!]_{\mathcal{I}} = \mathbf{T} \Leftrightarrow [\![\varphi]\!]_{\mathcal{I}} = \mathbf{T}$ *and* $[\![\psi]\!]_{\mathcal{I}} = \mathbf{T}$
- $[\![\bigvee_{i=\mathbf{a}}^{\mathbf{b}} \varphi_i]\!]_{\mathcal{I}} = \mathbf{T} \Leftrightarrow \exists\alpha \in \mathbb{Z}$ *such that* $[\![\mathbf{a}]\!]_{\mathcal{I}} \leq_{\mathbb{Z}} \alpha \leq_{\mathbb{Z}} [\![\mathbf{b}]\!]_{\mathcal{I}}$ *and* $[\![\varphi_i]\!]_{\mathcal{I}[\alpha/i]} = \mathbf{T}$
- $[\![\bigwedge_{i=\mathbf{a}}^{\mathbf{b}} \varphi_i]\!]_{\mathcal{I}} = \mathbf{T} \Leftrightarrow \forall\alpha \in \mathbb{Z}$, $[\![\mathbf{a}]\!]_{\mathcal{I}} \leq_{\mathbb{Z}} \alpha \leq_{\mathbb{Z}} [\![\mathbf{b}]\!]_{\mathcal{I}}$ *implies* $[\![\varphi_i]\!]_{\mathcal{I}[\alpha/i]} = \mathbf{T}$

In the above definition, by $[\![\varphi_i]\!]_{\mathcal{I}[\alpha/i]}$ we mean every occurrence of i in φ_i is replaced by α. A propositional schema φ is *valid* (respectively *satisfiable*) iff for all (exists an interpretation) interpretations $\mathcal{I}$ s.t. $[\![\varphi]\!]_{\mathcal{I}} = T$. $\mathcal{I}$ is called a *model* of φ, written as $\mathcal{I} \models \varphi$. Two schemata φ, ψ are equivalent (written $\varphi \equiv \psi$) iff $\mathcal{I} \models \varphi \Leftrightarrow \mathcal{I} \models \psi$. φ and ψ are sat-equivalent (written $\varphi \equiv_S \psi$) iff φ and ψ are both satisfiable or both unsatisfiable (not necessarily by the same model).

Definition 5 (Unrolling Iterations [4]). *The following set S of rewrite rules is used to unroll the iterations of a given schematic formula φ:*

$$S = \begin{cases} \bigvee_{i=a}^{b} \psi \longrightarrow \bot & a, b \in \mathbb{Z} \text{ and } b <_{\mathbb{Z}} a \\ \bigwedge_{i=a}^{b} \psi \longrightarrow \top & a, b \in \mathbb{Z} \text{ and } b <_{\mathbb{Z}} a \\ \bigvee_{i=a}^{b} \psi \longrightarrow \left(\bigvee_{i=a}^{b-1} \psi \right) \vee \psi\,[b/i] & a, b \in \mathbb{Z} \text{ and } a \leq_{\mathbb{Z}} b \\ \bigwedge_{i=a}^{b} \psi \longrightarrow \left(\bigwedge_{i=a}^{b-1} \psi \right) \wedge \psi\,[b/i] & a, b \in \mathbb{Z} \text{ and } a \leq_{\mathbb{Z}} b \end{cases} \tag{2}$$

By $\leq_{\mathbb{Z}}$ we are referring to the standard ordering over the integers.

Definition 6 (Regular Schemata (as written in [4])). *A propositional schema ϕ is regular if it has a unique parameter n and if it is flat, of bounded propagation and aligned on $[\alpha, n - \beta]$:*

1) *A schema is flat if every $\Pi_{i=a}^{b}\psi$ occurring in the schema ψ does not contain an iteration, , where $\Pi \in \{\bigvee, \bigwedge\}$.*
2) *A schema is of bounded propagation if every atom that occurs in an iteration $\Pi_{i=a}^{b}\psi$ is of the form $P_{i+\gamma}$ for some $\gamma \in \mathbb{Z}$, where $\Pi \in \{\bigvee, \bigwedge\}$.*
3) *A schema is aligned on $[c, d]$ if all iterations occurring in the schema are of the form $\Pi_{i=c}^{d}\psi$, where $\Pi \in \{\bigvee, \bigwedge\}$.*

Example 3. Consider the following schema:

$$\varphi = p_0 \wedge \left(\bigwedge_{i=0}^{n} \neg p_i \vee p_{i+1} \right) \wedge \neg p_n \wedge 0 \leq n \tag{3}$$

φ is a regular schemata.

2.2   Basics of STAB and the ST Procedure

We now overview the main ingredients of the ST decision procedure of the STAB framework introduced in Aravantinos et al. [4]. In this paper, we only rely on the existence of the ST procedure and the propositional tableaux extension rules to define an extended decision procedure for our newly defined classes of schemata.

Definition 7 (Tableau). *A tableau is a tree T s.t. each node N occurring in T is labelled by a set of schemata written $\Phi_T(N)$.*

Definition 8 (Extension Rules). *The extension rules of the STAB procedure are as follows:*

Propositional Rules

- $\varphi \wedge \psi \Rightarrow \varphi, \psi$
- $\varphi \vee \psi \Rightarrow \varphi \mid \psi$

Iteration Rules

- $\bigwedge_{i=a}^{b} \varphi \Rightarrow a \leq b, \varphi\,[b/i] \wedge \bigwedge_{i=a}^{b-1} \varphi \mid b < a$
- $\bigvee_{i=a}^{b} \varphi \Rightarrow a \leq b, \varphi\,[b/i] \vee \bigvee_{i=a}^{b-1} \varphi$

Closure Rule

- $p_a, \neg p_b \Rightarrow p_a, \neg p_b, a \neq b$

The way the STAB extension rules work is by extending currently constructed tableau with new leaves containing all the formulae of the prior node minus the formula φ on which the extension rule was applied. The parts of φ will be added to the leaves in accordance with the extension rule definitions. The symbol $\mid$ in the extension rules means that the constructed tableau branches when this rule is applied. The closure rule, rather than extending the constructed tableau, tells us that there is no need to extend the considered branch because it contains an unsatisfiable sub-branch.

Theorem 1. *There is a decision procedure for satisfiability testing of regular schemata (ST procedure) based on the STAB extension rules (Def. 8) and an additional rule to deal with looping, which terminates on every regular schema. The procedure is sound and complete for regular schemata.*

Example 4. We provide an example of the ST procedure producing a closed tableau for the regular schema of example 3. Note that not every available formula is passed down the constructed tableau in the diagram.

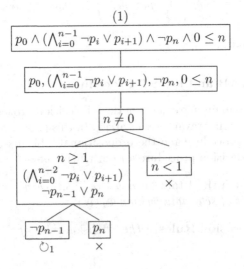

$$(1)$$

The symbol $\circlearrowleft_1$ at the bottom of the left-most branch represents the looping rule. Essentially it means that the branch at the denoted point is the same as the branch at (1) (the top of the tableau), but for $n-1$ instead of n. We will not delve deeper into the theory behind the looping rule as we only rely on the existence of such a rule for our procedure to work– for more details on the looping we refer to [4].

Finally, we recall the following two concepts over a set Φ of schemata: the interval constraints $(IC(\Phi))$ and the conjunction of arithmetic formulae in Φ (Φ_Z). The formula $IC(\Phi)$ is the conjunction of the arithmetic formulae $min_\phi(i) \leq i \wedge i \leq max_\phi(i)$ for each $\phi \in \Phi$ and for each bound variable in ϕ. We assume that all bound variables are distinct in Φ and $min_\phi(i)$ is defined as the minimal value that can be assigned to the bound variable i, whereas $max_\phi(i)$ is the maximum value that can be assigned to the bound variable i.

Definition 9 (Pure Literal). *A literal p_a (respectively $\neg p_a$) is pure in a set of schemata Φ iff for every occurrence of a literal $\neg p_b$ (respectively p_b) in Φ, the arithmetic formula $\Phi_Z \wedge IC(\Phi) \wedge a = b$ is unsatisfiable.*

This definition will be modified to formalize the class of pure overlap schemata.

3   Linked Schemata

The class of linked schemata is an extension of regular schemata based on the following observation:

$$\left(\bigwedge_{i=1}^{n} p_i\right) \wedge \left(\bigvee_{j=n+1}^{m} \neg p_i\right) \equiv_S \left(\bigwedge_{i=1}^{n} p_i\right) \wedge \left(\bigvee_{j=1}^{m} \neg q_i\right). \tag{4}$$

Simply, we choose the interpretations such that $[\![p_{n+k}]\!]_\mathcal{I} = [\![q_k]\!]_\mathcal{I}$ for $k \in [1, m]$. By the finiteness of the language, we can separate the integers into two distinct parts, those greater than n and those less than n. Thus, the propositional variable p in the interval $[1, n]$ is invariant to the labelling of the propositional variable in the interval $[n+1, m]$. They can share the same name or not, the assignment will not influence the interpretations which model the schema. This observation is similar to the reduction from monadic predicate logic with monadic function symbols to monadic predicate logic without monadic function symbols, as outlined in Sec. 6.2 of "The Classical Decision Problem" [10].

3.1   Construction

The simplest way to understand the construction of the class of linked schemata is that any regular schema consists of atoms (specifically, ones not contained in iterations) and iterations. We will refer to these "parts" as the *principal objects*, denoted by $\mathcal{P}(\phi)$ of a schema ϕ. We consider sets Φ of regular schemata, such that the propositional symbols are distinct with regards to the regular schemata in the set, i.e. if $\phi, \psi \in \Phi$ and ϕ contains a propositional variable using the symbol p then ψ cannot contain propositional variables using this symbol. We can compute $\bigcup_{\phi \in \Phi} \mathcal{P}(\phi)$ without any propositional symbols occuring in two iterations indexed by different free parameters. Using this set of "parts" and the propositional connectives $\neg$, $\vee$, and $\wedge$ we can construct new formulae. The rest of this section will be focused on the formalization of this concept.

Definition 10. *Let $p \in P$ be a propositional symbol and φ a formula schema, then $occ(p, \varphi) = 1$ iff p occurs in φ, otherwise it is $occ(p, \varphi) = 0$*

Definition 11 (principal Objects). *Given a schema φ we can construct the set of principal objects $\mathcal{P}(\varphi)$ using the following inductive definition:*

- $\mathcal{P}(P_a) \Rightarrow \{P_a\}$
- $\mathcal{P}(\bigvee_{i=a}^{b} \psi) \Rightarrow \left\{ \bigvee_{i=a}^{b} \psi \right\}$
- $\mathcal{P}(\bigwedge_{i=a}^{b} \psi) \Rightarrow \left\{ \bigwedge_{i=a}^{b} \psi \right\}$
- $\mathcal{P}(\neg\psi) \Rightarrow \mathcal{P}(\psi)$
- $\mathcal{P}(\phi \vee \psi) \Rightarrow \mathcal{P}(\phi) \cup \mathcal{P}(\psi)$
- $\mathcal{P}(\phi \wedge \psi) \Rightarrow \mathcal{P}(\phi) \cup \mathcal{P}(\psi)$

One can consider $\mathcal{P}(\varphi)$ as a specially constructed set of formula schema.

Example 5. Let use compute the set of principal objects of the following regular schema:

$$\varphi \equiv (0 \leq n) \wedge P_0 \wedge \bigwedge_{i=1}^{n} (\neg P_{i-1} \vee P_i) \wedge \neg P_n \tag{5}$$

We get $\mathcal{P}(\varphi) = \{(0 \leq n), P_0, \bigwedge_{i=1}^{n} (\neg P_{i-1} \vee P_i), P_n\}$

We will abbreviate the set of propositional connectives used as $\mathcal{O} = \{\wedge, \vee, \neg\}$. By $\psi \in cl_\mathcal{O}(\Phi)$, we mean that ψ can be constructed using the set of formula schema Φ and the logical connective set $\mathcal{O}$.

Example 6. Using the principal object set from Ex. 5 and the set of operators $\mathcal{O} = \{\wedge, \vee, \neg\}$, some of the formulae we can construct are:

$$\psi_1 = (0 \leq n) \wedge P_0 \wedge \bigwedge_{i=1}^{n} (\neg P_{i-1} \vee P_i) \wedge \neg P_n \tag{6}$$

$$\psi_2 = \left((0 \leq n) \wedge P_0 \wedge \bigwedge_{i=1}^{n} (\neg P_{i-1} \vee P_i) \wedge \neg P_n \right) \vee \left((0 \leq n) \wedge \neg P_0 \wedge \bigwedge_{i=1}^{n} (\neg P_{i-1} \vee P_i) \wedge P_n \right) \tag{7}$$

It is not necessary that the constructed formulae are valid, satisfiable, or unsatisfiable. One can check that both $\psi_1, \psi_2 \in cl_\mathcal{O}(\mathcal{P}(\varphi))$.

Lemma 1. *If φ is a regular schema, then all $\psi \in cl_\mathcal{O}(\mathcal{P}(\varphi))$ have the same aligned interval as φ.*

Proof. Assuming that φ has an aligned interval $[\alpha, n - \beta]$, then any, of its parts must have an aligned interval of at most $[\alpha, n - \beta]$ and are themselves regular schema. Thus, ψ is a boolean combination with the same aligned interval, implying that its aligned interval must be the same. $\square$

Using this simple result we will define the class of linked schemata, as follows.

Definition 12 (The class of Linked Schemata). *Let us consider the class Λ of all finite sets Φ of regular schemata such that for all propositional symbols p, we have that $\left(\sum_{\phi \in \Phi} occ(p, \phi) \right)$ is either 1 or 0, we define the class **LS** of linked schemata as*

$$\mathbf{LS} = \bigcup_{\Phi \in \Lambda} cl_{\mathcal{O}} \left(\bigcup_{\phi \in \Phi} \mathcal{P}(\phi) \right)$$

Lemma 2. *If φ is a regular schema, then it is a linked schema.*

Proof. By definition 12, we can consider the set $\Phi = \{\varphi\}$, also, $\varphi \in cl_{\mathcal{O}}(\mathcal{P}(\varphi))$, and thus, $\varphi \in \mathbf{LS}$. $\square$

Theorem 2. *The class of regular schemata is contained but not equal to the class of linked schemata.*

Proof. We prove this by providing an example, see Ex. 7, of a linked schema which is not a regular schema. $\square$

Example 7. Let us consider Φ containing the following three regular schemata. In what follows, we write $A \leftrightarrow B$ as an abbreviation for $(\neg A \vee B) \wedge (\neg B \vee A)$:

$$\varphi_1 = \bigvee_{i=1}^{k} \neg P_i \wedge \neg \bigvee_{i=1}^{k} \neg P_i \tag{8a}$$

$$\varphi_2 = \bigvee_{i=1}^{m} Q_i \wedge \bigvee_{i=1}^{m} R_i \wedge \bigwedge_{i=1}^{m} Q_i \leftrightarrow R_i \tag{8b}$$

$$\varphi_3 = \bigwedge_{i=1}^{n} M_i \tag{8c}$$

We can construct the following **LS** formula using Φ:

$$\left(\left(\bigvee_{i=1}^{k} \neg P_i \rightarrow \bigvee_{i=1}^{m} Q_i \right) \wedge \left(\bigvee_{i=1}^{m} R_i \rightarrow \bigwedge_{i=1}^{n} M_i \right) \wedge \bigwedge_{i=1}^{m} (Q_i \leftrightarrow R_i) \right) \rightarrow$$
$$\left(\bigvee_{i=1}^{k} \neg P_i \rightarrow \bigwedge_{i=1}^{n} M_i \right). \tag{9}$$

Formula 9 gives a formalization of the composition of certain boolean functions when one function's range has the same number of bits as another function's domain. This formula is obviously not regular, but it is linked. This concludes the proof of Thm. 2.

4 Pure Overlap Schemata

In this section we show how one can weaken the restriction that propositional symbols occur indexed by only one parameter. Consider the following formula schema ψ:

$$0 \leq n \wedge \left(\bigwedge_{i=0}^{n} p_i \right) \vee \left(\bigwedge_{i=0}^{m} \neg p_i \right) \wedge 0 \leq m \tag{10}$$

It is not a linked schema because p occurs indexed by two different parameters, however, using the tableaux extension rule for propositional $\vee$ we see that the occurrences are handled by two different branches, thus each parameter can be handled separately. It is also important to note that

$$0 \leq n \wedge \left(\bigwedge_{i=0}^{n} p_i \right) \vee \left(\bigwedge_{i=0}^{m} \neg q_i \right) \wedge 0 \leq m \tag{11}$$

Replacing p with q in Eqn. 10 results in Eqn. 11, which changes the formula from valid to satisfiable (only when $0 \leq n, m$). Thus, we cannot reduce this formula to linked schemata without changing its semantic properties. To deal with this problem we introduce relatively pure literals, based on the observation that if the negation of a literal occurs in the same branch indexed by a different parameter then the literal must not be of arithmetic importance. We then show that relatively pure literals can be dropped without effecting satisfiability of the considered pure overlap schemata.

4.1 Construction

We first introduce the notion of relatively pure literals and detail the construction of pure overlap schemata.

Definition 13 (Iteration Invariant DNF (IIDNF)). *The* Iteration Invariant disjunctive normal *form of a linked schema is a schema of the form:*

$$(\varphi_{1,1} \wedge \cdots \wedge \varphi_{1,n_1}) \vee \cdots \vee (\varphi_{m,1} \wedge \cdots \wedge \varphi_{m,n_m})$$

where $m, n_1, \cdots, n_m \in \mathbb{N}$ *(note they are not free parameters, but rather meta variables) and* $\varphi_{i,j}$ *is either an iteration, an atom, or negated atom. We will refer to the formula* $(\varphi_{i,1} \wedge \cdots \wedge \varphi_{i,n_i})$ *as clauses* C_i *for* $i \in [1, m]$, *That is, given a formula* φ *in IIDNF we will write* $C_i \in \varphi$ *as the* i^{th} *clause of* φ.

Lemma 3. *Given a set of regular schemata* Φ, *for all* $\psi \in cl_O \left(\bigcup_{\phi \in \Phi} P(\phi) \right)$ *there exists an IIDNF of* ψ.

Proof. Since, iterations are not unfolded in the creation of an IIDNF form of ψ, the problem reduces to showing that all propositional formulae have a DNF form, which is a well known result. Also, it is possible to put a regular schemata into *Negation Normal Form* (NNF) because negation can be passed over iterations, i.e $\neg \bigwedge_{i=a}^{b} \phi_i \equiv_\models \bigvee_{i=a}^{b} \neg \phi_i$ and $\neg \bigvee_{i=a}^{b} \phi_i \equiv_\models \bigwedge_{i=a}^{b} \neg \phi_i$. $\qquad \square$

Definition 14 (Relatively Pure Literal). *Given a set of regular schemata* Φ, *let* $\psi \in cl_{\mathcal{O}}\left(\bigcup_{\phi\in\Phi}\mathcal{P}(\phi)\right)$ *and* ψ' *be the IIDNF of* ψ. *A literal* p_a $(\neg p_a)$ *is relatively pure in* ψ *iff for every clause* $C \in \psi'$ *and for any two distinct regular schemata* $\varphi_1, \varphi_2 \in \Phi$ *used to construct* ψ, *where* $p_a \in C$ $(\neg p_a \in C)$, $\neg p_b \in C$ $(p_b \in C)$, $p_a \in \varphi_1$ $(\neg p_a \in \varphi_1)$ *and* $\neg p_b \in \varphi_2$ $(p_b \in \varphi_2)$, *the arithmetic formula* $\Phi_Z \wedge IC(\Phi) \wedge a = b$ *is unsatisfiable, where* $\Phi = \mathcal{P}(C)$.

Example 8. Consider the schemata:

$$\neg(5 < n) \wedge \left(\bigwedge_{i=0}^{n} p_i\right) \wedge \left(\bigwedge_{j=6}^{m} \neg p_i\right) \wedge 0 \leq m \tag{12}$$

The literal p_i $(\neg p_i)$ is relatively pure in this example.

We will refer to a schema as *relatively pure* if all the literals in the schema are either relatively pure or in the IIDNF of the schema they only occur in clauses being indexed by a single parameter. The non-IIDNF form of a relatively pure schema is also relatively pure. Given a set of regular schemata Φ, let $cl_{\mathcal{O}}^{rp}(\Phi)$ be the set of all schema which can be constructed using the logical connectives $\mathcal{O}$ such that they are relatively pure.

Definition 15 (The class of Pure Overlap Schemata). *Let us consider the class* Λ *of all finite sets* Φ *of regular schemata. We define the class of* pure overlap schemata *as*

$$\mathbf{POS} = \bigcup_{\Phi\in\Lambda} cl_{\mathcal{O}}^{rp}\left(\bigcup_{\phi\in\Phi}\mathcal{P}(\phi)\right)$$

It should be noted that even though the definition of relatively pure literals uses the IIDNF of a positional schema it is not the case that members of **POS** must be in IIDNF.

Example 9. Both Ex. 8 and Eqn. 10 are in the class of pure overlap schemata.

Lemma 4. *If* φ *is a linked schema, then it is a pure overlap schema.*

Proof. A linked schema is a pure overlap schema where each propositional variable is indexed by only one parameter. $\qquad\square$

Theorem 3. *The class of linked schemata is contained but not equal to the class of pure overlap schemata.*

Proof. Eqn. 10 is a pure overlap schema but not a linked schema. $\qquad\square$

5   A Decision Procedure for POS

We now introduce a decision procedure for the class POS of schemata, by using and extending results of [4], as follows.

Algorithm 1 (ST$^{\textbf{POS}}$ Procedure). *Given a schema $\varphi \in$ **POS** in negation normal form. The following algorithm, called the $ST^{\textbf{POS}}$ procedure, decides the satisfiability of φ:*

1) *Apply STAB propositional extension rules with highest priority until no more can be applied. This results in m sets of atoms and iterations referred to as $B_1, \ldots, B_m$.*
2) *For each B_i, we separate B_i into n (the number of parameters in B_i) sub-branches $B_{(i,1)}, \cdots B_{(i,n)}$, where each $B_{(i,j)}$ contains iterations and atoms indexed by a single parameter. Atoms without a free parameter in the indices can be added to every $B_{(i,j)}$. We will mark such a sub-branching with $\otimes_n$ where n is the number of parameters on the branch.*
3) *Run the ST procedure on the sub-branch $B_{(i,j)}$.*
4) *For any branch B_i, if one of its sub-branches $B_{(i,j)}$ has a closed tableau after following the ST procedure, then the branch B_i is closed.*

Let us make the following observation about the $ST^{\textbf{POS}}$ decision procedure. When it comes to constructing the interpretation for a formula in **POS** we specifically defined the class such that the procedure to construct the model would be precisely the procedure used for regular schemata, except the number of possible models would increase. For linked schemata this is obvious, the propositional symbols are distinct in every sub-branch. However, for pure overlap schemata two distinct sub-branches (of the same branch) can contain the same propositional symbol, but by Def. 14 the occurrences are distinct from each other arithmetically if one occurrence is negated and the other occurrence is not. Thus, when a propositional symbol occurs on two distinct sub-branches and the two occurrences are not arithmetically distinct, the two occurrences must be of the same polarity. In this case when one sub-branch forces the propositional variable using the positional symbol to be true (false in the case of a negated literal), the other sub-branches will also interpret this literal as true. In some sense one can consider it as a local tautology which can be removed from consideration when constructing the model.

Theorem 4. *The $ST^{\textbf{POS}}$ procedure terminates for **POS**.*

Proof. The key to the termination is that we only need to decompose the members of **POS** using the procedure outlined above. This decomposition process always terminates being that we are, up to this point, only applying propositional tableaux extension rules. When the formulae are completely decomposed we use the ST procedure on each sub-branch. The procedure is known to terminate for regular schemata [4] and each of the sub-branches is regular. □

In regards to the soundness and completeness of $ST^{\textbf{POS}}$ the procedure, it was shown that STAB is sound and complete for all propositional schemata [4] (Sec. 5.4). The propositional schemata we introduce in this paper are constructed using exactly the same language as in the work by Aravantinos et al. [4] Our extension of STAB with the sub-branching rule does not change the soundness and completeness results being that the sub-branching rule, rather than being an additional tableaux rule, is more a method to enforce termination. It essentially states that instead of considering the given branch as a whole we consider it in parts using the same tableaux rules introduced for STAB in prior work.

Theorem 5. *The $ST^{\mathbf{POS}}$ decision procedure is sound and complete for all propositional schemata $\psi \in \mathbf{POS}$.*

Example 10. We conclude this section by illustrating our $ST^{\mathbf{POS}}$ decision procedure on the following formula ψ:

$$p_0 \wedge \left(\left(\left(\bigwedge_{i=0}^{k} \neg q_i \right) \vee \left(\bigwedge_{i=0}^{n} \neg p_i \vee p_{i+1} \right) \wedge \neg p_{n+1} \right) \vee \left(\left(\bigwedge_{i=0}^{m} \neg p_{i-1} \vee p_i \right) \wedge p_{m+1} \wedge \neg q_{w+3} \right) \right)$$

Applying $ST^{\mathbf{POS}}$ on the above formula, we obtain the following branching tree (corresponding to the run of $ST^{\mathbf{POS}}$):

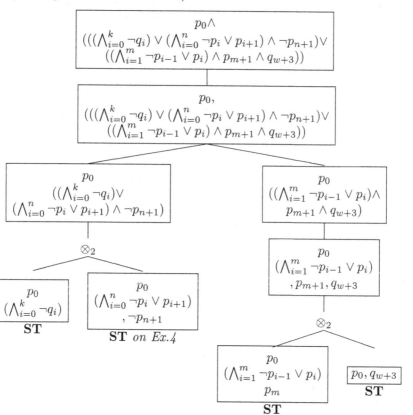

Interesting result of this derivation is that the assignment to w influences the interpretation modelling the formula. If an interpretation $\mathcal{I}$ assigns $q_{-1} = T, q_0 = F, p_0 = T, p_{-1} = F$, $p_5 = F$,$w \leftarrow -4$, $n \leftarrow -2$, $k \leftarrow 0$, and $m \leftarrow 5$ then $\mathcal{I} \models \psi$. But if $\mathcal{I}$ assigns to $n \leftarrow 0$ keeping the same propositional variable assignments, then $\mathcal{I} \not\models \psi$.

6   Conclusion and Future Work

In this work we have shown that the ST procedure of Aravantinos et al. [4] can be extended to handle more expressive classes of schemata which allow for restricted use of multiple free parameters. The two classes shown, though their construction is awkward, are simple to conceptually understand and work with. Also, neither requires the heavy machinery of normalized clause sets, nor do the classes require a conversion of the schemata into a clausal normal form. Also, the introduced decision procedure $ST^{\textbf{POS}}$ is sound, complete, and terminates for all propositional schemata in the class **POS**. Though an advantage normalized clause sets have over both of the introduced classes of schemata is that they can handle propositional variables being indexed by multiple free parameters without restriction. This is one of the significant advantages to separating the propositional part from the equational part, and using a levelled resolution calculus. When it is not required to have unrestricted usage of the propositional variables it suffices to use STAB. This also has the added value of compression being that it is possible for clausal form to result in an exponential increase in the size of the formula.

As for future work, further increase in expressivity by relaxing the purity constraint does not seem feasible as this would require two parameters to be active in the same branch. This is when the undecidability result for propositional schemata [4] stops us in our tracks. However, investigating how the new classes outlined here can interact with the class of *regular nested schemata* [3] could lead to new expressivity results. In particular, we are interested in the relationship between alternation-free μ-calculus [11] and such a class of schemata. Also, Aravantinos et al. [5] investigated the relationship between LTL and regular schemata. Being that pure overlap schemata are a super class of regular schemata it is quite possible that a more expressive temporal logic is related to pure overlap schemata or linked schemata. In either case this work has broaden the scope of application of propositional schemata.

Acknowledgements. I would like to give special thanks to Daniel Weller[1] and Alexander Leitsch[2] for their help with constructing a concise mathematical formalism, as well as Giselle Reis [3] for help with editing of the formalisms.

References

1. Aczel, P.: An introduction to inductive definitions. In: Barwise, J. (ed.) Handbook of Mathematical Logic. Studies in Logic and the Foundations of Mathematics, vol. 90, pp. 739–782. Elsevier (1977)
2. Aravantinos, V., Caferra, R., Peltier, N.: A schemata calculus for propositional logic. In: Giese, M., Waaler, A. (eds.) TABLEAUX 2009. LNCS, vol. 5607, pp. 32–46. Springer, Heidelberg (2009)

---

[1] http://www.logic.at/staff/weller/index.html
[2] http://www.logic.at/people/leitsch/
[3] http://www.logic.at/staff/giselle/

3. Aravantinos, V., Caferra, R., Peltier, N.: A decidable class of nested iterated schemata. In: Giesl, J., Hähnle, R. (eds.) IJCAR 2010. LNCS, vol. 6173, pp. 293–308. Springer, Heidelberg (2010)
4. Aravantinos, V., Caferra, R., Peltier, N.: Decidability and undecidability results for propositional schemata. J. Artif. Int. Res. 40(1), 599–656 (2011)
5. Aravantinos, V., Caferra, R., Peltier, N.: Linear temporal logic and propositional schemata, back and forth. In: Proceedings of the 2011 Eighteenth International Symposium on Temporal Representation and Reasoning, TIME 2011, pp. 80–87. IEEE Computer Society, Washington, DC (2011)
6. Aravantinos, V., Echenim, M., Peltier, N.: A resolution calculus for first-order schemata. Fundamenta Informaticae (2013)
7. Baaz, M.: Note on the generalization of calculations. Theoretical Computer Science 224(1-2), 3–11 (1999)
8. Baaz, M., Hetzl, S., Leitsch, A., Richter, C., Spohr, H.: Ceres: An analysis of Fürstenberg's proof of the infinity of primes. Theor. Comput. Sci. 403(2-3), 160–175 (2008)
9. Baaz, M., Zach, R.: Short proofs of tautologies using the schema of equivalence. In: Börger, E., Gurevich, Y., Meinke, K. (eds.) CSL 1993. LNCS, vol. 832, pp. 33–35. Springer, Heidelberg (1994)
10. Börger, E., Grädel, E., Gurevich, Y.: The Classical Decision Problem. Springer (1997)
11. Bradfield, J.C.: The modal mu-calculus alternation hierarchy is strict. In: Montanari, U., Sassone, V. (eds.) CONCUR 1996. LNCS, vol. 1119, pp. 233–246. Springer, Heidelberg (1996)
12. Comon, H.: Inductionless induction. In: Robinson, J.A., Voronkov, A. (eds.) Handbook of Automated Reasoning (in 2 volumes), pp. 913–962. Elsevier and MIT Press (2001)
13. Cooper, D.: Theorem proving in arithmetic without multiplication. Machine Intelligence (1972)
14. Corcoran, J.: Schemata: The concept of schema in the history of logic. Bulletin of Symbolic Logic (2), 219–240
15. Dunchev, C., Leitsch, A., Rukhaia, M., Weller, D.: Ceres for first-order schemata. CoRR, abs/1303.4257 (2013)
16. Gentzen, G.: Fusion of several complete inductions. In: Szabo, M.E. (ed.) The Collected Papers of Gerhard Gentzen. Studies in Logic and the Foundations of Mathematics, vol. 55, pp. 309–311. Elsevier (1969)
17. Giesl, J., Kapur, D.: Decidable classes of inductive theorems. In: Goré, R.P., Leitsch, A., Nipkow, T. (eds.) IJCAR 2001. LNCS (LNAI), vol. 2083, pp. 469–484. Springer, Heidelberg (2001)
18. Kapur, D., Subramaniam, M.: Extending decision procedures with induction schemes. In: McAllester, D. (ed.) CADE-17. LNCS, vol. 1831, pp. 324–345. Springer, Heidelberg (2000)
19. Krajíček, J., Pudlák, P.: The number of proof lines and the size of proofs in first order logic. Archive for Mathematical Logic 27(1), 69–84 (1988)
20. Orevkov, V.P.: Proof schemata in Hilbert-type axiomatic theories. Journal of Soviet Mathematics 55(2), 1610–1620 (1991)
21. Parikh, R.J.: Some results on the length of proofs. Transactions of the American Mathematical Society 177, 29–36 (1973)
22. Takeuti, G.: Proof Theory. Studies in logic and the foundations of mathematics, vol. 81. American Elsevier Pub. (1975)

Detecting Unknots via Equational Reasoning, I: Exploration

Andrew Fish[1] and Alexei Lisitsa[2]

[1] School of Computing, Engineering and Mathematics, University of Brighton
[2] Department of Computer Science, The University of Liverpool
Andrew.fish@brighton.ac.uk, A.Lisitsa@csc.liv.ac.uk

Abstract. We explore the application of automated reasoning techniques to unknot detection, a classical problem of computational topology. We adopt a two-pronged experimental approach, using a theorem prover to try to establish a positive result (i.e. that a knot is the unknot), whilst simultaneously using a model finder to try to establish a negative result (i.e. that the knot is not the unknot). The theorem proving approach utilises equational reasoning, whilst the model finder searches for a minimal size counter-model. We present and compare experimental data using the involutary quandle of the knot, as well as comparing with alternative approaches, highlighting instances of interest. Furthermore, we present theoretical connections of the minimal countermodels obtained with existing knot invariants, for all prime knots of up to 10 crossings: this may be useful for developing advanced search strategies.

1 Introduction

One of the most well-known and intriguing problems in computational topology is *unknot detection* (UKD): given a *knot*, which is a closed loop without self-intersection embedded in 3-dimensional Euclidean space $\mathbb{R}^3$, is it possible to deform $\mathbb{R}^3$ continuously such that the knot is transformed into a trivial unknotted circle without passing through itself? Knots are often studied as a diagrammatic system: (i) a knot diagram is a regular projection of the knot onto a plane, having a finite number of singularities, all of which are transverse double points annotated to indicate which strand is passing over and which is passing under at each crossing; (ii) knots are equivalent if and only if their diagrams differ by a finite sequence of Reidemesiter moves [20]. Figure 1 shows the diagrams of two knots with a negative a) and positive b) answers to the unknottedness question. All work and results stated assume that knots are tame (a common technical requirement which is generally imposed on knots, ruling out pathological cases such as permitting infinite sequences of trefoil-like knot pieces of decreasing sizes glued together within a knot); see [26] for more details, for instance.

The unknot detection (or unknot recognition) problem has attracted a lot of attention, but some of the fundamental questions about it still remain open. In particular, it is unknown whether it is possible to recognize unknots in PTIME. It is known, though, that the problem lies in NP ∩ coNP [12,16] (membership

S.M. Watt et al. (Eds.): CICM 2014, LNAI 8543, pp. 76–91, 2014.

a)   b)

Fig. 1. a) Non-trivial trefoil knot and b) trivial knot or unknot

in $coNP$ is subject to generalized Riemann hypothesis holding). There has been a slow but steady development of algorithms for unknot detection and their experimental evaluation. An early algorithm, presented by W. Haken in his proof of the decidability of UKD [11], was developed for theoretical purposes and was deemed to be impractical due to being too complex to attempt to implement it. Since then various algorithms for unknot detection have been proposed with various degrees of implementability and efficiency [7]. The algorithms based on *monotone simplifications* [7] provide practically fast recognition of unknots but do not necessarily yield a decision procedure. The algorithms based on *normal surface theory*, implemented in Regina system [3], provide efficient recognition of non-trivial knots. In particular, it is reported that every non-trivial knot with crossing number ≤ 12 is recognized as such by the procedure from [3] in under 5 minutes. There still are efficiency problems with the existing algorithms, which in the worst case are exponential, and it appears that establishing that a particular diagram with a few hundred (or even dozens of) crossings represents a non-trivial knot may well be out of reach of the available procedures. Thus the exploration of alternative procedures for unknot detection is an interesting and well-justified task.

In this paper we explore the following route to the efficient practical algorithms for unknot detection. The unknotedness property can be faithfully characterized by the properties of algebraic invariants associated with knot projections. We attempt to establish the properties of concrete invariants by using methods and procedures developed in the *automated reasoning* area. A key observation is that the task of unknot detection can be reduced to the task of (dis)proving a first-order formulae, and for this there are efficient generic automated procedures, notwithstanding the fact that generally first-order-order validity is undecidable.

2 Involutory Quandles and Unknot Detection

We provide relevant background definitions; for example, see [9,15,19,6] for further details.

Definition 1. *Let Q be a set equipped with a binary operation $\triangleright$ (product) such that the following hold:*

Q1 $x \triangleright x = x$ *for all* $x \in Q$.

Q2 *For all* $x, y \in Q$, *there is a unique* $z \in Q$ *such that* $x = z \triangleright y$.

Q3 *For all* $x, y, z \in Q$, *we have* $(x \triangleright y) \triangleright z = (x \triangleright z) \triangleright (y \triangleright z)$.

Then Q *is called a* quandle[1]. *If* Q *additionally satisfies the identity Q2' below, then* Q *is called an* involutory quandle:

Q2' $(x \triangleright y) \triangleright y = x$ *for all* $x, y \in Q$.

Remark 1. For a quandle Q, the unique element $z \in Q$ from axiom 2 is denoted by $z = x \triangleright^{-1} y$, and $\triangleright^{-1}$ also defines a quandle structure. However, for involutory quandles, we have $\triangleright = \triangleright^{-1}$, which can be taken as an equivalent definition of involutory; axiom 2' supersedes axiom 2.

Definition 2. *A function* $\phi : Q_1 \to Q_2$ *between quandles is a homomorphism if* $(a \triangleright b)\phi = (a)\phi \triangleright (b)\phi$ *for any* $a, b \in Q_1$.

Given a knot K (i.e. a circle embedded in $\mathbb{R}^3$), a well known invariant is the *knot group* of K, which is $\pi(K) = \pi_1(\mathbb{R}^3 - K)$, the fundamental group of the complement of the knot K in $\mathbb{R}^3$ (i.e. homotopy class of paths in the complement of the knot). One can compute a presentation of the knot group, in terms of generators and relations, from a knot diagram, following Wirtinger (e.g. see [26] for details). An analogous construction can be used to construct a presentation of the *quandle* of the knot, $Q(K)$ (e.g. see [19] for details). One acquires the presentation of the knot group from the presentation of the quandle by considering the generators and relations in the group, and imposing the quandle operation to be conjugation. Since we focus primarily on involutory quandles, we provide the simplified construction for these below; a method for generalising to (not involutory) quandles is to assign an orientation to the knot, which yields a sign for each crossing according to the relative orientations of the involved curves, and then the relation assigned to the crossing is either $a \triangleright b = c$ or $a \triangleright^{-1} b = c$ according to the sign of the crossing. Interpreting the quandle relation $\triangleright$ as conjugation in the knot group (i.e. $a \triangleright b = b^1 a b$), and $\triangleright^{-1}$ as its inverse (i.e. $a \triangleright^{-1} b = bab^1$) returns the well known Wirtinger presentation of the knot group.

Definition 3. *A presentation of the involutory knot quandle,* $IQ(K)$ *for a knot* K, *is obtained from a diagram* D *for* K *as follows: a solid arc of the diagram is an unbroken line of the diagram with an undercrossings at each of its ends; every solid arc of the diagram is labelled by an unique label; all labels of* D *form the set* G_D *of generators; to every crossing of* D *one associates a relation, as shown in Figure 2; denote the set of all such relations by* R_D. *Then the presentation* $\langle G_D \mid R_D \rangle$ *defines the involutory quandle* $IQ(D)$. *This is a quotient of the free involutory quandle modulo the equational theory defined by* R_D.

The three equalities $Q1, Q2'$ and $Q3$ form an equational theory of involutory quandles, which we denote by E_{iq}.

---

[1] A *rack* [9] is such a Q that satisfies Q2 and Q3 but not necessarily Q1.

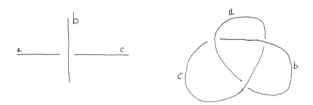

Fig. 2. (a) Left: A labelled crossing and its corresponding relation $a \triangleright b = c$; here a and c are the labels of the underarcs at this crossing, whilst b is the label of the overarc, and we often identify the arcs with their labels to simplify language in discussions. (b) Right: The trefoil knot diagram, with solid arcs a,b,c.

Example 1. Let D_{tr} be the diagram of the trefoil knot K shown in Figure 2. The involutory quandle of D_{tr} is defined by the presentation $IQ(D_{tr}) = \langle a, b, c \mid a \triangleright b = c, b \triangleright c = a, c \triangleright a = b \rangle$. For comparison, the quandle $Q(D_{tr})$ has presentation $Q(D_{tr}) = \langle a, b, c \mid a \triangleright^{-1} b = c, b \triangleright^{-1} c = a, c \triangleright^{-1} a = b \rangle$, whilst the knot group has presentation $G(K) = \langle a, b, c \mid bab^{-1} = c, cbc^{-1} = a, aca^{-1} = b \rangle$. In detail, consider the crossing of the diagram in which a and c are underarcs, whilst b is an overarc; i.e. match up the crossing locally with (a rotated version of) Figure 2(a). This gives rise to either the relation $a \triangleright^{-1} b = c$ or $a \triangleright b = c$, depending on whether the *sign* of the crossing is negative or positive, respectively. One method for reading off the sign is to choose an orientation of the knot (i.e. pick a direction on the knot, often depicted using an arrowhead) and if a is the approaching underarc, following orientation (one can traverse a knot, or part of a knot, intuitively being a walk around the knot along the arcs; then following orientation means that one is traversing the arc in the direction determined by the orientation), check if one turns left or right, respectively, when passing onto the overarc b, following orientation. In this case, all three crossings are negative; a mirror of the diagram (i.e. exchanging all over and under crossings) would have the above presentation for knot quandle, but with $\triangleright^{-1}$ replaced by $\triangleright$, and similarly for the knot group.

2.1   Overview of the Approach

The importance of involutory quandles, in the context of unknot detection, relies on the following properties [14,15,19]:

- Involutory quandle is a knot invariant, i.e. it does not depend on the choice of diagram;
  [Theorem 15.1 of [14] shows that the quandle $Q(K)$ of knot K is an invariant of the knot type of K, and the involutory quandle $IQ(K)$ is a homomorphic image of $Q(K)$]
- Involutory quandle $IQ(K)$ of a knot K is trivial (i.e. it contains a single element e with $e * e = e$) if and only if K is the *unknot*.
  [Theorem 5.2.5 of [19]].

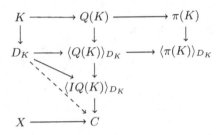

Fig. 3. An overview of the objects related to the unknot detection programme

These properties suggest the following approach to unknot detection. Given a knot diagram, one can try to decide whether its associated involutory quandle is trivial. Notice that an involutory quandle of a knot can be an infinite set [19]. Not much progress has been made towards the development of specific decision procedures for such a problem, apart of that presented in the thesis of S. Winker [19]; the diagrammatic method presented there, together with details and explanations, allows one to construct the involutory quandles for many knot diagrams, and in our opinion, is a very good starting point for developing algorithmic procedures directly dealing with the involutary quandles. In this paper, we take an alternative route and propose, instead of applying a specific involutary quandles decision procedure, to tackle unknot detection as follows:

- Given a knot diagram, compute its involutary quandle presentation;
- Convert the task of involutary quandle triviality detection into the task of proving a first-order equational formula;
- Concurrently, apply generic automated reasoning tools for first-order equational logic to tackle the (dis)proving task

Thus, we concurrently search for a proof and for a model to disprove the formula. After explaining the details, we apply these methods in parallel, present empirical data for many knots, and compare the countermodels with existing knot invariants that are encoded in the smallest homomorphic image of the involutory quandle of the knot. As an overview of the related objects, Figure 3 shows the following. In the top row, we have the knot K, fundamental quandle of the knot, $Q(K)$, and the projection from $Q(K)$ onto the fundamental group $\pi(K)$ obtained by forgetting information (peripheral subgroup and meridian, discussed later), and essentially setting the quandle operation to be conjunction. In the second row, the diagram D_K of the knot K, together with presentations of $Q(K)$ and $\pi(K)$ obtained from those diagrams. The third row shows the involutory quandle presentation $IQ(K)$ obtained by identifying the quandle operation with its inverse. Here C is a finite involutory quandle that is the homomorphic image of the involutory quandle of the knot. Then, if X is any quandle, one can ask if imposing the involutory condition on X can yield such a C.

2.2 Unknot Detection by Equational Reasoning

Given a knot diagram D, with n arcs, consider its involutory quandle representation $IQ(D) = \langle G_D \mid R_D \rangle$ with $G_D = \{a_1, \ldots, a_n\}$. Denote by $E_{iq}(D)$ an equational theory of $IQ(D)$, i.e. $E_{iq}(D) = E_{iq} \cup R_D$. It is known that the axioms of (involutory) quandles are algebraic counterparts of the Reidemeister moves (see further discussion of that in Section 6).

Proposition 1. *A knot diagram D is a diagram of the unknot if and only if $E_{iq}(D) \vdash \wedge_{i=1\ldots n-1}(a_i = a_{i+1})$, where $\vdash$ denotes derivability in the equational logic (or, equivalently in the first-order logic with equality).*

Proof. *(Sketch)* D is a diagram of unknot iff $IQ(D)$ is a trivial involutive quandle [19]. The proposition "$IQ(D)$ is a trivial involutory quandle iff $E_{iq}(D) \vdash \wedge_{i=1\ldots n-1}(a_i = a_{i+1})$" is an easy consequence of the soundness and completeness of equational logic (Birkhoff Theorem) [4]. See also Lemma 4.2.7 p. 30 of [19]. The case of first-order logic with equality follows from the conservativity of first-order logic with equality over equational logic for equational theories. □

Now, if $E_{iq}(D) \vdash \wedge_{i=1\ldots n-1}(a_i = a_{i+1})$ holds true, then this fact can be established by a proof of the formula $E_{iq}(D) \rightarrow \wedge_{i=1\ldots n-1}(a_i = a_{i+1})$ by a *complete* automated theorem prover for first-order logic with equality, of which there are many around, see e.g. [2]. By a complete theorem prover we mean an automated procedure which, given a valid formula, terminates with a proof of the formula.

For an introduction to automated theorem proving see e.g. [10]. In order to show that $E_{iq}(D) \vdash \wedge_{i=1\ldots n-1}(a_i = a_{i+1})$ does not hold, it suffices to disprove $E_{iq}(D) \rightarrow \wedge_{i=1\ldots n-1}(a_i = a_{i+1})$. We propose to do this by the application of generic finite model finding procedures [5,18] to find a finite countermodel to the formula, or equivalently a finite model for $E_{iq}(D) \wedge \neg \wedge_{i=1\ldots n-1}(a_i = a_{i+1})$. So, the unknot detection procedure P which we propose here consists of the parallel composition of

- automated proving $E_{iq}(D) \rightarrow \wedge_{i=1\ldots n-1}(a_i = a_{i+1})$, and
- automated disproving $E_{iq}(D) \rightarrow \wedge_{i=1\ldots n-1}(a_i = a_{i+1})$ by a finite model finder.

It is obvious that the parallel composition above provides with *at least* a semidecision algorithm for unknotedness. If D is a diagram of the unknot then the termination of the theorem proving is guaranteed by the completeness of a theorem prover. On the other hand, if D is a diagram of a non-trivial knot then the termination can be guaranteed only if a *finite* countermodel exists. In general, in the first-order logic, there are formulae which can only be refuted on *infinite* countermodels, so for arbitrary formulae the termination of the automated disproving cannot be guaranteed.

For the specific type of formulae $E_{iq}(D) \rightarrow \wedge_{i=1\ldots n-1}(a_i = a_{i+1})$ we conjecture that they have *finite countermodel property*, that is if there exists a countermodel for a formula of this form at all, then there is a finite countermodel too. Since a countermodel for $E_{iq}(D) \rightarrow \wedge_{i=1\ldots n-1}(a_i = a_{i+1})$ is a model for $E_{iq}(D) \wedge \neg \wedge_{i=1...n-1} (a_i = a_{i+1})$, it follows that: 1) such a countermodel is a *homomorphic* image of the involutory quandle $IQ(D)$ of D, by satisfaction of $E_{iq}(D)$; and 2) it is non-trivial involutory quandle, by satisfaction of $\neg \wedge_{i=1...n-1} (a_i = a_{i+1})$.

Thus the required *finite countermodel* property, for the programme to yield a decision procedure, is equivalent to the property of the involutory quandles of knots being *residually finite*, as formulated in the following conjecture.

Conjecture 1 (Involutory quandles are finitely residual). For any knot diagram D, if $IQ(D)$ is not trivial (i.e. consists of more than 1 element), then there is a finite non-trivial involutory quandle Q which is a homomorphic image of $IQ(D)$.

We remark on the conjecture. Following Hempel and Thurston, we know that knot groups are residually finite. Thurston [21] states that "It is a standard fact that a finitely generated subgroup of $GL_n(\mathbb{Q})$ (the general linear group with coefficients in the rationals) is residually finite. Using this, one easily sees that the fundamental group of any geometric 3-manifold is residually finite. After a certain amount of fussing, one can assemble finite quotients of the fundamental groups of pieces of a geometric decomposition of a 3-manifold to obtain finite quotients of the fundamental group of the entire manifold.". The question of whether the proof can be lifted to quandles is another matter. Since the knot quandle contains the same information as a group system which is the triple (G, P, m) consisting of the knot group G, a peripheral subgroup P, and a meridian m in P, one would require that the *group system* (see e.g. [19,14]) is somehow preserved, and the finite homomorphic image has an induced quandle structure.

Theorem 1. *The unknot detection procedure, P, given above, is a decision procedure if conjecture 1 holds and a semi-decision procedure otherwise.*

In the next section we illustrate the practical applicability of the proposed (semi)-decision procedure to various instances of unknot detection problem. In the experiments, we use an automated theorem prover Prover9 and a finite model finder Mace4, both by W. McCune [18]. We present some examples of the results of the approach on unknots with interesting properties.

3   Experiments: Detecting Unknots

Culprit Unknot. is shown in Figure 4. This is an interesting unknot which, during any untangling of it by Reidemeister moves, necessarily requires an increase in the number of crossings. The formula of the form $E_{iq}(D) \to \wedge_{i=1...n-1}(a_i = a_{i+1})$ for the culprit unknot diagram in the syntax of Prover9/Mace4 (with $*$ denoting involutory quandle operation $\triangleright$) is presented to the right of the figure. Prover9 proves the formula in 0.03 seconds demonstrating thereby that culprit is indeed the unknot. The entire proof can be found in [17].

Assumptions:

```
%Involutory quandle axioms
x * x = x.
(x * y) * y = x.
(x * z) * (y * z) = (x * y) * z.
%Culprit unknot
a1 = a9 * a7.
a3 = a1 * a2.
a2 = a3 * a4.
a5 = a2 * a10.
a6 = a5 * a4.
a7 = a6 * a1.
a8 = a7 * a4.
a10 = a8 * a9.
a4 = a10 * a3.
a9 = a4 * a8.
```

Goals:

```
(a1 = a2) & (a2 = a3) &
(a3 = a4) & (a4 = a5) &
(a5 = a6) & (a6 = a7) &
(a7 = a8) & (a8 = a9) &
(a9 = a10).
```

Fig. 4. Culprit Unknot

Haken's Gordian unknot diagram has 141 crossings, and is the one of the most well-known concrete, hard-to-detect, unknots; see Figure 4. Prover9 produces the proof of the formula of the form $E_{iq}(D) \to \wedge_{i=1...n-1}(a_i = a_{i+1})$ for this diagram in just under 15 seconds, demonstrating that indeed it is the unknot. The input, and the proof produced by the prover, can be found in [17].

The only alternative approach capable of detecting unknotedness of Haken's Gordian Unknot in practice, that we are aware of, is Dynnikov's algorithm based on *monotone simplifications* [7,8,1].

We have experimented also with the detection of other well-known hard unknots, such as Goerlitz unknot, Thistlethwaite unknot, Friedman's Twisted unknot. In all of these cases Prover9 was able to establish unknotedness in under a second. See further details in [17]. This can be compared with the times "in only a few seconds" required to detect unknotedness of these instances by the Heegaard tool, reported by the author in [27].

4   Experiments: Detecting Non-trivial Knots

We first provide an example of the output from the counter-model finder for a non-trivial knot. Next, we present a table containing the time taken for each of the prime knots in the knot tables, for up to 10 crossings. We will compare the output sizes with the known invariant of knots, called the determinant of the knot, later.

Trefoil. (Figure 1 a) is the simplest non-trivial knot; the countermodel found is:

Fig. 5. Haken's Gordian Unknot

```
interpretation( 3, [number=1, seconds=0], [
        function(a1, [ 0 ]),
        function(a2, [ 1 ]),
        function(a3, [ 2 ]),
        function(*(_,_), [
    0, 2, 1,
    2, 1, 0,
    1, 0, 2 ])
]).
```

The table of prime knots, the size of the minimal countermodel found, and the time taken to find, is given in the Appendix.

Comparisons. For the collection of prime non-trivial knots of up to 12 crossings from [22], the system in [3], based on linear programming in conjunction with normal surface theory, claims to massively improve on previous approaches, claiming to solve all cases efficiently (in under 5 minutes). Their time-based data is subsequent to highly optimised polynomial time pre-processing simplifications and uses an encoding based on triangulations of the complement of the knot. Bearing in mind the differences in encodings we have compared the performance of the Regina unknot detection algorithm with our approach using Mace4 on the knots with 10 crossings. For the five special cases $10_{83}, 10_{91}, 10_{92}, 10_{117}, 10_{119}$ for which our approach does not terminate in a reasonable time, Regina completed the work in under 3 minutes per knot. For the remaining cases the average time was 47 seconds for Regina, and 1230 seconds for our approach. In general our approach demonstrates much higher discrepancy in timing data: it was very efficient for the cases when small countermodels were found. For countermodel sizes of up to 15-17 our detection time is under a second – this holds in more than

70% of the instances, and here our approach outperforms Regina's algorithm. In a few cases with large countermodels (e.g $10_{88}, 10_{94}, 10_{115}$) our approach takes 40000-80000 seconds to complete the search. Further comparisons on the large sets of knots is a subject of ongoing and future work.

5   Countermodels and Knot Invariants

Any countermodel found is a finite quandle, C, which is a homomorphic image of $IQ(K)$. Thus it a homomorphic image of $Q(K)$ which factors through $IQ(K)$. Such finite quandles may be constructed via "involutising" quandles which are homomorphic images of $Q(K)$. The countermodel search process finds the smallest such finite quandle C, and we see that these do not all arise via the same known quandles. We present results which demonstrate that the majority, but not all, of the small alternating knots (of size up to 10) arise as quotients of the dihedral quandle. The question of how the other minimal size finite quandles arise is still open, but this demonstrates that the methodology is particularly interesting in that it is discovering the smallest size quandle invariant, over all such invariants, for each case, as opposed to computing each invariant in turn, as per the common approach using invariants.

Definition 4 (dihedral quandle). *Let R_n be the set of reflections in the dihedral group D_{2n} of order $2n$ (which one can regard as the symmetry group of the regular n-gon). Then R_n forms a quandle of order n, called the dihedral quandle of order n.*

Proposition 2. *For any knot K, with determinant not equal to ± 1, R_p is a finite non-trivial involutory quandle which is the image of the fundamental quandle of the knot, where p is smallest prime divisor of the determinant of the knot.*

Proof. For more details on racks and quandles, and this construction, see [9]. A homomorphic image of the fundamental quandle of the knot K into R_n may be given by colouring the arcs of any diagram of K with n colours $0, 1, \ldots, n - 1$ such that at each crossing if x_a, x_b, x_c are the three colours assigned to the arcs labelled a, b, c, with b the overarc, then $x_c \equiv 2x_b - x_a$ mod n. If n is prime, then it is well known that these equations have a non-constant solution if and only if n divides $det(K)$ the *determinant* of K (obtainable as the evaluation of the Alexander polynomial at $t = -1$, sometimes denoted $\Delta(-1)$). In general, a representation into any finite quandle can be interpreted as a suitable colouring scheme for the diagram. Since the elements of the dihedral quandle are all reflections, the quandle is involutory by definition.

The experimental computation, together with comparison of the determinant of the knot gives us the following, where as usual, the numbering convention is that generally adopted for prime knots in the knot tables; see [22] for example.

Proposition 3. *Out of the 251 prime, alternating knots of up to 10 crossings, from the knot tables, a smallest non-trivial involutory quandle which is a homomorphic image of the fundamental quandle of the knot is: of size 15 for 22 knots:*

$9_{22}, 9_{25}, 9_{30}, 9_{36}, 9_{44}, 9_{45}, 10_{46}, 10_{47}, 10_{49}, 10_{70}, 10_{72}, 10_{73}, 10_{79}, 10_{80}, 10_{93}, 10_{102},$ $10_{124}, 10_{126}, 10_{127}, 10_{148}, 10_{149}, 10_{153}$; *of size 28 for 11 knots:* $10_{50}, 10_{51}, 10_{52}, 10_{53},$ $10_{54}, 10_{55}, 10_{57}, 10_{131}, 10_{135}, 10_{150}, 10_{151}$; *of size 31 for 1 knot:* 10_{115}; *of size 32 for 1 knot:* 10_{118}; *of size 36 for 1 knot:* 10_{110}; *of size equal to the smallest prime divisor of the determinant of the knot for the remaining 213 knots.*

Corollary 1. *For 213 of the 251 prime, alternating knots of up to 10 crossings, there is no smaller non-trivial involutory quandle which is a homomorphic image of the fundamental quandle of the knot than the dihedral rack on p elements, where p is smallest prime divisor of the determinant of the knot. For the remaining 38 knots, there is a smaller such non-trivial involutory quandle.*

5.1 Discussion: Countermodels and Small Quandles

Winker [19] remarks that for a certain class of knots (technically, those which are the closure of 4-strand braids), the involutory quandle $IQ(K)$ is finite and has order equal to the knot determinant $|K|$ (or $det(K)$), citing [14], and that since every prime knot of 7 or fewer crossings is 4-strand it has finite involutory quandle. On the other hand, $IQ(8_{16})$ and $IQ(9_{35})$ are infinite, as are the (k, m, n)-pretzel knots (knots obtainable by a construction involving a certain process of k twists, m twists and n twists) when $1/k + 1/m + 1/n \leq 1$; knots $8_5 = K_{2,3,3}$ and $9_{35} = K_{3,3,3}$ are examples of such pretzel knots. There does not appear to be an existing complete classification of involutory quandles. We observe that for any knot K which has a finite involutory quandle $IQ(K)$, this involutory quandle itself would be a countermodel, as would the projection onto the quandle arising from each of the colouring numbers. Whilst we we may find the homomophic image of the quandle with size the smallest prime divisor of $det(K)$, corresponding to the smallest colouring number, this does not, a priori, rule out smaller homomorphic images of involutory quandles that arise in other ways. Furthermore, there are some prime knots K with $det(K) = \pm 1$. The knots 10_{124} and 10_{153} are the only such prime alternating knots with up to 10 crossings. For both of these knots we find a smallest homomorphic image of involutory quandle of size 15. In [9], they observe that the representation in a reflection rack, whose elements are the edges of a dodecahedron, can be used to distinguish knots which have determinant ± 1 and so have no non trivial representation to R_n, giving 10_{124} an an example. Similarly, in [14], they present the knot 10_{124} with determinant 1, indicating that $AbQ_2(K)$ is trivial (this is the abelian, involutory quandle on K, where abelian means that $(w \triangleright x) \triangleright (y \triangleright z) = (w \triangleright y) \triangleright (x \triangleright z)$). But the involutory quandle $IQ(K) = Q_2(K)$ is non-trivial and has order 30, and may be faithfully represented on a sphere as the 30 midpoints of the edges of a dodecahedron projected onto the sphere.

In [23], they use the Library for Automated Deduction Research in the endeavour of identifying isomorphism classes of small quandles, and they present several families of quandles; as well as considering the dihedral quandle, they also refer to linear quandles, the Alexander quandles, and transposition quandles. These are candidate classes to consider in the identification of the countermodels. For instance, the transposition quandle, T_n, has size $n(n-1)/2$, and so these are candidates to explain our countermodels arising at sizes 15, 28 and 36. The identification and understanding of exactly which quandle families have smallest homomorphic image is an intriguing open problem to be explored in future work, with guidance from the countermodel finder approach adopted. Furthermore, Proposition 11.2 of [14] says that every involutory quandle is representable as an involutory quandle with geodesics, and so the construction of the smallest such involutory quandle with geodesics for a knot K will, in fact, correspond to our search for minimal countermodel.

Winker [19] states that the involutory quandle of a knot or link is either finite or "not too infinite", and gives examples to show that knots can have different knot groups but the same involutory quandle (e.g. the Figure of Eight knot and the $(5,2)$ torus knot), and that the involutory quandle of a particular prime link (the Borromean rings) is infinite.

6   Equational Reasoning and Untangling Unknots

Recall Proposition 1: a knot diagram D is a diagram of the unknot if and only if $E_{iq}(D) \vdash \wedge_{i=1...n-1}(a_i = a_{i+1})$, where $\vdash$ denotes derivability in the equational logic (or, equivalently in the first-order logic with equality). We adopt the abbreviation $TRIV \equiv \wedge_{i=1...n-1}(a_i = a_{i+1})$ for the generators $a_1, \ldots, a_n$. Then the condition above will be rewritten as $E_{iq} \vdash ?TRIV$. The axioms of involutory quandles can be seen as algebraic counterparts of the Reidemeister moves:

1. $x \triangleright x = x$ for all $x \in Q$ ($\sim RM_1$)
2. $(x \triangleright y) \triangleright y = x$ for all $x, y \in Q$ ($\sim RM_2$)
3. $(x \triangleright z) \triangleright (y \triangleright z) = (x \triangleright y) \triangleright z$ for all $x, y, z \in Q$ ($\sim RM_3$)

For $I \subseteq \{1, 2, 3\}$, denote by E_{iq}^I an equational theory formed by the corresponding subset of the axioms $1-3$ given above. In particular $E_{iq}^{\{1,2,3\}} = E_{iq}$. Reidemeister's theorem [20] says that a diagram D is a diagram of an unknot if and only if D can be transformed to a trivial diagram D_U by a finite sequence of Reidemeister moves. Denote by $D \rightarrow^I D'$ the fact that D can be transformed to D' using the Reidemeister moves drawn only from I. In this section we explore possible connections between equational proofs and Reidemeister transformations. The following proposition expresses the fact that the equational proof can simulate simplifications by Reidemeister moves.

Proposition 4. *For any non-empty $I \subseteq \{1, 2, 3\}$, if $D \rightarrow^I D^U$ then $E_{iq}^I \vdash TRIV$. Furthermore an equational proof can be constructively built by a simple procedure from the untangling sequence of Reidemeister moves.*

Proof. (Sketch) Consider the case of $I = \{1, 2, 3\}$. Assume that for a diagram D we have $D \rightarrow^I D^U$. That is, there is a sequence of diagrams $D = D_1, \ldots, D_i, \ldots, D_n = D^U$ such that every diagram in the sequence is obtained from the previous one by a single application of a Reidemeister move. Let

$IQ(D) = \langle G_D \mid R_D \rangle$ be a presentation of the involutory quandle of D with the set of generators G_D and set of relators R_D. Denote by $\mathcal{T}(D)$ the set of all terms built upon the set of constants, identified with the generators G_D, together with the involutory quandle operation $\triangleright$ as the only term construct. Denote by $A(D_i)$ the set of solid arcs of the diagram D_i. A *labelling* L of a diagram D_i is a mapping $L : A(D_i) \to \mathcal{T}(D)$. Now we demonstrate the inductive construction of a sequence of pairs $(E(D_i), L_i)$, associating with each D_i a set of equations $E(D_i)$ and a labelling $L_i : A(D_i) \to \mathcal{T}(D)$ satisfying the following properties:

1. $E(D_i) \subseteq E(D_{i+1})$;
2. $E(D_i) \vdash E(D_{i+1})$;
3. L_i is consistent w.r.t. involutory quandle labelling rules on solid arcs of $A(D_i)$, meaning $E(D_i) \vdash L_i(a) \triangleright L_i(b) = L_i(c)$ for all $a, b, c \in A(D_i)$ positioned as shown in Figure 2 (a);
4. If $t \in \mathcal{T}(D)$ is in $L_i(A(D_i))$ but not in $L_{i+1}(A(D_{i+1}))$, and $|A(D_i)| > 1$ then there exist an s in $L_{i+1}(A(D_{i+1}))$ and $t = s$ in $E(D_{i+1})$.

The intuition is that the equation set grows, adding statements of equality which are derived from the axioms, according to the progress in the unknotting sequence. The condition on the size of the arc set in Property 4 is required since one can still apply RM moves to untangle a diagram which has only one arc, but there is no more equational rewriting to perform.

Assume that the above four properties are satisfied. Then, in any untangling sequence of diagrams the last diagram D^U is a trivial diagram of the unknot. Then D^U has just one arc with a label τ, say. Let D_k be the last diagram in the sequence which has this property of having just one arc with label τ. Then, by Property 4, the label ρ of the last removed arc (in the diagram D_{k-1} preceding D_k in the sequence) is provably equal to τ, that is, $E(D_{k-1}) \vdash (\tau = \rho)$, and $(\tau = \rho) \in E(D_{k-1})$. Unwinding the process backwards (and formally applying induction) we obtain that the labels of all of the removed arcs are provably equal to each other, including all of the generators. Thus, the required $n - 1$ pairwise equalities of the n generators are derivable from $E(D^U)$, and the result follows. The details of the construction of the sequence $(E(D_i), L_i)$ can be found in the Appendix of the extended version of this paper in [17]. For any I which is a proper subset of $\{1, 2, 3\}$ the proof follows the same route using the property that the construction of $(E(D_{i+1}), L_{i+1})$ depends only on $(E(D_i), L_i)$ and a type of RM used to transform D_i into D_{i+1}.

The approach can be used to investigate which of the Reidemeister moves are required in a proof of unknottedness.

Proposition 5. *Culprit unknot (see above) needs all three Reidemeister moves to untangle.*

Proof. For $I = \{2, 3\}, \{1, 3\}, \{1, 2\}$ one can disprove $E_{iq}^I \vdash TRIV$ by finding countermodels by Mace4 automatically of sizes 2,3,4, respectively.

An interesting question: is it possible to make a simulation in the opposite direction, that is, to extract an untangling sequences of Reidemeister moves from equational proofs? Although we don't have a definite answer here, in some simple cases one can indeed extract the moves from the proofs. We leave the development of systematic Reidemester move extraction procedures for future work.

7 Conclusion

We presented the basis for a new method for unknot detection, based on parallel application of theorem prover and (counter)-model finder. It appears interesting, in that it has different abilities to existing approaches. In particular, the countermodel finder is producing the smallest non-trivial homomorphic image of the involutory quandle of the knot; thus it is, in some sense, finding the smallest invariant which distinguishes it from the unknot. Furthermore, the approach lends itself to new avenues of research, such as the exploration of the correlations between the equational proofs provided by the theorem provers and the corresponding sequences of labelled diagrams in an unknotting sequence. Thus, whilst we have provided some interesting examples of unknot detection, exploring the whole spectrum of unknots and the relative difficulty of their detection, in comparison with other methodologies, will be an interesting avenue to explore. Furthermore, developing any correspondences between quandle-labelled diagram transformations and the unknotting proofs produced may provide interesting insights, potentially leading to more advanced tailored reasoning strategies.

Another direction is to explore automated deduction approach using different knot invariants such as knot groups and (non-involutory) quandles. In terms of groups, unknotedness corresponds to *commutativity* deciding which can also be reduced to the equational theorem (dis)proving. We have early indications that using involutory quandles, as explored in this paper, might be more efficient than using knot groups; for example, for a torus knot $T_{3,5}$ disproving using involutory quandles took less than a second and produced a countermodel of size 15, whilst disproving commutativity of the group of $T_{3,5}$ did not finish in 500 seconds and no countermodels of size less than 120 were found.

References

1. Andreeva, M., Dynnikov, I., Koval, S., Polthier, K., Taimanov, I.: Book Knot Simplifier, http://www.javaview.de/services/knots/doc/description.html (accessed March 14, 2014)
2. The CADE ATP System Competition, The World Championship for Automated Theorem Proving, http://www.cs.miami.edu/~tptp/CASC/ (accessed June 07, 2013)
3. Burton, B.A., Olzen, M.: A fast branching algorithm for unknot recognizion with experimental polynomial-time behaviour. arXiv:1211.1079 [math.GT]
4. Birkhoff, G.: On the structure of abstract algebras. Proc. Cambridge Philos. Soc. 31, 433–454 (1935)
5. Caferra, R., Leitsch, A., Peltier, N.: Automated Model Building. Applied Logic Series, vol. 31. Kluwer (2004)

6. Scott Carter, J.: A survey of quandle ideas. arXiv:1002.4429 [math.GT]
7. Dynnikov, I.A.: Recognition algorithms in knot theory. Uspekhi Mat. Nauk 58(6(354)), 45–92 (2003)
8. Dynnikov, A.: Three-page link presentation and an untangling algorithm. In: Proc. of the International Conference Low-Dimensional Topology and Combinatorial Group Theory, Chelyabinsk, Kiev, July 31-August 7, pp. 112–130 (1999, 2000)
9. Fenn, R., Rourke, C.: Racks and Links in Codimension two. J. Knot Theory Ramifications 01, 343 (1992)
10. Goubault-Larrecq, J., Mackie, I.: Proof Theory and Automated Deduction. Applied Logic Series, vol. 6. Kluwer (2001)
11. Haken, W.: Theorie der Normal achen. Acta Math. 105, 245–375 (1961)
12. Hass, J., Lagarias, J.C., Pippenger, N.: The computational complexity of knot and link problems. J. Assoc. Comput. Mach. 46(2), 185–211 (1999)
13. Hempel, J.: Residual finiteness for 3-manifolds. In: Combinatorial Group Theory and Topology (Alta, Utah, 1984). Ann. of Math. Stud., vol. 111, pp. 379–396. Princeton Univ. Press, Princeton (1987)
14. Joyce, D.: A Classifying Invariant of Knots, the Knot Quandle. Journal of Pure and Applied Algebra 23, 37–65 (1982)
15. Joyce, D.: Simple Quandles. Journal of Algebra 79, 307–318 (1982)
16. Kuperberg, G.: Knottedness is in NP, modulo GRH, Preprint, arXiv:1112.0845 (November 2011)
17. Unknot detection by equational reasoning, http://www.csc.liv.ac.uk/~alexei/unknots/ (accessed April 14, 2014)
18. McCune, W.: Prover9 and Mace4, http://www.cs.unm.edu/~mccune/mace4/
19. Winker, S.N.: Quandles, Knot Invariants and the N-fold Branching Cover. PhD Thesis, University of Illinois at Chicago (1984)
20. Reidemeister, K.: Elementare Begründung der Knotentheorie. Abh. Math. Sem. Univ. Hamburg 5, 24–32 (1926)
21. Thurston, W.P.: Three-dimensional manifolds, Kleinian groups and hyperbolic geometry. Bull. Amer. Math. Soc. (N.S.) 6(3), 357–381 (1982)
22. Cha, J.C., Livingston, C.: KnotInfo: Table of knot invariants, http://www.indiana.edu/~knotinfo (accessed January 2014)
23. Wu, Z.S.: Computable Invariants for Quandles. Thesis, Bard College, New York (2012)
24. Wallace, S.D.: Homomorphic images of link quandles. MA thesis, Houston, Texas (2004)
25. Manoim, B.: Toward an Online Knowledgebase for Knots and Quandles. Technical report, http://asclab.org/asc/sites/default/files/docs/ Online%20database%20knots%20%26%20quandles.pdf
26. Rolfsen, D.: Knots and Links. AMS Chelsea Publishing (2004)
27. http://mathoverflow.net/questions/144158/what-is-the-state-of-the-art-for-algorithmic-knot-simplification

Appendix

The table below shows the experimental data obtained so far for the knots in the standard knot tables of at most 10 crossings. For each case, we present the size of the minimal countermodel to unknottedness which is found, together with the time taken to find this countermodel. In the majority of these cases the size of the countermodel is the smallest prime divisor of the determinant of the knot. In a few cases, the process was manually terminated after a certain time had elapsed, as indicated.

Table 1. Experimental data for the time taken to find the minimal countermodels for the knots in the standard knot tables of at most 10 crossings.

Knot	3_1	4_1	5_1	5_2	6_1	6_2	6_3	7_1	7_2	7_3	7_4	7_5	7_6	7_7	8_1	8_2
Size	3	5	5	7	3	11	13	7	11	13	3	17	19	3	13	17
Time	0.0	0.01	0.01	0.0	0.0	0.01	0.01	0.01	0.03	0.03	0.0	0.08	0.16	0.01	0.06	0.11

Knot	8_3	8_4	8_5	8_6	8_7	8_8	8_9	8_{10}	8_{11}	8_{12}	8_{13}	8_{14}	8_{15}	8_{16}	8_{17}	8_{18}
Size	17	19	3	23	23	5	5	3	3	29	29	31	3	5	37	3
Time	0.12	0.17	0.01	0.41	0.47	0.01	0.01	0.01	0.01	1.31	2.11	2.06	0.01	0.01	37.66	0.03

Knot	8_{19}	8_{20}	8_{21}	9_1	9_2	9_3	9_4	9_5	9_6	9_7	9_8	9_9	9_{10}	9_{11}	9_{12}	9_{13}
Size	3	3	3	3	3	19	3	23	3	29	31	31	3	3	5	37
Time	0.00	0.01	0.01	0.00	0.00	0.12	0.00	0.45	0.00	1.53	3.77	1.50	0.00	0.01	0.01	3.77

Knot	9_{14}	9_{15}	9_{16}	9_{17}	9_{18}	9_{19}	9_{20}	9_{21}	9_{22}	9_{23}	9_{24}	9_{25}	9_{26}	9_{27}	9_{28}	9_{29}
Size	37	3	3	3	41	41	41	43	15	3	3	15	3	3	3	3
Time	4.28	0.00	0.01	0.00	5.97	12.0	122.73	20.83	0.11	0.00	0.00	0.11	13.20	0.00	0.03	0.00

Knot	9_{30}	9_{31}	9_{32}	9_{33}	9_{34}	9_{35}	9_{36}	9_{37}	9_{38}	9_{39}	9_{40}	9_{41}	9_{42}	9_{43}	9_{44}	9_{45}
Size	15	5	59	61	3	3	15	3	3	5	3	7	7	13	15	15
Time	0.09	0.00	1498.44	670	0.00	0.00	0.11	0.03	0.05	0.01	0.01	0.01	0.01	0.06	0.08	0.08

Knot	9_{46}	9_{47}	9_{48}	9_{49}	10_1	10_2	10_3	10_4	10_5	10_6	10_7	10_8	10_9	10_{10}	10_{11}	10_{12}
Size	3	3	3	5	17	23	5	3	3	37	43	29	3	3	43	47
Time	0.03	0.01	0.03	0.05	0.09	0.72	0.03	0.01	0.03	8.38	16.94	22.52	0.01	0.01	554	21.50

Knot	10_{13}	10_{14}	10_{15}	10_{16}	10_{17}	10_{18}	10_{19}	10_{20}	10_{21}	10_{22}	10_{23}	10_{24}	10_{25}	10_{26}	10_{27}	10_{28}
Size	53	3	43	47	41	5	3	5	3	7	5	5	5	61	71	53
Time	90.53	0.03	16.22	11.84	18.14	0.01	0.03	0.00	0.00	0.01	144.42	0.01	0.00	1000	153	73.41

Knot	10_{29}	10_{30}	10_{31}	10_{32}	10_{33}	10_{34}	10_{35}	10_{36}	10_{37}	10_{38}	10_{39}	10_{40}	10_{41}	10_{42}	10_{43}	10_{44}
Size	3	67	3	3	5	37	7	3	53	59	61	3	71	3	79	79
Time	0.01	298.11	0.03	0.00	0.03	11.38	0.03	0.03	36.98	157.08	76.25	0.00	554.25	0.01	728.36	968

Knots	10_{45}	10_{46}	10_{47}	10_{48}	10_{49}	10_{50}	10_{51}	10_{52}	10_{53}	10_{54}	10_{55}	10_{56}	10_{57}	10_{58}	10_{59}	10_{60}
Size	89	15	15	7	15	28	28	28	28	28	28	5	28	5	3	5
Time	2629.41	0.11	0.11	0.01	0.14	4.12	6.33	4.76	5.94	3.28	8.53	0.03	5.51	0.01	0.03	0.01

Knot	10_{61}	10_{62}	10_{63}	10_{64}	10_{65}	10_{66}	10_{67}	10_{68}	10_{69}	10_{70}	10_{71}	10_{72}	10_{73}	10_{74}	10_{75}	10_{76}
Size	3	3	3	3	3	3	3	3	3	15	7	15	15	3	3	3
Time	0.01	0.01	0.03	0.01	0.03	0.01	0.01	0.01	0.01	0.09	0.00	0.12	0.09	0.03	0.01	0.00

Knots	10_{77}	10_{78}	10_{79}	10_{80}	10_{81}	10_{82}	$\mathbf{10_{83}}$	10_{84}	10_{85}	10_{86}	10_{87}	10_{88}	10_{89}	10_{90}	$\mathbf{10_{91}}$	$\mathbf{10_{92}}$
Size	3	3	15	15	5	3	**≥ 42**	3	3	5	3	101	3	7	**≥ 34**	**≥ 62**
Time	0.03	0.01	0.12	0.14	0.00	0.01	**≥ 21170**	0.00	0.00	0.00	0.03	63849	0.00	0.03	**≥ 20828**	**≥ 62848**

Knot	10_{93}	10_{94}	10_{95}	10_{96}	10_{97}	10_{98}	10_{99}	10_{100}	10_{101}	10_{102}	10_{103}	10_{104}	10_{105}	10_{106}	10_{107}	10_{108}
Size	15	71	7	3	3	3	3	5	5	15	3	7	7	3	3	3
Time	0.62	40879.72	0.01	0.00	0.01	0.01	0.01	0.01	0.01	1.01	0.00	0.01	0.03	0.03	0.00	0.00

Knot	10_{111}	10_{112}	10_{113}	10_{114}	10_{115}	10_{116}	$\mathbf{10_{117}}$	10_{118}	$\mathbf{10_{119}}$	10_{120}	10_{121}	10_{122}	10_{123}	10_{124}	10_{125}	10_{126}
Size	7	3	3	3	31	5	**≥ 37**	32	**≥ 41**	3	5	11	11	15	11	15
Time	0.00	0.00	0.00	0.03	85098	0.01	**≥ 150410**	28	**≥ 261100**	0.00	0.00	0.03	0.12	0.08	0.03	0.11

Knot	10_{127}	10_{128}	10_{129}	10_{130}	10_{131}	10_{132}	10_{133}	10_{134}	10_{135}	10_{136}	10_{137}	10_{138}	10_{139}	10_{140}	10_{141}	10_{142}
Size	15	11	5	17	28	5	19	23	28	3	5	3	3	3	3	3
Time	0.11	0.03	0.01	0.28	6.5	0.03	0.62	1.50	3.80	0.03	0.00	0.03	0.01	0.03	0.03	0.01

Knots	10_{143}	10_{144}	10_{145}	10_{146}	10_{147}	10_{148}	10_{149}	10_{150}	10_{151}	10_{152}	10_{153}	10_{154}	10_{155}	10_{156}	10_{157}	10_{158}
Size	3	3	3	3	3	15	15	28	28	11	15	13	5	5	7	3
Time	0.00	0.01	0.01	0.01	0.00	0.12	0.11	6.72	6.36	0.06	0.12	0.08	0.00	0.01	0.03	0.01

Knot	10_{159}	10_{160}	10_{161}	10_{162}	10_{163}	10_{164}	10_{165}									
Size	3	3	5	5	3	3	3									
Time	0.01	0.03	0.01	0.03	0.03	0.01	0.01									

Applying Machine Learning to the Problem of Choosing a Heuristic to Select the Variable Ordering for Cylindrical Algebraic Decomposition

Zongyan Huang[1], Matthew England[2], David Wilson[2], James H. Davenport[2], Lawrence C. Paulson[1], and James Bridge[1]

[1] University of Cambridge Computer Laboratory, Cambridge CB3 0FD, U.K.
[2] University of Bath, Department of Computer Science, Bath, BA2 7AY, U.K.
{zh242,lp15,jpb65}@cam.ac.uk,
{J.H.Davenport,M.England,D.J.Wilson}@bath.ac.uk

Abstract. Cylindrical algebraic decomposition(CAD) is a key tool in computational algebraic geometry, particularly for quantifier elimination over real-closed fields. When using CAD, there is often a choice for the ordering placed on the variables. This can be important, with some problems infeasible with one variable ordering but easy with another. Machine learning is the process of fitting a computer model to a complex function based on properties learned from measured data. In this paper we use machine learning (specifically a support vector machine) to select between heuristics for choosing a variable ordering, outperforming each of the separate heuristics.

Keywords: machine learning, support vector machine, symbolic computation, cylindrical algebraic decomposition, problem formulation.

1 Introduction

Cylindrical algebraic decomposition (CAD) is a key tool in real algebraic geometry. It was first introduced by Collins [18] to implement quantifier elimination over the reals, but has since been applied to applications including robot motion planning [49], programming with complex valued functions [22], optimisation [28] and epidemic modelling [15]. Decision methods for real closed fields are of great use in theorem proving [25]. METITARSKI [1], for example, decides the truth of statements about special functions using CAD and rational function bounds.

When using CAD, we often have a choice over which variable ordering to use. It is well known that this choice is very important and can dramatically affect the feasibility of a problem. In fact, Brown and Davenport [14] presented a class of problems in which one variable ordering gave output of double exponential complexity in the number of variables and another output of a constant size. Heuristics have been developed to help with this choice, with Dolzmann et al. [23] giving the best known study. However, in CICM last year [8], it was shown that even the best known heuristic could be misled. Although that paper provided

S.M. Watt et al. (Eds.): CICM 2014, LNAI 8543, pp. 92–107, 2014.

an alternative heuristic, this had its own shortcomings, and it now seems likely that no one heuristic is suitable for all problems.

Our thesis is that the best heuristic to use is dependent upon the problem considered. However, the relationship between the problems and heuristics is far from obvious and so we investigate whether machine learning can help with these choices. Machine learning is a branch of artificial intelligence. It uses statistical methods to infer information from supplied data which is then used to make predictions for previously unseen data [2]. We have applied machine learning (specifically a support vector machine) to the problem of selecting a variable ordering for both CAD itself and quantifier elimination by CAD, using the nlsat dataset [50] of fully existentially quantified problems. Our results show that the choices made by machine learning are on average superior to both any individual heuristic and to picking a heuristic at random. The results also provide some new insight on the heuristics themselves. This appears to be the first application of machine learning to problem formulation for computer algebra, although it follows recent application to theorem proving [10, 31].

We conclude the introduction with background theory on CAD and machine learning. Then in Sections 2, 3 and 4 we describe our experiment, its results and how they may be extended in the future. Finally in Section 5 we give our conclusions and ideas for future work.

1.1   Quantifier Elimination and CAD

Let $Q_i \in \{\exists, \forall\}$ be quantifiers and ϕ be some quantifier free formula. Then given

$$\Phi(x_1, \ldots, x_k) := Q_{k+1} x_{k+1} \ldots Q_n x_n \, \phi(x_1, \ldots, x_n),$$

quantifier elimination (QE) is the problem of producing a quantifier free formulae $\psi(x_1, \ldots, x_k)$ equivalent to Φ. In the case $k = 0$ this reduces to the *decision problem*, is Φ true? Tarski proved that QE was possible for semi-algebraic formulae (polynomials and inequalities) over $\mathbb{R}$ [47]. However, the complexity of Tarski's method is non-elementary (indescribable as a finite tower of exponentials) and so CAD was a major breakthrough when introduced, despite complexity doubly exponential in the number of variables. For some problems QE is possible through algorithms with better complexity (see for example the survey by Basu [5]), but CAD implementations remain the best general purpose approach.

Collins' algorithm [3] works in two stages. First, *projection* calculates sets of projection polynomials S_i in variables $(x_1, \ldots, x_i)$. This is achieved by repeatedly applying a projection operator onto a set of polynomials, producing a set with one variable fewer. We start with the polynomials from ϕ and eliminate variables this way until we have the set of univariate polynomials S_1.

Then in the *lifting* stage, decompositions of real space in increasing dimensions are formed according to the real roots of those polynomials. First, the real line is decomposed according to the roots of the polynomials in S_1. Then over each cell c in that decomposition, the bivariate polynomials S_2 are taken at a sample point and a decomposition of $c \times \mathbb{R}$ is produced according to their roots. Taking the union gives the decomposition of $\mathbb{R}^2$ and we proceed this way to a decomposition of $\mathbb{R}^n$. The decompositions are cylindrical (projections of any two cells onto their first i coordinates are either identical or disjoint) and each cell is a semi-algebraic set (described by polynomial relations). Collins' original algorithm used a projection operator which guaranteed CADs of $\mathbb{R}^n$ on which the polynomials in ϕ had constant sign, and thus ϕ constant truth value, on each cell. Hence only a single sample point from each cell needed to be tested and the equivalent quantifier free formula ψ could be generated from the semi-algebraic sets defining the cells in the CAD of $\mathbb{R}^k$ for which Φ is true.

Since the publication of the original algorithm, there have been numerous improvements, optimisations and extensions of CAD (with a summary of the first 20 years given by Collins [19]). Of great importance is the improvement to the projection operator used. Hong [29] proved that a refinement of Collins' operator was sufficient and then McCallum [37] presented a further refinement which could only be used for input that was *well-oriented* and was in turn improved by Brown [11]. Further refinements are possible by removing the need for sign-invariance of polynomials while maintaining truth-invariance of a formula, with McCallum [38] presenting an operator for use when an equational constraint is present (an equation logically implied by a formula) and Bradford *et al.* [7] extending this to the case of multiple formulae. Collins and Hong [20] described Partial CAD for QE, where lifting over a cell is aborted if there already exists sufficient information to determine the truth of ϕ on that cell. Other recent CAD developments of particular note include the use of symbolic-numeric techniques in the lifting stage [33, 45] and the alternative to projection and lifting offered by decompositions of complex space via regular chains technology [17].

When using CAD we have to assign an ordering to the variables (the labels i on the x_i in the discussion above). This dictates the order in which the variables are eliminated during projection and thus the sub-spaces for which CADs are produced en route to a CAD of $\mathbb{R}^n$. For some applications this order is fixed but for others there may be a free or constrained choice. When using CAD for QE we must project quantified variables before unquantified ones. Further, the quantified variables should be projected in the order they occur, unless successive ones have the same quantifier in which case they may be swapped. The ordering can have a big effect on the output and performance of CAD [8, 14, 23].

1.2   Machine Learning

Machine learning [2] deals with the design of programs that can learn rules from data. This is often a very attractive alternative to manually constructing them when the underlying functional relationship is very complex. Machine learning techniques have been widely used in many fields, such as web searching [6], text categorization [42], robotics [44], expert systems [27] and many others.

Various machine learning techniques have been developed. McCulloch and Pitts [39] created the first computational model for *neural networks* called *threshold logic*. Following that, Rosenblatt [40] proposed the *perceptron* as an iterative algorithm for supervised classification of an input into one of several possible non-binary outputs. A later development was the *decision tree* [2], which is a simple representation for classifying examples. The main idea here is to apply serial classifications which refine the output state. At the same time as the *decision tree* was being developed, the *multi-layer perceptron* [30] was explored. It is a modification of the standard linear perceptron and can distinguish data that are non-linearly separable.

In the last decade, the use of machine learning has spread rapidly following the invention of the *Support Vector Machine* (SVM) [41]. This was a development of the perceptron approach and gives a powerful and robust method for both classification and regression. *Classification* refers to the assignment of input examples into a given set of classes (the output being the class labels). *Regression* refers to a supervised pattern analysis in which the output is real-valued. The SVM technology can deal efficiently with high-dimensional data, and is flexible in modelling diverse sources of data. The standard SVM classifier takes a set of input data and predicts one of two possible classes from the input. Given a set of examples, each marked as belonging to one of two classes, an SVM training algorithm builds a model that assigns new examples into one of the classes. The examples used to fit the model are called training examples.

An important concept in the SVM theory is the use of a kernel function [43], which maps data into a high dimensional kernel-defined feature space and then separates samples in the transformed space. Kernel functions enable operations in feature space without ever computing the coordinates of the data in that space. Instead they simply compute the inner products between all pairs of data vectors. This operation is generally computationally cheaper than the explicit computation of the coordinates.

The machine learning experiment described in this paper uses SVM-LIGHT (see Joachims [34]) which is an implementation of SVMs in C. The SVM-LIGHT software consists of two programs: SVM LEARN and SVM CLASSIFY. SVM LEARN fits the model parameters based on the training data and user inputs (such as the kernel function and the parameter values). SVM CLASSIFY uses the generated model to classify new samples. It calculates a hyperplane of the n-dimensional transformed feature space, which is an affine subspace of dimension $n-1$ dividing the space into two corresponding to the two distinct classes. SVM CLASSIFY outputs margin values which are a measure of how far the sample is from this separating hyperplane. Hence the margins are a measure of the confidence in a correct prediction. A large margin represents high confidence in a correct prediction. The accuracy of the generated model is largely dependent on the selection of the kernel functions and parameter values.

2 Methodology

2.1 CAD Implementation and Heuristics

For the machine learning experiment we decided to focus on a single CAD implementation, QEPCAD [12]. We note that other CAD implementations are available, as discussed further in Section 4.

QEPCAD is an interactive command line program written in C for performing **Q**uantifier **E**limination with **P**artial **CAD**. It was chosen as it is a competitive implementation of both CAD and QE that also allows the user some control and information during its execution. We used QEPCAD with its default settings which implement McCallum's projection operator [37] and partial CAD [20]. It can also makes use of an equational constraint automatically (via the projection operator [38]) when one is explicit in the formula, (where *explicit* means the formula is a conjunction of the equational constraint with a sub-formula).

In the experiment we used three existing heuristics for picking a CAD variable ordering:

Brown: This heuristic chooses a variable ordering according to the following criteria, starting with the first and breaking ties with successive ones:
  (1) Eliminate a variable first if it has lower overall degree in the input.
  (2) Eliminate a variable first if it has lower (maximum) total degree of those terms in the input in which it occurs.
  (3) Eliminate a variable first if there is a smaller number of terms in the input which contain the variable.
  It is labelled after Brown who suggested it [13].

sotd: This heuristic constructs the full set of projection polynomials for each permitted ordering and selects the ordering whose corresponding set has the lowest sum of total degrees for each of the monomials in each of the polynomials. It is labelled sotd for *sum of total degree* and was suggested by Dolzmann, Seidell and Sturm [23], whose study found it to be a good heuristic for both CAD and QE by CAD.

ndrr: This heuristic constructs the full set of projection polynomials for each ordering and selects the ordering whose set has the lowest number of distinct real roots of the univariate polynomials within. It is labelled ndrr for *number of distinct real roots* and was suggested by Bradford *et al.* [8]. Ndrr was shown to assist with examples where sotd failed.

Brown's heuristic has the advantage of being very cheap, since it acts only on the input and checks only simple properties. The ndrr heuristic is the most expensive (requiring real root isolation), but is the only one to explicitly consider the real geometry of the problem (rather than the geometry in complex space).

All three heuristics may identify more than one variable ordering as a suitable choice. In this case we took the heuristic's choice to be the first of these after they had been ordered lexicographically.[1]

---

[1] This final choice may depend on the convention used for displaying the variable ordering. QEPCAD and the notes where Brown introduces his heuristic [13] use the convention of ordering variables from left to right so that the last one is projected first. On the other hand, MAPLE and the papers introducing sotd and ndrr [8, 23] use the opposite convention. The heuristics were implemented in MAPLE and so ties were broken by picking the first lexicographically on the second convention. This corresponds to picking the first under a reverse lexicographical order under the QEPCAD convention. The important point is that all three heuristics had ties broken under the same convention and so were treated fairly.

2.2   Problem Data

Problems were taken from the nlsat dataset [50], chosen over more traditional CAD problem sets (such as Wilson *et al.* [48]) as these did not have sufficient numbers of problems for machine learning. 7001 three-variable CAD problems were extracted for our experiment. The number of variables was restricted for two reasons. First to make it feasible to test all possible variable orderings and second to avoid the possibility that QEPCAD will produce errors or warnings related to well-orientedness with the McCallum projection [37].

Two experiments were undertaken, applying machine learning to CAD itself and to QE by CAD. QE is clearly very important throughout engineering and the sciences, but increasingly CAD has been applied outside of this context, as discussed in the introduction. We performed separate experiments since for quantified problems QEPCAD can use the partial CAD techniques to stop the lifting process early if the outcome is already determined, while the full process is completed for unquantified ones and the two outputs can be quite different.

The problems from the nlsat dataset are all fully existential (satisfiability or SAT problems). A second set of problems for the quantifier free experiment was obtained by simply removing all quantifiers. An example of the QEPCAD input for a SAT problem is given in Figure 1 with the corresponding input for the unquantified problem in Figure 2. Of course, for such quantified problems there are better alternatives to building a CAD (see for example the work of Jovanovic and de Moura [36]). However, our decision to use only SAT problems was based on availability of data rather than it being a requirement of the technology, and so we focus on CAD only here and discuss how we might generalise our data in Section 4. For both experiments, the problems were randomly split into training sets (3545 problems in each), validation sets (1735 problems in each) and test sets (1721 problems in each) [2].

2.3   Evaluating the Heuristics

Since each problem has three-variables and all the quantifiers are the same, all six possible variable orderings are admissible. For each ordering we had QEPCAD build a CAD and measured the number of cells. The best ordering was defined as the one resulting in the smallest cell count, (and if more than one ordering gives the minimal both orderings are considered the best). The decision to focus on cell counts (rather than say computation time) was made so that our experiment could validate the use of machine learning to CAD theory, rather than just the QEPCAD implementation. Further, it is usually the case that cell counts and timings are strongly correlated.

The heuristics (Brown, sotd and ndrr) have been implemented in MAPLE (as part of the freely available `ProjectionCAD` package [26]) and for each problem the orderings suggested by the heuristics were recorded and compared to the cell

---

[2] The data is available at `http://www.cl.cam.ac.uk/~zh242/data`.

```
(x0,x1,x2)
0
(Ex0)(Ex1)(Ex2)[[((x0 x0) + ((x1 x1) + (x2 x2))) = 1]].
go
go
go
d-stat
go
finish
```

Fig. 1. Sample QEPCAD input for a quantified problem

```
(x0,x1,x2)
3
[[((x0 x0) + ((x1 x1) + (x2 x2))) = 1]].
go
go
d-proj-factors
d-proj-polynomials
go
d-fpc-stat
go
```

Fig. 2. Sample QEPCAD input for a quantifier free problem

counts produced by QEPCAD [3]. Note that all three heuristics do not discriminate on the structure of any quantifiers. As discussed above, some heuristics are more expensive than others. However, since none of the costs were prohibitive for our data set they are not considered here.

Machine learning was applied to predict which of the three heuristics will give an *optimal* variable ordering for a given problem, where *optimal* means the lowest cell count of the selected CADs. Note that in the quantified case QEPCAD can collapse stacks when sufficient truth values for the constituent cells have been discovered to determine a truth value for the base cell. Hence, since our problems are all fully existential, the output for all quantified problems is always a single cell: true or false. Therefore, in these cases it was not the number of cells in the output that was used but instead the number of cells constructed during the process (hence the statistics commands in Figures 1 and 2 differ).

---

[3] When comparing care must be taken when changing between the different variable ordering conventions (see Footnote 1).

2.4   Problem Features

To apply machine learning, we need to identify features of the CAD problems that might be relevant to the correct choice of the heuristics. A feature is an aspect or measure of the problem that may be expressed numerically. Table 1 shows the 11 features that we identified, where (x_0, x_1, x_2) are the three variable labels used in all our problems. The number of features is quite small, compared to other machine learning experiments. They were chosen as easily computable features of the problems which could affect the performances of the heuristics. Other features were considered (such as the maximum coefficient and the proportion of constraints that were equations) but were not found to be useful. Further investigation into feature selection may be a topic of our future work.

Table 1. Description of the features used. The proportion of a variable occurring in polynomials is the number of polynomials containing the variable divided by total number of polynomials. The proportion of a variable occurring in monomials is the number of terms containing the variable divided by total number of terms in polynomials.

Feature number	Description
1	Number of polynomials.
2	Maximum total degree of polynomials.
3	Maximum degree of x_0 among all polynomials.
4	Maximum degree of x_1 among all polynomials.
5	Maximum degree of x_2 among all polynomials.
6	Proportion of x_0 occurring in polynomials.
7	Proportion of x_1 occurring in polynomials.
8	Proportion of x_2 occurring in polynomials.
9	Proportion of x_0 occurring in monomials.
10	Proportion of x_1 occurring in monomials.
11	Proportion of x_2 occurring in monomials.

Each feature vector in the training set was associated with a label, $+1$ (positive examples) or -1 (negative examples), indicating in which of two classes it was placed. To take Brown's heuristic as an example, a corresponding training set was derived with each problem labelled $+1$ if Brown's heuristic suggested a variable ordering with the lowest number of cells, or -1 otherwise.

The features could all be easily calculated from the problem input using MAPLE. For example. if the input formula is defined using the set of polynomials

$$\{-6x_0^2 - x_2^3 - 1, \quad x_0^4 x_2 + 9x_1, \quad x_0 + x_0^2 - x_2 x_0 - 5\}$$

then the problem will have the feature vector

$$\left[3, 5, 4, 1, 3, 1, \frac{1}{3}, 1, \frac{5}{9}, \frac{1}{9}, \frac{1}{3}\right].$$

After the feature generation process, the training data (feature vectors) were normalized so that each feature had zero mean and unit variance across the set. The same normalization was then also applied to the validation and test sets.

2.5 Parameter Optimization

SVM-LIGHT was used to do the classification for this experiment. As stated in Section 1.2, SVMs use kernel functions to map the data into higher dimensional spaces where the data may be more easily separated. SVM-LIGHT has four standard kernel functions: linear, polynomial, sigmoid tanh and radial basis function. For each kernel function, there are associated parameters which must be set. An earlier experiments applying machine learning to an automated theorem prover [9] found the radial basis function (RBF) kernel performed well in finding a relation between the simple algebraic features and the best heuristic choice. Hence the same kernel was selected for this experiment (other kernel functions may be tested in future work). The RBF function is defined as:

$$K(x, x\prime) = \exp\left(-\gamma||x - x\prime||^2\right)$$

where K is the kernel function, x and $x\prime$ are feature vectors. There is a single parameter γ in the RBF kernel function. Besides the parameter γ, two other parameters are involved in the SVM fitting process. The parameter C governs the trade-off between margin and training error, and the cost factor j is used to correct imbalance in the training set and we set it equal to the ratio between negative and positive samples. Given a training set, we can easily compute the value of parameter j by looking at the sign of the samples. However, it is not that trivial to find the optimal values of γ and C.

In machine learning, *Matthew's correlation coefficient* (MCC) [4] is often used to evaluate the performance of the binary classifications. It takes into account true and false positives and negatives:

$$\text{MCC} = \frac{\text{TP} * \text{TN} - \text{FP} * \text{FN}}{\sqrt{(\text{TP} + \text{FP})(\text{TP} + \text{FN})(\text{TN} + \text{FP})(\text{TN} + \text{FN})}}$$

In this equation, TP is the number of true positives, TN is the number of true negatives, FP is the number of false positives and FN is the number of false negatives. The denominator is set to 1 if any sum term is zero. This measure has the value 1 if perfect prediction is attained, 0 if the classifier is performing as a random classifier, and -1 if the classifier exactly disagrees with the data.

A grid-search optimisation procedure was used with the training and validation set, involving a search over a range of (γ, C) values to find the pair which would maximize MCC. We tested a commonly used range of value of γ (varied between $2^{-15}, 2^{-14}, 2^{-13}, \ldots, 2^3$) and C (varied between $2^{-5}, 2^{-4}, 2^{-3}, \ldots, 2^{15}$) in our grid search process [32]. Following the completion of the grid-search, the values for kernel function and model parameters giving optimal MCC results were selected for each individual CAD heuristic classifier. We also performed a similar calculation, selecting parameters to maximise the F_1-score [35], but the results using MCC were superior.

The classifiers with optimal (γ, C) were applied to the test set to output the margin values [21]. In an ideal case, only one classifier would return a positive result for any problem, where selecting a best heuristic is just a case of observing which classifier returns a positive result. However, in practice, more than one classifier will return a positive result for some problems, while no classifiers may return a positive for others. Thus, instead we used the relative magnitudes of the classifiers in our experiment. The classifier with most positive (or least negative) margin was selected to indicate the best decision procedure for the selection.

3   Results

The experiment was run as described in Section 2. We use the number of problems for which a selected variable ordering is optimal to measure the efficacy of each heuristic separately, and of the heuristic selected by machine learning.

Table 2 breaks down the results into a set of mutually exclusive outcomes that describe all possibilities. The column headed 'Machine Learning' indicates the heuristic selected by the machine learned model with the next three columns indicating each of the fixed heuristics tested. For each of these four heuristics, we may ask the question "Did this heuristic select the optimal variable ordering?" A 'Y' in the table indicates yes and an 'N' indicates no, with each of the 13 cases listed covering all possibilities. Note that at least one of the fixed heuristics must have a 'Y' since, by definition, the optimal ordering is obtained by at least one heuristic while if they all have a Y it is not possible for machine learning to fail. For each of these cases we list the number of problems for which this case occurred for both the quantifier free and quantified experiments.

For many problems more than one heuristic selects the optimal variable ordering and the probability of a randomly selected heuristic giving the optimal ordering depends on how many pick it. For example, a random selection would be successful 1/3 of the time if one heuristic gives the optimal ordering or 2/3 of the time if two heuristics do so.

In Table 2, case 1 is where machine learning cannot make any difference as all heuristics are equally optimal. We compare the remaining cases pairwise. For each pair, the behaviour of the fixed heuristics are identical and the difference is whether or not machine learning picked a winning heuristic (one of the ones with a Y). We see that in each case machine learning succeeds far more often than fails. For each pair we can compare with a random heuristic selection. For example, consider cases 2 and 3 where sotd and ndrr are successful heuristics and Brown is not. A random selection would be successful 2/3 of the time. For the quantifier free examples, machine learned selection is successful $146/(146+39)$ or approximately 79% of the time, which is significantly better.

We repeated this calculation for the quantified case and the other pairs, as shown in Table 3. In each case the values have been compared to the chance of success when picking a random heuristic, and so there are two distinct sets in

Table 2. Categorising the problems into a set of mutually exclusive cases characterised by which heuristics were successful

Case	Machine Learning	sotd	ndrr	Brown	Quantifier Free	Quantified
1	Y	Y	Y	Y	399	573
2	Y	Y	Y	N	146	96
3	N	Y	Y	N	39	24
4	Y	Y	N	Y	208	232
5	N	Y	N	Y	35	43
6	Y	N	Y	Y	64	57
7	N	N	Y	Y	7	11
8	Y	Y	N	N	106	66
9	N	Y	N	N	106	75
10	Y	N	Y	N	159	101
11	N	N	Y	N	58	89
12	Y	N	N	Y	230	208
13	N	N	N	Y	164	146

Table 3. Proportion of examples where machine learning picks a successful heuristic

sotd	ndrr	Brown	Quantifier Free	Quantified
Y	Y	N	79% (>67%)	80% (>67%)
Y	N	Y	86% (>67%)	84% (>67%)
N	Y	Y	90% (>67%)	84% (>67%)
Y	N	N	50% (>33%)	47% (>33%)
N	Y	N	73% (>33%)	53% (>33%)
N	N	Y	58% (>33%)	59% (>33%)

Table 3: those where only one heuristic was optimal and those where two are. We see that machine learning did better for some classes of problems than others. For example in quantifier free examples, when only one heuristic is optimal machine learning does considerably better if that one is ndrr, while if only one is not optimal machine learning does worse if is Brown. Nevertheless, the machine learning selection is better than random in every case in both experiments.

By summing the numbers in Table 2 in which Y appears in a row for the machine learned selection and each individual heuristic, we get Table 4. This compares, for both the quantifier free and quantified problem sets, the learned selection with each of the CAD heuristics on their own.

Of the three heuristics, Brown seems to be the best, albeit by a small margin. Its performance is a little surprising, both because the Brown heuristic is not so

Table 4. Total number of problems for which each heuristic picks the best ordering

	Machine Learning	sotd	ndrr	Brown
Quantifier free	1312	1039	872	1107
Quantified	1333	1109	951	1270

well known (having never been formally published) and because it requires little computation (taking only simple measurements on the input).

For the quantifier free problems there were 399 problems where every heuristic picked the optimal, 499 where two did and 823 where one did. Hence for this problem set the chances of picking a successful heuristic at random is

$$\frac{100}{1721} \left(399 + 499 * \tfrac{2}{3} + 823 * \tfrac{1}{3} \right) \simeq 58\%$$

which compares with $100 * 1312/1721 \simeq 76\%$ for machine learning. For the quantified problems the figures are 64% and 77%. Hence machine learning performs significantly better than a random choice in both cases. Further, if we were to use only the heuristic that performed the best on this data, the Brown heuristic, then we would pick a successful ordering for approximately 64% of the quantifier free problems and 74% of the quantified problems. So we see that a machine learned choice is also superior to using any one heuristic.

4   Possibilities for Extending the Experiment

Although a large data set of real world problems was used, we note that in some ways the data was quite uniform. A key area of future work is experimentation on a wider data set to see if these results, both the benefit of machine learning and the superiority of Brown's heuristic, are verified more generally. An initial extension would be to relax the parameters used to select problems from the nlsat dataset, for example by allowing problems with more variables.

One key restriction with this dataset is that all problems have one block of existential quantifiers. Note that our restriction to this case followed the availability of data rather than any technical limitation of the machine learning. Possible ways to generalise the data include randomly applying quantifiers to the the existing problems, or randomly generating whole problems. However, this would mean the problems no longer originate from real applications, and it has been noted in the past that random problems for CAD can be unrepresentative.

We do not suggest SVM as the only suitable machine learning method for this experiment, but overall a SVM with the RBF kernel worked well here. It would be interesting to see if other machine learning methods could offer similar or even better selections. Further improvements may also come from more work on the feature selection. The features used here were all derived from the polynomials involved in the input. One possible extension would be to consider also the type of relations present and how they are connected logically (likely to be particularly beneficial if problems with more variables or more varied quantifiers are allowed).

A key extension for future work will be the testing of other heuristics. For example the greedy sotd heuristic [23] which chooses an ordering one variable at a time based on the sotd of new projection polynomials or combined heuristics, (where we narrow the selection with one and then breaking the tie with another). We also note that there are other questions of CAD problem formulation besides variable ordering [8] for which machine learning might be of benefit.

Finally, we note that there are other CAD implementations. In addition to QEPCAD there is ProjectionCAD [26], RegularChains [17] and SyNRAC [33] in MAPLE, MATHEMATICA [46] and Redlog [24] in REDUCE. Each implementation has its own intricacies and often different underlying theory so it would be interesting to test if machine learning can assist with these as it does with QEPCAD.

5   Conclusions

We have investigated the use of machine learning for making the choice of which heuristic to use when selecting a variable ordering for CAD, and quantifier elimination by CAD. The experimental results confirmed our thesis, drawn from personal experience, that no one heuristic is superior for all problems and the correct choice will depend on the problem. Each of the three heuristics tested had a substantial set of problems for which they were superior to the others and so the problem was a suitable application for machine learning.

Using machine learning to select the best CAD heuristic yielded better results than choosing one heuristic at random, or just using any of the individual heuristics in isolation, indicating there is a relation between the simple algebraic features and the best heuristic choice. This could lead to the development of a new individual heuristic in the future.

The experiments involved testing heuristics on 1721 CAD problems, certainly the largest such experiment that the authors are aware of. For comparison, the best known previous study on such heuristics [23] tested with six examples. We observed that Brown's heuristic is the most competitive for our example set, and this is despite it involving less computation than the others. This heuristic was presented during an ISSAC tutorial in 2004 (see Brown [13]), but does not seem to be formally published. It certainly deserves to be better known.

Finally, we note that CAD is certainly not unique amongst computer algebra algorithms in requiring the user to make such a choice of problem formulation. More generally, computer algebra systems (CASs) often have a choice of possible algorithms to use when solving a problem. Since a single formulation or algorithm is rarely the best for the entire problem space, CASs usually use *meta-algorithms* to make such choices, where decisions are based on some numerical parameters [16]. These are often not as well documented as the base algorithms, and may be rather primitive. To the best of our knowledge, the present paper appears to be the first applying machine learning to problem formulation for computer algebra. The positive results should encourage investigation of similar applications in the field of symbolic computation.

Acknowledgements. This work was supported by the EPSRC grant: EP/J003247/1 and the China Scholarship Council (CSC). The authors thank the anonymous referees for useful comments which improved the paper.

References

1. Akbarpour, B., Paulson, L.: MetiTarski: An automatic theorem prover for real-valued special functions. Journal of Automated Reasoning 44(3), 175–205 (2010)
2. Alpaydin, E.: Introduction to machine learning. MIT Press (2004)
3. Arnon, D., Collins, G., McCallum, S.: Cylindrical algebraic decomposition I: The basic algorithm. SIAM Journal of Computing 13, 865–877 (1984)
4. Baldi, P., Brunak, S., Chauvin, Y., Andersen, C.A., Nielsen, H.: Assessing the accuracy of prediction algorithms for classification: an overview. Bioinformatics 16(5), 412–424 (2000)
5. Basu, S.: Algorithms in real algebraic geometry: A survey (2011),
   www.math.purdue.edu/~sbasu/raag_survey2011_final.pdf
6. Boyan, J., Freitag, D., Joachims, T.: A machine learning architecture for optimizing web search engines. In: AAAI Workshop on Internet Based Information Systems, pp. 1–8 (1996)
7. Bradford, R., Davenport, J., England, M., McCallum, S., Wilson, D.: Cylindrical algebraic decompositions for boolean combinations. In: Proc. ISSAC 2013, pp. 125–132. ACM (2013)
8. Bradford, R., Davenport, J.H., England, M., Wilson, D.: Optimising problem formulation for cylindrical algebraic decomposition. In: Carette, J., Aspinall, D., Lange, C., Sojka, P., Windsteiger, W. (eds.) CICM 2013. LNCS (LNAI), vol. 7961, pp. 19–34. Springer, Heidelberg (2013)
9. Bridge, J.P.: Machine learning and automated theorem proving. University of Cambridge Computer Laboratory Technical Report UCAM-CL-TR-792 (2010),
   http://www.cl.cam.ac.uk/techreports/UCAM-CL-TR-792.pdf
10. Bridge, J., Holden, S., Paulson, L.: Machine learning for first-order theorem proving. Journal of Automated Reasoning, 1–32 (2014)
11. Brown, C.: Improved projection for cylindrical algebraic decomposition. Journal of Symbolic Computation 32(5), 447–465 (2001)
12. Brown, C.: QEPCAD B: A program for computing with semi-algebraic sets using CADs. ACM SIGSAM Bulletin 37(4), 97–108 (2003)
13. Brown, C.: Companion to the Tutorial: Cylindrical algebraic decomposition. Presented at ISSAC 2004 (2004), www.usna.edu/Users/cs/wcbrown/
    research/ISSAC04/handout.pdf
14. Brown, C., Davenport, J.: The complexity of quantifier elimination and cylindrical algebraic decomposition. In: Proc. ISSAC 2007, pp. 54–60. ACM (2007)
15. Brown, C., Kahoui, M.E., Novotni, D., Weber, A.: Algorithmic methods for investigating equilibria in epidemic modelling. Journal of Symbolic Computation 41, 1157–1173 (2006)
16. Carette, J.: Understanding expression simplification. In: Proc. ISSAC 2004, pp. 72–79. ACM (2004)
17. Chen, C., Maza, M.M., Xia, B., Yang, L.: Computing cylindrical algebraic decomposition via triangular decomposition. In: Proc. ISSAC 2009, pp. 95–102. ACM (2009)

18. Collins, G.E.: Quantifier elimination for real closed fields by cylindrical algebraic decomposition. In: Brakhage, H. (ed.) GI-Fachtagung 1975. LNCS, vol. 33, pp. 134–183. Springer, Heidelberg (1975)
19. Collins, G.: Quantifier elimination by cylindrical algebraic decomposition – 20 years of progress. In: Quantifier Elimination and Cylindrical Algebraic Decomposition. Texts & Monographs in Symbolic Computation, pp. 8–23. Springer (1998)
20. Collins, G., Hong, H.: Partial cylindrical algebraic decomposition for quantifier elimination. Journal of Symbolic Computation 12, 299–328 (1991)
21. Cristianini, N., Shawe-Taylor, J.: An introduction to support vector machines and other kernel-based learning methods. Cambridge University Press (2000)
22. Davenport, J., Bradford, R., England, M., Wilson, D.: Program verification in the presence of complex numbers, functions with branch cuts etc. In: Proc. SYNASC 2012, pp. 83–88. IEEE (2012)
23. Dolzmann, A., Seidl, A., Sturm, T.: Efficient projection orders for CAD. In: Proc. ISSAC 2004, pp. 111–118. ACM (2004)
24. Dolzmann, A., Sturm, T.: REDLOG: Computer algebra meets computer logic. SIGSAM Bulletin 31(2), 2–9 (1997)
25. Dolzmann, A., Sturm, T., Weispfenning, V.: Real quantifier elimination in practice. In: Algorithmic Algebra and Number Theory, pp. 221–247. Springer (1998)
26. England, M.: An implementation of CAD in Maple utilising problem formulation, equational constraints and truth-table invariance. University of Bath Department of Computer Science Technical Report 2013-04 (2013), http://opus.bath.ac.uk/35636/
27. Forsyth, R., Rada, R.: Machine learning: Applications in expert systems and information retrieval. Halsted Press (1986)
28. Fotiou, I., Parrilo, P., Morari, M.: Nonlinear parametric optimization using cylindrical algebraic decomposition. In: 2005 European Control Conference on Decision and Control, CDC-ECC 2005, pp. 3735–3740 (2005)
29. Hong, H.: An improvement of the projection operator in cylindrical algebraic decomposition. In: Proc. ISSAC 1990, pp. 261–264. ACM (1990)
30. Hornik, K., Stinchcombe, M., White, H.: Multilayer feedforward networks are universal approximators. Neural Networks 2(5), 359–366 (1989)
31. Huang, Z., Paulson, L.: An application of machine learning to rcf decision procedures. In: Proc. 20th Automated Reasoning Workshop (2013)
32. Hsu, C., Chang, C., Lin, C.: A practical guide to support vector classification (2003)
33. Iwane, H., Yanami, H., Anai, H., Yokoyama, K.: An effective implementation of a symbolic-numeric cylindrical algebraic decomposition for quantifier elimination. In: Proc. SNC 2009, pp. 55–64 (2009)
34. Joachims, T.: Making large-scale SVM learning practical. In: Advances in Kernel Methods - Support Vector Learning, pp. 169–184. MIT Press (1999)
35. Joachims, T.: A support vector method for multivariate performance measures. In: Proc. 22nd Intl. Conf. on Machine Learning, pp. 377–384. ACM (2005)
36. Jovanović, D., de Moura, L.: Solving non-linear arithmetic. In: Gramlich, B., Miller, D., Sattler, U. (eds.) IJCAR 2012. LNCS, vol. 7364, pp. 339–354. Springer, Heidelberg (2012)
37. McCallum, S.: An improved projection operation for cylindrical algebraic decomposition. In: Quantifier Elimination and Cylindrical Algebraic Decomposition. Texts & Monographs in Symbolic Computation, pp. 242–268. Springer (1998)
38. McCallum, S.: On projection in CAD-based quantifier elimination with equational constraint. In: Proc. ISSAC 1999, pp. 145–149. ACM (1999)

39. McCulloch, W.S., Pitts, W.: A logical calculus of the ideas immanent in nervous activity. The Bulletin of Mathematical Biophysics 5(4), 115–133 (1943)
40. Rosenblatt, F.: The perceptron: a probabilistic model for information storage and organization in the brain. Psychological Review 65(6), 386 (1958)
41. Schölkopf, B., Tsuda, K., Vert, J.-P.: Kernel methods in computational biology. MIT Press (2004)
42. Sebastiani, F.: Machine learning in automated text categorization. ACM Computing Surveys (CSUR) 34(1), 1–47 (2002)
43. Shawe-Taylor, J., Cristianini, N.: Kernel methods for pattern analysis. Cambridge University Press (2004)
44. Stone, P., Veloso, M.: Multiagent systems: A survey from a machine learning perspective. Autonomous Robots 8(3), 345–383 (2000)
45. Strzeboński, A.: Cylindrical algebraic decomposition using validated numerics. Journal of Symbolic Computation 41(9), 1021–1038 (2006)
46. Strzeboński, A.: Solving polynomial systems over semialgebraic sets represented by cylindrical algebraic formulas. In: Proc. ISSAC 2012, pp. 335–342. ACM (2012)
47. Tarski, A.: A decision method for elementary algebra and geometry. In: Quantifier Elimination and Cylindrical Algebraic Decomposition. Texts and Monographs in Symbolic Computation, pp. 24–84. Springer (1998)
48. Wilson, D., Bradford, R., Davenport, J.: A repository for CAD examples. ACM Communications in Computer Algebra 46(3), 67–69 (2012)
49. Wilson, D., Davenport, J., England, M., Bradford, R.: A "piano movers" problem reformulated. In: Proc. SYNASC 2013. IEEE (2013)
50. The benchmarks used in solving nonlinear arithmetic. New York University (2012), http://cs.nyu.edu/~dejan/nonlinear/

Hipster: Integrating Theory Exploration in a Proof Assistant

Moa Johansson, Dan Rosén, Nicholas Smallbone, and Koen Claessen

Department of Computer Science and Engineering,
Chalmers University of Technology
{jomoa,danr,nicsma,koen}@chalmers.se

Abstract. This paper describes Hipster, a system integrating theory exploration with the proof assistant Isabelle/HOL. Theory exploration is a technique for automatically discovering new interesting lemmas in a given theory development. Hipster can be used in two main modes. The first is *exploratory mode*, used for automatically generating basic lemmas about a given set of datatypes and functions in a new theory development. The second is *proof mode*, used in a particular proof attempt, trying to discover the missing lemmas which would allow the current goal to be proved. Hipster's proof mode complements and boosts existing proof automation techniques that rely on automatically selecting existing lemmas, by inventing new lemmas that need induction to be proved. We show example uses of both modes.

1   Introduction

The concept of theory exploration was first introduced by Buchberger [2]. He argues that in contrast to automated theorem provers that focus on proving one theorem at a time in isolation, mathematicians instead typically proceed by exploring entire theories, by conjecturing and proving layers of increasingly complex propositions. For each layer, appropriate proof methods are identified, and previously proved lemmas may be used to prove later conjectures. When a new concept (e.g. a new function) is introduced, we should prove a set of new conjectures which, ideally, "completely" relates the new with the old, after which other propositions in this layer can be proved easily by "routine" reasoning. Mathematical software should be designed to support this workflow. This is arguably the mode of use supported by many interactive proof assistants, such as Theorema [3] and Isabelle [17]. However, they leave the generation of new conjectures relating different concepts largely to the user. Recently, a number of different systems have been implemented to address the conjecture synthesis aspect of theory exploration [15,13,16,5]. Our work goes one step further by integrating the discovery and proof of new conjectures in the workflow of the interactive theorem prover Isabelle/HOL. Our system, called Hipster, is based on our previous work on HipSpec [5], a theory exploration system for Haskell programs. In that work, we showed that HipSpec is able to automatically discover many of the kind of equational theorems present in, for example, Isabelle/HOL's libraries for

S.M. Watt et al. (Eds.): CICM 2014, LNAI 8543, pp. 108–122, 2014.

natural numbers and lists. In this article we show how similar techniques can be used to speed up and facilitate the development of new theories in Isabelle/HOL by discovering basic lemmas automatically.

Hipster translates Isabelle/HOL theories into Haskell and generates equational conjectures by testing and evaluating the Haskell program. These conjectures are then imported back into Isabelle and proved automatically. Hipster can be used in two ways: in *exploratory mode* it quickly discovers basic properties about a newly defined function and its relationship to already existing ones. Hipster can also be used in *proof mode*, to provide lemma hints for an ongoing proof attempt when the user is stuck.

Our work complements Sledgehammer [18], a popular Isabelle tool allowing the user to call various external automated provers. Sledgehammer uses *relevance filtering* to select among the available lemmas those likely to be useful for proving a given conjecture [14]. However, if a crucial lemma is missing, the proof attempt will fail. If theory exploration is employed, we can increase the success rate of Isabelle/HOL's automatic tactics with little user effort.

As an introductory example, we consider the example from section 2.3 of the Isabelle tutorial [17]: proving that reversing a list twice produces the same list. We first apply structural induction on the list xs.

```
theorem rev_rev : "rev(rev xs) = xs"
apply (induct xs)
```

The base case follows trivially from the definition of rev, but Isabelle/HOL's automated tactics simp, auto and sledgehammer all fail to prove the step case. We can simplify the step case to:

$$rev(rev\ xs) = xs \implies rev((rev\ xs)\ @\ [x]) = x\#xs$$

At this point, we are stuck. This is where Hipster comes into the picture. If we call Hipster at this point in the proof, asking for lemmas about rev and append (@), it suggests and proves three lemmas:

```
lemma lemma_a:   "xs @ [] = xs"
lemma lemma_aa : "(xs @ ys) @ zs = xs @ (ys @ zs)"
lemma lemma_ab : "(rev xs) @ (rev ys) = rev (ys @ xs)"
```

To complete the proof of the stuck subgoal, we need lemma ab. Lemma ab in turn, needs lemma a for its base case, and lemma aa for its step case. With these three lemmas present, Isabelle/HOL's tactics can take care of the rest. For example, when we call Sledgehammer in the step case, it suggests a proof by Isabelle/HOL's first-order reasoning tactic metis [11], using the relevant function definitions as well as lemma_ab:

```
theorem rev_rev : "rev(rev xs) = xs"
apply (induct xs)
apply simp
sledgehammer
by (metis rev.simps(1) rev.simps(2) app.simps(1) app.simps(2) lemma_ab)
```

The above example shows how Hipster can be used interactively in a stuck proof attempt. In exploratory mode, there are also advantages of working in an interactive setting. For instance, when dealing with large theories that would otherwise generate a very large search space, the user can instead incrementally explore different relevant sub-theories while avoiding a search space explosion. Lemmas discovered in each sub-theory can be made available when exploring increasingly larger sets of functions.

The article is organised as follows: In section 2 we give a brief overview of the HipSpec system which Hipster uses to generate conjectures, after which we describe Hipster in more detail in section 3, together with some larger worked examples of how it can be used, both in proof mode and exploratory mode. In section 4 we describe how we deal with partial functions, as Haskell and Isabelle/HOL differ in their semantics for these. Section 5 covers related work and we discuss future work in section 6.

2   Background

In this section we give a brief overview of the HipSpec system which we use as a backend for generating conjectures, and of Isabelle's code generator which we use to translate Isabelle theories to Haskell programs.

2.1   HipSpec

HipSpec is a state-of-the-art inductive theorem prover and theory exploration system for Haskell. In [5] we showed that HipSpec is able to automatically discover and prove the kind of equational lemmas present in Isabelle/HOL's libraries, when given the corresponding functions written in Haskell.

HipSpec works in two stages:

1. Generate a set of conjectures about the functions at hand. These conjectures are equations between terms involving the given functions, and have not yet been proved correct but are nevertheless extensively tested.
2. Attempt to prove each of the conjectures, using already proven conjectures as assumptions. HipSpec implements this by enumerating induction schemas, and firing off many proof obligations to automated first-order logic theorem provers.

The proving power of HipSpec comes from its capability to automatically discover and prove lemmas, which are then used to help subsequent proofs.

In Hipster we can not directly use HipSpec's proof capabilities (stage (2) above); we use Isabelle/HOL for the proofs instead. Isabelle is an LCF-style prover which means that it is based on a small core of trusted axioms, and proofs must be built on top of those axioms. In other words, we would have to reconstruct inside Isabelle/HOL any proof that HipSpec found, so it is easier to use Isabelle/HOL for the proofs in the first place.

The part of HipSpec we directly use is its conjecture synthesis system (stage (1) above), called QuickSpec [6]), which efficiently generates equations about a given set of functions and datatypes.

QuickSpec takes a set of functions as input, and proceeds to generate all type-correct terms up to a given limit (usually up to depth three). The terms may contain variables (usually at most three per type). These parameters are set heuristically, and can be modified by the user. QuickSpec attempts to divide the terms into equivalence classes such that two terms end up in the same equivalence class if they are equal. It first assumes that all terms of the same type are equivalent, and initially puts them in the same equivalence class. It then picks random ground values for the variables in the terms (using QuickCheck [4]) and evaluates the terms. If two terms in the same equivalence class evaluate to different ground values, they cannot be equal; QuickSpec thus breaks each equivalence class into new, smaller equivalence classes depending on what values their terms evaluated to. This process is repeated until the equivalence classes stabilise. We then read off equations from each equivalence class, by picking one term of that class as a representative and equating all the other terms to that representative. This means that the conjectures generated are, although not yet proved, fairly likely to be true, as they have been tested on several hundred different random values. The confidence increases with the number of tests, which can be set by the user. The default setting is to first run 200 tests, after which the process stops if the equivalence classes appear to have stabilised, i.e. if nothing has changed during the last 100 tests. Otherwise, the number of tests are doubled until stable.

As an example, we ask QuickSpec to explore the theory with list append, @, the empty list, [], and three list variables xs, ys, zs. Among the terms it will generate are (xs @ ys) @ zs, xs @ (ys @ zs), xs @ [] and xs. Initially, all four will be assumed to be in the same equivalence class. The random value generator for lists from QuickCheck might for instance generate the values: xs ↦ [], ys ↦ [a] and zs ↦ [b], where a and b are arbitrary distinct constants. Performing the substitutions of the variables in the four terms above and evaluating the resulting ground expressions gives us:

	Term	Ground Instance	Value
1	(xs @ ys) @ zs	([] @ [a]) @ [b]	[a,b]
2	xs @ (ys @ zs)	[] @ ([a] @ [b])	[a,b]
3	xs @ []	[] @ []	[]
4	xs	[]	[]

Terms 1 and 2 evaluate to the same value, as do terms 3 and 4. The initial equivalence class will therefore be split in two accordingly. After this, whatever variable assignments QuickSpec generates, the terms in each class will evaluate to the same value. Eventually, QuickSpec stops and the equations for associativity and right identity can be extracted from the resulting equivalence classes.

2.2   Code Generation in Isabelle

Isabelle/HOL's code generator can translate from Isabelle's higher-order logic to code in several functional programming languages, including Haskell [9,8]. Isabelle's higher-order logic is a typed λ-calculus with polymorphism and type-classes. Entities like constants, types and recursive functions are mapped to corresponding entities in the target language. For the kind of theories we consider in this paper, this process is straightforward. However, the code generator also supports user-given *code lemmas*, which allows it to generate code from non-executable constructs, e.g. by replacing sets with lists.

3   Hipster: Implementation and Use

We now give a description of the implementation of Hipster, and show how it can be used both in theory exploration mode and in proof mode, to find lemmas relevant for a particular proof attempt. An overview of Hipster is shown in figure 1. The source code and examples are available online[1].

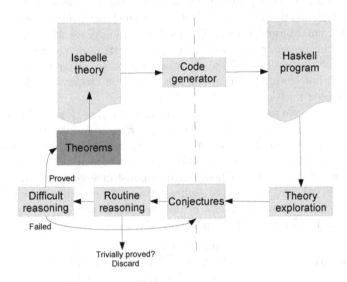

Fig. 1. Overview of Hipster

Starting from an Isabelle/HOL theory, Hipster calls Isabelle/HOL's code generator [9,8] to translate the given functions into a Haskell program. In order to use the testing framework from QuickCheck, as described in the previous section, we also post-process the Haskell file, adding *generators*, which are responsible for producing arbitrary values for evaluation and testing. A generator in Haskell's QuickCheck is simply a function which returns a random ground value for a

---

[1] https://github.com/moajohansson/IsaHipster

particular datatype [4]. In our case, the generators pick a random constructor and then recursively generate its arguments. To ensure termination, the generators are parametrised by a size; the generator reduces the size when it invokes itself recursively and when the size reaches zero, it always picks a non-recursive constructor.

Another important issue in the translation is the difference in semantics for partial functions between Isabelle/HOL and Haskell. In order for HipSpec not to miss equations that hold in Isabelle/HOL, but not in Haskell, we have to translate partial functions specially. This is explained in more detail in section 4.

Once the Haskell program is in place, we run theory exploration and generate a set of equational conjectures, which HipSpec orders according to generality. More general equations are preferred, as we expect these to be more widely applicable as lemmas. In previous work on HipSpec, the system would at this stage apply induction on the conjectures and send them off to some external prover. Here, we instead import them back into Isabelle as we wish to produce checkable LCF-style proofs for our conjectures.

The proof procedure in Hipster is parametrised by two tactics, one for easy or *routine reasoning* and one for *difficult reasoning*. In the examples below, the routine reasoning tactic uses Isabelle/HOL's simplifier followed by first-order reasoning by Metis [11], while the difficult reasoning tactic performs structural induction followed by simplification and first-order reasoning. Metis is restricted to run for at most one second in both the routine and difficult tactic. If there are several possible variables to apply induction on, we may backtrack if the first choice fails. Both tactics have access to the theorems proved so far, and hence get stronger as the proof procedure proceeds through the list of conjectures.

As theory exploration produces rather many conjectures, we do not want to present them all to the user. We select the most interesting ones, i.e. those that are difficult to prove, and filter out those that follow only by routine reasoning. Depending on the theory and application we can vary the tactics for routine and difficult reasoning to suit our needs. If we want Hipster to produce fewer or more lemmas, we can choose a stronger or weaker tactic, allowing for flexibility.

The order in which Hipster tries to prove things matters. As we mentioned, it will try more general conjectures first, with the hope that they will be useful to filter out many more specific routine results. Occasionally though, a proof will fail as a not-yet-proved lemma is required. In this case, the failed conjecture is added back into the list of open conjectures and will be retried later, after at least one new lemma has been proved. Hipster terminates when it either runs out of open conjectures, or when it can not make any more progress, i.e. when all open conjectures have been tried since it last proved a lemma.

Below we give two typical use cases for Hipster. In both examples, Hipster has been instantiated with the routine and difficult reasoning tactics that we described above.

3.1 Exploring a Theory of Binary Trees

This example is about a theory of binary trees, with data stored at the leaves:

```
datatype 'a Tree =
    Leaf 'a
  | Node "'a Tree" "'a Tree"
```

Let us define some functions over our trees: `mirror` swaps the left and right subtrees everywhere, and `tmap` applies a function to each element in the tree.

```
fun mirror :: 'a Tree => 'a Tree
where
  mirror (Leaf x) = Leaf x
| mirror (Node l r) = Node (mirror r) (mirror l)

fun tmap :: ('a => 'b) => 'a Tree => 'b Tree
where
  tmap f (Leaf x) = Leaf (f x)
| tmap f (Node l r) = Node (tmap f l) (tmap f r)
```

Now, let us call theory exploration to discover some properties about these two functions. Hipster quickly finds and proves the two expected lemmas:

```
lemma lemma_a [thy_expl]: "mirror (tmap x y) = tmap x (mirror y)"
by (tactic {* Hipster_Tacs.induct_simp_metis . . .*})

lemma lemma_aa [thy_expl]: "mirror (mirror x) = x"
by (tactic {* Hipster_Tacs.induct_simp_metis . . . *})
```

The output produced by Hipster can be automatically pasted into the proof script by a mouseclick. Recall that Hipster discards all lemmas that can be proved by routine reasoning (here, without induction). The tactic `induct_simp_metis` appearing in the proof script output is the current instantiation of "difficult reasoning". Note that the lemmas discovered are tagged with the attribute `thy_expl`, which tells Hipster which lemmas it has discovered so far. If theory exploration is called several times, we can use these lemmas in proofs and avoid rediscovering the same things. The user can also inspect what theory exploration has found so far by executing the Isabelle command `thm thy_expl`.

Next, let us also define two functions extracting the leftmost and rightmost element of the tree:

```
fun rightmost :: 'a Tree => 'a
where
  rightmost (Leaf x) = x
| rightmost (Node l r) = rightmost r

fun leftmost :: 'a Tree => 'a
where
  leftmost (Leaf x) = x
| leftmost (Node l r) = leftmost l
```

Asking Hipster for lemmas about all the functions defined so far, it provides one additional lemma, namely:

```
lemma lemma_ab [thy_expl]: "leftmost (mirror x2) = rightmost x2"
by (tactic {* Hipster_Tacs.induct_simp_metis . . . *})
```

Finally, we define a function flattening trees to lists:

```
fun flat_tree :: 'a Tree => 'a list
where
  flat_tree (Leaf x) = [x]
| flat_tree (Node l r) = (flat_tree l) @ (flat_tree r)
```

We can now ask Hipster to explore the relationships between the functions on trees and the corresponding functions on lists, such as rev, map and hd. Hipster produces four new lemmas and one open conjecture:

```
lemma lemma_ac [thy_expl]: "flat_tree (tmap x y) = map x (flat_tree y)"
by (tactic {* Hipster_Tacs.induct_simp_metis . . . *})

lemma lemma_ad [thy_expl]: "map x (rev xs) = rev (map x xs)"
by (tactic {* Hipster_Tacs.induct_simp_metis . . . *})

lemma lemma_ae [thy_expl]: "flat_tree (mirror x) = rev (flat_tree x)"
by (tactic {* Hipster_Tacs.induct_simp_metis . . . *})

lemma lemma_af [thy_expl]: "hd (xs @ xs) = hd xs"
by (tactic {* Hipster_Tacs.induct_simp_metis . . . *})

lemma unknown: "hd (flat_tree x) = leftmost x"
oops
```

Lemmas ad and af are perhaps not of much interest, as they only relate functions on lists. In fact, lemma ad is already in Isabelle/HOL's list library, but is not labelled as a simplification rule, which is why Hipster rediscovers it. Lemma af is a variant of a conditional library-lemma: xs ≠ [] ⟹ hd(xs @ ys) = hd xs. Observe that lemma af holds due to the partiality of hd. Hipster can not discover conditional lemmas, so we get this version instead.

In addition to the four lemmas which have been proved, Hipster also outputs one interesting conjecture (labelled **unknown**) which it fails to prove. To prove this conjecture, we need a lemma stating that, as the trees store data at the leaves, flat_tree will always produce a non-empty list: flat_tree t ≠ []. As this is not an equation, it is not discovered by Hipster.

This example shows that Hipster can indeed find most of the basic lemmas we would expect in a new theory. The user has to provide the constants Hipster should explore, and the rest is fully automatic, thus speeding up theory development. Theory exploration in this example takes just a few seconds, no longer than it takes to run tools like Sledgehammer. Even if Hipster fails to prove some properties, they may still be interesting, and the user can choose to prove them interactively.

Exploring with different Tactics. To illustrate the effects of choosing a slightly different tactic for routine and difficult reasoning, we also experimented with an instantiation using only Isabelle/HOL's simplifier as routine reasoning and induction followed by simplification as difficult reasoning. The advantage of this instantiation is that the simplifier generally is faster than Metis, but less powerful. However, for this theory, it turns out that the simplifier is sufficient to prove the same lemmas as above. Hipster also suggests one extra lemma, namely `rightmost (mirror x) = leftmost x`, which is the dual to lemma `ab` above. When we used Metis, this lemma could be proved without induction, by routine reasoning, and was thus discarded. Using only the simplifier, difficult reasoning and induction is required to find a proof, and the lemma is therefore presented to the user.

3.2 Proving Correctness of a Small Compiler

The following example is about a compiler to a stack machine for a toy expression language[2]. We show how theory exploration can be used to unblock a proof on which automated tactics otherwise fail due to a missing lemma.

Expressions in the language are built from constants (`Cex`), values (`Vex`) and binary operators (`Bex`):

```
type_synonym 'a binop = 'a => 'a => 'a
```

```
datatype ('a, 'b) expr =
    Cex 'a
  | Vex 'b
  | Bex "'a binop" "('a,'b) expr" "('a,'b) expr"
```

The types of values and variables are not fixed, but given by type parameters `'a` and `'b`. To evaluate an expression, we define a function `value`, parametrised by an environment mapping variables to values:

```
fun value :: ('b => 'a) => ('a,'b) expr => 'a
where
      value env (Cex c) = c
  | value env (Vex v) = env v
  | value env (Bex f e1 e2) = f (value env e1) (value env e2)
```

A program for our stack machine consists of four instructions:

```
datatype ('a, 'b) program =
    Done
  | Const 'a "('a, 'b) program"
  | Load 'b "('a, 'b) program"
  | Apply "'a binop" "('a, 'b) program"
```

A program is either empty (`Done`), or consists of one of the instructions `Const`, `Load` or `Apply`, followed by the remaining program. We further define a function `sequence` for combining programs:

---

[2] This example is a slight variation of that in §3.3 in the Isabelle tutorial [17].

```
fun sequence :: ('a, 'b) program => ('a, 'b) program => ('a, 'b) program
where
    sequence Done p = p
  | sequence (Const c p) p' = Const c (sequence p p')
  | sequence (Load v p) p' = Load v (sequence p p')
  | sequence (Apply f p) p' = Apply f (sequence p p')
```

Program execution is modelled by the function exec, which given a store for variables and a program, returns the values on the stack after execution.

```
fun exec :: ('b => 'a) => ('a,'b) program => 'a list => 'a list
where
    exec env Done stack = stack
  | exec env (Const c p) stack = exec env p (c#stack)
  | exec env (Load v p) stack = exec env p ((env v)#stack)
  | exec env (Apply f p) stack =
    exec env p ((f (hd stack) (hd(tl stack)))#(tl(tl stack)))
```

We finally define a function compile, which specifies how expressions are compiled into programs:

```
fun compile :: ('a,'b) expr => ('a,'b) program
  where
    compile (Cex c) =  Const c Done
  | compile (Vex v) =  Load v Done
  | compile (Bex f e1 e2) =
    sequence (compile e2) (sequence (compile e1) (Apply f Done))"
```

Now, we wish to prove correctness of the compiler, namely that executing a compiled expression indeed results in the value of that expression:

```
theorem "exec env (compile e) [] = [value env e]"
```

If we try to apply induction on e, Isabelle/HOL's simplifier solves the base-case but neither the simplifier or first-order reasoning by Sledgehammer succeeds in proving the step-case. At this stage, we can apply Hipster's theory exploration tactic. It will generate a set of conjectures, and interleave proving these with trying to prove the open sub-goal. Once Hipster succeeds in finding a set of lemmas which allow the open goal to be proved by routine reasoning, it presents the user with a list of lemmas it has proved, in this case:

```
Try first proving lemmas:

lemma lemma_a: "sequence x Done = x"
by (tactic {* Hipster_Tacs.induct_simp_metis . . . *})

lemma lemma_aa: "exec x y (sequence z x1) xs = exec x y x1 (exec x y z xs)"
by (tactic {* Hipster_Tacs.induct_simp_metis . . . *})

lemma lemma_ab: "exec x y (compile z) xs = value x y z # xs"
by (tactic {* Hipster_Tacs.induct_simp_metis . . . *})
```

Our theorem is a trivial instance of `lemma_ab`, whose proof depends on `lemma_aa`. Hipster takes about twenty seconds to discover and prove the lemmas. Pasting them into our proof script we can try Sledgehammer on our theorem again. This time, it succeeds and suggests the one-line proof:

```
theorem "exec env (compile e) [] = [value env e]"
by (metis lemma_ab)
```

4   Dealing with Partial Functions

Isabelle is a logic of total functions. Nonetheless, we can define apparently partial functions, such as `hd`:

```
fun hd :: 'a list => 'a where
  hd (x#xs) = x
```

How do we reconcile `hd` being partial with Isabelle functions being total? The answer is that in Isabelle, `hd` is total, but the behaviour of `hd` `[]` is unspecified: it returns some arbitrary value of type `'a`. Meanwhile in Haskell, `head` is partial, but the behaviour of `head` `[]` is specified: it crashes. We must therefore translate *partially defined* Isabelle functions into *total but underspecified* Haskell functions.

Hipster uses a technique suggested by Jasmin Blanchette [1] to deal with partial functions. Whenever we translate an Isabelle function that is missing some cases, we need to add a default case, like so:

```
hd :: [a] -> a
hd (x:xs) = x
hd [] = ???
```

But what should we put for the result of `hd` `[]`? To model the notion that `hd` `[]` is unspecified, whenever we evaluate a test case we will pick a *random* value for `hd` `[]`. This value will vary from test case to test case but will be consistent within one run of a test case. The idea is that, if an equation involving `hd` in Haskell always holds, for all values we could pick for `hd` `[]`, it will also hold in Isabelle, where the value of `hd` `[]` is unspecified.

Suppose we define the function `second`, which returns the second element of a list, as

```
second (x#y#xs) = y
```

It might seem that we should translate `second`, by analogy with `hd`, as

```
second :: [a] -> a
second (x:y:xs) = y
second _ = ???
```

and pick a random value of type `a` to use in the default case. But this translation is wrong! If we apply our translated `second` to a single-element list, it will give the same answer regardless of which element is in the list, and HipSpec will discover the lemma `second [x] = second [y]`. This lemma is certainly not true of our Isabelle function, which says nothing about the behaviour of `second` on single-element lists, and Hipster will fail to prove it.

We must allow the default case to produce a different result for different arguments. We therefore translate `second` as

```
second :: [a] -> a
second (x:y:xs) = y
second xs = ??? xs
```

where `???` is a random *function* of type `[a] -> a`. (QuickCheck can generate random functions.) As before, whenever we evaluate a test case, we instantiate `???` with a new random function[3]. This second translation mimics Isabelle's semantics: any equation that holds in Haskell no matter how we instantiate the `???` functions also holds in Isabelle.

In Hipster, we first use Isabelle/HOL's code generator to translate the theory to Haskell. Then we transform *every* function definition, whether it is partial or not, in the same way we transformed `second` above. If a function is already total, the added case will simply be unreachable. This avoids having to check functions for partiality. The extra clutter introduced for total functions is not a problem as we neither reason about nor show the user the generated program.

5   Related Work

Hipster is an extension to our previous work on the HipSpec system [5], which was not designed for use in an interactive setting. HipSpec applies structural induction to conjectures generated by QuickSpec and sends off proof obligations to external first-order provers. Hipster short-circuits the proof part and directly imports the conjectures back into Isabelle. This allows for more flexibility in the choice of tactics employed, by letting the user control what is to be considered routine and difficult reasoning. Being inside Isabelle/HOL provides the possibility to easily record lemmas for future use, perhaps in other theory developments. It gives us the possibility to re-check proofs if required, as well as increased reliability as proofs have been run through Isabelle's trusted kernel. As Hipster uses HipSpec for conjecture generation, any difference in performance (e.g. speed, lemmas proved) will depend only on what prover backend is used by HipSpec and what tactic is used by Hipster.

There are two other theory exploration systems available for Isabelle/HOL, IsaCoSy [13] and IsaScheme [16]. They differ in the way they generate conjectures, and both discover similar lemmas as HipSpec/Hipster. A comparison between HipSpec, IsaCoSy and IsaScheme can be found in [5], showing that all

---

[3] To avoid having to retranslate the Isabelle theory every time we evaluate a test case, in reality we parametrise the generated program on the various `???` functions. That way, whenever we evaluate a test case, we can cheaply change the default cases.

three systems manage to find largely the same lemmas on theories about lists and natural numbers. HipSpec does however outperform the other two systems on speed. IsaCoSy ensures that terms generated are non-trivial to prove by only generating irreducible terms, i.e. conjectures that do not have simple proofs by equational reasoning. These are then filtered through a counter-example checker and passed to IsaPlanner for proof [7]. IsaScheme, as the name suggests, follows the scheme-based approach first introduced for algorithm synthesis in Theorema [3]. IsaScheme uses general user-specified schemas describing the shape of conjectures and then instantiates them with available functions and constants. It combines this with counter-example checking and Knuth-Bendix completion techniques in an attempt to produce a minimal set of lemmas.

Unfortunately, the counter-example checking in IsaCoSy and IsaScheme is often too slow for use in an interactive setting. Unlike IsaCoSy, Hipster may generate reducible terms, but thanks to the equivalence class reasoning in QuickSpec, testing is much more efficient, and conjectures with trivial proofs are instead quickly filtered out at the proof stage. None of our examples takes more than twenty seconds to run.

Neither IsaCoSy or IsaScheme has been used to generate lemmas in stuck proof attempts, but only in fully automated exploratory mode. Starting from stuck proof attempts allows us to reduce the size of the interesting background theory, which promises better scalability.

Proof planning critics have been employed to analyse failed proof attempts in automatic inductive proofs [12]. The critics use information from the failure in order to try to speculate a missing lemma top-down, using techniques based on rippling and higher-order unification. Hipster (and HipSpec) takes a less restricted approach, instead constructing lemmas bottom-up, from the symbols available. As was shown in our previous work [5], this succeeds in finding lemmas that the top-down critics based approach fails to find, at the cost of possibly also finding a few extra lemmas as we saw in the example in section 3.2.

6   Further Work

The discovery process is currently limited to equational lemmas. We plan to extend the theory exploration to also take side conditions into account. If theory exploration is called in the middle of a proof attempt, there may be assumptions associated with the current sub-goal, which could be a useful source of side conditions. For example, if we are proving a lemma about sorting, there will most likely be an assumption involving the "less than" operator; this suggests that we should look for equations that have "less than" in a side condition. Once we have a candidate set of side conditions, it is easy to extend QuickSpec to find equations that assume those conditions.

The parameters for Hipster, e.g. the number of QuickSpec generated tests, the runtime for Metis and so on, are largely based on heuristics from development and previous experience. There is probably room for fine-tuning these heuristics and possibly adapt them to the theory we are working in. We plan to add and experiment with additional automated tactics in Hipster. Again, we expect that different tactics will suit different theories.

Another interesting area of further work in theory exploration is reasoning by analogy. In the example in section 3.1, theory exploration discovers lemmas about `mirror` and `tmap` which are analogous to lemmas about lists and the functions `rev` and `map`. Machine learning techniques can be used to identify similar lemmas [10], and this information could then be used to for instance suggest new combinations of functions to explore, new connections between theories or directly suggest additional lemmas about trees by analogy to those on lists.

7   Summary

Hipster integrates lemma discovery by theory exploration in the proof assistant Isabelle/HOL. We demonstrated two typical use cases of how this can help and speed up theory development: by generating interesting basic lemmas in *exploration mode* or as a lemma suggestion mechanism for a stuck proof attempt in *proof mode*. The user can control what is discovered by varying the background theory, and by varying Hipster's "routine reasoning" and "difficult reasoning" tactics; only lemmas that need difficult reasoning (e.g. induction) are presented.

Hipster complements tools like Sledgehammer: by discovering missing lemmas, more proofs can be tackled automatically. Hipster succeeds in automating inductive proofs and lemma discovery for small, but non-trivial, equational Isabelle/HOL theories. The next step is to increase its scope, to conditional equations and to bigger theories: our goal is a practical automated inductive proof tool for Isabelle/HOL.

Acknowledgements. The third author's research was supported by the EU FP7 Collaborative project *PROWESS*, grant number 317820.

References

1. Blanchette, J.C.: Personal communication (2013)
2. Buchberger, B.: Theory exploration with Theorema. Analele Universitatii Din Timisoara, ser. Matematica-Informatica 38(2), 9–32 (2000)
3. Buchberger, B., Creciun, A., Jebelean, T., Kovacs, L., Kutsia, T., Nakagawa, K., Piroi, F., Popov, N., Robu, J., Rosenkranz, M., Windsteiger, W.: Theorema: Towards computer-aided mathematical theory exploration. Journal of Applied Logic 4(4), 470–504 (2006), Towards Computer Aided Mathematics
4. Claessen, K., Hughes, J.: QuickCheck: a lightweight tool for random testing of Haskell programs. In: Proceedings of ICFP, pp. 268–279 (2000)
5. Claessen, K., Johansson, M., Rosén, D., Smallbone, N.: Automating inductive proofs using theory exploration. In: Bonacina, M.P. (ed.) CADE 2013. LNCS, vol. 7898, pp. 392–406. Springer, Heidelberg (2013)
6. Claessen, K., Smallbone, N., Hughes, J.: QUICKSPEC: Guessing formal specifications using testing. In: Fraser, G., Gargantini, A. (eds.) TAP 2010. LNCS, vol. 6143, pp. 6–21. Springer, Heidelberg (2010)

7. Dixon, L., Fleuriot, J.D.: Higher order rippling in ISAPLANNER. In: Slind, K., Bunker, A., Gopalakrishnan, G.C. (eds.) TPHOLs 2004. LNCS, vol. 3223, pp. 83–98. Springer, Heidelberg (2004)
8. Haftmann, F., Bulwahn, L.: Code generation from Isabelle/HOL theories (2013), http://isabelle.in.tum.de/doc/codegen.pdf
9. Haftmann, F., Nipkow, T.: Code generation via higher-order rewrite systems. In: Blume, M., Kobayashi, N., Vidal, G. (eds.) FLOPS 2010. LNCS, vol. 6009, pp. 103–117. Springer, Heidelberg (2010)
10. Heras, J., Komendantskaya, E., Johansson, M., Maclean, E.: Proof-pattern recognition and lemma discovery in ACL2. In: McMillan, K., Middeldorp, A., Voronkov, A. (eds.) LPAR-19. LNCS, vol. 8312, pp. 389–406. Springer, Heidelberg (2013)
11. Hurd, J.: First-order proof tactics in higher-order logic theorem provers. In: Design and Application of Strategies/Tactics in Higher Order Logics (STRATA), number NASA/CP-2003-212448 in NASA Technical Reports, pp. 56–68 (2003)
12. Ireland, A., Bundy, A.: Productive use of failure in inductive proof. Journal of Automated Reasoning 16, 79–111 (1996)
13. Johansson, M., Dixon, L., Bundy, A.: Conjecture synthesis for inductive theories. Journal of Automated Reasoning 47(3), 251–289 (2011)
14. Kühlwein, D., Blanchette, J.C., Kaliszyk, C., Urban, J.: MaSh: Machine learning for Sledgehammer. In: Blazy, S., Paulin-Mohring, C., Pichardie, D. (eds.) ITP 2013. LNCS, vol. 7998, pp. 35–50. Springer, Heidelberg (2013)
15. McCasland, R.L., Bundy, A., Smith, P.F.: Ascertaining mathematical theorems. Electronic Notes in Theoretical Computer Science 151(1), 21–38 (2006)
16. Montano-Rivas, O., McCasland, R., Dixon, L., Bundy, A.: Scheme-based theorem discovery and concept invention. Expert Systems with Applications 39(2), 1637–1646 (2012)
17. Nipkow, T., Paulson, L.C., Wenzel, M.T.: Isabelle/HOL. LNCS, vol. 2283. Springer, Heidelberg (2002)
18. Paulson, L.C., Blanchette, J.C.: Three years of experience with Sledgehammer, a practical link between automatic and interactive theorem provers. In: IWIL 2010 (2010)

Formalization of Complex Vectors in Higher-Order Logic

Sanaz Khan Afshar, Vincent Aravantinos, Osman Hasan, and Sofiène Tahar

Dept. of Electrical & Computer Engineering, Concordia University
1455 de Maisonneuve W., Montreal, Quebec, H3G 1M8, Canada
{s_khanaf,vincent,o_hasan,tahar}@ece.concordia.ca
http://hvg.ece.concordia.ca

Abstract. Complex vector analysis is widely used to analyze continuous systems in many disciplines, including physics and engineering. In this paper, we present a higher-order-logic formalization of the complex vector space to facilitate conducting this analysis within the sound core of a theorem prover: HOL Light. Our definition of complex vector builds upon the definitions of complex numbers and real vectors. This extension allows us to extensively benefit from the already verified theorems based on complex analysis and real vector analysis. To show the practical usefulness of our library we adopt it to formalize electromagnetic fields and to prove the law of reflection for the planar waves.

1 Introduction

Vector analysis is one of the most useful branches of mathematics; a highly scientific field that is used in practical problems arising in engineering and applied sciences. Not only the real vectors but the *complex* vectors are a convenient tool to describe natural and physical phenomena, including waves. They are thus used in fields as diverse as optics, electromagnetics, quantum mechanics, nuclear physics, aerospace, and communications. Therefore, a concrete development of (real and complex) vector analysis by formal methods can be a huge asset in the development and application of formal methods in connection with a variety of disciplines, e.g., physics, medicine, control and signal processing.

The theory of vector analysis is formalized in many theorem provers, e.g., HOL Light [7], Isabelle/HOL [3], PVS [8], and Coq [12]. However, these works are either limited to real vector analysis or provide very abstract formalizations which are useful for the formalization of mathematics but lack many notions required for applied sciences. For instance, Coq [12] provides a general, axiomatic, formalization of linear algebras. However, this work lacks many notions that are useful for applications. In PVS [8], real vector spaces are formalized but the work does not support complex vectors. Another example is the Isabelle/HOL [3] with a library of real vector analysis ported from the formalization of multivariate analysis available in the HOL Light theorem prover [7]. To the best of our knowledge, the only work addressing complex vectors is [10], where the

S.M. Watt et al. (Eds.): CICM 2014, LNAI 8543, pp. 123–137, 2014.

authors present a formalization of complex vectors, using the concept of complex function spaces in HOL Light. However, their formalization is very abstract and is focused on infinite dimension linear spaces: those definitions and properties which are only meaningful in a finite dimension are not addressed in the formalization presented in [10], e.g., cross product.

In this paper we present an alternate formalization of complex vectors using the HOL Light theorem prover. HOL Light is an interactive theorem prover which is based on classical higher-order logic with axioms of infinity, extensionality, and choice in the form of Hilbert's ϵ operator [6]. HOL Light uses functional programming language Objective CAML (OCaml) as both the implementation and interaction language. This theorem prover has been particularly successful in verifying many challenging mathematical theorems by providing formal reasoning support for different mathematical theories, including real analysis, complex analysis and vector calculus. The main motivation behind choosing HOL Light for the formalization of complex vector analysis is the availability of rich libraries of multivariate analysis, e.g., complex analysis [5] and Euclidean space [7].

Our formalization of complex vectors is inspired by the concept of bivectors, originally introduced by the famous American physicist J. William Gibbs, in his famous pamphlet "Elements of Vector Analysis" [4]. We adopt a great part of definition and properties of vector analysis and extend them to our formalization of complex vectors. Then, we prove many of complex vectors properties by introducing componentwise operators, and inheriting properties of complex analysis from HOL Light multivariate libraries [5]. Our formalization thus overcomes the limitations of [10] by providing the support of finite vector spaces, in two of the most widely used representations of vectors in applied sciences: vector of complex numbers and bivectors. In general, the subject of vector analysis can be sub-divided into three distinct parts [4], i.e, the algebra of vectors, the theory of the linear vector function, and the differential and integral calculus of vector functions. In this paper, we mainly focus on the first two parts.

The rest of the paper is organized as follows: In Section 2 we introduce two sets of operators, which are extensively used in our formalization to make use of the multivariate libraries of HOL Light. Section 3 presents our formalization of complex vector algebra followed by the formalization of linearity and infinite summation of complex vector functions. Finally, Section 4 provides an application that illustrates the usage of our current development of complex vectors by the formalization of some basics of electromagnetics: In particular, the law of reflection and the law of plane of incidence for plane electromagnetic waves have been verified.

All the source codes corresponding to the work presented in this paper are available at: http://hvg.ece.concordia.ca/projects/optics/cxvec/.

2   Complex Vectors vs. Bivectors

Complex vectors, by definition, are vectors whose components are complex numbers. One approach to formalize complex vectors is as a pair of two real vectors, very similar to the definition of bivectors by J. William Gibbs [4]. Adopting this approach, first, we instantly inherit all the topological and analytic apparatus of Euclidean Space, described in [7], for real and imaginary parts of complex vectors. Next, we can adopt the approach used for developing complex analysis based on real analysis [5] to extend our libraries for complex vector analysis.

However, in many analytical problems, we need to have access to each element of complex vectors as a complex number. For instance, in the standard definition of complex vector derivative, derivative exists if and only if a function of complex variable is complex analytic, i.e., if it satisfies the Cauchy-Riemann equations [9]. However, this condition is very strong and in practice, many systems, which are defined as a non-constant real-valued functions of complex variables, are not complex analytic and therefore are not differentiable in the standard complex variables sense. In these cases, it suffices for the function to be only differentiable in directions of the co-ordinate system, i.e., in case of cartesian co-ordinate systems, for the x, y, and z. In these cases, it is preferred to define the complex vector as a vector of complex numbers.

In order to have the advantages of both approaches, we define a set of operators which makes a one to one mapping between these two representations of complex vectors.

First, we formalize the concept of vector operators, with no restriction on the data type[1]:

Definition 1 (Unary and Binary Componentwise Operators)

$\vdash \forall(\mathtt{k} : \mathtt{A}).\ \mathtt{vector\_const}\ \mathtt{k} =\ \mathtt{lambda}\ \mathtt{i.}\ \mathtt{k}\ :\mathtt{A}^{\mathtt{N}}$

$\vdash \forall(\mathtt{f} : \mathtt{A} \to \mathtt{B})(\mathtt{v} : \mathtt{A}^{\mathtt{N}}).\ \mathtt{vector\_map}\ \mathtt{f}\ \mathtt{v} =\ \mathtt{lambda}\ \mathtt{i.}\ \mathtt{f}(\mathtt{v\$i}) : \mathtt{B}^{\mathtt{N}}$

$\vdash \forall(\mathtt{f} : \mathtt{A} \to \mathtt{B} \to \mathtt{C})(\mathtt{v1} : \mathtt{A}^{\mathtt{N}})(\mathtt{v2} : \mathtt{B}^{\mathtt{N}}).$
$\qquad\qquad\qquad \mathtt{vector\_map2}\ \mathtt{f}\ \mathtt{v1}\ \mathtt{v2} =\ \mathtt{lambda}\ \mathtt{i.}\ \mathtt{f}(\mathtt{v1\$i})(\mathtt{v2\$i}) : \mathtt{C}^{\mathtt{N}}$

where $\mathtt{v\$i}$ returns the ith component of $\mathtt{v}$ and $\mathtt{lambda\ i.\ f(v\$i)}$ applies the unary operator $\mathtt{f}$ on each component of $\mathtt{v}$ and returns the vector which has $\mathtt{f(v\$n)}$ as its nth component, i.e., $\mathtt{f(v\$1)}$ as its first component, $\mathtt{f(v\$2)}$ as its second component, and so on.

In Section 3, we will show how to verify all the properties of "componentwise" operations of one (complex) vector space by its *counterpart field* (complex numbers) using Definition 1. For instance, after proving that:

$\vdash \forall \mathtt{i}\ \mathtt{f}\ \mathtt{v}.\ (\mathtt{vector\_map}\ \mathtt{f}\ \mathtt{v})\mathtt{\$i} =\ \mathtt{f(v\$i)}$

it is trivial to prove that real vector negation is $\mathtt{vector\_map}$ of real negation:

$\vdash (--) : \mathtt{real}^{\mathtt{N}} \to \mathtt{real}^{\mathtt{N}} = \mathtt{vector\_map}((--) : \mathtt{real} \to \mathtt{real}).$

Next, we extract the real and imaginary parts of a complex vector with type $\mathtt{complex}^{\mathtt{N}}$ as two $\mathtt{real}^{\mathtt{N}}$ vectors, by $\mathtt{cvector\_re}$ and $\mathtt{cvector\_im}$, respectively,

---

[1] In order to improve readability, occasionally, HOL Light statements are written by mixing HOL Light script notations and pure mathematical notations.

and also import the real and imaginary parts of a complex vector into its original format by `complex_vector`, as follows:

Definition 2 (Mapping between $\texttt{complex}^N$ and $\texttt{real}^N \times \texttt{real}^N$)

$\vdash$ `cvector_re = vector_map Re`

$\vdash$ `cvector_im = vector_map Im`

$\vdash \forall$`v1 v2. complex_vector (v1, v2) =`
$\qquad\qquad$ `vector_map2` $(\lambda$ `x y.Cx x + ii * Cx y) v1 v2`

Finally, we formally define a bijection between complex vectors and real vectors with even size, as follows:

Definition 3 (Flatten and Unflatten)

$\vdash \forall$`v. flatten (v : complex`^{N}`) : real`$^{(N,N)\texttt{finite\_sum}}$ `=`
$\qquad\qquad\qquad$ `pastecart (cvector_re v) (cvector_im v)`

$\vdash \forall$`v. unflatten (v : real`$^{(N,N)\texttt{finite\_sum}}$`) : complex`^{N} `=`
$\qquad\qquad\qquad$ `complex_vector (fstcart v, sndcart v)`

where type : $\texttt{real}^{(N,N)\texttt{finite\_sum}}$ refers to a real vector with size $N + N$. All three functions `pastecart`, `fstcart`, and `sndcart` are HOL Light functions. The function `flatten`, takes a complex vector with size N and returns a real vector with size $N + N$ in which the first N elements provide the real part of the original vector and the second N elements provide the imaginary part of the original complex vector. The `unflatten` is an inverse function of `flatten`. The most important properties of `flatten` and `unflatten` are presented in Table 1. The first two properties in Table 1, i.e., Inverse of Flatten and Inverse of Unflatten, guarantee that `flatten` and `unflatten` are bijective. The second two properties ensure

Table 1. Mapping Complex Vectors and Real Vectors

Property	Formalization
Inv. of Flatten	$\vdash$ `unflatten o flatten = I : complex`^{N} $\to$ `complex`^{N}
Inv. of Unflatten	$\vdash$ `flatten o unflatten = I : real`$^{(N,N)\texttt{finite\_sum}}$ $\to$ `real`$^{(N,N)\texttt{finite\_sum}}$
Flatten map	$\vdash \forall$`f g. f = vector_map g` $\Rightarrow$ $\forall$`x. flatten(vector_map f x) = vector_map g (flatten x)`
Flatten map2	$\vdash \forall$`f g. f = vector_map2 g` $\Rightarrow$ $\forall$`x y. flatten(vector_map2 f x y) = vector_map2 g (flatten x)(flatten y)`

that as long as an operator is a map from $\texttt{complex}^N$ to $\texttt{real}^N$, applying this operator in the $\texttt{complex}^N$ domain, and flattening the result will give the same result as flatten operand(s) and applying the operation in the $\texttt{real}^N$ domain. This property is very helpful to prove componentwise properties of complex vectors from their counterparts in real vectors analysis.

3   Complex Vector Algebra

The first step towards the formalization of complex vector algebra is to formalize *complex vector space*. Note that the HOL Light built-in type of vectors does not represent, in general, elements of a vector space. Vectors in HOL Light are basically lists whose length is explicitly encoded in the type. Whereas a vector space is a set S together with a field F and two binary operators that satisfy eight axioms of vector spaces[13] (Table 2). Therefore, we have to define these operators for complex^N and prove that they satisfy the linear space axioms.

Vector Space

Now, it is easy to formally define "componentwise" operations of complex vectors using their counterparts in complex field, including addition and scalar multiplication:

Definition 4 (Arithmetics over complex^N)

$\vdash$ $\text{cvector\_add} = \text{vector\_map2}\ (+ : \text{complex} \to \text{complex})$
$\vdash$ $\text{cvector\_mul}\ (a : \text{complex}) = \text{vector\_map}\ ((* : \text{complex} \to \text{complex})\ a)$

Table 2. Vector Space Axioms for Complex Vectors

Property	Formalization
Addition associativity	$\vdash \forall\ u\ v\ w.\ u + v + w = (u + v) + w$
Addition commutativity	$\vdash \forall\ u\ v.\ u + v = v + u$
Addition unit	$\vdash \forall\ u.\ u + \text{cvector\_zero} = u$
Addition inverse	$\vdash \forall\ u.\ u + (--u) = \text{cvector\_zero}$
Vector distributivity	$\vdash \forall\ a\ u\ v.\ a\ \%\ (u + v) = a\ \%\ u + a\ \%\ v$
Scalar distributivity	$\vdash \forall\ a\ b\ u.\ (a + b)\ \%\ u = a\ \%\ u + b\ \%\ u$
Mul. associativity	$\vdash \forall\ a\ b\ u.\ a\ \%\ b\ \%\ u = (a * b)\ \%\ u$
Scalar mul. unit	$\vdash \forall\ u.\ \text{Cx}(\&1)\ \%\ u = u$

We developed a tactic, called `CVECTOR_ARITH_TAC`, which is mainly adapted from `VECTOR_ARITH_TAC` [7] and is able to prove simple arithmetics properties of complex vectors automatically. Using this tactic we prove all eight axioms of vector spaces, indicated in Table 2, plus many other basic but useful facts. In Table 2, the symbol (&) and the function (Cx) are HOL Light functions for typecasting from `integer` to `real` and from `real` to `complex`, respectively. The symbol (%) and (--) are overloaded by scalar multiplication and complex vector negation, respectively, and `cvector_zero` is a complex null vector. The negation and `cvector_zero` are formalized using componentwise operators `vector_map` and `vector_const` (Definition 1), respectively.

Vector Products

Two very essential notions in vector analysis are *cross product* and *inner product*.

The cross product between two vectors u and v of size 3 is classically defined as $(u_2v_3 - u_3v_2, u_3v_1 - u_1v_3, u_1v_2 - u_2v_1)$. This is formalized as follows[2]:

Definition 5 (Complex Cross Product)

$\vdash \forall$u v : complex^3. u $\times$ v : complex^3 =
  vector [u\$2 * v\$3 $-$ u\$3 * v\$2; u\$3 * v\$1 $-$ u\$1 * v\$3; u\$1 * v\$2 $-$ u\$2 * v\$1]

where `vector` is a HOL Light function taking a list as input and returning a vector. Table 3 presents some of the properties we proved about the cross product.

Table 3. Cross Product Properties

Property	Formalization
Left zero	$\vdash \forall$ u. cvector_zero $\times$ u = cvector_zero
Right zero	$\vdash \forall$ u. u $\times$ cvector_zero = cvector_zero
Irreflexivity	$\vdash \forall$ u. u $\times$ u = cvector_zero
Asymmetry	$\vdash \forall$ u v. $--$(u $\times$ v) = v $\times$ u
Left-distributivity over addition	$\vdash \forall$ u v w. (u + z) $\times$ w = u $\times$ w + z $\times$ w
Right-distributivity over addition	$\vdash \forall$ u v w. u $\times$ (v + w) = u $\times$ v + u $\times$ w
Left-distributivity over scalar mul.	$\vdash \forall$ a u v. (a%u) $\times$ v = a%(u $\times$ v)
Right-distributivity over scalar mul.	$\vdash \forall$ a u v. u $\times$ (a%v) = a%(u $\times$ v)

The inner product is defined for two complex vectors u and v of dimension N as $\sum_{i=1}^{N} u_i \overline{v}_i$, where $\overline{x}$ denotes the complex conjugate of x. This is defined in HOL Light as follows:

Definition 6 (Complex Vector Inner Product)

$\vdash \forall$(u : complex^N) (v : complex^N).
              u cdot v = vsum (1..dimindex(: N)) (λi. u\$i * cnj(v\$i))

where `cnj` denotes the complex conjugate in HOL Light, and `vsum s f` denotes $\sum_{x \in s}$ f x, `dimindex` s is the number of elements of s, and (: N) is the set of all inhabitants of the type N (the "universe" of N, also written UNIV).

Proving properties for the *inner product space*, presented in Table 4, are quite straightforward except for the positive definiteness, which involves inequalities. The inner product of two complex vector is a complex number. Hence, to prove the positive definiteness, we first prove that $\vdash \forall$x. real(x cdot x), where `real` is a HOL Light predicate which returns true if the imaginary part of a complex number is zero. Then, by introducing function `real_of_complex`, which

---

[2] The symbol "$\times$" indicates `ccross` in our codes.

returns a complex number with no imaginary part as a real number, we formally prove the positive definiteness. This concludes our formalization of complex inner product spaces.

Table 4. Inner Product Space

Property	Formalization
Conjugate Symmetry	$\vdash \forall x\ y.\ x$ cdot $y = $ cnj$(y$ cdot $x)$
Linearity (scalar multiplication)	$\vdash \forall c\ x\ y.\ (c\ \%\ x)$ cdot $y = c * (x$ cdot $y)$
Linearity (vector addition)	$\vdash \forall x\ y\ z.\ (x + y)$ cdot $z = (x$ cdot $z) + (y$ cdot $z)$
Zero length	$\vdash \forall x.\ x$ cdot $x = $ Cx$(\&0) \Leftrightarrow x = $ cvector_zero
Positive definiteness	$\vdash \forall x.\ \&0 \leq$ real_of_complex$(x$ cdot $x)$

Norm, orthogonality, and the angle between two vectors are mathematically defined using the inner product. Norm is defined as follows:

Definition 7 (Norm of Complex Vectors)
$\vdash$ cnorm $=$ sqrt o cnorm2

where $\vdash \forall x.$ cnorm2 $x = $ real_of_complex$(x$ cdot $x)$. Then the norm is overloaded with our new definition of cnorm.

We also define the concept of orthogonality and collinearity, as follows:

Definition 8 (Orthogonality and Collinearitiy of Complex Vectors)
$\vdash \forall x\ y.$ corthogonal $x\ y \Leftrightarrow x$ cdot $y = $ Cx$(\&0)$
$\vdash \forall x\ y.$ collinear_cvectors $x\ y\ \Leftrightarrow\ \exists a.\ (y = a\%x)\ \lor\ (x = a\%y)$

Next, the angle between two complex vectors is formalized just like its counterpart in real vectors. Obviously, defining the vector angle between any complex vector and cvector_zero as Cx$(pi/\&2)$ is a choice, which is widely used and accepted in literature. Note that in a very similar way we define Hermitian angle and the real angle [11].

Definition 9 (Complex Vector Angle)

$\vdash \forall x\ y : $ complex^N.
cvector_angle $x\ y = $ if $x = $ cvector_zero $\lor\ y = $ cvector_zero
        then Cx$(pi/\&2)$
        else cacs$((x$ cdot $y)/$Cx$($norm $x\ *$ norm $y))$

In Table 5, some of the properties related to norm and analytic geometry are highlighted.

We also define many basic notions of linear algebra, e.g., the canonical basis of the vector space, i.e., the set of vectors $(1, 0, 0, 0, \dots)$, $(0, 1, 0, 0, \dots)$, $(0, 0, 1, 0, \dots)$, and so on. This is done as follows:

Table 5. Highlights of properties related to vector products

Property	Formalization
Cross product & collinearity	$\vdash \forall$x y. x $\times$ y $=$ cvector_zero $\Leftrightarrow$
	collinear_cvectors x y
Cauchy-Schwarz inequality	$\vdash \forall$x y. norm(x cdot y) $\leq$ norm x $*$ norm y
Cauchy-Schwarz equality	$\vdash \forall$x y. collinear_cvectors x y $\Leftrightarrow$
	norm(x cdot y) $=$ norm x $*$ norm y
Triangle inequality	$\vdash \forall$x y. norm(x $+$ y) $\leq$ norm x $+$ norm y
Pythagorean theorem	$\vdash \forall$x y. corthogonal x y $\Leftrightarrow$
	cnorm2(x $+$ y) $=$ cnorm2 x $+$ cnorm2 y
Dot product and angle	$\vdash \forall$x y. x cdot y $=$ Cx(norm x $*$ norm y)$*$
	ccos(cvector_angle x y)
Vector angle range	$\vdash \forall$x y. $\neg$collinear_cvectors x y $\Rightarrow$
	&0 $<$ Re(cvector_angle x y) $\wedge$
	Re(cvector_angle x y) $<$ pi

Definition 10 $\vdash$ $\forall$k. cbasis k $=$ vector_to_cvector (basis k) : complex^{N}

With this definition cbasis 1 represents $(1, 0, 0, 0, \ldots)$, cbasis 2 represents $(0, 1, 0, 0, \ldots)$, and so on.

Another essential notion of vector spaces is the one of *matrix*. Matrices are essentially defined as vectors of vectors: a $M \times N$ matrix is formalized by a value of type (complex^{N})^{M}. Several arithmetic notions can then be formalized over matrices, again using the notions of operators, presented in Definition 1. Table 6 presents the formalization of complex matrix arithmetics. The formal definition of arithmetics in matrix is almost identical to the one of complex vectors, except for the type. Hence, in Table 6, we specify the type of functions, where $\mathbb{C}$ represents the type complex.

Table 6. Complex Matrix Arithmetic

Operation	Types	Formalization
Negation	: $\mathbb{C}^{NM} \to \mathbb{C}^{NM}$	$\vdash$ cmatrix_neg $=$ vector_map $(--)$
Conjugate	: $\mathbb{C}^{NM} \to \mathbb{C}^{NM}$	$\vdash$ cmatrix_cnj $=$ vector_map (cvector_cnj)
Addition	: $\mathbb{C}^{NM} \to \mathbb{C}^{NM} \to \mathbb{C}^{NM}$	$\vdash$ cmatrix_add $=$ vector_map2 $(+)$
Scalar Mul.	: $\mathbb{C} \to \mathbb{C}^{NM} \to \mathbb{C}^{NM}$	$\vdash$ cmatrix_smul $=$ vector_map o $(\%)$
Mul.	: $\mathbb{C}^{NM} \to \mathbb{C}^{PN} \to \mathbb{C}^{PM}$	$\vdash \forall$m$_1$ m$_2$. cmatrix_mul $=$
		lambda i j.vsum(1..dimindex(: N)) (λk.m$_1$\$i\$k $*$ m$_2$\$k\$j)

Finally, we formalize the concepts of summability and infinite summation, as follows:

Definition 11 (Summability and Infinite Summation)

$\vdash \forall (s : \texttt{num} \to \texttt{bool})(f : \texttt{num} \to \texttt{complex}^N).$
    $\texttt{csummable s f} \Leftrightarrow \texttt{summable s}(\texttt{cvector\_re o f}) \wedge$
                    $\texttt{summable s } (\texttt{cvector\_im o f})$

$\vdash \forall (s : \texttt{num} \to \texttt{bool})(f : \texttt{num} \to \texttt{complex}^N).$
    $\texttt{cinfsum s f} = \texttt{vector\_to\_cvector } (\texttt{infsum s } (\lambda x.\texttt{cvector\_re } (f\ x)))$
                    $+ \texttt{ii\%vector\_to\_cvector}(\texttt{infsum s } (\lambda x.\texttt{cvector\_im } (f\ x)))$

We, again, prove that summability and infinite summation can be addressed by their counterparts in real vector analysis. Table 7 summarizes the key properties of summability and infinite summation, where the predicate `clinear f` is true if and only if the function $f : \texttt{complex}^M \to \texttt{complex}^N$ is linear. As it can be observed in Table 7, to extend the properties of summability and infinite summation of real vectors to their complex counterparts, the two functions of `flatten` and `unflatten` are used. When we `flatten` a complex vector, the result would be a real vector which can be accepted by `infsum`. Since summation is a componentwise operand and, as presented in Table 1, `unflatten` is the inverse of `flatten`, `unfalttening` the result of `infsum` returns the desired complex vector.

Table 7. Linearity and summability of complex vector valued functions

Properties	Formalization
Linearity	$\vdash \forall f.$ `clinear f` $\Leftrightarrow$ `linear (flatten o f o unflatten)`
Summability	$\vdash \forall s\ f.$ `csummable s f` $\Leftrightarrow$ `summable s (flatten o f)`
Infinite Summation	$\vdash \forall s\ f.$ `csummable s f` $\Rightarrow$
	(`cinfsum s f` $=$ `unflatten (infsum s (flatten o f))`)

In summary, we successfully formalized 30 new definitions and proved more than 500 properties in our libraries of complex vectors. The outcome of our formalization is a set of libraries that is very easy to understand and to be adopted for the formalization of different applications, in engineering, physics, etc. Our proof script consists of more than 3000 lines of code and took about 500 man-hours for the user-guided reasoning process.

Infinite Dimension Complex Vector Spaces

As mentioned earlier in Section 1, in [10], the infinite-dimension complex vector spaces are formalized using function spaces. However, this definition brings unnecessary complexity to the problems in the finite space. As a result, we chose to develop our library of complex vectors then prove that there exist an isomorphism between our vector space and a finite vector space developed based on the work presented in [10].

The complex-valued functions, in [10], are defined as functions from an arbitrary set to `complex`, i.e., values of type $\texttt{cfun} = A \to \texttt{complex}$, where A is a type variable. In general, complex vector spaces can be seen as a particular case of complex function spaces. For instance, if the type A in A → complex is instantiated with a finite type, say a type with three inhabitants one, two, three, then functions mapping each of one, two, three to a complex number can be seen as 3-dimension complex vectors. So, for a type t with n inhabitants, there is a bijection between complex-valued functions of domain t and complex vectors of dimension n.

This bijection is introduced by Harrison [7], defining the type constructor (A)finite_image which has a finite set of inhabitants. If A is finite and has n elements then (A)finite_image also has n elements. Harrison also defines two inverse operations, as follows:

$$\vdash \text{mk\_cart} : ((A)\text{finite\_image} \to B) \to B^A$$
$$\vdash \text{dest\_cart} : B^A \to ((A)\text{finite\_image} \to B)$$

where B^A is the type of vectors with as many coordinates as the number of elements in A.

By using the above inverse functions, we can transfer complex vector functions, for instance, cfun addition to the type complex^N as follows:

$$v +_{\text{complex}^N} w = \text{mk\_cart} (\text{dest\_cart} \ v +_{\text{cfun}} \text{dest\_cart} \ w)$$

where the indices to the + provides the type information. This definition basically takes two vectors as input, transforms them into values of type cfun, computes their cfun-addition, and transforms back the result into a vector. This operation can actually be easily seen to be equivalent to cvector_add in Definition 4. Therefore, not only we have a bijection between the types $(N)\text{finite\_image} \to \text{complex}$ and complex^N, but also an *isomorphism* between the structures $((N)\text{finite\_image} \to \text{complex}, +_{\text{cfun}})$ and $(\text{complex}^N, +_{\text{complex}^N})$.

We use these observations to develop a framework proving that there exists an isomorphism between the two type A → complex and complex^N when A has N elements. This development results in a uniform library addressing both finite and infinite dimension vector spaces. The details of this development can be found in [1].

In order to show the effectiveness of our formalization, in the next section we formally describe monochromatic light waves and their behaviour at an interface between two mediums. Then we verify the laws of incidence and reflection. We intentionally developed our formalization very similar to what one can find in physics textbooks.

4   Application: Monochromatic Light Waves

In the electromagnetic theory, light is described by the same principles that govern all forms of electromagnetic radiations. The reason we chose this application is that vector analysis provides an elegant mathematical language in which electromagnetic theory is conveniently expressed and best understood. In fact, in the literature, the two subjects, electromagnetic theory and complex vector analysis, are so entangled, that one is never explained without an introduction to the other. Note that, although physically meaningful quantities can only be represented by real numbers, in analytical approaches, it is usually more convenient to introduce the complex exponential rather than real sinusoidal functions.

An electromagnetic radiation is composed of an electric and a magnetic field. The general definition of a field is "a physical quantity associated with each point of space-time". Considering electromagnetic fields ("EMF"), the "physical quantity" consists of a 3-dimensional complex vector for the electric and the magnetic field. Consequently, both those fields are defined as complex vector valued functions $E(r,t)$ and $H(r,t)$, respectively, where r is the position and t is the time. Points of space are represented by 3-dimensional real vectors, so we define the type `point` as an abbreviation for the type `real`^{3}. Instants of time are considered as real so the type `time` represents type *real* in our formalization. Consequently, the type `field` (either magnetic or electric) is defined as `point` $\rightarrow$ `time` $\rightarrow$ `complex`^{3}. Then, since an EMF is composed of an electric and a magnetic field, we define the type `emf` to represent `point` $\rightarrow$ `time` $\rightarrow$ `complex`^{3} $\times$ `complex`^{3}. The electric and magnetic fields are therefore complex vectors. Hence their formalization and properties make use of the complex vectors theory developed in Section 3.

One very important aspect in the formalization of physics is to make sure that all the postulates enforced by physics are formalized. We call these sets of definitions as "constraints". For instance, we define a predicate which ensures that the electric and magnetic field of an electromagnetic field are always orthogonal, as follows:

Constraint 1 (Valid Electromagnetic Field)
$\vdash \forall$emf. is_valid_emf emf $\Leftrightarrow$
$$(\forall r\ t.\ \text{corthogonal}\ (\text{e\_of\_emf emf r t})\ (\text{h\_of\_emf emf r t}))$$

where e_of_emf and h_of_emf are two helpers returning the electric field and magnetic field of an emf, respectively.

An electromagnetic plane wave can be expressed as $U(r,t) = a(r)e^{j\phi(r)}e^{j\omega t}$, where U can be either the electric or magnetic field at point r and time t, $a(r)$ is a complex vector called the *amplitude* of the field, and $\phi(r)$ is a complex number called its *phase*. In monochromatic plane waves (i.e., waves with only one *frequency* ω), the phase $\phi(r)$ has the form $-k \cdot r$, where "$\cdot$" denotes the inner product between real vectors. We call k the *wavevector* of the wave; intuitively, this vector represents the propagation direction of the wave. This yields the following definition:

Definition 12 (Monochromatic Plane Wave)
$\vdash$ plane_wave (k : real^3) (ω : real) (E : complex^3) (H : complex^3) : emf
$$= \lambda(r : \text{point})\ (t : \text{time}).\ (e^{-\text{ii}(k\cdot r - \omega t)}E,\ e^{-\text{ii}(k\cdot r - \omega t)}H)$$

where, again we accompany this physical definition with its corresponding Constraint 2, is_valid_wave, which ensures that a plane wave U is indeed an electromagnetic field and the wavevector is indeed representing the propagation direction of the wave. This former condition will be satisfied, if and only if, the wavevector k, electric field, and magnetic field of the wave are all perpendicular.

Constraint 2 (Valid Monochromatic Wave)

⊢ ∀emf. is_valid_wave wave ⇔
    (is_valid_emf wave ∧
    (∃k w e h.
      &0 < w ∧ ¬(k = vec 0) ∧ wave = plane_wave k w e h ∧
      corthogonal e (vector_to_cvector k) ∧
      corthogonal h (vector_to_cvector k))

where, vector_to_cvector is a function from $real^N$ to $complex^N$, mapping a real vector to a complex vector with no imaginary part.

Now, focusing on electromagnetic optics, when a light wave passes through a medium, its behaviour is governed by different characteristics of the medium. The *refractive index*, which is a real number, is the most dominant among these characteristics, therefore the type medium is defined simply as the abbreviation of the type *real*. The analysis of an optical device is essentially concerned with the passing of light from one medium to another, hence we also define the plane *interface* between the two mediums, as medium × medium × plane × $real^3$, i.e., two mediums, a plane (defined as a set of points of space), and an orthonormal vector to the plane, indicating which medium is on which side of the plane.

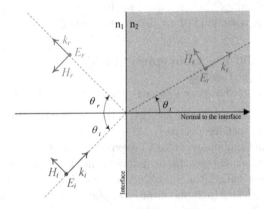

Fig. 1. Plane Interface Between Two Mediums

In order to show how the effectiveness of our formalization, we prove some properties of waves at an interface (e.g., the law of reflection, i.e., a wave is reflected in a symmetric way to the normal of the surface) derived from the *boundary conditions* on electromagnetic fields. These conditions state that the projection of the electric and magnetic fields shall be equal on both sides of the interface plane. This can be formally expressed by saying that the cross product between those fields and the normal to the surface shall be equal:

Definition 13 (Boundary Conditions)

⊢ boundary_conditions emf$_1$ emf$_2$ n p t ⇔
  n × e_of_emf emf$_1$ p t = n × e_of_emf emf$_2$ p t ∧
  n × h_of_emf emf$_1$ p t = n × h_of_emf emf$_2$ p t

We then formalize a plane interface between two mediums, in the presence of a plane wave, shown in Fig. 1, with the following predicate:

Constraint 3 (Plane Wave and a Plane Interface)

⊢ is_plane_wave_at_interface i emf$_i$ emf$_r$ emf$_t$ ⇔
  is_valid_interface i ∧ is_plane_wave emf$_i$ ∧
  is_plane_wave emf$_r$ ∧ is_plane_wave emf$_t$ ∧
  let $(n_1, n_2, p, n) = $ i in
  let $(k_i, k_r, k_t) = $ map_triple k_of_w (emf$_i$, emf$_r$, emf$_t$) in
  let $(e_i, e_r, e_t) = $ map_triple (norm ∘ e_of_w) (emf$_i$, emf$_r$, emf$_t$) in
  let $(h_i, h_r, h_t) = $ map_triple (norm ∘ k_of_w) (emf$_i$, emf$_r$, emf$_t$) in
  $0 \le$ (k$_i$ · norm_of_plane p) ∧ (k$_r$ · norm_of_plane p) ≤ 0 ∧
  $0 \le$ (k$_t$ · norm_of_plane p) ∧
  (∀pt. pt ∈ p ⇒
  ∀t. boundary_conditions (emf$_i$ + emf$_r$) emf$_t$ n pt t) ∧
  ∃k$_0$. norm k$_i$ = k$_0$n$_1$ ∧ norm k$_r$ = k$_0$n$_1$ ∧ norm k$_t$ = k$_0$n$_2$ ∧
  ∃η_0. h$_i$ = e$_i$n$_1$/η_0 ∧ h$_r$ = e$_r$n$_1$/η_0 ∧ h$_t$ = e$_t$n$_2$/η_0 ∧ e$_i$ ≠ 0 ∧ e$_r$ ≠ 0

where map_triple f $(x, y, z) = ($f x, f y, f z$)$, norm denotes the norm of a complex vector (defined by using the inner product), e_of_w (shorthand for "electric field of wave") is a helper function allowing to retrieve the electric amplitude of a wave, and k_of_w allows us to retrieve the wavevector of a wave. The predicate of Constraint 3 takes an interface i and three EMFs e$_i$, e$_r$, and e$_t$, intended to represent the incident wave, the reflected wave, and the transmitted wave, respectively. The first four atoms of the predicate ensure that the arguments are well-formed, i.e., i is a valid interface and the three input fields exist and are plane waves (we refer to the implementation for details about these predicates). From this predicate, which totally describes the interface in Fig. 1, we can prove several foundational properties of plane waves. For instance, the incident, reflected, and transmitted waves all lie in the same plane (the so-called "plane of incidence"):

Theorem 1 (Law of Plane of Incidence)

⊢ ∀i emf$_i$ emf$_r$ emf$_t$ x y z.
    is_plane_wave_at_interface i emf$_i$ emf$_r$ emf$_t$ ∧
    is_incident_basis (x, y, z) emf$_i$ i ⇒
    k_of_w emf$_i$ · x = 0 ∧ k_of_w emf$_r$ · x = 0 ∧ k_of_w emf$_t$ · x = 0

where is_incident_basis (x, y, z) emf$_i$ i asserts that (x, y, z) is a basis of the incident plane, i.e., the plane characterized by the wavevector of emf$_i$ and the normal to i (note that if these two vectors are collinear then there is an infinity of planes of incidence). Another non-trivial consequence is the fact that the reflected wave is symmetric to the incident wave with respect to the normal to the surface:

Theorem 2 (Law of reflection)

$\vdash \forall i$ emf$_i$ emf$_r$ emf$_t$.
      is_plane_wave_at_interface i emf$_i$ emf$_r$ emf$_t$ $\Rightarrow$
                are_sym_wrt $(-($k_of_w emf$_i))$ (k_of_w emf$_r$)
                (norm_of_plane (plane_of_interface i))

where **are_sym_wrt** u v w formalizes the fact that u and v are symmetric with respect to w (this is easily expressed by saying that $v = 2*(u \cdot w)w - u$). Referring to Fig. 1, Theorem 2 just means that $\theta_i = \theta_r$, which is the expression usually found in textbooks.

The proofs of these results make an extensive use of the formalization of complex vectors and their properties presented in Section 3: the arithmetic of complex vectors is used everywhere as well as the dot and cross product. Our proof script for the application consists of approximately 1000 lines of code. Without availability of the formalization of complex vectors, this reasoning would not have been possible, which indicates the usefulness of our work.

5 Conclusion

We presented formalization of complex vectors in HOL Light. The concepts of linear spaces, norm, collinearity, orthogonality, vector angles, summability, and complex matrices which are all elements of more advanced concepts in complex vector analysis are formalized. An essential aspect of our formalization is to keep it *engineering* and *applied science*-oriented. Our libraries developed originally upon two different representations of complex vectors: vector of complex numbers (complex^N) and bivectors ($\text{real}^N \times \text{real}^N$), which were verified to be isomorphic. We favoured the notions and theorems that are useful in applications, and we provided tactics to automate the verification of most commonly encountered expressions. We then illustrated the effectiveness of our formalization by formally describing electromagnetic fields and plane waves, and we verified some classical laws of optics: the law of the plane of incidence and of reflection.

Our future works is extended in three different directions. The first is to enrich the libraries of complex vector analysis. We have not yet addressed differential and integral calculus of vector functions. Second, we plan to formally address practical engineering systems. For instance, we already started developing electromagnetic applications, particularly in the verification of optical systems [2]. Finally, to have our formalization to be used by engineers, we need to develop more tactics and introduce automation to our libraries of complex vectors.

References

1. Afshar, S.K., Aravantinos, V., Hasan, O., Tahar, S.: A toolbox for complex linear algebra in HOL Light. Technical Report, Concordia University, Montreal, Canada (2014)
2. Afshar, S.K., Siddique, U., Mahmoud, M.Y., Aravantinos, V., Seddiki, O., Hasan, O., Tahar, S.: Formal analysis of optical systems. Mathematics in Computer Science 8(1) (2014)
3. Chaieb, A.: Multivariate Analysis (2014), http://isabelle.in.tum.de/repos/isabelle/file/tip/src/HOL/Multivariate_Analysis
4. Gibbs, J.W.: Elements of Vectors Analysis. Tuttle, Morehouse & Taylor (1884)
5. Harrison, J.: Formalizing basic complex analysis. In: From Insight to Proof: Festschrift in Honour of Andrzej Trybulec. Studies in Logic, Grammar and Rhetoric, pp. 151–165. University of Białystok (2007)
6. Harrison, J.: HOL Light: An Overview. In: Berghofer, S., Nipkow, T., Urban, C., Wenzel, M. (eds.) TPHOLs 2009. LNCS, vol. 5674, pp. 60–66. Springer, Heidelberg (2009)
7. Harrison, J.: The HOL Light theory of Euclidean space. Journal of Automated Reasoning 50(2), 173–190 (2013)
8. Herencia-Zapana, H., Jobredeaux, R., Owre, S., Garoche, P.-L., Feron, E., Perez, G., Ascariz, P.: PVS linear algebra libraries for verification of control software algorithms in C/ACSL. In: Goodloe, A.E., Person, S. (eds.) NFM 2012. LNCS, vol. 7226, pp. 147–161. Springer, Heidelberg (2012)
9. LePage, W.R.: Complex variables and the Laplace transform for engineers. Dover Publications (1980)
10. Mahmoud, M.Y., Aravantinos, V., Tahar, S.: Formalization of infinite dimension linear spaces with application to quantum theory. In: Brat, G., Rungta, N., Venet, A. (eds.) NFM 2013. LNCS, vol. 7871, pp. 413–427. Springer, Heidelberg (2013)
11. Scharnhorst, K.: Angles in complex vector spaces. Acta Applicandae Mathematica 69(1), 95–103 (2001)
12. Stein, J.: Linear Algebra (2014), http://coq.inria.fr/pylons/contribs/view/LinAlg/trunk
13. Tallack, J.C.: Introduction to Vector Analysis. Cambridge University Press (1970)

A Mathematical Structure for Modeling Inventions

Bernd Wegner and Sigram Schindler

Mathematical Institute, TU Berlin, TELES Patent Rights International GmbH, Germany

Abstract. The paper is the first of several ones [14,17] describing a mathematical structure developed in the FSTP project, mathematically modeling Substantive Patent Law ("SPL") and its US Highest Courts' precedents primarily for emerging technologies inventions. Chapter 2 presents this mathematical structure comprising particularly, 3 abstraction levels - each comprising "inventive concepts", their "subset coverings", "concept transformations", "induced concept relations', and "refinements". Chapters 3 and 4 explain its practical application in describing an invention respectively testing it by an Innovation Expert System (IES) for its satisfying SPL.

Using the notion of "inventive concepts" for precisely describing emerging technologies inventions has been introduced into SPL precedents by the US Supreme Court during its ongoing "SPL initiative" - marked by its KSR/Bilski/ Mayo/ Myriad decisions. It induced, into the FSTP project, a rigorous mathematical analysis of allegedly new problems caused by these Highest Courts' SPL decisions about emerging technologies inventions. This analysis proved extremely fertile by enabling not only clarifying/removing obscurities in such problems but also developing powerful "patent technology" in the FSTP project.

1 Introduction

This FSTP project paper addresses the community of mathematicians not that much interested in the most recent problems in SPL precedents about emerging technologies inventions, but in contributing to its scientification in an unquestionable manner, i.e. in exerting rigorous mathematical scrutiny to it. It hence deals with providing a mathematical fundament - established by a sophisticated mathematical structure - for a very topical area of greatest socio/economic importance for the world's wealthy countries, as controlling the flow of annually several 100 billions of US Dollars [15], just as for supporting the transfer into emerging countries the know-how about innovativity enabling this wealth.

The FSTP project comprises, on the one side, a range of publications about developing the "patent/innovation technology" for a cutting edge prototype, the "Innovation Expert System (IES)", as technologically today possible. In the future such IESes will be indispensable for the efficiency of the everyday professional activities of the hundred thousands of patent/innovation professionals of all kinds - in particular researchers/inventors, their R&D managers, R&D investors, patent/license lawyers, patent examiners, judges, product managers, marketing managers, ..., in all kinds of emerging technologies, be it telecommunications/nano/genetics/drugs/....../business/ sports/... technologies. These publications deal with leveraging on the capability of

S.M. Watt et al. (Eds.): CICM 2014, LNAI 8543, pp. 138–152, 2014.

precisely describing emerging technologies inventions, as required by the US Supreme Court's SPL precedents during its ongoing "SPL initiative", i.e. its famous line of KSR/Bilski/Mayo/Myriad decisions. It thus induced, on the FSTP project, a rigorous mathematical analysis of the thereby arising new decision problems in SPL precedents. Its scrutiny proved extremely fertile and enabled clarifying and removing the notional and legal obscurities in SPL stirred up by emerging technologies, i.e. having been lingering in SPL since ever, as well as developing "patent/innovation technology" based on that scientific approach to creativity/innovativity.

On the other side, there are FSTP publications just as this one, elaborating on the mathematical foundation of this new technology for ascertaining its well- definedness, e.g. for excluding allegedly correct legal statements about an invention but evidently contradicting each other - as it recently repeatedly occurred in the CAFC. The research results published by them deal with groundbreaking mathematical/theoretical modeling issues by Advanced IT for enabling these practical developments in the area of stimulating/supporting/protecting/commercializing such emerging technology innovations.

This paper, in particular, reports about a sophisticated mathematical structure enabling deciding in an unquestionable as mathematically assured - and hence consistent and predictable - way about the patent eligibility and patentability of a developing/claimed/reexamined/licensed/infringed/... invention represented by a patent/ application /contract/..... To this end, this new mathematical structure has been designed such as to facilitate precisely modeling the interpretation of SPL by the Highest Courts and their respective precedents as to emerging technologies inventions, just as the technical/factual needs arising from such inventions.

The established classical interpretations of SPL in all industrial nations are strongly depending on the tangibility/visibility of inventions and hence prove vastly deficient when dealing with emerging technologies inventions, as these are virtually always intangible and invisible, i.e. "model based". They hence may be "indefinite" and/or "preemptive" as their scopes of patent monopoly are ambiguous resp. potentially comprise inventions not yet known at all when granting them - or just socially unwanted, e.g. as felt threatening - all these being new reasons of their non-patent-eligibility.

But the main purpose the here presented mathematical structure must serve, for emerging technologies inventions just as for classical ones, is to facilitate modeling the metes and bounds of any patent analysis and achieving its objectives, namely to support deciding correctly and semi-automatically whether it satisfies SPL. In its above mentioned SPL initiative, more precisely in its Mayo decision, the US Supreme Court introduced into SPL precedents the key notion of "inventive concepts" and required using them for construing, for any claimed emerging technologies invention its (thus refined) claim construction, being a shorter way than saying "using its inventive concepts in checking whether it meets the requirements stated by SPL" - as explained in legal terms in [8,9] and will mathematically be presented in detail in this and the following mathematical paper(s). [17]

To conclude and summarize these introductory remarks: The aim of this paper is to present in mathematical rigor the mathematical structure developed in the FSTP project as fundament of any IES, not just the present IES prototype, and to briefly indicate how this mathematical frame work is motivated by and related to the everyday legal business of patent professionals. All publications about the FSTP project and its IES prototype are available on www.fstp-expert-system.com.

2 Concepts and Mathematical Structure

In principle, an inventive concept, in-C, is a pair of a legal concept and a technical alias factual concept, i.e. in-C = (le-C, cr-C), the latter one supposed to be creative, hence its name. In this Chapter totally focused on mathematical structures, only their creative concept are considered, and for further simplification they are temporarily called just concepts - in Section 9 these simplifications are elaborated on. In any practical application this bisection of inventive concepts must be preserved, for sake of the quite different beings of the notions of creative and legal concepts, i.e. for thereby establishing a legal and factual clarity often disregarded hitherto.

1. As a first step we define the mathematical structure for a description of a concept:

Definition 1: A *concept* C is given by a triple of data

$$C = (DC, VC, UC),$$

where DC is a non-empty set called the *domain* of C, VC is a non-empty finite set of non-empty sets $vC \in VC$, VC being called the *set of value sets* vC of C and where a vC is called a *value set* of C, UC is a map from DC to VC, represented by a non-empty relation from DC to VC, i.e., $UC \subseteq DC \times VC$, each $d \in DC$ is related to some $vC \in VC$ and no $d \in DC$ is related to two different value sets in VC. UC is called the *universe* of C.

A concept is called *binary concept* if VC has exactly two value sets, one being identified with $T = True$ and the other with $F = False$. This is abbreviated by $VC = \{T, F\}$. In the following only **binary concepts** will be considered.

2. A binary concept has three equivalent descriptions:

i) The usual one as a triple $C = (DC, \{T, F\}, UC)$, where DC is the domain of the concept and UC is a relation between DC and $\{T, F\}$. As defined above UC has to satisfy the following requirements: a) each $d \in DC$ is related to some $vC \in \{T, F\}$ and b) no $d \in DC$ is related to T and F.

ii) A binary concept may be expressed by a triple $C = (DC, MC, \{T, F\})$, where $MC : DC \longrightarrow \{T, F\}$ is the map connected with UC by $UC = \{(d, MC(d)) \mid d \in DC\}$.

iii) The third description is given by the separation of the domain DC into the *truth-set* $TS(C)$, being the complete preimage of T under MC, and the *false-set* $FS(C)$ being the complete preimage of F under MC, i.e. $C = (DC, TS(C), FS(C))$, where $DC = TS(C) \cup FS(C)$ and $TS(C) \cap FS(C) = \emptyset$.

These descriptions are linked by the following equations:

$$UC = (TS(C) \times \{T\}) \cup (FS(C) \times \{F\}),$$

where the universe UC now also is represented as a set, and the sets $TS(C) \times \{T\}$ and $FS(C) \times \{F\}$ represent a partition of UC into two disjoint subsets, MC decomposes into two constant maps $MC|TS(C)$ having the value T and $MC|FS(C)$ having the value F, and $DC = p_1(UC)$, p_1 denoting the projection of $DC \times \{T, F\}$ onto its first factor DC.

3. Today, everyday business of a patent practitioner deals with only information in natural language or simple graphics representation, which the posc considers to be lawfully disclosed by the document providing it (whereby the posc represents the fictional "person of ordinary skill and creativity" pertinent alias familiar to the subject area of the patent preparation/application/prosecution/licensing - at issue; i.e. is the perfect representative of the pertinent ordinary skill and creativity, and hence of no extraordinary qualification). Advanced IT (e.g. Semantics, Natural Language, Knowledge Representation, Compiler/Interpreter research) tells that on this information representation "level" precise statements are impossible. For deriving from this imprecise original information representation to a precise one, two preciseness increasing steps are indispensable in FSTP alias Patent Technology today to be performed by the posc:

• Firstly, transforming the original and NL syntax based information representation, e.g. of a patent specification, into a NL terms but some FOL syntax based information, and

• secondly transforming the latter representation using compound inventive concepts as "alphabet" (consisting of these NL terms) - by preserving its exact FOL syntax - into a refined knowledge representation, again using NL terms, but refined ones such as to achieve that anyone of these non- refined NL terms is logically equivalent to the conjunction of certain ones of these refined NL terms (in total defining the refined alphabet).

In order to link to each other these three different kinds/representations of inventive concepts of (often) different preciseness of an invention - as generated by the posc by its refinement process - we need transformations between the sets of such associated binary concepts. These transformations are given by bijections between underlying subset coverings of the union of the universes for each concept set. Subset coverings are defined by the following:

Definition 2: Let $Cset$ be a finite set of concepts. A finite *subset collection* of the union of the universes of the concepts belonging to $Cset$ is given by a finite sequence $SSU = (ssU_1, \ldots, ssU_\Lambda)$ of Λ non-empty sets

$$ssU_\lambda \subseteq \bigcup \{UC \mid C \in Cset\}$$

whereby the union is considered as (concept-wise) disjoint. Such a subset collection is called a finite *subset covering* $SSCov = (ssCov_1, \ldots, ssCov_\Lambda)$ of the union of the universes of the concepts in $Cset$, iff the following covering condition is satisfied:

$$\bigcup \{ssCov_\lambda \mid 1 \leq \lambda \leq \Lambda\} = \bigcup \{UC \mid C \in Cset\}.$$

The choice of a finite sequence of subsets for the collection instead of just choosing a finite number of subsets is motivated by the fact that for the transformation to be defined below some subsets may have to be considered several times, which requires to distinguish several copies of the same set formally for these cases. We may describe a subset covering in the form $SSCov = \{ssCov_1, \ldots, ssCov_\Lambda\}$, if multiple copies of the same set do not occur.

4. Every element $ssCov_\lambda$ of a subset covering $SSCov = (ssCov_1, \ldots, ssCov_\Lambda)$ of the union of the universes of the concepts in $Cset$ can be interpreted as the universe of a binary concept $CssCov_\lambda := (DCssCov_\lambda, \{T, F\}, ssCov_\lambda)$, where $DCssCov_\lambda$ still has to be defined. The following equations and inclusions shows that $ssCov_\lambda$ is a relation between $\bigcup \{DC \mid C \in Cset\}$ and $\{T, F\}$:

$$ssCov_\lambda = ssCov_\lambda \cap (\bigcup \{UC \mid C \in Cset\}) = \bigcup \{ssCov_\lambda \cap UC \mid C \in Cset\} \subseteq \bigcup \{DC \times \{T, F\} \mid C \in Cset\} = (\bigcup \{DC \mid C \in Cset\}) \times \{T, F\}.$$

For the domain, which has not been defined so far, we get

$$DCssCov_\lambda := \bigcup \{p_1(ssCov_\lambda \cap UC) \mid C \in Cset\},$$

where p_1 denotes the projection onto the first factor of a product of sets. The partition of $DCssCov_\lambda$ into a truth- and a false-set is given by

$$TS(CssCov_\lambda) := \bigcup \{p_1(ssCov_\lambda \cap UC) \cap TS(C) \mid C \in Cset\} \text{ and}$$
$$FS(CssCov_\lambda) := \bigcup \{p_1(ssCov_\lambda \cap UC) \cap FS(C) \mid C \in Cset\}.$$

5. **Definition 3:** Let $Cset$ and $Cset'$ be two sets of concepts equipped with subset coverings $SSCov$ and $SSCov'$ of their unions of universes. A *concept transformation* between $Cset$ and $Cset'$ is given by a bijection Z from $SSCov$ to $SSCov'$. Hence we have the consequences:

    i) The lengths Λ and Λ' of $SSCov$ and $SSCov'$ are the same.
    ii) There is a permutation σ of the natural numbers from 1 to Λ such that $Z(ssCov_\lambda) = ssCov'_{\sigma(\lambda)}$ for all $1 \leq \lambda \leq \Lambda$.

Remark: i) Each concept transformation has an inverse given by $Z^{-1}(ssCov'_\lambda) = ssCov_{\sigma^{-1}(\lambda)}$.

6. **Definition 4:** A concept transformation Z between a concept set $Cset$ and a concept set $Cset'$ leads to an *induced concept relation* $Ind(Z)$ from $Cset$ to $Cset'$ as follows: $(C, C') \in Ind(Z) \subseteq Cset \times Cset'$, if and only if there exists a λ, $1 \leq \lambda \leq \Lambda$, such that $UC \cap ssCov_\lambda \neq \emptyset$ and $UC' \cap Z(ssCov_\lambda) = UC' \cap ssCov'_{\sigma(\lambda)} \neq \emptyset$.

Remarks: i) The induced concept relation for Z^{-1} is given by the inverse relation of $Ind(Z)$, $Ind(Z^{-1}) = (Ind(Z))^T$.

    ii) For the next paragraph we have to investigate the case that the **induced relation** $Ind(Z)$ on the concept level **is a map**. This implies extra properties for Z. The uniqueness of the values of the map $Ind(Z)$ leads to the condition, that for each $C \in Cset$ and each λ, $1 \leq \lambda \leq \Lambda$, satisfying $UC \cap ssCov_\lambda \neq \emptyset$, we have

$$UC' \cap Z(ssCov_\lambda) = UC' \cap ssCov'_{\sigma(\lambda)} = \emptyset \text{ for all } C' \in Cset' \setminus \{(Ind(Z))(C)\}.$$

This implies that $ssCov'_{\sigma(\lambda)}$ is a subset of the universe of $(Ind(Z))(C)$. Hence $SSCov'$ consists of subsets of the universes of the concepts in $Cset'$ and splits up into mutually disjoint subsequences, each of them consisting of subsets of the universe of the same concept in $Cset'$. The covering condition for $SSCov'$ implies, that the subsequence of those $ssCov'_\tau$, which are contained in the universe UC' of a fixed $C' \in Cset'$, are covering UC'.

Another consequence is that $UC_0 \cap ssCov_\lambda \neq \emptyset$ implies $UC \cap ssCov_\lambda = \emptyset$ for all $\{C \in Cset \mid (Ind(Z))(C) \neq (Ind(Z))(C_0)\}$, i.e., $SSCov$ splits up into mutually disjoint subsequences, such that the sets in each subsequence are subsets of the union of the universes of those concepts in $Cset$, which have the same image under $Ind(Z)$. Hence the concept transformation given by Z^{-1} may be called a *refining transformation* and $Cset$ may be called a *refinement* of $Cset'$.

Definition 5: Let $Cset$ and $Cset'$ be two sets of concepts equipped with subset coverings $SSCov$ and $SSCov'$ of their unions of universes, Z a concept transformation between $Cset$ and $Cset'$, given by a bijection Z from $SSCov$ to $SSCov'$, let $G_\lambda : DCssCov_\lambda \longrightarrow DCssCov'_{\sigma(\lambda)}, 1 \leq \lambda \leq \Lambda$, be a system of maps between the domains of the covering sets, where $Z(ssCov_\lambda) = ssCov'_{\sigma(\lambda)}$ and where the sets of the coverings are interpreted as concepts (see paragraph 4). Then the transformation given by Z and the system of maps $G_\lambda, 1 \leq \lambda \leq \Lambda$, are called *truth-preserving*, if

$$G_\lambda(TS(CssCov_\lambda)) = TS(CssCov'_{\sigma(\lambda)}), 1 \leq \lambda \leq \Lambda, \text{ and}$$
$$G_\lambda(FS(CssCov_\lambda)) = FS(CssCov'_{\sigma(\lambda)}), 1 \leq \lambda \leq \Lambda.$$

These conditions imply that G_λ is surjective for all $1 \leq \lambda \leq \Lambda$.

8. In connection with a concept transformation Z between a concept set $Cset$ and a concept set $Cset'$ we have the possibility to enhance the concepts in $Cset$ by parts of concepts in $Cset'$. Such an extension is constructed as follows:

i) As a first step the induced relation $Ind(Z)$ of a concept transformation is used to enhance each concept $C = (DC, \{T, F\}, UC)$ in $Cset$ to an *extended concept* $C_e = (DC_e, \{T, F\}, UC_e)$ in the following way:

$$DC_e = DC \cup (\bigcup \{p_1(UC' \cap ssCov'_{\sigma(\lambda)}) \mid 1 \leq \lambda \leq \Lambda \text{ and } (C, C') \in Ind(Z)\}),$$
$$UC_e = UC \cup (\bigcup \{UC' \cap ssCov'_{\sigma(\lambda)} \mid 1 \leq \lambda \leq \Lambda \text{ and } (C, C') \in Ind(Z)\}).$$

ii) Having extended all concepts in $Cset$ in this way, we get the *extension* $Cset_e$ of $Cset$. Simultaneously we get the extended subset covering $SSCov_e$ for $Cset_e$ defined by $(ssCov_e)_{sly\lambda} = ssCov_\lambda \cup ssCov'_{\sigma(\lambda)}$ for all $1 \leq \lambda \leq \Lambda$. The extended concept transformation Z_e is given by just selecting the second set from the disjoint union, i.e. $Z_e((ssCov_e)_\lambda) = ssCov'_{\sigma(\lambda)} = Z((ssCov)_\lambda)$.

Everything just mentioned also holds, if the above restriction "... by parts of concepts in $Cset$" is left away. Then it becomes evident that the simplification performed at the beginning of Section 1, of considering of inventive concepts just their therein embedded creative concepts, is easily reversed: Simply by extending any inventive concept - wherever it occurs - in the way just described by its legal concept component. An alternative way of looking at that phenomenon is to ignore the just quoted initial simplification as well as that an inventive concept is a pair of a creative concept and a legal concept - as further going explained in Chapter 3.

To put it in other words: Anyone of the 3 components DC, VC, UC of an inventive concept is a pair of a technical/factual space and a legal space. For facilitating grasping the elaborations of this Chapter, we initially abstracted and may abstract also in what follows from the inventive concepts being bifid - i.e. mentally just leave their bisection aside - though in the end both spaces are indispensable for matching the needs of patent jurisprudence, in particular Highest Courts' SPL precedents.

9. Important for these knowledge representations and their transformations is also, what restriction of concept sets and concept transformations are induced by the needs arising from the mathematical structures representing a patent, more precisely: the legally and technically in-depth description of an invention/innovation - also as explained in Chapter 3. Let Z be a concept transformation between the concept set $Cset$ and the concept set $Cset'$ given as a bijection from a subset covering $SSCov$ for $Cset$ to a subset covering $SSCov'$ for $Cset'$. Let $Cset'_r$ be a subset of $Cset'$. We assume without loss of generality that

$$(ssCov')_\lambda \cap \bigcup \{UC' \mid C' \in Cset'_r\} \neq \emptyset \text{ for all } 1 \leq \lambda \leq \Lambda_r \text{ and}$$
$$(ssCov')_\lambda \cap \bigcup \{UC' \mid C' \in Cset'_r\} = \emptyset \text{ for all } 1 + \Lambda_r \leq \lambda \leq \Lambda.$$

Then the sets $(ssCov'_r)_\lambda \cap \bigcup \{UC' \mid C' \in Cset'_r\}$, $1 \leq \lambda \leq \Lambda_r$ provide a subset covering $SSCov'_r = \{(ssCov'_r)_\lambda \mid 1 \leq \lambda \leq \Lambda_r\}$ for $Cset'_r$, called the *restriction* of $SSCov'$ to $Cset'_r$. Applying the inverse induced concept relation $(Ind(Z))^T$ to $Cset'_r$, i.e. looking for all $C \in Cset$ being in relation $Ind(Z)$ to concepts in $Cset'_r$, we get the *preliminary restriction* $Cset_{pr} := (Ind(Z))^T(Cset'_r)$ of $Cset$ corresponding to $Cset'_r$. Finally we get with σ as in Definition 4 the restriction $SSPCov_r$ of $SSCov$ by setting $(ssPCov_r)_\lambda := ssCov_{\sigma^{-1}(\lambda)} \cap \bigcup \{UC \mid C \in Cset_r\}$ for all $1 \leq \lambda \leq \Lambda_r$. The definition of $Ind(Z)$ implies that $(ssPCov_r)_\lambda \neq \emptyset$ for all $1 \leq \lambda \leq \Lambda_r$ and even more: For every $C \in Cset_r$ there exists a λ, $1 \leq \lambda \leq \Lambda_r$ such that $(ssPCov_r)_\lambda \cap UC \neq \emptyset$. Setting $Z_r((ssPCov_r)_\lambda) := (ssCov'_r)_\lambda$ for $1 \leq \lambda \leq \Lambda_r$ we get a bijection Z_r from $SSPCov_r$ to $SSCov'_r$.

In order to guarantee that $SSPCov_r$ is a covering we have to reduce the universes of the concepts in $Cset_{pr}$ as follows:

$$UCnew = UC \cap (\bigcup \{(ssPCov_r)_\lambda \mid 1 \leq \lambda \leq \Lambda_r\})$$

for all $C \in Cset_{pr}$. Then $SSPCov_r$ will be a subset covering of the discrete union of the new universes. If we want the universes of the new concepts to be surjective on the domains, which means that the corresponding maps are defined on the whole domains, then the domains should be modified as follows: $DCnew = p_1(UCnew)$ for all $C \in Cset_{pr}$. Defining $Cset_r := \{Cnew \mid C \in Cset_{pr}\}$, where $Cnew$ has the universe $UCnew$ and the domain $DCnew$, the bijection Z_r from $SSPCov_r$ to $SSCov'_r$ represents a concept transformation from $Cset_r$ to $Cset'_r$. Truth- and false-sets for the $Cnew$ are obtained by $TS(Cnew) = DCnew \cap TS(C)$ and $FS(Cnew) = DCnew \cap FS(C)$.

3   Construing an Invention's Mathematical Structure

As mathematically introduced by Chapter 2, three levels of abstraction for the representations of all inventive concepts of an invention are defined, as usual in Semantics [10]. Next, the representation of these semantics and their preciseness associated with these three levels is explained in some more detail. In Chapter 4 then also is indicated, how subsets of an invention's concept set must be limited for being meaningful as well as how to define different interpretations of a claimed invention, defining also the scope of this claim claiming it [1].

But, let's put it simple. First, by the below 3 bullet points colloquially describe the 3 levels of granularity of the notional resolution of their inventive concepts alias "inCs" of the invention at issue. Thereafter, Sections 10-12 describe the 3 level, once more and more precisely, in terms of the mathematical structure introduced in Chapter 2. For any level trivially holds that a concept is inventive iff skill does not know a set of concepts and how to combine them such that this combination factually is equivalent to the former concept.

- *OCset* is defined to denote and comprise the invention's inventive concepts, the O-inCs, disclosed for the posc by reading the original documents containing them and grasping the technical teaching they infer within it. I.e.: The layman does not exist in this context - its understanding of the invention or of the specification or of the claim or of the terms therein is completely irrelevant, though it may coincide with that of the posc. As explained elsewhere, e.g. [5,6,16] and outlined in Chapter 4, the meaning of O-inCs in isolation is often principally not definable precisely, as it may depend on the invention's set of concepts selected on the BED level, as indicated in Chapter 2 by Section 3 - but even for a claim(ed invention) with only a single interpretation the meanings of compound O-inCs are, due to natural language deficiencies, often blurring/imprecise/indefinite.
- *BADset* is defined to denote and comprise the invention's binary aggregated disclosed inventive concepts, the BAD-inCs. Due to the limitations imposed on their general expressiveness explained in Sect 3, and in any SPL test case to be approved by the posc, the refined meanings of BAD-inCs (logically modeling their resp. O-inCs)

---

[1] It is important to see already here - this has been presented in [13,17] and will be mathematically elaborated on in detail in [18] - that a patent may comprise several independent inventions, anyone claimed by an independent claim of the patent. In particular in patents dealing with emerging technology inventions - being just model based, as explained above - such a claim(ed invention) may have several different interpretations, anyone identified by its BID-inC generating it. These different interpretations of this claim(ed invention) are called isomorphic iff their respective generative BID-inCs are isomorphic. For any claim(ed invention)'s interpretation exactly one scope is defined, thus also identified by its generative BID-inC.

This raises for emerging technology inventions several important new questions - actually, these question existed for classical technologies inventions, too, but due to their tangibility/visibility never became virulent or at least noticed - such as: How to prevent granting pre-emptive claim(ed invention)s and/or how to separate patent-eligible from non-patent-eligible issues in a claim(ed invention) and/or what makes different but isomorphic interpretations of a claim(ed invention) patent-eligible and patentable separately from each other (i.e. treat them as non-isomorphic), . . .?

are precise/definite - at least, if their interpretation dependency just mentioned may be disregarded, otherwise only the next step of refining them (then e.g. interpretation dependent or otherwise dependent on other BAD-inCs, as outlined in Chapter 4 and FIG 1) will achieve their preciseness/definiteness, as required anyway for the claim(ed invention)'s SPL test.

• *BEDset* is defined to denote and comprise the invention's binary elementary disclosed inventive concepts, the BED-inCs. As to their refinement explained in Sect 3 holds that the refined meanings of BED-inCs (conjunctions of which logically model the BAD-inCs) are precise/definite - possibly achieved by defining for components of BAD-inCs different BED-inCs, e.g. being interpretation specific (see footnote 1). But the fundamental requirement to be met by the claim(ed invention)'s inventive concepts on this level of notional refinements is that they are identified/defined such they show the maximal number of distinctions as to concepts known by skill (i.e. a further refining of a BED-inC into a set of BEDBED-inCs is either factually not possible, at all, or all these BEDBED-inCs were known by skill already just as how to combine them such as to be factually equivalent to BED-inC, which contradicts BED-inC being inventive by the definition of in-Cs).

Next is shown, by Sections 10-12, that the mathematical structure defined in Chapter 2 assures that the total information represented on the 3 levels always is the same and that the respective various chunks of this information on the 3 levels are properly mapped onto each other by concept transformations, as defined by Sections 3-9.

10. *OCset*: The concepts in *OCset* are based on the *mark-up items*, *MUIs*, taken in accordance with the legal requirements from the patent application doc_0. The legal requirements provide the rules from which parts of doc_0 the *MUIs* can (must) be taken: Denoting by *SMUI* the set of all *MUIs* selected from doc_0, sets $SSMUI_h$, $h = 1, \ldots, H$, of subsets of *SMUI* are identified as clusters for the domains DOC_h of the O-level concepts OC_h, $h = 1, \ldots, H$. If in a cluster more than one copy of the same set of *MUIs* will be of relevance, because this set will have different interpretations in *BADset*, then we have to distinguish these copies by adding labels. Having selected all $SSMUI_h$ there may be a non-empty remainder

$$RM := SMUI \setminus \bigcup \{ \bigcup \{sMUI \mid sMUI \in SSMUI_h\} \mid h = 1, \ldots, H\}.$$

Then OC_h is defined as follows: $TS(OC_h)$ is given by the finite set of subsets in $SSMUI_h$ having a meaning with respect to the corresponding concept $BADC_h$ and $FS(OC_h) := \{RM\} \cup (SSMUI_h \setminus TS(OC_h))$. The rules for selecting the sets in $TS(OC_h)$ will be made more precise in connection with the related concept $BADC_h$ in *BADset*. Having defined the partition for $D(OC_h)$ in a truth- and a false-set, all other data for OC_h can be concluded easily as has been explained in paragraph 2.

11. *BADset* and *OCset* to *BADset*: The concepts in *BADset* are set up in bijective correspondence $M_{OAD} : OCset \longrightarrow BADset$ with those in *OCset*. $BADC_h = M_{OAD}(OC_h)$ is the conceptual reference set of a (possibly) aggregated statement making the corresponding O-level concepts more precise. The elements of the truth-set $TS(BADC_h)$ of the domain of $BADC_h$ are given by sets of *MUIs* belonging to $SSMUI_h$, each of them being combined with a meaning or technical notion they

are related with. These sets of $MUIs$ are in bijective correspondence $M_{OAD}D_h$: $TS(OC_h) \longrightarrow TS(BADC_h)$ with the sets of $MUIs$ in $TS(OC_h)$, or more precisely, each $d \in TS(OC_h)$ is exactly the set of $MUIs$ defining $M_{OAD}D_h(d) \in TS(BADC_h)$. This closes the gap left in the setup of $OCset$, because the selection of the subsets of $SSMUI_h$ for $TS(OC_h)$ is delegated to the selection of sets of $MUIs$ for $TS(BADC_h)$ related with some meaning or technical notion. For completing the partition of $D(BADC_h)$ into a truth- and a false-set, we set $FS(BADC_h) := \{RM\} \cup (SSMUI_h \setminus TS(OC_h))$.

All other data for $BADC_h$ can be concluded easily as we have explained in paragraph 2. $OCset$ and $BADset$ are connected by the following bijections:

$$M_{OAD} : OCset \longrightarrow BADset \quad \text{and} \quad M_{OAD}D_h : D(OC_h) \longrightarrow D(BADC_h),$$

$h = 1, \ldots, H$, where $M_{OAD}D_h$ is extended to the false-sets by the corresponding identity map $M_{OAD}D_h(d) := d$ for all $d \in \{RM\} \cup (SSMUI_h \setminus TS(OC_h))$. Hence $M_{OAD}D_h$ is truth-preserving for all $h = 1, \ldots, H$.

12. *BEDset* and *BADset* to *BEDset*: The concepts (more precisely concept representations) in $BEDset$ (binary elementary disclosed) are obtained by disaggregating those in $BADset$ into elementary concepts, which could not be disaggregated further in a reasonable way, still represent the properties of the invention and are formulated in a non-ambiguous or definite way.

This procedure is represented as a concept transformation Z_{BAD} between $BADset$ and $BEDset$ by a bijection $Z_{BAD} : SSCBAD \longrightarrow SSCBED$ from a subset covering $SSCBAD$ of $BADset$ to a special subset covering $SSCBED$ of $BEDset$. $SSCBED$ is simple in the sense that it is given by the set of universes of the concepts in $BEDset$, i.e. each $ssCBED$ in $SSCBED$ is the universe of a concept in $BEDset$, there are no duplications and all universes of concepts in $BEDset$ are covered. This implies that the length of the sequence $SSCBAD$ is given by the number of concepts in $BEDset$. Furthermore $Ind(Z_{BAD}^{-1})$ has to be a map on the concept level. Hence according to the remarks in paragraph 6 the disaggregation of the concepts in $BADset$ is based on a concept wise subdivision of the union of the universes of the concepts in $BADset$. Hence we have a mapping $Ref : \{1, \ldots \Lambda\} \longrightarrow BADset$ such that $ssCBAD_\lambda \subseteq Ref(\lambda)$. This leads to a mapping from $BEDset$ to $BADset$ given by $Ref \circ Z_{BAD}^{-1}$. The covering condition for $SSCBAD$ implies the covering condition

$$UC = \bigcup \{ssCBAD_\lambda \mid Ref(\lambda) = C\} \text{ for all } C \in BADset.$$

Remark. Starting in the way described above Z_{BAD} does not comprise any information whether or how the $ssCBAD$ are mapped to the $U(BEDC)$. If in each case there is an underlying mapping, which may be a bijection in addition, the condition of being truth-preserving can be imposed on Z_{BAD}, like it has been explained in paragraph 7.

Remark: The condition for a concept to be elementary needs a confirmation. The same applies to the condition that the set of concepts still describes the invention. Though some automatized (and still to be developed) support from semantics may apply, this decision more or less depends on the person of posc.

13. *BIDset*, and *BEDset* to *BIDset*: $BEDset$ as constructed in paragraph 12 may contain almost similar concepts or concepts having parts, which represent parts of

concepts in the same collection. In the next chapter this will be formulated more precisely by the notion of a dependent set of concepts. If the creative parts of $BIDset$ should pass the independency test being part of the 10 tests representing SPL in the final chapter, concepts leading to dependencies in $BEDset$ have to be removed from $BEDset$. In addition, in order that $BIDset$ will be able to pass the novelty/nonobviousness test, the creative parts of $BIDset$ should not include concepts, which are anticipated by all $doc.i$ representing the prior art under consideration. Hence the creative parts of $BIDset$, called $BIDset$ for short in the following considerations, are obtained by removing concepts from $BEDset$ until we arrive at a set of binary independent, not totally anticipated disclosed concepts. This procedure will be explained in the next chapter. For the following considerations the starting point is just that $BIDset$ is a subset of $BEDset$. Applying the restriction procedure introduced in paragraph 9 with respect to $BIDset$ to Z_{BAD} and then to a still to be defined concept (representation) transformation Z_{OC} from the O-representation level to the BAD-representation level, we will be able to establish similar transformations for the new concept sets, as we initially had for $OCset$, $BADset$ and $BEDset$. This will be described explicitly in the next paragraphs.

14. Adjusting $BADset$ to $BIDset$: As a next step we are using the restriction construction given in paragraph 9 for reducing $BADset$ to $BADset_r$, such that the restriction $Z_{BAD,r}$ of the concept transformation Z_{BAD} is a concept transformation between $BADset_r$ and $BIDset$. According to paragraph 12 Z_{BAD}^{-1} induces a map $Ind(Z_{BAD}^{-1})$ from $BEDset$ to $BADset$. Hence the concepts of the restriction $BADset_r$ resulting from the restriction of $BEDset$ to $BIDset$ is given by modifications of the concepts in $Ind(Z_{BAD}^{-1})(BIDset)$ as follows: The universes of the concepts in $BADset_r$ are obtained from universes of the concepts in $(Ind(Z_{BAD}^{-1}))(BIDset)$ by removing all covering subsets mapped by Z_{BAD} to the universes of concepts in $BEDset \setminus BIDset$. $Z_{BAD,r}$ is just the restriction of Z_{BAD} to the remaining covering subsets. By definition of the induced concept relation none of the universes obtained in this way can be empty.

15. Adjusting $OCset$ to $BADset_r$: This is quite obvious. We only have to restrict the inverse maps of the bijections
$M_{OAD} : OCset \longrightarrow BADset$ and $M_{OAD}D_h : D(OC_h) \longrightarrow D(BADC_h)$
to the restricted sets in $BADset_r$, which finally leads to $OCset_r$. This also can be used to transfer the covering sets of $BADset_r$ to $OCset_r$, leading in an obvious way to a concept transformation Z_{OC} between $OCset_r$ and $BADset_r$.

16. So far we have ignored the so-called *elements* $X_{in}, 0 \leq i \leq I, 1 \leq n \leq N$, describing roughly said the general properties of the patent application, derived from doc_0 in terms of aggregated concepts $X_{0n}, 1 \leq n \leq N$ combined with mirror FOL predicates $\underline{X}_{0n}, 1 \leq n \leq N$, and the general properties of the prior art, derived from the documents $doc_i, 1 \leq i \leq I$, in terms of aggregated concepts $X_{in}, 1 \leq i \leq I, 1 \leq n \leq N$ combined with mirror FOL predicates $\underline{X}_{in}, 1 \leq i \leq I, 1 \leq n \leq N$. Every FOL predicate $\underline{X}_{0n}$ can be represented by the conjunction of mirror predicates of a uniquely determined subset of concepts in $BADset_r$, where these subsets of $BADset_r$ are mutually disjoint and provide a covering of $BADset_r$. Hence we have a surjective map

$El_{BAD} : BADset_r \longrightarrow \{1, \ldots, N\}$, separating $BADset_r$ into N mutually disjoint subsets $BADset_{r,n} = El_{BAD}^{-1}(n)$, $1 \leq n \leq N$. Composing El_{BAD} with $M_{OAD,r}$ and $Ind(Z_{BAD,r}^{-1})$ we get the same kind of decompositions $OCset_{r,n}$ and $BIDset_n$, $1 \leq n \leq N$, for $OCset_r$ and $BIDset$ respectively. It is easy to see that the maps $M_{OAD,r}$ and $Ind(Z_{BAD,r}^{-1})$ and the chain given by the maps Z_{OC} and $Z_{BAD,r}$ decompose element wise in accordance with the decompositions of the sets of concepts.

4   The Usefulness of This Mathematical Structure

"Claim construction" is a key notion of US SPL precedents. Yet, as to emerging technologies inventions, this classical notion of claim construction has proven to be deficient: In a whole series of CAFC decisions these notional deficiencies led to situations, which its Chief Judge recently - in a case remanded to it by the Supreme Court for reconsideration in the light of Mayo - called irreconcilable within the CAFC. A "refined claim construction", as implicitly required by the Supreme Court's Mayo decision - for mathematically modeling of which the mathematical structure of Chapter 2 has been developed - completes the established/classical notion of claim construction in the sense describable as follows:

A claimed invention passes its SPL test
$\Longleftrightarrow$ *it passes the FSTP-Test*
$\Longleftrightarrow$ *the refined claim construction is construable for it*
$\Longleftrightarrow$ *the "refined mathematical structure" is construable for it.*

The conclusion is: *A claimed invention passes its SPL test* $\Longleftrightarrow$ *the "refined mathematical structure" is construable for it.*

The next paragraphs outline, by what extensions of the mathematical structure of an invention (as defined in Chapter 2 and established for the invention in Chapter 3) it becomes the refined mathematical structure, which will be mathematically presented in [17]. Yet, the above 3 equivalences indicate already here: If the mathematical structure modeling a claimed invention is a substantial part of its refined mathematical structure, it also models a substantial part of its SPL test - and therefore is already very useful [2], though it will unfold its full usefulness only as refined mathematical structure.

What the mathematical structure of Chapter 2 of an invention requires for becoming its refined mathematical structure - i.e. for modeling its refined claim construction, i.e. for modeling its SPL test - is outlined by the next two bullet points: It must

• extend its analysis (in Chapters 2 and 3) of solely an invention respectively its technical teaching ("TT.0") to the analysis of a PTR, being defined to be a **"pair of TT.0 over RS"**. Thereby a "reference set" RS is a finite set of prior art documents, doc.i's, the TT.i's of which allegedly anticipate TT.0 or make it obvious over a combination of them, and

---

[2] It is important to see that, for most inventions, the seemingly plausible implication *Its refined mathematical structure models its refined claim construction* $\Rightarrow$ *Its mathematical structure models its claim construction* is principally wrong; the reason being that the (classical) claim construction principally has no in-Cs, at all. Practically, any limitation of a claim(ed invention) is the cr-C of one of its in-Cs, but it may have more different in-Cs than such limitations [13].

• complete this extended analysis - and hence the correspondingly extended mathematical structure, as defined by the preceding bullet point - by the subtests FSTP test.o, $2 \leq o \leq 10$, of the FSTP-Test (see the list given by **FIG 1**), as the extended mathematical structure for this invention performs only FSTP test.1.

The FSTP-Test in **FIG 1**, is here simply quoted from [12] and hence it cannot be understood completely - in [14,17] it will be described in the mathematical style as the above description of the mathematical structure. Yet its principle of working may be figured out already, here, by footnote 2 and the following 3 hints:

• Anyone of the finitely many "set of inCs, SoI" identifies a different interpretation of the claim(ed invention) (see footnote 2), for which the execution of all 10 FSTP test.o is attempted to complete, $2 \leq o \leq 10$. The {SoI} of the claim(ed invention) is assumed to be determined prior to starting the FSTP-Test. On any prompt by the FSTP-Test, the user must input into it the information it prompted for. If for a specific claim interpretation alias SoI the execution of one of the 10 test.o cannot be completed, because the user cannot provide the input prompted for by test.o, this SoI is abandoned and another SoI is tested. In any case all SoIs are executed, completely or partially only.
• Any input provided by the user may be augmented by its correctness confirmation by the posc.
• The problem $P.0^{SoI}$ of the NAIO test is an EPC notion, in the US SPL to be replaced by the total usefulness modeled by the generative {inCs} (see footnote 2) of the claim(ed invention) identified by SoI.

FIG 1: The FSTP-Test,
whereby several of the FSTP test.o are solely indicated by their headlines.

test.1: The FSTP-Test is executed for all claim interpretations (see footnote 2), with the **posc justified definite disaggregation** of the compound inventive concepts, after the **posc justified these as definite** for the set of interpretations, SoI, selected in 2) and 3), comprising the steps: It

1. prompts the user for the claim(ed invention)'s and prior art's docs with their **"marked-up items, MUIs"**;
2. prompts for all SoI and for any SoI's $\forall$
   $BAD^{SoI} - \underline{X}in ::= \bigwedge_{1 \leq SoI.in \leq SoI.IN} BAD - crCin^{SoI.in}$ in doc.i-MUI's, $0 \leq i \leq I, 1 \leq n \leq N$;
3. prompts for the posc's definiteness justification of $\forall$ compound inCs in SoI, i.e. of $\forall AD - crCin^{SoI.in}$;
4. prompts to disaggregate $\forall AD - crCin^{SoI.in} \forall 0 \leq i \leq I, 0 \leq n \leq N$ into $\{BED - cr\underline{C}ink^{SoI.in} \mid 1 \leq k^{SoI.in} \leq K^{SoI.IN}\}$:
   $BAD - crCin^{SoI.in} =$
   $\bigwedge_{1 \leq kSoI.in \leq KSoI.IN} BED - cr\underline{C}ink^{SoI.in} \wedge BED - cr\underline{C}ink^{SoI.in} \neq$
   $BED - cr\underline{C}ink^{SoI.in'} \forall k^{SoI.in} \neq k^{SoI.in'}$;
5. prompts for the posc's definiteness justification of its disaggregation in 4);
6. Set $K^{SoI} ::= \sum_{1 \leq 0n \leq 0N} K^{0n}$, $S^{SoI} ::= \{BED - cr\underline{C}0nk^{SoI.0n} \mid 1 \leq k^{0n} \leq K^{0N}\}$, with $K^{SoI} = |\{BED - cr\underline{C}0nk^{SoI.0n} \mid 1 \leq k^{0n} \leq K^{0N}\}|$;

test.2: Prompts for justifying $\forall$ BED-crCs in S^{Sol}: Their **lawful disclosures**;

test.3: Prompts for justifying $\forall$ BED-inCs in S^{Sol}: Their **definiteness** under §112.6;

test.4: Prompts for justifying $\forall$ BED-inCs in S^{Sol}: Their **enablement**;

test.5: Prompts for justifying $\forall$ BED-inCs in S^{Sol}: Their **independence**;

test.6: Prompts for justifying $\forall$ BED-inCs in S^{Sol}: Their **posc-nonequivalence**:

1. if $|RS| = 0$ then $BED^* - inC0k ::=$ "dummy";
2. else performing **c-f** $\forall\ 1 \leq i \leq |RS|$;
3. It prompts to disaggregate $\forall\ BAD - \underline{X}in$ into $\bigwedge_{1 \leq kn \leq Kn} BED - in\underline{C}ik^n$;
4. It prompts to define $BED^* - inCik^n ::=$
   either $BED - inC0k^n$ iff
   $BED - inCik^n = BED - inC0k^n \wedge$ disclosed $\wedge$ definite $\wedge$ enabled,
   else "dummy(ik^n)";
5. It prompts for $JUS^{posc}(BED^* - inCik^n)$.

test.7: Prompts for justifying by NAIO test (see i) below) on $(S^{Sol} : P.0^{Sol})$: TT.0 is **not an abstract idea only**;

test.8: Prompts for justifying $\forall$ BED-inCs in S^{Sol}: TT.0 is **not natural phenomena solely**;

test.9: Prompts for justifying $\forall$ BED-inCs on $(S^{Sol} : P.0^{Sol})$: TT.0 is **novel and nonobvious** by NANO test (see ii) below) on the pair $(S,\ if\ |RS| = 0\ then\ \{BED^* - inC0k | 1 \leq k \leq K\}\ else\ \{BED^* - inCik | 1 \leq k \leq K, 1 \leq i \leq |RS|\})$;

test.10: Prompts for justifying $\forall$ BED-inCs in S^{Sol}: TT.0 is **not idempotent** by NANO test (see ii) below) on the pair $S' \subseteq S$.

i) The **"Not an Abstract Idea Only, NAIO"** test basically comprises 4 steps [4, 6, 7, 8, 9, 25], ignoring RS:

1. verifying the specification discloses a problem, $P.0^{Sol}$, to be solved by the claim(ed invention) as of S^{Sol};
2. verifying, using the inventive concepts of S^{Sol}, that the claimed invention solves $P.0^{Sol}$;
3. verifying that $P.0^{Sol}$ is not solved by the claim(ed invention), if a BED-inC of S^{Sol} is removed or relaxed;
4. if all verifications 1)-3) apply, then this pair (claim(ed invention), S^{Sol}) is "not an abstract idea only".

ii) The **"Novel And Not Obvious, NANO"** test basically comprises 3 steps, checking all "binary anticipation combinations, $BAC^{Sol}s$" derivable from the prior art documents in RS for the invention defined by S^{Sol}:

1. generating the ANC^{Sol} matrix, its lines representing for any prior art document.i, $i = 1, 2, \ldots, I$, the relations between its $invention^{i.Sol}$'s BED-inCs to their peers of $TT.0^{Sol}$, represented by its columns;
2. automatically deriving from the ANC^{Sol} matrix the set of $\{AC^{Sol}s\}$ with the minimal number $Q^{plcs/SoI}$;
3. automatically delivering $(Q^{plcs/SoI}, \{AC^{Sol}\})$, indicating the creativity of the pair (claim(ed invention), SoI).

References

[1] "Advanced IT" denotes IT research areas such as AI, Semantics, KR, DL, NL

[2] Brachmann, R., Levesque, H.: Knowledge Representation & Reasoning. Elsevier (2004)

[3] The Description Logic Handbook. Cambridge UP (2010)

[4] Schindler, S.: Mathematically Modeling Substantive Patent Law (SPL) Top- Down vs. Bottom-Up, Yokohama, JURISIN 2013 (2013) ∗)

[5] SSBG pat. appl.: THE FSTP EXPERT SYSTEM ∗)

[6] SSBG pat. appl.: AN INNOVATION EXPERT SYS., IES, & ITS DATA STRUC., PTR-DS ∗)

[7] SSBG's Amicus Brief to the CAFC in LBC (2013) ∗)

[8] SSBG Amicus Brief to the Supreme Court in CLS (October 07, 2013) ∗)

[9] SSBG Amicus Brief to the Supreme Court in WildTangent (September 23, 2013) ∗)

[10] Schindler, S., Paschke, A., Ramakrishna, S.: Formal Legal Reasoning that an Invention Satisfies SPL, Bologna, JURIX 2013 (2013) ∗)

[11] Schindler, S.: Substantive Trademark Law (STL), Substantive Copyright Law (SCL), and SPL - STL Tests Are True SCL Subtests, Which Are True SPL Subtests (in prep.)

[12] Schindler, S.: Boon and Bane of Inventive Concepts and Refined Claim Construction in the Supreme Court's New Patent Precedents, Hawaii, IAM-2014 (2014) ∗)

[13] SSBG's Amicus Brief to the Supreme Court as to its (In)Definiteness Questions (March 3, 2014) ∗)

[14] Wegner, B., Schindler, S.: Unabbreviated Version of "A Mathematical Structure for Modeling Inventions, Coimbra, CICM-2014", http://www.fstp-expert-system.com

[15] Fiacco, B.: Amicus Brief to the CAFC in VERSATA v. SAP & USPTO (March 24, 2014) ∗)

[16] Schindler, S.: The Supreme Court's 'SPL Initiative': Scientizing Its SPL Interpretation Clarifies Three Initially Evergreen SPL Obscurities (submitted for publ., 2014) ∗)

[17] Wegner, B., Schindler, S.: A Refined Mathematical Structure for Modeling Inventions (in prep.)

[18] Schindler, S., Shipley, H.: Petition for Certiorari to the Supreme Court in the 902 case

∗) available at www.fstp-expert-system.com

Search Interfaces for Mathematicians

Andrea Kohlhase

Jacobs University Bremen and FIZ Karlsruhe, Germany

Abstract. Access to mathematical knowledge has changed dramatically in recent years, therefore changing mathematical search practices. Our aim with this study is to scrutinize professional mathematicians' search behavior. With this understanding we want to be able to reason why mathematicians use which tool for what search problem in what phase of the search process. To gain these insights we conducted 24 repertory grid interviews with mathematically inclined people (ranging from senior professional mathematicians to non-mathematicians). From the interview data we elicited patterns for the user group "mathematicians" that can be applied when understanding design issues or creating new designs for mathematical search interfaces.

1 Introduction

Mathematical practices are changing due to the availability of mathematical knowledge on the Web. This paper deals with the question whether mathematicians have special needs or preferences when accessing this knowledge and if yes, what are those? In particular, we focus on how mathematicians think of search on the Web: what are their cognitive categories, what kinds of searches do they distinguish, and which attributes do they associate with tools for math access?

The usability study [10] conducted interviews with mathematicians and essentially stated that mathematicians didn't know how to use the offerings of mathematical search interfaces. To get a better understanding we wanted to dig deeper. In [27] ZHAO concentrates on user-centric and math-aware requirements for math search. The former are based on mathematicians' specific information needs and search behaviors, the latter are the needs for structured indizes by the system. In contrast, we focus on eliciting attributions of existing math search interfaces by mathematicians versus non-mathematicians. We hope to learn what exactly sets mathematicians apart, since from this knowledge we can deduce implications for future mathematical designs.

We decided on using repertory grid interviews as main methodology to elicit evaluation schemes with respect to selected math search interfaces ("**mSI**") and to understand how mathematicians classify those mSIs. The main advantage of the method is its semi-empirical nature. On the one hand, it allows to get deep insights into the topic at hand through deconstruction and intense discussion of each subject's idiosyncratic set of constructs and their resp. mapping to the set of mSIs. On the other hand, the grids produced in such RGI sessions can be analyzed with a General Procrustes Analysis to obtain statistically significant

S.M. Watt et al. (Eds.): CICM 2014, LNAI 8543, pp. 153–168, 2014.

correlations between the elicited constructs or the chosen mSIs. We used the Idiogrid [3] and the OpenRepGrid [21] software for this.

Information search is not a single act, but a process with many strategies and options: *"In fact, we move fluidly between models of ask, browse, filter, and search without noting the shift. We scan feeds, ask questions, browse answers, and search again."* [19, p.7]. Therefore, we can consider the term "search" as an umbrella term for (at least) the following approaches:

Finding = already knowing what one is looking for ([20, 23] call it "fact-finding")

Browsing = getting an overview over a topic or an idea of a concept ([20] calls it "exploration of availability")

Surfing = surrendering to the links, drifting from one to another (see [26])

Solving/Information Gathering = creating a search plan, i.e., specifying a sequence of actions that achieves the solution of a problem (see [22, 65ff.], [8])

Asking = posing a question to find an answer (see [25])

Our question here is, what search approach is used with which assessment attributes for what kind of math search tool? The answer could enable us to design specifically for more math search approaches by learning from the used ones.

We start out in Section 2 with a description of the RGI study. In Section 3 we present the elicited interview data and note the patterns that emerge from this data. The patterns state interesting, prototypical attributions of mathematicians, which separate the data gathered from the group of mathematicians from the one of non-mathematicians. To demo the utility of such patterns, we apply them in a discussion of an interesting, confusing evaluation of two specific mSIs in Section 4. We conclude in Section 5 by hinting at general design implications for mathematical (search) interfaces based on the found set of patterns.

2   The Study

The aim of our study was to find out what distinguishes mathematicians from non-mathematicians when using a web interface for searching relevant content, here math content. From the outset it was clear that observational methods wouldn't work as the working context of a mathematician is typically neither restrained to certain locations nor time slots. Surveys (or structured interviews) were out of question as the answers require a deep insight of subjects into their own math search behavior, which cannot be assumed in general. Unstructured interviews could have been made use of to get such deep insights, but we would either have to do too many to be able to soundly interpret them or too few to draw general conclusions. Finally, the option of semi-structured interviews as methodology was discarded, since it became clear in the first pilot study trials that mathematicians tend to describe "truths" and "falsities". In particular, they try to scrutinize the interview or interviewer and manipulate the outcome towards what they think is the correct answer. Thus, the interviewer has to trade her observational stance with a continuously sparring stance, which hinders the process of gaining deep insights.

In the end, we opted for the methodology of repertory grid interviews, as they allow a semi-empirical analysis, and interviewees understand quickly that they are not asked to decide on rights or wrongs. The **Repertory Grid Interview (RGI) Technique** [4, 7, 9] explores personal constructs, i.e., how persons perceive and understand the world around them. It has been used as a usability/user experience method to research users' personal constructs when interacting with software artifacts (see [5, 6, 24] for examples). RGI has the advantage that it can deliver valuable insights into the perception of users even with relative low numbers of study subjects (seeo [12] for more details).

Table 1. The RGI Elements in the Study

Element Name	Short Description	URL
zbMathNew	"an *abstracting and reviewing service* in pure and applied mathematics"	zbMath.org
zbMathOld	the former interface of `zbMathNew`	not available
MathSciNet	"searchable *database of reviews, abstracts and bibliographic information* for much of the mathematical sciences literature"	ams.org/mathscinet
Google-Scholar	"search of *scholarly literature* across many disciplines and sources"	scholar.google.com
Google	"Search the *world's information*, including webpages, images, videos and more"	google.com
myOffice	the personal *office* as math search interface	—
TIB	The *online catalogue* of the Uni Hannover Library	tib.uni-hannover.de
vifamath	"The Virtual Library of Mathematics" - a *meta online catalogue*	vifamath.de
myLibrary	a *physical library* known by the subject	—
arXiv	"*Open e-print archive* with over [...] 10000 [articles] in mathematics"	arxiv.org
ResearchGate	"a *network* dedicated to science and research"	researchgate.net
mathoverflow	"a *question and answer site* for professional mathematicians"	mathoverflow.net
myColleagues	personal *colleagues* as math search interface	—
MSC-Map	"accessing math via *interactive maps*" based on an MSC metric	map.mathweb.org
arXiv-Catchup	an interface for *catching up* with the newest articles in math	arxiv.org/catchup
FormulaSearch	"allows to search for *mathematical formulae* in documents indexed in zbMath"	zbmath.org/formulae
Bibliography	a *bibliography* as math search interface	

2.1   The RGI Elements

As we want to cover a broad range of different types of math search interfaces we opted for a set of 17 mSIs as **RGI elements** – ranging from standard mSIs like

"Zentralblatt Mathematik (zbMath)" or "MathSciNet" via social media platforms like "mathoverflow" to scientific prototypes like the "MSC map" interface (MSC = Math Subject Classification, see [17]). To avoid being limited to digital mSIs, we included traditional search situations like asking colleagues or personal office spaces as well. Table 1 summarizes the 17 elements used in the RGIs and gives short descriptions – the ones from their websites where available – and their web addresses if applicable. Note that wikis (e.g., "Wikipedia" or "PlanetMath") were excluded as the tension between searching for articles versus encyclopedia entries was perceived problematic in the pilot study, so we opted for the former. As we were only interested in the search behavior of mathematicians we disregarded mathematical software whose main task is computation or verification.

2.2   The RGI Set-Up

At the beginning of each interview the interviewer introduced the interviewee to all mSIs based on print-outs. Both the home page with its search facilities and the search result pages were discussed. The front page of each print-out presented the homepage initialized with the phrase "Cauchy sequence" in the search box if applicable. The back page displayed the search result wrt to this query. For mSIs with special features extra pages were attached. For FormulaSearch the LATEX query corresponding to $?a_{?n} \in \mathbb{N}$ was used.

An RGI interview iterates the following process until the interviewee's individual construct space seems to be exhausted:

i. The interviewee randomly chooses three RGI elements.

ii. He declares which two of the three elements seem more similar.

iii. He determines the aspect under which these two are more similar and the aspect under which the third one is different. Those aspects are the **"poles"** of an interviewee-dependent evaluation dimension, the so-called **"construct"**.

To get a sense of what the users consider important properties of mSIs, we extended this set-up by encouraging most interviewees to judge the "fitness" of each mSI for mathematical search. As is typical with RGIs, the interviews were very intense. Therefore, the findings are not only based on the actual data elicited in the RGI but also on the deep discussions taking place during each interview.

2.3   The RGI Data

We conducted interviews with 24 people, all of which were interested in accessing math on the web. Out of these, 18 had a degree in mathematics. For the final analysis we decided to use 22 RGIs: interviews with a group of 11 professional mathematicians working in a scientific environment ("inMATH"), a group of 5 content experts for mathematical information ("infoMATH"), and a group of 6 non-mathematicians ("noMATH"). Only 3 of the participants were female.

Each interview took between 1.75 and 3 hours, in which an average of 4 constructs were elicited. The inMATH group created 50 constructs, infoMATH reported 28 constructs and NOMATH 29 constructs. The rating scale for these 107 elicited constructs was a 7-point Likert scale.

3 Findings

As already mentioned, the RGI method is semi-empirical. This means that there will be a quantitative and a qualitative analysis of the data gathered. Due to space limitations we will focus on presenting and interpreting the most interesting, statistically significant quantitative results in form of dendrograms and qualitative results in form of patterns. Note that here, the theory emerges from the data, thus, it provides us with patterns but not with proofs.

With the **Generalized Procrustes Analysis (GPA)** method (see [2]) 3-dimensional data matrices can be analyzed with a multivariate statistical technique. In particular, in our RGI we can compare the individual (dim 1) natural language constructs (dim 2) rated on our fixed set of mSIs (dim 3). We conducted a GPA with Idiogrid for each data set and refer to [11] for a detailed description of an analoguous GPA procedure. To provide a shared set of (virtual) standard constructs on which the individual ratings of the RGI elements of each interviewee can be compared, the GPA method produces "**abstract constructs**" of the form "Con_i - ConOpo_i" with poles "Con_i" and "ConOpo_i".

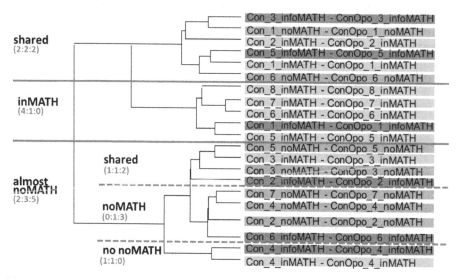

Fig. 1. Dendogram of the Abstract Construct Clusters (wrt. Euclidean distance and Ward clustering) of INMATH, INFOMATH and NOMATH: we can clearly discern a "common" cluster, which is equally shared by all three, a strong INMATH cluster and a fairly strong NOMATH cluster.

Based on a pre-study we suspected a distinction of the interviewee group not only into mathematicians and non-mathematicians, but into research mathematicians, mathematics practitioners and non-mathematicians. Therefore we compared the element evaluations of the INMATH, INFOMATH, and NOMATH group. We subjected the union of the group-specific sets of abstract constructs to a cluster analysis run by OpenRepGrid resulting in the dendrogram in Fig. 1. Recall that **dendrograms** are a visual representation of correlation data. Two constructs in Fig. 1 are closely correlated, if their scores on the RGI elements are similar. The distance to the next upper level of two constructs/groups of constructs *indicates* this relative closeness. Please note that we left out the scale in the dendrograms, as we are not interested in the absolute numbers, only in their relative groupings. This also means, that we won't use arguments in our discussion of findings based on this scale. Nevertheless, we can for example, conclude from Fig. 1 that Con_6_INMATH and Con_7_INMATH are the most correlated constructs. For the conversion of Idiogrid data to OpenRepGrid data we developed the according software.

The interview data seen in Fig. 1 indeed suggest a difference between how people in the INMATH, INFOMATH and the NOMATH group think about mSIs. The INFOMATH interviewees' point of view lies between the one of INMATH and NO-MATH subjects. In particular, there are INFOMATH abstract constructs in every cluster and there is no cluster dominated by the INFOMATH abstract constructs. As this user group dilutes possible similarities or dissimilarities wrt the user group in focus – the professional mathematicians – we further on only analyzed the INMATH and NOMATH data in depth. From here on we will call INMATH members "mathematicians" and NOMATH members "non-mathematicians".

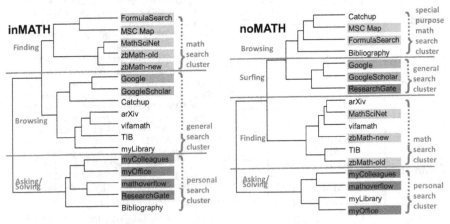

Fig. 2. Cluster Dendrograms of mSI Elements for INMATH and NOMATH

Fig. 2 gives a visualization of the element clusters of group INMATH resp. NOMATH as dendrograms. The difference between the clusters is evident; we will elaborate the interpretations in the next paragraphs.

There are three main element clusters for INMATH in Fig. 2. Clearly, one of these contains the mSI elements whose main purpose it is to find mathematical content ("**math search cluster**"). In the math search cluster both, Formula-Search and MSC-Map, are innovative mathematical services, nevertheless they are identified as being most similar to the standard mSIs `zbMathNew`, `MathSciNet` and `zbMathOld`. This shows that

> **Pattern 1:** *"Mathematicians do not assess mSIs based on familiarity."*

Another cluster includes all mSI elements that provide a personal touch in the search process (**"personal search cluster"**). Here, the term "personal" labels the interactive adaptation and customization of the search or search results in a process driven by human interactions. In the interviews it became quite clear, that anything involving human beings or communities was highly distinctive and predominantly highly appreciated. So

> **Pattern 2:** *"Mathematicians trust human and community resources."*

Note that we don't mean a naive trust here, but a trust given the sensible precautions. Even though the element clusters of the NOMATH interviewees also include a personal search cluster (see Fig. 2), the elements `Bibliography` and `ResearchGate` are missing and replaced by `myLibrary`. The INMATH participants explicitly commented that they don't have confidence in the librarians' expertise in math. Interestingly, mathematicians showed a lot of skepticism wrt `Research-Gate` but not because they could not rely on the links the `ResearchGate` members would provide, but rather because they mistrusted `ResearchGate`'s competence in judging the relevance of links. An indication of this is also given by the well-known observation that mathematicians like anecdotes about fellow mathematicians like no other community of practice.

The third cluster groups the remaining elements. Noticeably `Google` and `Google-Scholar`, which mathematicians nowadays use heavily for mathematical searches, are in this cluster. Nevertheless, these elements are not specific to math search, therefore we label this cluster as the **"general search cluster"**.

According to ZHAO's usability study in [27], mathematicians use *"three main approaches: general keyword search, browsing math-specific resources and personal contact."* This can also be seen in our three clusters for the INMATH group.

For the NOMATH element clusters we only want to point out that the clusters are indeed very different from the ones in the INMATH dendrogram. For example, for mathematicians the mSIs `MathSciNet`, `zbMathNew` and `zbMathOld` correlate the highest, whereas for non-mathematicians each of them correlates more with a different element than with each other. The only similarities seem to be the obvious correlation between `Google` and `Google-Scholar`, and the same very high correlation distance between the personal search cluster and the others.

For a more precise qualitative analysis consider the dendrogram in Fig. 3. First we decided on fitting categories/subcategories for each cluster. We looked, for instance, at the first main cluster and decided on the category "fit for math". Then we elaborated on its four subclusters, e.g., for the fourth cluster we selected "preconditions for search" as a subcategory. Note that there are blue-colored abstract constructs "Con_i - ConOpo_i" among the constructs. We can interpret them now as characteristic constructs of the corresponding major subcluster, so we associate each abstract construct with its subcategory. Out of convenience,

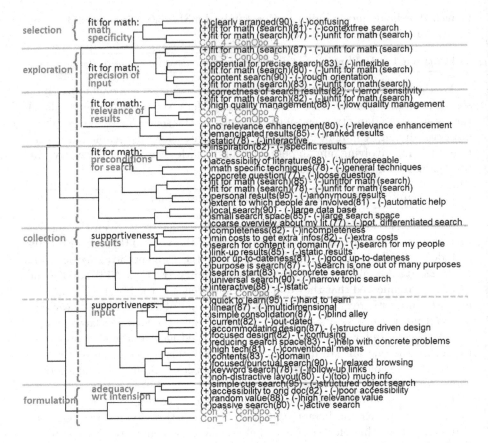

Fig. 3. Cluster Dendrogram of Construct Clusters in INMATH. The first two levels of the dendrogram were contracted for a more readable image. Moreover, the numbers in parentheses in each construct encode the individual interviewee issuing it.

we call them by their explicit pole name together with the corresponding data set, thus we say for example, "Con_4_INMATH 'means' math specificity".

According to KUHLTAU ET AL. in [13, 14] the information search process can be described by a six-phase framework consisting of *initiation* (prompting a search), *selection* (identifying information needs), *exploration* (pondering available tools and thus search strategies), *formulation* (formalizing search queries), *collection* (gathering information and goal-oriented cherry picking in search results), and *search closure* (giving up on the search). In our study we are not interested in the entire search process, but in the interactions with the user interface, so we focus on the iterative acts of selection, exploration, formulation and collection. In these phases a user seeking information translates a search intension into a query or series of queries optimizing for the relevance of the final collection of search results. Interestingly, the four phases are mirrored in the construct clusters of the INMATH group (see Fig. 3 on the left).

Meaning		PC_1	PC_2	PC_3	Length
relev. of results: driven by user	Con_6_inMATH	-0.61	0.52	0.52	0.96
relev. of results: driven by data	Con_7_inMATH	-0.72	0.09	0.63	0.96
precision of input	Con_5_inMATH	-0.59	-0.61	-0.37	0.93
adequacy: search philosophy	Con_1_inMATH	0.03	0.70	-0.59	0.92
math specificity	Con_4_inMATH	-0.81	-0.20	-0.35	0.90
adequacy: trustfulness	Con_3_inMATH	0.05	0.72	-0.03	0.72
supportiveness: result	Con_2_inMATH	-0.36	0.50	-0.34	0.70
preconditions for search	Con_8_inMATH	0.34	-0.05	0.49	0.60

Fig. 4. Abstract Construct Ranking in INMATH via Structure Coefficients

To obtain a **ranking for the abstract constructs** consider the structure coefficients of the abstract constructs (wrt their ratings on the three main principal components PC_i) for each interviewee group in Fig. 4 and 5. The Euclidean **length** of the resp. 3-dimensional construct vector indicates its construct's relevance. To distinguish between two abstract constructs that are in the same subcluster in Fig. 3, we compare their structure coefficients. For any distinctive deviation we take a closer look in the biplots for the resp. PC-dimension and elaborate on its meaning.

It is obvious that the rankings of the INMATH group are distinct from the ones of the NOMATH group. What mathematicians care about the most is the relevance of the search result with respect to their search intension. So they seek interfaces and databases that allow them to formulate precisely that in accordance with the respective search philosophy they want to apply (Con_6-_INMATH , Con_7_INMATH , Con_5_INMATH , Con_1_INMATH). As this describes a search process that enables the user to find exactly what he is looking for, we have

> **Pattern 3:** *"Finding is the primary mathematical search task."*

Note that the math search cluster of the INMATH group in Fig. 2 also has a clear focus on "finding". For the NOMATH math search cluster this is much less clear, e.g., the vifamath mSI, which concentrates on collecting mathematical information (from legacy math articles to images of mathematicians), but not on its findability, thus mimicking a physical math library without noticeable presence of other people. The interviewer observed that interviewees aligned the distinct kinds of search like finding, browsing or solving with the clusters, but that the evaluation of search activities was different for mathematicians and non-mathematicians. The former had a clear preference for finding, followed by browsing and solving/asking, and even a hint of rejection for surfing. In contrast, the NOMATH participants indicated a preference for browsing and surfing, followed by solving/asking and finally finding. Note that the position of "finding" in this ranking may be well due to the fact, that only one participant in the NOMATH group worked in a scientific environment.

It is conspicuous that even though there was an obvious mSI cluster with respect to "people" (the personal search cluster) for the INMATH group in Fig. 3, there is no appreciation of "socialness" in their ranked list of constructs in Fig. 4.

In particular, mathematicians distinguish certain mSIs, i.e., the tools, as socially driven, but as professionals they do not appreciate "socialness" as a value per se in their evaluation schemes. In the theory of "Communities of Practice (**CoP**)" [15], **practices** are not only typical customs shared within a community, but they are tools that define the community. Whereas in other CoPs social interaction is a tool for achieving *social* bindings, in the mathematical CoP, social interaction is a tool for doing mathematics, i.e., it is a mathematical practice. Therefore, we note that

---

Pattern 4: *"Mathematicians appreciate social interaction as a* mathematical *tool. In particular, it is a mathematical practice to collaborate and exchange feedback."*

In this sense, we confirm BROWN's dictum in [1] that mathematicians may rely more heavily on their social network than other disciplines.

Meaning		PC_1	PC_2	PC_3	Length
input design	Con_1_noMATH	0.11	0.61	0.71	0.94
math specificity	Con_5_noMATH	-0.90	0.21	0.15	0.94
flex.: what and when	Con_2_noMATH	-0.92	-0.01	0.07	0.92
adequacy	Con_3_noMATH	-0.54	-0.70	-0.15	0.90
flex.: what	Con_4_noMATH	-0.10	-0.42	0.76	0.87
flex.: where and how	Con_7_noMATH	-0.80	0.19	0.01	0.82
usability and interactivity	Con_6_noMATH	0.28	-0.56	0.43	0.76

Fig. 5. Abstract Construct Ranking in NoMATH via Structure Coefficients

Let us recall from Fig. 1, that some mSI elements were in a subcluster shared by all three user groups. That is, with respect to these constructs the INMATH, INFOMATH as well as NOMATH interviewees agreed on the evaluation of the given mSIs. In particular, mSI scores correlate on Con_3_INFOMATH ("usability"), Con_1-_NOMATH ("input design"), Con_5_INFOMATH ("simple design"), Con_6_NOMATH ("usability and interactivity"), and Con_2_INMATH ("supportiveness:result"), Con_1-_INMATH ("adequacy: search philosophy").

We note the different flavor of the non-INMATH constructs versus the INMATH constructs. Where the former aim for design aspects, the latter are only concerned with fitness of the mSI for achieving the search intension. It becomes even clearer if we consider the phrasing "usability" in the non-INMATH group and "supportiveness" in the INMATH group: Whereas usability is a neutral measure for all kinds of qualities while using an object, supportiveness is a task-oriented requirement in the use-flow of a human person. The media-theoretic difference is that the first doesn't tell us anything about whether the user adopts a mSI as a mere tool or as a medium (in the sense of MCLUHAN as *"any extension of the human body [...] as a side-effect of a technology"* [18, p. 564], i.e., a technology that empowers its users):

Pattern 5: *"Mathematicians aim at adopting a search tool as a medium."*

One consequence is that once they have adopted it as a medium, they won't easily change to other media. Not surprisingly, this shared construct cluster also supports a long-standing belief that

Pattern 6: *"Mathematicians appreciate function over form."*

Even though the mSI elements' scores were highly correlated in the shared cluster, their respective conceptualization can still disagree. To understand the conceptualization, we look at the meanings of the distinct constructs and the location of an element wrt constructs. For instance, for mathematicians Google-Scholar enables a top-down approach (as search philosophy =Con_1_INMATH) by using a very general technique of ranking the search results (high supportiveness for presenting search results =Con_2_INMATH), but offers a very textual input design (=Con_1_NOMATH) and a medium-rated effectiveness (as part of usability = Con_6_NOMATH) for non-mathematicians. Here, note that the evaluation by the mathematicians concerns the outcome, whereas the non-mathematicians rather assess it by the input. This argument can be made more generally, as the resp. construct clusters for the INMATH resp. NOMATH groups favor the result resp. the input. Interestingly one cluster category of the INMATH group didn't make it into the consensus grid. In particular, the category "supportiveness of input" has no representative among the abstract constructs of INMATH. We conclude

Pattern 7: *"Mathematicians care more for the outcome than the input."*

This also means that mathematicians seem to be willing to trade input hardships (like more complex interfaces) for output satisfaction (i.e., having perfect **precision** – all found results fit the search intension– and **recall** – all fitting results were found).

In Fig. 1 we observe that the constructs Con_5_INMATH, Con_6_INMATH, Con_7-_INMATH, and Con_8_INMATH are part of an abstract construct cluster only containing INMATH constructs; they are the enhanced (yellow colored) constructs in Fig. 4. Here, the math interfaces scored similarly according to the attributes "relevance of results: driven by user", "relevance of results: driven by data", "precision of input", and "preconditions of search". Thus, we can interpret that a mathematical search interface that empowers the user by enabling him to fine-tune the search query is considered to strongly improve the relevance of the result. This interpretation is supported by Pattern 7, thus we note that

Pattern 8: *"Mathematicians want to be empowered in the search process."*

Moreover, mathematicians obviously realize that this precision comes at a cost: the underlying data have to be structured enough. Therefore, if the data do not allow such a fine-tuning right away, they are willing to iteratively refine their query themselves. A direct consequence seems to be that mathematicians want as much support in formulating a search query as they can get. Whereas non-mathematicians will agree that Pattern 7 is different from their own approach, wrt the above consequence their attitude might be different as the pattern describes the disregard of input facilities by mathematicians and the latter

the total investment of time and energy towards satisfying the search intension. We already observed that this is a cluster of elements marked by mathematicians only. That is, this kind of evaluation scheme didn't occur to non-mathematicians, thus it isn't a dominant one.

For Pattern 3 we argued with the abstract construct ranking within Fig. 4. Interestingly, three of the four first ranked items in that list belong to the uniquely mathematical cluster. The fourth one (Con_8_INMATH , "preconditions for search") occurred in the INMATH group only, that is, it discriminates between mathematicians' and non-mathematicians' search behavior ever more. As mathematicians take the preconditions for search into account in the exploration phase of an information search process, they value their anticipation of the search outcome. This has two consequences:

> **Pattern 9:** *"Mathematicians base their information search process on transparency of the search result."*

Additionally, if they put a lot of thought into the exploration phase, they expect to be rewarded by a good search result. So we hold

> **Pattern 10:** *"Mathematicians expect to find meaningful information in the search result."*

In the interviews, it was striking how much awe `Google` evoked. Pattern 10 solves this riddle: Considering the low amount of work to be invested in the exploration phase, the expectations towards the search results are really low. Therefore, the relevance of `Google` searches amazes mathematicians tremendously.

4   Understanding the Mathematical Perspective on mSIs: an Example

To see the utility of the elicited patterns, we will now discuss the mSIs `MathSci-Net` and `zbMathNew` under a mathematical perspective, which is informed by our elicited patterns.

Let us start with `MathSciNet` as seen in Fig. 6 and `zbMathNew` shown in Fig. 7. From above we know that mathematicians don't discern between `MathSciNet` and `zbMathNew`. This immediately raises the question why this might be the case. Evidently both layouts use a lot of vacuity to focus the users' attention and use bright colors sparingly. But we know because of Pattern 6, that the

Fig. 6. mSI of `MathSciNet`

form is not important to mathematicians, so the reason for their alignment cannot stem from these observations. Unfortunately, at first glance the similarity of the start page already ends here: zbMathNew provides a simple search, i.e., a one-step search, MathSciNet a multi-dimensional, structured search. Moreover, zbMathNew offers innovative extra services like mathematical software search and formula search, MathSciNet an extra citation service. The former offers inline search fields to specify the search. The latter provides social media connections. If we look at the search result page of each, we will find that there are as many differences.

Now let us take a closer look, for example, at the difference between zbMath-New's simple search and MathSciNet's structured search. We know because of Pattern 7, that mathematicians value the outcome higher than the input. Therefore, as zbMathNew offers not only the functionality of MathSciNets multi-dimensional search via inline search fields in the simple search but also choicewise a link to a structured search, the functionality seems to be the same for mathematicians. The input inefficiencies can be neglected, the potential outcome is the same.

Fig. 7. mSI of zbMathNew

What about the clear differences in functionality in these mSIs? Note that the social media links weren't recognized once in the interviews with the mathematicians, which also fits Pattern 4, stating that they appreciate the communities' practices, but not the links to a community themselves. Then, it seems rather evident that zbMathNew offers *more* functionality, as MathSciNet's extra functionality consists only of the citations index. So shouldn't Pattern 8 kick in and lead to a distinctive perception of both systems?

We can counter-argue with two patterns. On the one hand, Pattern 3 tells us that finding is the major kind of search a mathematician is conducting. The additional services zbMathNew provides the user with are essentially no services that support finding, they rather support browsing. This is clear for the mathematical software search. The facetted search with its abilities to refine a search in the process also supports browsing behavior explicitly. In contrast, the formula search feature was designed for finding, but in the interviews, mathematicians indicated that they simply don't believe in the finding capability of the software (unfair as it is). In [27], interestingly, a similar phenomenon was observed. The underlying reason for this disbelief could lie in Pattern 5, namely that they have adopted zbMathNew as a medium, and that uses string search. Therefore, their conceptualization of this service doesn't fit yet and is a challenge to change.

On the other hand, `zbMathNew` is rather new. The older version didn't have as many relevant extra features as this new one. Thus, Pattern 5 strikes again. Quite a few interviewees reported that they use `MathSciNet` and even when they became aware that `zbMathNew` has more to offer now, they didn't mention any intention to change over.

We can summarize that the patterns help us understand the perception of mathematicians much better. This new-found understanding in turn triggers new design challenges and ultimately better, more math-oriented designs.

5 Conclusion

We have presented an RGI study that was concerned with mathematical search interfaces. To be able to understand the idiosyncracies of mathematicians, we interviewed mathematicians as well as non-mathematicians, with a focus on the former. From the quantitative data and its qualitative interpretation several patterns emerged:

P 1 *"Mathematicians do not assess mSIs based on familiarity."*

P 2 *"Mathematicians trust human and community resources."*

P 3 *"Finding is the primary mathematical search task."*

P 4 *"Mathematicians appreciate social interaction as a mathematical tool. In particular, it is a mathematical practice to collaborate and exchange feedback."*

P 5 *"Mathematicians aim at adopting a search tool as a medium."*

P 6 *"Mathematicians appreciate function over form."*

P 7 *"Mathematicians care more for the outcome than the input."*

P 8 *"Mathematicians want to be empowered in the search process."*

P 9 *"Mathematicians base their information search process on transparency of the search result."*

P 10 *"Mathematicians expect to find meaningful information in the search result."*

With these patterns many design issues for mSIs can be understood and elaborated on much deeper now.

For instance, LIBBRECHT posed in[16] the question whether (mathematical) search queries may become too precise (so that the search result becomes too small). But this question does only make sense for browsing queries not for finding queries. The Pattern 8 suggests that the solutions should be finetuned to the distinct kind of searches. If that is not possible, the default case should be "finding" because of Pattern 3.

Pattern 5 indicates that a change from one tool to another is not easily done by mathematicians. In particular, a change of media will only occur if the innovation is disruptive, a mere incremental innovation won't suffice. Therefore, phrasing a major change (like the one from `zbMathOld` to `zbMathNew`) as a mere update won't convince mathematicians to switch, and because of Pattern 7, neither will an announcement of change that essentially points to the new Google-like layout of the homepage.

Moreover, our data suggest that the search approach "finding" is used by mathematicians predominantly when interacting with elements from the math search cluster, "browsing" when interacting with mSIs in the general search cluster and "solving/asking" when using elements in the personal search cluster. Thus, we can look for the properties of the resp. cluster to extend mSIs by more search approaches. Note that `Google` is best-known for its browsing qualities, only for specific kinds of queries it is now also successful in finding. Under this aspect `Google` is also often used by mathematicians.

Our future work is concerned with general design implications based on the foundational work conducted in this paper. For example, one simple consequence concerns the development process of math user interface development: Specify the user group of your math service beforehand and appreciate the credo of "participatory design" that strongly admonishes developers to acknowledge the fact that *"You are not the user!"*. In our study, e.g., the mathematics practitioners turned out to be different from the professional mathematicians. Another consequence might be that we should make mathematicians more effective by supporting their *interventions* in formulating a search query.

Finally, we like to note that the RGI methodology – even though strenuous at times – seems to be a worthy methodology for use with mathematicians.

Acknowledgement. I thank all my interviewees for their motivation and patience with the RGI method. Moreover, I appreciated the supportive work environment at zbMath, especially discussions with Wolfram Sperber. This work has been funded by the Leibniz association under grant SAW-2012-FIZ.

References

[1] Brown, C.M.: Information Seeking Behavior of Scientists in the Electronic Information Age: Astronomers, Chemists, Mathematicians, and Physicists. JASIS 50(10), 929–943 (1999)

[2] Gower, J.: Generalized procrustes analysis. Psychometrika 40, 33–51 (1975) ISSN: 0033-3123

[3] Grice, J.W.: Idiogrid: Software for the management and analysis of repertory grids. Behavior Research Methods, Instruments, & Computers 34, 338–341 (2002)

[4] Hassenzahl, M., Wessler, R.: Capturing Design Space From a User Perspective: The Repertory Grid Technique Revisited. International Journal of Human-Computer Interaction. 3rd ser. 12, 441–459 (2000) ISSN: 1044-7318

[5] Heidecker, S., Hassenzahl, M.: Eine gruppenspezifische Repertory Grid Analyse der wahrgenommenen Attraktivität von Universitätswebsites. In: Gross, T. (ed.) Mensch & Computer, pp. 129–138. Oldenbourg Verlag (2007)

[6] Hertzum, M., Clemmensen, T.: How do usability professionals construe usability? Int. J. Hum.-Comput. Stud. 70(1), 26–42 (2012)

[7] Jankowicz, D.: The Easy Guide to Repertory Grids. Wiley (2003) ISBN: 0470854049

[8] Kellar, M., Watters, C.R., Shepherd, M.A.: A field study characterizing Web-based information-seeking tasks. JASIST 58(7), 999–1018 (2007)

[9] Kelly, G.: A Brief Introduction to Personal Construct Theory. In: International Handbook of Personal Construct Technology, pp. 3–20. John Wiley & Sons (2003)

[10] Kitchen, T.: The European Digital Mathematics Library: Usability Study, https://wiki.eudml.eu/eudml-w/images/D6.1.pdf (visited on March 14, 2014)

[11] Kohlhase, A.: Framings of Information: Readers' Perception of Information Sources in Spreadsheets. Tech. rep. 30. Jacobs University (March 2013), kwarc.info/ako/pubs/tr_hsi_2013.pdf

[12] Kohlhase, A.: Human-Spreadsheet Interaction. In: Kotzé, P., Marsden, G., Lindgaard, G., Wesson, J., Winckler, M. (eds.) INTERACT 2013, Part IV. LNCS, vol. 8120, pp. 571–578. Springer, Heidelberg (2013)

[13] Kuhlthau, C.C., Heinström, J., Todd, R.J.: The information search process revisited: is the model still useful? IR Information Research 13(4) (2008)

[14] Kuhlthau, C.C.: Seeking meaning. A process approach to library and information services, 2nd edn., XVII, 247 p. Libraries Unlimited (2004)

[15] Lave, J., Wenger, E.: Situated Learning: Legitimate Peripheral Participation (Learning in Doing: Social, Cognitive and Computational Perspectives S.). Cambridge University Press (1991)

[16] Libbrecht, P.: Escaping the Trap of Too Precise Topic Queries. In: Carette, J., Aspinall, D., Lange, C., Sojka, P., Windsteiger, W. (eds.) CICM 2013. LNCS (LNAI), vol. 7961, pp. 296–309. Springer, Heidelberg (2013), http://ceur-ws.org/Vol-1010

[17] Mathematics Subject Classification (MSC) SKOS (2012), http://msc2010.org/resources/MSC/2010/info/ (visited on August 31, 2012)

[18] McLuhan, M.: Understanding Media: The Extensions of Man (Critical Edition), edited by W. Terrence Gordon. Gingko Press, Corte Madera (1964) (2003)

[19] Morville, P., Callender, J.: Search Patterns: Design for Discovery. O'Reilly Media (2010) ISBN: 9781449383060

[20] Navarro-Prieto, R., Scaife, M., Rogers, Y.: Cognitive strategies in web searching. In: Proceedings of the 5th Conference on Human Factors & the Web (1999)

[21] OpenRepGrid.org, http://www.openrepgrid.org (visited on March 14, 2014)

[22] Russell, S.J., Norvig, P.: Artificial Intelligence — A Modern Approach. Prentice Hall, Upper Saddle River (1995)

[23] Shneiderman, B., Byrd, D., Croft, W.B.: Clarifying Search: A User-Interface Framework for Text Searches (1997) ISSN: 1082-9873

[24] Tan, F.B., Gordon Hunter, M.: The Repertory Grid Technique: A Method for the Study of Cognition in Information Systems. English. MIS Quarterly 26(1), 39–57 (2002) ISSN: 02767783

[25] Taylor, R.S.: The process of Asking Questions. American Documentation 13(4), 391–396 (1962), doi:10.1002/asi.5090130405

[26] Wise, K., Kim, H.J., Kim, J.: The effect of searching versus surfing on cognitive and emotional responses to online news. Journal of Media Psychology: Theories, Methods, and Applications 21(2), 49–59 (2009), doi:10.1027/1864-1105.21.2.49

[27] Zhao, J., Kan, M.-Y., Theng, Y.L.: Math Information Retrieval: User Requirements and Prototype Implementation. In: Proceedings of the 8th ACM/IEEE-CS Joint Conference on Digital Libraries, JCDL 2008, pp. 187–196. ACM, Pittsburgh (2008) ISBN: 978-1-59593-998-2

A Data Model and Encoding for a Semantic, Multilingual Terminology of Mathematics

Michael Kohlhase

Computer Science, Jacobs University Bremen
http://kwarc.info/kohlhase

Abstract. To understand mathematical language we have to understand the words of mathematics. In particular, for machine-supported knowledge management and digital libraries, we need machine-actionable terminology databases (termbases). However, terminologies for Mathematics and related subjects differ from vocabularies for general natural languages in many ways. In this paper we analyze these and develop a data model for SMGloM the Semantic, Multilingual Glossary of Mathematics and show how it can be encoded in the OMDoc/MMT theory graph model. This structured representation naturally accounts for many of the terminological and ontological relations of a semantic terminology (aka. glossary). We also demonstrate how we can account for multilinguality in this setting.

1 Introduction

Text-based information systems for mathematics and the linguistics of mathematics are still in their infancy due to the inherent complexity of mathematical documents, domains, and knowledge. One issue of particular importance is the problem of dealing with mathematical vocabularies, since they are intimately linked with both the underlying domain of mathematical knowledge and the linguistic structures that make up the particular documents. In general natural language processing, the establishment of machine-actionable terminology databases has kick-started so many applications and systems that the field is unthinkable without such resources. The SMGloM (<u>S</u>emantic <u>M</u>ultilingual <u>Glo</u>ssary for <u>M</u>athematics) is an attempt to jump-start similar applications.

The SMGloM system [SMG] builds on the MMT API [Rab13] and MathHub. info [CICM1414] for archiving and editing support. It supplies glossary-oriented web services that answer termbase queries, e.g. for terminological relations, definitions, or translations and generates glossaries for sub-corpora. The current glossary contains

- ca. 150 glossary entries from elementary mathematics, to provide a basis for further development and
- ca. 350 are special concepts from number theory to explore the suitability of the SMGloM for more advanced areas of mathematics.

In this paper we analyze the special needs of terminologies for Mathematics and related subjects and develop a data model for the SMGloM. This structured

S.M. Watt et al. (Eds.): CICM 2014, LNAI 8543, pp. 169–183, 2014.

representation naturally accounts for many of the terminological and ontological relations of a semantic terminology (aka. glossary).

Let us briefly recap the relevant linguistic and epistemological issues involved in terminological databases to ground our discussion of the special case of mathematical terminologies.

Glossaries Traditionally, a glossary consists of a list of technical/non-standard terms with short definitions ordered alphabetically or in the chronology of the document it illustrates. Figure 1 shows an example from Mathematics.

---

braid ...
branch has multiple meanings:
1. In complex analysis, a **branch** (also called a **sheet**) is a portion of the range of a multivalued function over which the function is single-valued.
2. In a directed graph $G = \langle V, E \rangle$ we call E the set of **edges** or **branches** in G.
3. If $T = \langle V, E \rangle$ is a tree and $u \in V$, then the **branch** at u is the maximal subtree with root u (Harary 1994, p. 35).
4. ...
branch curve ...

Fig. 1. A Glossary Entry for Mathematics

Terminologies Modern glossaries are usually generated from **terminologies** or **termbases** – i.e. special ontologies that organize terms and their definitions by terminological relations and/or the inherent structure of the underlying domain.

Terms are words and compound words that in specific contexts are given specific meanings. These may deviate from the meaning the same words have in other contexts and in everyday language. More specifically, we consider terms as **lexemes** which summarize the various inflectional variants of a word or compound word into a single unit of lexical meaning. Lexemes are usually referenced by their **lemma** (or **citation form**) – a particular form of a lexeme that is chosen by convention to represent a canonical form of a lexeme. Grammatical information about a lexeme is represented in a **lexicon** – a listing of the lexemes of a language or sub-language organized by lemmata.

Terminological relations are semantic relations between terms[1]. The ones commonly used in terminologies are the following:

synonymy two terms are synonymous, if they have the same meaning, i.e. they are interchangeable in a context without changing the truth value of the proposition in which they are embedded.

---

[1] In linguistics, these relations are usually called "semantic relations", but in the context of this note, the term "semantic" is so convoluted that we will highlight the fact that they are relations between terms.

hypernymy term Y is a hypernym of term X if every X is a (kind of) Y.

hyponymy the converse relation of hypernymy

meronomy term Y is a meronym of term X if Y is a part of X

holonymy the converse relation of meronomy

homonymy two terms are homonyms if they have the same pronunciation and spelling (but different meanings).

antonomy two terms are antonyms, if they have opposite meanings: one is the antithesis of the other.

We will call a termbase **semantic**, if it contains terminological relations and/or a representation of the domain relations.

The paradigmatic example of a termbase organized along terminological relations is WordNet [Fel98; WN]. In WordNet the synonymy relation is treated specially: the set of synonyms – called a **synset** – is taken to represent a specific entity in the world – a **semantic object** – and forms the basic representational unit of digital vocabularies. Indeed, all other terminological relations are inherited between synonyms, so it is sensible to quotient out the synonymy relation and use synsets.

Semantic terminologies are very useful linguistic resources: WordNet been used as the basis for many different services and components in information systems, including word sense disambiguation, information retrieval, automatic text classification, automatic text summarization, and machine translation. Note that WordNet and related lexical resources do not model the relations of the objects the terms describe other than via the terminological relations above. For instance, WordNet is ignorant of the fact that a "son" is "male"[2] and a "child" of another "man". In particular, definitions are not first-class citizens in WordNet-like resources, they are included into the data set for the purposes documentation, primarily so that human lexicographers can delineate the synsets. But to fully "understand" terms in their contexts – e.g. to automate processing of documents that involve such terms, and drawing inferences from them – domain relations like the ones above are crucial.

Domain Relations. Semantic glossaries and digital vocabularies usually make some relations between entries explicit, so that they can be used for reasoning and applications. Linguistically, the domain relations – i.e. the relations between the (classes of) objects denoted by words – come into play in the form of **semantic roles** – the thematic relations that express the role that a noun phrase plays with respect to the action or state described by a sentence's verb. The basic idea is that one cannot understand the meaning of a single word without access to all the essential knowledge that relates to that word.

Prominent examples of termbases with semantic roles include FrameNet [FN10; FN] and PropBank [PKG05; PB]. The former collects the semantic roles into **frames** like `Being_born` with a role `Child`, and additional roles like `Time`, `Place`, `Relatives`, etc. Such resources allow additional natural language processing steps

---

[2] Do not confuse that with the grammatical gender of the word "son" is masculine or the fact that "man" is a hypernym of "son".

like "semantic role labeling", which in turn allow the extraction of facts from texts, e.g. in the form of RDF triples which can then be used for textual entailment queries, question answering, etc.; see e.g. [Leh+13] for applications and references.

A Semantic, Multilingual Termbase for Mathematics. For the SMGloM data model we will essentially start with the intuitions from term bases above, but adapt them to the special situation of **mathematical vernacular**, the everyday language used in writing mathematics in textbooks, articles, and to blackboards. This is a mixture of natural language, formulae, and diagrams[3] all of which utilize special, domain-dependent, and dynamically extensible vocabularies. SMGloM differs from resources like FrameNet in the domain representation: we will reuse the OMDoc/MMT format for representing mathematical domains.

2   A Data Model for SMGloM

The data model of SMGloM is organized as a semantic term base with strong terminological relations and an explicit and expressive domain ontology. The terms are used as "named mathematical entities" in the sense that they are rigid designators in Kripke's sense, rather than univalent descriptions.

2.1   Components of Terminology in Mathematics

Whereas in general natural language word meanings are grounded in the perceived world, the special vocabularies used in mathematics are usually grounded by (more or less rigorous) definitions of the mathematical objects and concepts they denote: We have learned to reliably and precisely recognize an object as a "chair" even though we have a hard time when asked to give a precise definition[4] of what constitutes a "chair", but we cannot directly experience a "symplectic group" and are left only with its definition to determine its meaning. In both cases, the word references an object or a set of objects that are uniform in some way so they can be subsumed under a concept; we will consider both as **semantic objects**. As mathematical objects can still have multiple "names" with which designate them, we will use the definitions themselves as the representatives of the respective semantic objects. Every definition will have an identifier which we call the **symbol** and use it for identifying the semantic object.

Note that even though the symbol name will in practice usually be derived from (the lemma of) the definiens of the definition, they are not (conceptually) the same. The technical terms normally found in glossaries arise as "verbalizations" (see Section 2.6) of symbols in diverse languages. In general there is a many-to-many relationship between terms and symbols: several terms pointing

---

[3] Even though diagrams and their structural and lexical components are very interesting subject of study, we leave them to future work.

[4] Arguably such definitions exist – take for instance Wikipedia's page on chairs, but they are usually post-hoc and have little to do with our day-to-day use of the word and its meaning derived from this practice.

to the same definition, as well as several definitions communicated via the same term. In this way, symbols roughly correspond to synsets in WordNet.

But mathematical vernacular also contains formulae as special phrasal structures. We observe that formulae are complex expressions that describe mathematical objects in terms of symbols. In fact, they can be "read out" into equivalent verbal phrases, e.g. for visually impaired recipients. In this transformation, specific and characteristic parts of the formulae correspond to the symbols involved. We call these their **notations**, they act as an additional lexical component. Finally, we have the terminological and domain relations as above, only that we have to re-interpret them to the more rigorous and structured domain of mathematical knowledge.

For the purposes of SMGloM a glossary entry consists of five kinds of information, which we will describe in the rest of this section.

1. a *symbol* identified by a definition (see Section 2.2)
2. its *verbalizations* (common names; see Section 2.6)
3. its various *notations* (formula representations; see 2.5)
4. *terminological relations* to other glossary entries. (see 3.5)
5. *domain relations* to other glossary entries. (see 3.6)

2.2   Symbols and their Definitions

A **definition** consists of a **definiendum** – the term introduced in the definition – and a **definiens** – a text fragment that gives the definiendum its meaning. In the simplest of all cases, the definiens is an expression or formula that does not contain the definiendum and we can directly associate a symbol for the definiendum with the definition as an identifier. We call this case a **simple definition**.

Definition: A **directed graph** (or **digraph**) is a pair $\langle V, E \rangle$ such that V is a set of **vertices** (or **nodes**) and $E \subseteq V \times V$ is the set of its **edges**.

Fig. 2. A Definition for multiple concepts

We will rely on the reader's mathematical experience and forego a classification of definitional forms here, but note that definitions of structured mathematical objects often naturally define more than one term. Take, for instance, the definition of a graph in Figure 2. This introduces three concepts: "directed graph", "vertex", and "edge", which we take as symbols and the synonyms "digraph" for "directed graph" and "node" for "vertex". We can allow such definitions in SMGloM without losing the principal one-definition-one-symbol invariant if we understand them as aggregated forms. The one in Figure 2 is an aggregation of the three definitions (one per symbol) in Figure 3. But the separation of the definitions in Figure 3 is awkward and artificial, and arguably readers would prefer to see the single definition in Figure 2 in a glossary over one of the ones in Figure 3.

> *Def*: A **directed graph** (or **digraph**) is a pair $\langle V, E \rangle$ of sets, such that $E \subseteq V \times V$.
> *Def*: Let $G = \langle V, E \rangle$ be a digraph, then V is the set of **vertices** (or **nodes**) of G.
> *Def*: Let $G = \langle V, E \rangle$ be a digraph, then we call E the set of **edges** of G.

Fig. 3. The Definition from Figure 2 separated into Simple Definitions

2.3 Glossary Modules

To further support grouping symbols into semantic fields, SMGloM provides **modules**: groups of definitions that belong together conceptually. SMGloM modules are conceptually similar to OPENMATH content dictionaries [Bus+04] (CDs), and we follow the lead of OPENMATH and identify symbols by their module name (c) and their symbol name s (and their CD base g, the base URI of the CDs) and write this as $g?c?s$ following MMT conventions [RK13].

Note that there is a non-trivial design decision in taking the definitions as representatives of mathematical semantic objects in SMGloM as there are often multiple, equivalent ways of defining the "same" mathematical objects. For instance, a group can be defined as a base set with a binary i) associative operation $\circ$ that admits a unit and inverses or ii) cancellative operation $/$. These two definitions are logically equivalent, since we can define a/b as $a \circ b^{-1}$ and $a \circ b$ as $a/(b/(b/b))$. As this example already shows, logical equivalence can be non-trivial, and in many cases is only discovered a long time after the definition of the mathematical objects themselves. Therefore different definitions receive different glossary entries in SMGloM with different symbols.

In our example the two definitions give rise to two symbols `group1` and `group2`, and we do not consider them synonyms (they are in different synsets), but **homonyms** words that have different "meanings" (which are logically equivalent in this case). In a sense, the two symbols model how an objects appears to the observer, similarly to the "evening star" and the "morning star" which both refer to the planet Venus. It seems reasonable to conserve this level of modeling in a linguistic/semantic resource like SMGloM.

2.4 Symbols and Multilinguality

Another SMGloM design decision is that we allow mathematical vernacular for definitions. As written/spoken mathematical language is tied to a particular natural language, we abstract from this arbitrary choice by allowing **translations** of the definition in different languages, which we consider "indistinguishable" for a SMGloM module.

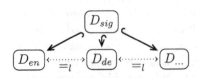

Fig. 4. Language Equality

In Figure 4 we see a situation where the content of a glossary entry D_{sig} is characterized as the equivalence class of definitions in specific languages D_* that are translations of each other – we call the translation relation **language equality** and we depict it by $=_l$; see [KK06] for an in-depth discussion on language-equality and related issues.

Concretely, a glossary module is represented as $n+1$ **glossary components**:
- one for the language-independent part (called the **module signature**, it introduces the symbols, their dependencies, and notations, since they are largely independent of the natural language), and
- n **language bindings**, which introduce the definitions – they are written in a particular mathematical vernacular – and the language-specific verbalizations of symbols. We could imagine "language bindings" for different logical systems as a possible future extension of the SMGloM, which adds formalizations. These would behave just like the regular language bindings, only that they are fully formal.

The reason for this construction is that the vocabulary of mathematics is language-independent, because it is grounded in definitions, which can be translated – unlike general natural language vocabularies where semantic fields do not necessarily coincide.

2.5   Notations

Many mathematical objects have special symbols or formula fragments that identify them. For instance, Euler's number is written as e and the imaginary unit of complex numbers is written as i (in mathematics, in electrical engineering it is written as j; "standard" notations vary with the community). Parameterized or functional mathematical objects, often have complex notations, e.g. the n-th Bernoulli number is written as B_n and the special linear group of degree n over a field F is traditionally written as $\mathrm{SL}(n, F)$. In SMGloM, we treat notations as mathematical objects themselves and reify them into notation definitions, since we want to model them in glossary components. Notation definitions are pairs $\langle \mathbb{C}, \mathbb{P} \rangle$, where $\mathbb{C}$ is a content schema (a representation of a formula with metavariables – here indicated by ?x) paired with a presentation $\mathbb{P}$ of the same schema. For instance, the notation for a functional symbols like the special linear group above, the head is a pattern of the form @(slg; ?n, ?f)[5] and the body is the formula $\mathrm{SL}(?n, ?f)$. Notation definitions are useful in two ways: used left-to-right (i.e. given a content representation) they can be used for styling, i.e. transforming content representations (here Content MathML) to presentations (here Presentation MathML). In the other direction, they can be used for notation-based parsing – i.e. context-sensitive parsing with a dynamic (formula) lexicon.

2.6   Verbalizations

Abstract mathematical concepts (named mathematical entities; NMEs) may have multiple names – at least one per language, e.g. the English nouns "vertex" and "node" in the example in Figure 2 and the corresponding German noun "Knoten". We specify this symbol-to-phrase relation via **verbalization definitions**, which are symbol-phrase pairs. As the NMEs are often not part of

---

[5] We will use @(a; l) to denote a content MathML application of a function a to an argument list l.

the regular lexicon of a language, we often need to specify syntactic/semantic information about the phrases. We do this in the form of verbalization definitions. Similarly to a notation definition, a **verbalization definition** is a pair $\langle \mathbb{C}, \mathbb{N} \rangle$, where the **head** $\mathbb{C}$ is a content schema and the **body** $\mathbb{N}$ is a natural language phrase schema, i.e. a phrase with metavariables. For simple cases like the verbalization "node" for the symbol `vertex` the verbalization definition is rather simple, it is just the pair $\langle \text{vertex}, \text{node} \rangle$. For functional symbols like the special linear group above, the head is a pattern of the form @(slg; ?n, ?f) and the body is the text schema

[special linear group][of degree ?n][over the field ?f]

where phrases are delimited by square brackets. Note that verbalization definitions can be used in both directions like notation definitions. We use them as a linguistic resource for parsing, but also for the generation of standard glossaries or wikifiers. We abstract from grammatical information here and reduce terms and phrases to their lemmata, assuming a suitable lexicon component that manages information about inflection and aggregation schemata. For instance, with suitable notation and verbalization definitions we can generate or parse aggregated declarations like "$SL(n, \mathbb{R})$ and $SL(m, \mathbb{C})$ are the special linear groups of orders n and m over the fields $\mathbb{R}$ and $\mathbb{C}$".

3 Implementing the Data Model in OMDoc/MMT

We (re)-interpret the data model introduced in the last section in terms of the OMDoc/MMT theory graph (see [RK13] for a discussion of MMT theory graphs, the formal core of OMDoc). A **theory graph** is a graph, where the nodes are theories and the edges are theory morphisms: truth-preserving mappings from expressions in the source theory to expressions in the target theory. OMDoc/MMT theories are essentially collections of

- **concept declarations**, together with
- **axioms** (in particular definitions) that state what properties the concepts have, and
- **notation definitions** that specify the presentation of symbols.

Theory morphisms come in four forms:

- **structures** which define their target theory to be an extension of the source theory; **inclusions** are those structures whose mapping is the identity,
- **views** which interpret the mathematical objects of the source theory as such of the target theory (for instance, the natural numbers with addition can be interpreted as a monoid if we interpret 0 as the unit element).
- **metatheory-relations** which import the symbols of the meta-language into a theory.

Note that the notion theory morphism is rather strong in OMDoc/MMT, as it allows renaming of concepts. Structures and the meta theory relation are truth/meaning-preserving by virtue of the extension property, essentially the target theory is defined so that they are: all symbols and axioms are in the target after translation. To establish a view, we need to prove all the source axioms (after translation) in the target theory.

3.1 Glossary Components as OMDoc/MMT Theories

We can implement the SMGloM data model directly in OMDoc/MMT theory graphs – indeed the MMT API drives the SMGloM system. Note that the setup in Figure 4 can directly represented by giving theories for the module signature and its language bindings and interpreting the dependencies as OMDoc/MMT inclusions. There is however one problem we still need to solve: the module signatures introduce the symbols of the glossary module, but their meaning is specified in the language bindings, which include them. Therefore we need to extend MMT with a new feature: **adoptions**, i.e. views from the language bindings to the module signatures that are definitional – they establish the meaning of the symbols in the module signature by postulating that definitions in the language bindings hold there. Note that in this sense, adoptions are similar to structures – only that meaning travels in the reverse direction. Like these, adoptions do not induce proof obligations. The lower half of Figure 5 shows the situation, the double squiggly arrows are the adoptions. Note that the adoptions can only work, if the definitions in the various language bindings are translations – indeed the adoptions postulate them, but we cannot check them in the SMGloM system.

But we can use the theory graph to even more advantage in SMGloM, if we take the MMT meta-level into account. We can model the fact that e.g. the language binding D_{en} is written in English by specifying the theory MV_{en} (English mathematical vernacular) as its meta-theory. In Figure 5, we find the module/bindings construction of Figure 4 at the bottom layer, and their vernaculars in the layer above. These, inherit from generic language theories L_* and a

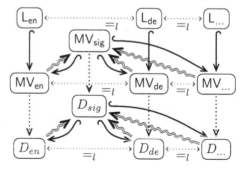

Fig. 5. The Language Metalevel

module signature MV for mathematical vernacular[6]. Note that the mathematical vernacular meta-level (the middle layer in Figure 5) is structurally isomorphic to the domain level. In particular, we can think of MV as a signature of mathematical vernacular: it contains symbols for meta-mathematical concepts like quantification, connectives, definitional equality, etc. In future extensions of the SMGloM by formal content, this is the spaces, where the logics would live – see [Cod+11].

The third level in Figure 5 contains the generic (i.e. non-mathematical) vocabularies of the respective natural languages. They are just stubs in SMGloM

---

[6] Actually, what we have depicted as a single theory here is a whole theory graph of inter-dependent theories.

that can be coupled with non-math-specific linguistic/lexical resources in the future.

3.2 Multilingual Theory Morphisms

In the SMGloM, where glossary items are structured, multilingual modules (see Figure 4), theory morphisms are similarity structured. Consider the situation on the right, where we have a module MS for metric spaces, and

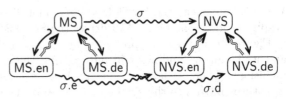

Fig. 6. A multilingual view

ule MS for metric spaces, and another (NVS) for metric vector spaces. It is well-known that a normed vector space $\langle V, \| \cdot \| \rangle$ induces a distance function $d(x, y) := \|x - y\|$ and thus a metric space $\langle V, d \rangle$. The OMDoc/MMT views that make up this structured relation between glossary modules is represented by the three wavy arrows in Figure 6. Here σ is the translation that assigns the base set V to itself and $d(x, y)$ to $\|x - y\|$. The two OMDoc/MMT views σ.e and σ.d include σ and add the proofs (in English and German respectively) for the proof obligations induced by the metric space axioms.

3.3 Notations and Verbalizations

We employ OMDoc notation definitions which directly implement the content/presentation pairs in XML syntax (see [Koh10] for details). It turns out that the for the structurally similar verbalization definitions introduced in Section 2.6, we re-use the OMDoc/MMT notation definitions mechanism, only that the "presentation" component is not presentation MathML, but in natural language phrase structures (in the respective languages).

3.4 Synsets: Direct Synonymy

We have two forms of "synonyms" in SMGloM: **direct synonyms** that are directly given in definitions, and induced ones (see below). For example, the definition in Figure 2 introduces the terms "vertex" and "node" as direct synonyms. Indeed, the definiendum markup gives rise to the verbalization definitions $\langle$dgraph?vertex, vertex$\rangle$ and $\langle$dgraph?vertex, node$\rangle$ respectively, i.e. the lemmata "vertex" and "node" refer to the symbol vertex in the theory dgraph. In essence we use symbol-synchronization for the representation of direct synonyms, and thus we can use the symbols as representations of synsets of the SMGloM term base. Note that this interpretation also sees translations as special cases of synonyms, as they also refer to the same (language-independent) symbol. In SMGloM we identify synsets with symbols and thus model terminological relations as relations between symbols. This allows us to model them as theory morphisms and use the OMDoc/MMT machinery to explain their contributions and properties. For the moment we restrict ourselves to inclusions and leave structures and views to Section 3.6.

3.5 Direct Terminological Relations

In OMDoc/MMT theory graphs, we often have a systematic dualism between the theory T as a structured object and the mathematical structure[7] it introduces, we call it the **primary object** and denote it with $\overline{T}$, all other symbols are called **secondary**. Consider for instance the case of directed graphs above, where the theory has secondary symbols for vertices and edges; and incidentally, the primary object of the glossary module in Figure 2 is the concept a digraph, i.e. the structure $\langle V, E \rangle$ which consists of (sets of) vertices and edges (both secondary concepts). Similarly, the theory of groups has a primary object made up of its secondary objects: the base set, the operation, the unit, and the inverse operation.

In our experience, secondary symbols mostly (all?) seem to be functional objects whose first argument is the primary symbol. For instance the "edges of" a graph. This makes the setup of SMGloM modules very similar to classes in object-oriented classes, where the secondary objects correspond to methods, and (more importantly for a linguistic resource like SMGloM) to frames in FrameNet, where the secondary symbols correspond to the semantic roles. We will conduct a survey on this on the SMGloM corpus once its more mature.

Hyper/Hyponomy. For the hyponomy and hyperonymy relations, we employ the notion of theory morphisms from OMDoc/MMT. If there is an import from S to T, then $\overline{T}$ is a hypernym of $\overline{S}$ and that a hyponym of S. Consider for instance, the notion of a "tree" as a digraph with special properties (a unique initial node and in-degree 1 on all others). Extending the digraph glossary module to one for trees naturally gives rise to an inclusion morphism that maps the principal symbol `digraph` to the new principal symbol `tree`. Thus the term "tree" is a hypernym of "digraph" (and "directed graph", since that is a direct synonym).

For the secondary symbols we have a related effect. They are usually inherited along theory morphisms together with the primary symbols, but they keep their meaning, only that their domain is restricted to the more specialized primary symbol. This relation which we tentatively call **domain restriction** is related to the notion of **selectional restriction** in lexical semantics – cf. [Ash14] for a recent contribution that seems compatible with the SMGloM data model.

Meronomy. Note that the inclusion relation we have encountered above is very naturally a theory morphism by construction: all objects and their properties of the source theory are imported into the target theory. As the imports relation is invoked whenever a mathematical object is referenced (used) in the definiens of another, we interpret the inclusion relation as the SMGloM counterpart of

---

[7] We have an unfortunate name clash with MMT "structures" here we mean the mathematical object, e.g. the pair $\langle V, E \rangle$ in Figure 2.

the meronymy relation: if there is an import from theory S to theory T, then $\overline{S}$ is part of $\overline{T}$. Take for instance a definition of a ring $\overline{\text{ring}} = \langle R, +, 0, -, *, 1 \rangle$ via an inclusion from a commutative group $\overline{\text{grp}} = \langle R, +, 0, - \rangle$ and a monoid $\overline{\text{mon}} = \langle R, *, 1 \rangle$. This directly gives us two meronomy relations: The monoid $\overline{\text{mon}}$ and the commutative group $\overline{\text{grp}}$ are both "parts of" the $\overline{\text{ring}}$. As a consequence, inclusions where the primary symbol of the source is not mapped to the primary symbol of the target theory give rise to meronomy relations between the primary symbols.

3.6 Induced Terminological Relations

We now turn to the other kind of theory morphisms: structures and views and their contribution to terminological relations. We first observe that structures and views bridge a greater conceptual distance than inclusions and adoptions, since the induced mapping is not the identity. Note that the distinction made here between inclusions and structures is a gradual one based on the complexity of the mapping. In particular, structures with injective symbol mappings may seem closer to inclusions than to structures that map to complex terms. Moreover, while inclusions and structures are definitional (their targets are defined in terms of them), views carry proof obligations that show their truth-preserving nature; this translates into an even greater cognitive distance of the **induced terminological relations**.

Homonymy. Logical equivalence of glossary modules – i.e. homonymy of the terms that verbalize the primary symbols – is just a case of theory isomorphism. In the example with the two groups from Section 2.3 we have two SMGloM modules which are represented OMDoc/MMT theories. Their equivalence can be encoded by a theory isomorphism: two views which compose to the identity. As any logical equivalence can be expressed as theory isomorphisms (given suitable glossary modules), homonymy is conservative over OMDoc/MMT theory graphs.

View-Induced Hyponomy (aka. Examples). We have already seen that theory inclusions induce hyponomy (the "isa relation") between the principal symbols, e.g. a group "is a" monoid. The "induced hyponomy relation" – e.g. $\langle \mathbb{N}, + \rangle$ "is a" monoid if we interpret 0 as the unit element is very salient in mathematics: we consider $\langle \mathbb{N}, +, 0 \rangle$ as an example of a monoid. The proof obligations of the underlying view verify that this is indeed true. Giving examples – and counter-examples – from other mathematical areas is an important mathematical practice necessary for fully understanding mathematical concepts and fostering intuitions about applications. Regular hyponyms are usually not considered good examples, since they are too direct.

Induced Synonymy. In a graph definition in [Har69] we find the terms "0-simplex" for the nodes and "1-simplex" for the edges of a graph. We interpret such "synonyms" as metaphoric. Given a definition

Definition: A k-**simplex** is a k-dimensional polytope which is the convex hull of $k+1$ affinely independent points in k-space.

Harary's definition makes sense if we map nodes to 0-simplices and edges to 1-simplices. In SMGloM we would model this via a glossary module for simplices and a view from the graph module to that. Then we can understand Harary's names as synonyms induced by this view. Note that in order for these to be the "synonyms" in the sense of this paper, we also need a (partial) view back from simplices to edges (that is defined on them), but that is also easy to do. The view directly accounts for the metaphoric character. We "borrow" terms for graphs from a related (via the view) field of simplices.

As the conceptual gap covered by views can vary greatly – the identity endomorphism covers none – the distinction between direct- and view-induced synonyms is flexible (and in the mind of the beholder). A first delineation could be whether the analogy mappings that give rise to the (originally metaphoric) names are inner-mathematical or extra-mathematical. If they are inner-mathematical then we should state the views, if they are not, then we cannot really. An example of synonyms introduced by an extra-mathematical (from plant anatomy) view is the junction/branch metaphor for vertices/edges in graphs. Given these criteria, it becomes debatable whether to interpret the synonyms point/line for vertex/edge via a view into point/line geometry.

A very positive effect of interpreting synonyms via views is that this also gives an account of the coordination of synonyms. We observe that verbalizations are coordinated in "conceptual systems". In particular, we will seldom find "mixed metaphors" in Math, where people use the word "point" for the concept of a vertex and "branch" for an edge in the same situation. Requiring the existence a view that maps the whole situation into a coherent glossary module explains this observation nicely. Similar consideration should hold for notations, but we will leave their study to future work.

4   Conclusion

We have presented a data model model for a mathematical termbase. As mathematical terminologies are based in definitions not in perceptions of the physical world, modeling the mathematical domain becomes as important as modeling the terminological relations for a machine-actionable resource. We integrate both aspects by modeling glossary terms by OMDoc/MMT symbols, glossary modules (semantic fields of terms correlated by their meaning) as theories, and terminological relations by theory morphisms, so that we can make use of the OMDoc/MMT machinery – and even implementation – for a glossary system.

Our treatment of multilinguality has some similarlity to the GF Mathematical Library (MGF [SX11]), but that concentrates on grammatical aspects where we focus on terminological relations and definitions. An extension of SMGloM by GF-based grammatical information seems an attractive avenue of future research as the OMDoc/MMT domain model is an extension of the OpenMath CDs used in MGF and the GF framework is based on a type theory that is similar to the meta-level of OMDoc/MMT.

Currently, the SMGloM only contains a handful of views to establish the concept and serve as examples. As we have seen above, views give rise to interesting semantic/linguistic phenomena, so this is where we have to invest most of the curation efforts.

Eventually, we will support multiple surface syntaxes for OMDoc, but initially, we use sTEX, a semantical variant of LATEX; see [Koh08; sTeX].

An feature of mathematical domain modeling which we have not included in the SMGloM is the assignment of sorts/types to mathematical concepts. This is probably the most immediate next step after consolidating the initial corpus to the data model described in this paper: Sortal and type-restrictions are important cognitive devices in semantic domains and representing them significantly enhances the expressivity and adequacy of lexical/linguistic as well as logical modeling. But the integration of linguistic and logical constraints – in particular selectional restrictions of verbs and adjectives – into a universal sort/type system for mathematics is no small feat, therefore we leave it to future work. But we conjecture that the SMGloM data model of a mathematical term base with a theory graph structure is the right setting to investigate **selectional restriction** in lexical semantics. We plan to use all "unary predicate symbols" in SMGloM as possible types and study what this means for the selection restrictions taking [Ash14] into account as a departure for this work.

Acknowledgements. Work on the concepts presented here has been partially supported by the Leibniz association under grant SAW-2012-FIZ_KA-2 and the German Research Foundation (DFG) under grant KO 2428/13-1. The development of the data model has profited from discussions with Deyan Ginev, Wolfram Sperber, and Mihnea Iancu.

References

[Ash14]   Asher, N.: Selectional Restrictions, Types, and Categories. Journal of Applied Logic 12(2), 75–87 (2014)

[Bus+04]   Buswell, S., et al.: The Open Math Standard, Version 2.0. Tech. rep. The OpenMath Society (2004), http://www.openmath.org/standard/om20

[CICM1414]   Iancu, M., Jucovschi, C., Kohlhase, M., Wiesing, T.: System Description: MathHub.info. In: Watt, S.M., Davenport, J.H., Sexton, A.P., Sojka, P., Urban, J. (eds.) CICM 2014. LNCS, vol. 8543, pp. 436–439. Springer, Heidelberg (2014),
             http://kwarc.info/kohlhase/submit/cicm14-mathhub.pdf

[Cod+11]   Codescu, M., Horozal, F., Kohlhase, M., Mossakowski, T., Rabe, F.: Project Abstract: Logic Atlas and Integrator (LATIN). In: Davenport, J.H., Farmer, W.M., Urban, J., Rabe, F. (eds.) Calculemus/MKM 2011. LNCS (LNAI), vol. 6824, pp. 289–291. Springer, Heidelberg (2011)

[Fel98]   Fellbaum, C. (ed.): WordNet: An Electronic Lexical Database. MIT Press (1998)

[FN]   Frame Net, https://framenet.icsi.berkeley.edu (visited on February 06, 2014)
[FN10]   Ruppenhofer, J., et al.: FrameNet II: Extended Theory and Practice (2010), https://framenet2.icsi.berkeley.edu/docs/r1.5/book.pdf
[Har69]   Harary, F.: Graph Theory. Addison Wesley (1969)
[KK06]   Kohlhase, A., Kohlhase, M.: An Exploration in the Space of Mathematical Knowledge. In: Kohlhase, M. (ed.) MKM 2005. LNCS (LNAI), vol. 3863, pp. 17–32. Springer, Heidelberg (2006), http://kwarc.info/kohlhase/papers/mkm05.pdf
[Koh08]   Kohlhase, M.: Using LaTeX as a Semantic Markup Format. Mathematics in Computer Science 2(2), 279–304 (2008), https://svn.kwarc.info/repos/stex/doc/mcs08/stex.pdf
[Koh10]   Kohlhase, M.: An Open Markup Format for Mathematical Documents OMDoc [Version 1.3]. Draft Specification (2010), https://svn.omdoc.org/repos/omdoc/branches/omdoc-1.3/doc/spec/main.pdf
[Leh+13]   Lehmann, J., et al.: DBpedia - A Large-scale, Multilingual Knowledge Base Extracted from Wikipedia. Semantic Web Journal, 29 (2013), http://www.semantic-web-journal.net/system/files/swj558.pdf
[PB]   Proposition Bank, http://verbs.colorado.edu/ mpalmer/projects/ace.html (visited on February 06, 2014)
[PKG05]   Palmer, M., Kingsbury, P., Gildea, D.: The Proposition Bank: An Annotated Corpus of Semantic Roles. Computational Linguistics 31(1), 71–106 (2005), doi:10.1162/0891201053630264
[Rab13]   Rabe, F.: The MMT API: A Generic MKM System. In: Carette, J., Aspinall, D., Lange, C., Sojka, P., Windsteiger, W. (eds.) CICM 2013. LNCS (LNAI), vol. 7961, pp. 339–343. Springer, Heidelberg (2013)
[RK13]   Rabe, F., Kohlhase, M.: A Scalable Module System. Information & Computation (230), 1–54 (2013), http://kwarc.info/frabe/Research/mmt.pdf
[SMG]   SMGloM Glossary, http://mathhub.info/mh/glossary (visited on April 21, 2014)
[sTeX]   Semantic Markup for LaTeX. Project Homepage, http://trac.kwarc.info/sTeX/ (visited on February 22, 2011)
[SX11]   Saludes, J., Xambó, S.: The GF Mathematics Library. In: Quaresma, P., Back, R.-J. (eds.) THedu. EPTCS, vol. 79, pp. 102–110 (2011), doi:10.4204/EPTCS.79.6
[WN]   WordNet: A lexical database for English, https://wordnet.princeton.edu/ (visited on May 26, 2013)

PDF/A-3u as an Archival Format
for Accessible Mathematics

Ross Moore

Macquarie University, Sydney, Australia
ross.moore@mq.edu.au

Abstract. Including LATEX source of mathematical expressions, within the PDF document of a text-book or research paper, has definite benefits regarding 'Accessibility' considerations. Here we describe three ways in which this can be done, fully compatibly with international standards ISO 32000, ISO 19005-3, and the forthcoming ISO 32000-2 (PDF 2.0). Two methods use embedded files, also known as 'attachments', holding information in either LATEX or MathML formats, but use different PDF structures to relate these attachments to regions of the document window. One uses structure, so is applicable to a fully 'Tagged PDF' context, while the other uses /AF tagging of the relevant content. The third method requires no tagging at all, instead including the source coding as the /ActualText relacement of a so-called 'fake space'. Information provided this way is extracted via simple Select/Copy/Paste actions, and is available to existing screen-reading software and assistive technologies.

1 Introduction

PDF/A is being adopted by publishers and Government agencies for the long-term preservation of important documents in electronic form. There are a few variants, which pay more or less regard to Accessibility considerations; i.e., 'a' for *accessible*, 'b' for *basic*, 'u' for (presence of) *unicode mappings* for all font characters. Later versions [3,4] of this ISO standard [2] allow for other file attachments in various data formats. In particular, the PDF/A-3u variant allows the inclusion of *embedded files* of arbitrary types, to convey supplementary descriptions of technical portions of a document's contents.

'Accessibility' is more relevant for reports and text-books than for research outputs. In fact in some countries it is a legal requirement that when a visually-impaired student enrols in unit of study for which a text-book is mandated as 'Required', then a fully accessible version of the contents of that book *must* be made available. Anecdotally, visually-impaired students of mathematics and related fields much prefer mathematical material to be made available as LATEX source, to any other format. With a Braille reader, this is text-based and sufficiently compact that expressions can be read and re-read with ease, until a full understanding has been achieved. This is often preferable to having an audio version [13,14], which is less-easy to navigate. Of course having *both* a well-structured audio version, as well as textual source, is even more useful. The

S.M. Watt et al. (Eds.): CICM 2014, LNAI 8543, pp. 184–199, 2014.
© Springer International Publishing Switzerland 2014

PDF example [12] accompanying this paper[1] in fact has both, though here we concentrate on how the latter is achievable within PDF documents.

Again anecdotally, the cost of reverse-engineering[2] all the mathematical expressions within a complete textbook is typically of the order of £10,000 or AUD 30,000 or CAD 10,000. This cost would have been dramatically reduced if the PDF had originally been created to include a LATEX or MathML description of each expression[3], attached or embedded for recovery by the PDF reader or other assistive technology. How to do this in PDF is the purpose of this paper.

The method of *Associated Files*, which is already part of the PDF/A-3 standard [4], is set to also become part of the ISO 32000-2 (PDF 2.0) standard [6], which should appear some time in 2014 or 2015. In Sect. 3 this mechanism is discussed in more detail, showing firstly how to include the relevant information as attachments, which can be extracted using tools in the PDF browser. The second aspect is to relate the attachments to the portion of content as seen on-screen, or within an extractable text-stream. This can be specified conveniently in two different ways. One way requires structure tagging to be present (i.e., a 'Tagged PDF' document), while the other uses direct tagging with an /AF key within the content stream. In either case a PDF reader needs to be aware of the significance of this /AF key and its associated embedded files.

With careful use of the /ActualText attribute of tagged content, LATEX (or other) source coding of mathematical expressions can be included within a PDF document, virtually invisibly, yet extractable using normal Select/Copy/Paste actions. A mechanism, using very small space characters inserted before and after each mathematical expression, is discussed in Sect. 4. This is applicable with *any* PDF file, not necessarily PDF/A. It is important that these spaces not interfere with the high-quality layout of the visual content in the document, so we refer to them as 'fake spaces'.

The various Figures in this paper illustrate the ideas and provide a look at the source coding of a PDF document[1] that includes all the stated methods, thus including the LATEX source of each piece of mathematical content. (Where explicit PDF coding is shown, the whitespace may have been massaged to conserve space within the pages of this paper.) Indeed the example document includes as many as 7 different representations of each piece of mathematical content:

- the visual form, as typically found in a PDF document;
- the LATEX source, in two different ways; i.e, an attachment associated with a /Formula structure tag and also associated directly to the (visual) content, and as the /ActualText replacement of a 'fake space'.
- a MathML version as an attachment, also associated to the /Formula structure tag and also associated directly to the (visual) content;

---

[1] ... should be attached prior to the 'References', else downloadable online; see [12].

[2] ... with prior permission granted by the publisher ...

[3] This is distinct from including the complete LATEX source of the whole document. There are many reasons why an author, and hence the publisher, might not wish to share his/her manuscript; perhaps due to extra information commented-out throughout the source, not intended for general consumption.

- a MathML representation through the structure tagging;
- words for a phonetic audio rendering, to be spoken by 'Read Out Loud';
- the original LATEX source of the complete document, as a file attachment associated with the document as a whole.

In practice not all these views need be included to satisfy 'Accessibility' or other requirements. But with such an array of representations, it is up to the PDF reading software to choose those which it wants to support, or which to extract according to particular requirements of end-users. It is remarkable that a single document can be so enriched, yet still be conforming with a standard such as PDF/A-3u, see [4]. Indeed, with all content being fully tagged, this document[1] would also validate for the stricter PDF/A-3a standard, apart from the lack of a way to specify the proper rôle of MathML structure tagging, so that tags and their attributes are preserved under the 'Save As Other ... XML 1.0' export method when using Adobe's 'Acrobat Pro' software. This deficiency will be addressed in PDF 2.0 [6].

Methods used to achieve the structure tagging in the example document[1] have been the subject of previous talks and papers [10,9] by the author. It is not the intention here to promote those methods, but rather to present the possibilities for mathematical publishing and 'Accessibility' that have been opened up by the PDF/A-3 and PDF/UA standards [4,7], and the 'fake spaces' idea. The example document [12] is then just a 'proof-of-concept' to illustrate these possibilities.

Since the PDF/A-3 standard [4] is so recent, and with PDF 2.0 [6] yet to emerge, software is not yet available that best implements the 'Associated Files' concept. The technical content of the Figures is thus intended to assist PDF software developers in building better tools in support of accessible mathematics. It details (i) exactly what kind of information needs to be included; (ii) the kind of structures that need to be employed; and (iii) how the information and structures relate to each other. For those less familiar with PDF coding, the source snippets have been annotated with high-lighting[4] and extra words indicating the ideas and intentions captured within each PDF object. Lines are used to show relationships between objects within the same Figure, or 'see Fig. Xx' is used where the relationship extends to parts of coding shown within a different Figure. Section 2 is supplied to give an overview of the PDF file structure and language features so that the full details in the Figures can be better understood and their rôle appreciated.

2   Overview of the PDF File Format

PDF files normally come employing a certain amount of compression, to reduce file-size, so appear to be totally intractable to reading by a human. Software techniques exist to undo the compression, or the PDF file may have been created without using any. The example document[1] was created without compression, so can be opened for reading in most editing software.

The overall structure of an uncompressed PDF file consists of:

---

[4] ... with consistent use of colours, in the PDF version of this paper ...

(a) a collection of numbered **objects**: written as <*num*> 0 obj ... endobj where the '...' can represent many, many lines of textual (or binary) data starting on a new line after obj and with endobj on a line by itself. The numbering need not be sequential and objects may appear in any order. An **indirect reference** sequence of the form <*num*> 0 R[5] is used where data from one object is required when processing another. A *cross-reference table* (described next), allows an object and its data to be located precisely. Such indirect references are evident throughout the coding portions of Figures 1–5.

(b) the **cross-reference table**: listing of byte-offsets to where each numbered object occurs within the uncompressed PDF file, together with a linked listing of unused object numbers. (Unused numbers are available for use by PDF editing software.)

(c) the *trailer*, including: (i) total number of objects used; (ii) reference to the document's /Catalog, see Fig. 3c; (iii) reference to the /Info dictionary, containing file properties (i.e., basic metadata); (iv) byte-offset to the cross-reference table; (v) encryption and decryption keys for handling compression; (vi) end-of-file marker.

Thus the data in a PDF file is contained within the collection of objects, using the cross-reference table to precisely locate those objects. A PDF browser uses the /Catalog object (e.g., object 2081 in Fig. 3c) to find the list of /Page objects (e.g., object 5 in Fig. 3b), each of which references a /Contents object. This provides each page's **contents stream** of graphics commands, which give the details of how to build the visual view of the content to be displayed. A small portion of the page stream for a particular page is shown in Figures 1b, 3a, 5a.

Character strings are used in PDF files in various ways; most commonly for ASCII strings, in the form (...); see Figures 1a, 1b, 2b, 2c, 3a, 3c, and 5a. Alternatively, a hexadecimal representation with byte-order mark <FEFF...> can be used, as in Figures 1b, 3a, 5a. This is required particularly for Unicode characters above position 255, with 'surrogate pairs' used for characters outside the basic plane, as with the k variable name in those figures. Below 255 there is also the possibility of using 3-byte octal codes within the (...) string format; see footnote 11 in Sect. 4. For full details, see §7.3.4 of PDF Specifications [1,5].

PDF names of the form /⟨*name*⟩, usually using ordinary letters, have a variety of uses, including (i) **tag-names** in the content stream (Figures 1b, 3a, 5a); (ii) identifiers for **named resources** (Fig. 3b within object 20 and in the /AF tagging shown in Fig. 3a); and extensively as (iii) dictionary **keys** (in all the Figures 1, 2, 3, 5) and frequently as dictionary **values** (see below).

Other common structures used within PDF objects are as follows.

(i) **arrays**, represented as [⟨*item*⟩ ⟨*item*⟩ ... ⟨*item*⟩], usually with similar kinds of ⟨*item*⟩, (see e.g., Figures 1a, 3b, 3c) or alternating kinds (e.g., the filenames array of Fig. 2b).

---

[5] The '0' is actually a *revision number*. In a newly constructed PDF this will always be 0; but with PDF editing software, higher numbers can result from edits.

(ii) **dictionaries** of *key–value pairs*, similar to alternating arrays, but represented as <<⟨*key*₁⟩ ⟨*value*₁⟩ ⟨*key*₂⟩ ⟨*value*₂⟩ ... >>. The ⟨*key*⟩ is always a *PDF name* whereas the ⟨*value*⟩ may be any other element (e.g., string, number, name, array, dictionary, indirect reference). The *key–value pairs* may occur in any order, with the proviso that if the same ⟨*key*⟩ occurs more than once, it is the first instance whose ⟨*value*⟩ is used. A /Type key, having a *PDF name* as value, is not always mandatory; but when given, one refers to the dictionary object as being of the type of this name. See Figures 1a, 2b, 2c, 3b, 3c and 5b for examples.

(iii) **stream objects** consist of a dictionary followed by an arbitrarily-long delimited *stream* of data, having the form << ... >> stream ... endstream, with the stream and endstream keywords each being on a separate line by themselves (see objects 26 and 28 in Fig. 2c). The dictionary must include a /Length key, whose value is the integer number of bytes within the data-stream. With the length of the data known, between the keywords on separate lines, there is no need for any escaping or special encoding of any characters, as is frequently needed in other circumstances and file-formats. See §7.3.8 of [1,5] for more details; e.g., how compression can be used.

(iv) **graphics operators** which place font characters into the visual view occur inside a page contents stream, within portions delimited by BT ... ET (abbreviations for Begin/End Text); see Figures 1b, 3a, 5a. These include coding /⟨*fontname*⟩ ⟨*size*⟩ Tf for selecting the (subsetted) font, scaled to a particular size, and [⟨*string*⟩]TJ for setting the characters of the string with the previously selected font. See §9.4 of [1,5] for a complete description of the available text-showing and text-positioning operators.

Dictionaries and arrays can be nested; that is, the ⟨*value*⟩ of a dictionary item's ⟨*key*⟩ may well be another dictionary or array, as seen in objects 20 and 90 within Fig. 3b. Similarly one or more ⟨*item*⟩s in an array could well be a dictionary, another array, or an indirect reference (regarded as a 'pointer' to another object).

With the use of *PDF names*, *objects*, and *indirect references* a PDF file is like a self-contained web of interlinked information, with names chosen to indicate the kind of information referenced or how that information should be used.

The use of objects, dictionaries (with key–value pairs) and indirect references makes for a very versatile container-like file format. If PDF reader software does not recognise a particular key occurring within a particular type of dictionary, then both the key and its value are ignored. When that value is an indirect reference to another object, such as a *stream object*, then the data of that stream may never be processed, so does not contribute to the view being built. Thus PDF producing or editing software may add whatever objects it likes, for its own purposes, without affecting the views that other PDF reading software wish to construct. This should be contrasted with HTML and XML when a browser does not recognise a custom tag. There that tag is ignored, together with its attributes, but any content of that tag *must still be handled*.

It is this feature of the PDF language which allows different reader software to support different features, and need not use all of the information contained within a PDF file. For example, some browsers support attachments; others do not. A PDF format specification now consists mostly of saying which tags and dictionary keys *must* be present, what others are allowed, and how the information attached to these keys and tags is intended to be used. Hence the proliferation of different standards: PDF/A, PDF/E, PDF/VT, PDF/UA, PDF/X, perhaps with several versions or revisions, intended for conveying different kinds of specialised information most relevant within specific contexts.

2.1 Tagging within PDF Documents

Two types of tagging can be employed within PDF files. 'Tagged PDF' documents use both, with content tags connected as leaf-nodes of the structure tree.

Tagging of content is done as /⟨tag⟩ ⟨dict⟩ BDC ... EMC within a contents stream. Here the BDC and EMC stand for 'Begin Dictionary Content' and 'End Marked Content' respectively, with the ⟨dict⟩ providing key-value pairs that specify 'properties' of the **marked content**, much like 'attributes' in XML or HTML tagging[6]. The ⟨tag⟩ can in principle be any *PDF name*; however, in §14.6.1 of the specifications [1,5] it stipulates that "All such tags shall be registered with Adobe Systems (see Annex E) to avoid conflicts between different applications marking the same content stream." Thus one normally uses a standard tag, such as /Span, or in the presence of structure tagging (see below) choose the same tag name as for the parent structure node. Figures 1b, 3a, 5a show the use of Presentation-MathML content tag names, which are expected to be supported in PDF 2.0 [6]. Typical attributes are the /ActualText and /Alt strings, which allow replacement text to be used when content is extracted from the document using Copy/Paste or as 'Accessible Text' respectively. The /MCID attribute allows *marked content* to be linked to document structure, as discussed below. A variant of this tagging uses a *named resource* for the ⟨dict⟩ element. This is illustrated with /AF content tagging in Sect. 3.2.

Tagging of structure. requires building a tree-like structural description of a document's contents, in terms of Parts, Sections, Sub-sections, Paragraphs, etc. and specialised structures such as Figures, Tables, Lists, List-items, and more [1,5, §14.8.4]. Each **structure node** is a dictionary of type /StructElem having keys /S for the structure type, /K an array of links to any child nodes (or Kids) including *marked content* items, and /P an indirect reference to the parent node. Optionally there can be a /Pg key specifying an indirect reference to a /Page dictionary, when this cannot be deduced from the parent or higher ancestor. Also, the /A key can be used to specify attributes for the structure tag when the document's contents are exported in various formats; e.g., using 'Save As Other ... XML 1.0' export from Adobe's 'Acrobat Pro' browser/editor. Fig. 1a shows the MathML tagging of some inline mathematical content. The tree structure is indicated with lines connecting nodes to their kids; reverse links to parents are

---

[6] Henceforth we use the term 'attribute', rather than 'property'.

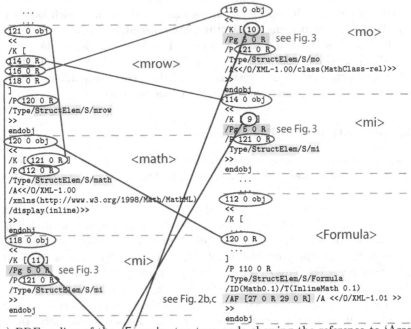

(a) PDF coding of the /Formula structure node showing the reference to 'Associated Files' via the /AF key in object 112. The indirect references (27 and 29) correspond to /Filespec dictionaries, as shown in Fig. 2. (In the coding '...' indicates parts omitted due to not being relevant to this structure; these portions are discussed in Sect. 4.) The corresponding *marked content* is specified via the /K [...] numbers (9, 10, 11) in the child structure nodes; i.e., objects 114 (<mi>), 116 (<mo>) and 118 (<mi>), which are children of object 121 (<mrow>) under object 120 (<math>).

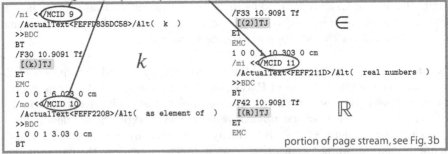

(b) Portion of the PDF page content stream showing the /MCID numbers (9, 10, 11) of the actual content portions of the mathematical expression. These correspond to leaf-nodes of the structure tree as presented in part (a).

Fig. 1. PDF coding for portions of (a) the structure tree and (b) the page content stream, corresponding to the mathematics shown as selected in Fig. 1a. It is through the /MCID numbers that the association is made to the /Formula structure tag for the corresponding piece of mathematical content.

not drawn, as this would unduly clutter the diagram. Other keys, such as /ID and /T can provide an identifier and title, for use primarily in editing software to locate specific nodes within appropriately ordered listings.

The link between structure and *marked content* (as leaf-nodes to the structure tree, say) is established using the /MCID number attribute. A numeric integer entry in the /K Kids array corresponds to an /MCID number occurring within the contents stream for that page specified via a /Pg entry, either of the structure node itself or the closest of its ancestors having such a key. Fig. 1b shows this linking via /MCID with lines drawn to the corresponding structure nodes shown in Fig. 1a. The interplay of structure with content was addressed in the author's paper [10], with Figure 1 of that paper giving a schematic view of the required PDF structural objects.

3 'Associated Files', Carrying LATEX and MathML Views of Mathematical Content

There are several ways in which file attachments may be associated with specific portions of a PDF document, using the 'Associated Files' technique [4, Annex E]. The file is embedded/attached and then associated, by a method, either to:

(i)   the document as a whole [4, §E.3], [6, §14.13.2] — e.g. the full LATEX source, or preamble file used when converting snippets of mathematical content into a MathML presentation of the same content;

(ii) a specific page within the document [4, §E.4], [6, §14.13.3] or to a (perhaps larger) logical document part using PDF 2.0 [6, §14.13.7];

(iii) graphic objects in a content stream [4, §E.5], [6, §14.13.4] — when structure is available, this is not the preferred method[7];

(iv) a structure node [4, §E.7], [6, §14.13.5] such as /Figure, /Formula, /Div, etc.

(v)   an /XObject [4, §E.6], [6, §14.13.6] such as an included image of a formula or other mathematical/technical/diagrammatic content;

(vi) an annotation [4, §E.8] — but this method can be problematic with regard to validation for PDF/A [4, §6.3], and PDF/UA [7, §7.18] standards[8].

Fig. 2a shows how attachments are presented within a separate panel of a browser window, using information from an array of filenames; see Fig. 2b. This is independent of the page being displayed, so the array must be referenced from the

---

[7] In the PDF/A-3 specifications [4, §E.5] the final paragraph explicitly states "When writing a PDF, the use of structure (and thus associating the /AF with the structure element, see [4, §E.7]) is preferred instead of the use of explicit marked content." with a corresponding statement also in [6, §14.13.4].

[8] The method of indicating an attachment with a 'thumb tack' annotation located at a specific point within a document, is deprecated in the PDF/A-3 standard, as it does not provide a proper method to associate with the portion of content. Besides, the appearance of such thumb-tacks all over paragraphs containing inline mathematics is, well, ... downright ugly.

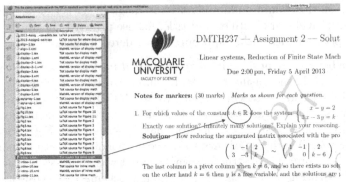

(a) Listing of attachments, indicating how one is associated to some inline content.

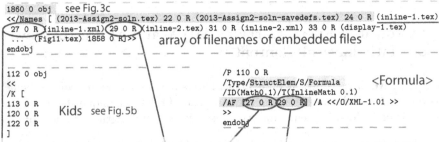

(b) A portion of the PDF coding of the /Names array (upper) for embedded files, and coding of the /Formula structure element (lower), showing the /AF key with array of two 'Associated' files given as indirect references. The /Names array is used to produce the listing in (a) and provides indirect links to /Filespec entries, as shown in (c).

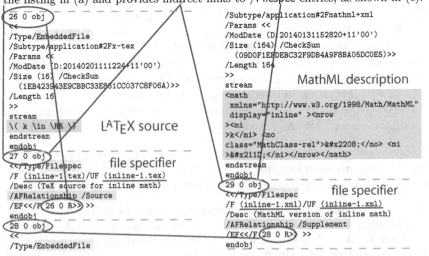

(c) PDF source of /Filespec dictionaries and /EmbeddedFile objects which hold the streams of L^ATEX and MathML coding for the mathematical content indicated in (a).

Fig. 2. Embedded files associated with a /Formula structure element

document level. This is seen in Fig. 3c using the /Names key of the /Catalog dictionary, which references object 2080, whose /EmbeddedFiles key then references the filenames array (object 1860 in Fig. 2b). One can also see in Fig. 2b how each filename precedes an indirect reference to the /Filespec dictionary [6, §7.11.3, Tables 44 and 45] for the named file; see Fig. 2c. This dictionary contains a short description (/Desc) of the type of content as well as the filename to use on disk, and a link via the /EF key to the actual EmbeddedFile stream object.

That a file is 'Associated' is indicated by the /AFRelationship key, whose value is a *PDF name* indicating how the file is related to visible content. Options here are /Source as used with the LATEX source coding, or /Supplement as used with the MathML description. Other possibilities are /Data (e.g., for tabular data) and /Alternative for other representations such as audio, a movie, projection slides or anything else that may provide an alternative representation of the same content. /Unspecified is also available as a non-specific catch-all.

Not all attachments need be 'Associated' and conversely not all 'Associated Files' need be displayed in the 'Attachments' panel, so there is another array (object 1859) as shown in Fig. 3c, linked to the /MarkInfo sub-dictionary of the /Catalog dictionary. Files associated with the document as a whole, as in method (i) above, link via the /AF key in the /Catalog dictionary (see Fig. 3c).

For the LATEX source of a mathematical expression method (iv) is preferred, provided structure tagging is present within the PDF. This is discussed below in Sect. 3.1. Method (iii) also works, provided the expression is built from content confined to a single page. This is described in Sect. 3.2.

As 'Associated Files' have only been part of published PDF/A standards [4] since late 2012, it may be some time before PDF readers provide a good interface for 'Associated Files', beyond using the 'Attachments' pane. This ought to include interfaces to view the contents of attached files, do searching within the files, and make the file's contents available to assistive technology. One possible way to display this association is apparent in earlier work [11], whereby a bounding rectangle appears as the mouse enters the appropriate region.

3.1 Embedded Files Associated with Structure

With an understanding of how structure tagging works, as in Sect. 2.1, then associating files to structure is simply a matter of including an /AF key in the structure node's dictionary, as shown in Figures 1a and 2b. The value for this key is an array of indirect references to /Filespec objects for the relevant files.

There is nothing in the content stream in Fig. 1b to indicate that there is a file associated with this structure node. Rather the browser, knowing that 'Associated' files are present, needs to have gone through some pre-processing to first locate the node (if any) to which it is associated, then trace down the structure tree to the deepest child nodes (objects 114, 116, 118). From their /K entries (viz., 9, 10, 11 resp.), the relevant *marked content* in the page's contents stream is located using these /MCID numbers.

```
/AF /inline-1 BDC
1 0 0 1 158.485 0 cm
...
/mi <</MCID 9        see Fig. 1b
    /ActualText<FEFFD835DC58>/Alt( k )
>>BDC
...                        k
EMC
...
1 0 0 1 10.303 0 cm
```

```
/mi <</MCID 11    see Fig. 1b
    /ActualText<FEFF211D>/Alt( real numbers )
>>BDC
BT
/F42 10.9091 Tf
    [(R)]TJ              ℝ
ET
EMC
...
EMC              portion of page stream, see Fig. 3b
```

(a) A portion of the PDF content stream associating content with embedded files for the mathematical expression indicated in Fig. 2a, using a *marked content* tag /AF to refer to a *named resource* /inline-1. All content down to the final EMC is associated.

dictionary of named resources — associated file arrays
```
20 0 obj
<< /inline-1 [27 0 R 29 0 R] /inline-2 [31 0 R 33 0 R] /display-1 [35 0 R 37 0 R]
   /inline-3 [39 0 R 41 0 R] /inline-4 [43 0 R 45 0 R] /inline-5 [47 0 R 49 0 R]
   ... /inline-31 [1475 0 R 1477 0 R] >>
```

```
90 0 obj
<<
/Properties 20 0 R
/Font << /F75 97 0 R /F79 100 0 R
   /F77 102 0 R /F45 105 0 R /F78 106 0 R
   ... >>
/XObject << /Im1 25 0 R >>
/ProcSet [ /PDF /Text ]
>>
endobj              Page Resources
```

```
5 0 obj
<<
/Type /Page
/Contents 91 0 R       reference to page stream
/Resources 90 0 R
/MediaBox [0 0 595.276 841.89]
/Tabs/S
/Parent 773 0 R
/StructParents 0
>>              Page dictionary
endobj
```

(b) A portion of the /Properties dictionary (upper, object 20) which is linked to a /Page object (lower right, object 5) via its /Resources key (see lower left, object 90). Thus a name (such as /inline-1) is associated with an array of /Filespec references (viz. [27 0 R 29 0 R]), which lead to the LaTeX and MathML files seen in Fig. 2c.

```
2080 0 obj    dictionary of name types
<<
/Dests 2079 0 R
/EmbeddedFiles 1860 0 R    see Fig. 2b
>>
endobj
2081 0 obj
<<
/Type /Catalog
/Pages 773 0 R         Document Catalog
/Names 2080 0 R
/ViewerPreferences <</DisplayDocTitle true >>
/OutputIntents [ << /Type /OutputIntent
```

```
/S/GTS_PDFA1 /DestOutputProfile 1 0 R
/OutputConditionIdentifier
  (sRGB_IEC61966-2-1_no_black_scaling)  /Info
  (sRGB IEC61966 v2.1 without black scaling) >> ]
/Metadata 2 0 R/Lang (en-US)
/PageMode/UseOutlines
/MarkInfo <</Marked true /AF 1859 0 R>>
/AF [ 22 0 R 24 0 R]
/PageLabels<</Nums[0<</P(1)>> ... ]>>
/OpenAction 4 0 R
/StructTreeRoot 95 0 R
>>
endobj
```

array of references to Associated embedded files
```
1859 0 obj
[ 27 0 R 29 0 R 31 0 R 33 0 R 35 0 R 37 0 R 39 0 R 41 0 R 43 0 R 45 0 R 47 0 R 49 0 R 51 0 R
  53 0 R 55 0 R 57 0 R 59 0 R 61 0 R 63 0 R 65 0 R 67 0 R 69 0 R 71 0 R 73 0 R 75 0 R 77 0 R
  ... 1850 0 R]
```

(c) The document's /Catalog (object 2081) indicates presence of embedded files via the /Names key (object 2080). This references the array (object 1860 in Fig. 2b), to establish the correspondence between filenames and /Filespec dictionaries. Embedded files which are 'Associated' to content portions are also listed in an array (object 1859) referenced from the /AF key in the /MarkInfo dictionary.

Fig. 3. Embedded files associated with specific content

3.2 Embedded Files Associated with Content

With an understanding of how content tagging works, as in Sect. 2.1, and the fact that *marked content* operators may be nested, then associating files to content is also quite simple. One simply uses an /AF tag within the page's content stream with BDC ... EMC surrounding the content to be marked, as shown in Fig. 3a. This employs the *named resource* variant (here /inline-1) to indicate the array of 'Associated' files. Fig. 3b shows how this name is used as a key (in dictionary object 20) having as value an array of indirect references to /Filespec objects (27 and 29). These resources can be specific to a particular page dictionary (object 5), but in the example document[1] the *named resources* are actually made available to all pages, since this accords with not including multiple copies of files when a mathematical expression is used repeatedly.

Finally Fig. 3c shows the coding required when embedded files, some of which may also be associated to content or structure, are present within a PDF document. One sees that the array (object 1859) of indirect object references in the lower part of Fig. 3c refer to the same /Filespec objects (27 and 29) as the *named resources* (object 20) in the upper part of Fig. 3b. These are the same references using /AF keys seen in Fig. 1b and Fig. 2b to the objects themselves in Fig. 2c.

This mechanism makes it easier for a PDF reader to determine that there are files associated to a particular piece of content, by simply encountering the /AF tag linked with a *named resource*. This should work perfectly well with a PDF file that is not fully tagged for structure. However, if the content is extended (e.g., crosses a page-boundary) then it may be harder for a PDF writer to construct the correct content stream, properly tagging two or more portions.

4 Access-Tags: Attaching LATEX Source to 'Fake' Spaces

A third method allows inclusion of the LATEX source of mathematics so that it may be readily extracted, using just the usual Select/Copy/Paste actions. This works with some existing PDF reader applications, including the freely available 'Adobe Reader'. It is achieved by making use of the /ActualText attribute [5, §14.9.4] for a piece of *marked content*, whether or not structure tagging is present. It can be done by existing PDF-writing software that supports tagging of content, as in Sect. 2.1, and specification of a value for the /ActualText attribute.

Fig. 4 shows how this works, by tagging a 'fake space' character immediately before mathematical content, and another immediately afterwards. By selecting (see Fig. 4a) then Copy/Paste the content into text-editing software, the result should be similar to Fig. 4b. The PDF reader must recognise[9] /Actual-Text and replace the copied content (e.g. a single font character) with its value. The *PDF name* /AccessTag tags a single 'space' character [( )]Tf as *marked content*, having replacement text in the /ActualText attribute; see Fig. 5a. An

---

[9] Adobe's 'Reader' and 'Acrobat Pro' certainly do, along with other software applications, but Apple's standard PDF viewers currently do not support /ActualText.

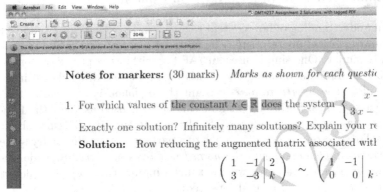

(a) Selection across mathematical content

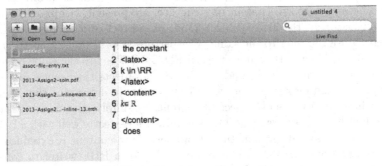

(b) Pasted text from the selection

(c) Access-tags selected in the 'Tags' tree, to show 'fake spaces'.

Fig. 4. This shows how the selection in (a), when copied and pasted into a text file, recovers the LATEX source (b) that was used to specify the visual appearance of the mathematical content. In (c) we see structure within a /Formula node, (see also Fig. 5b) with leaf-nodes of /accesstag structure nodes being *marked content* of type /AccessTag. This consists of a single space character carrying an /ActualText attribute which holds the replacement text; as seen explicitly in the coding shown in Fig. 5a. The 'fake spaces' are very narrow; when selected they can be seen very faintly in (c) within the ovals indicated, at the outer edge of the the bounding rectangles of the outermost math symbols.

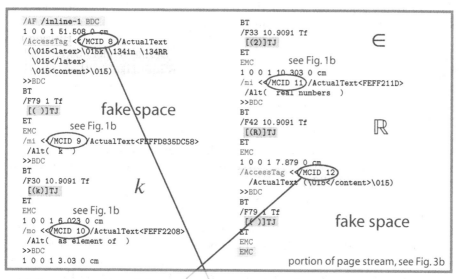

(a) Complete portion of the content stream corresponding to the mathematics shown as selected in Fig. 4a. This is the same content as in Figures 1b and 3a but with the '...' parts there now showing the /AccessTag coding of a 'fake space' with /ActualText attribute. The /AF ... BDC ... EMC wrapping of Fig. 3a is also shown. Being part of the document's content, these space characters are also assigned /MCID numbers to be linked to structure nodes, as in (b) below.

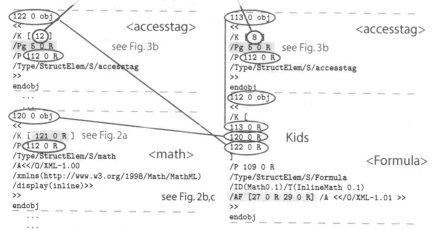

(b) Portion of the structure tree as in Fig. 1a, but now showing how the 'fake spaces' can be linked to structure nodes, here /accesstag. The missing portions of Fig. 1a, indicated there by '...' are now filled-in, but leaving out other parts whose purpose has already been explained. Fig. 4c, shows the tagging opened out within the 'Tags' navigation panel, with the /accesstag structure nodes selected.

Fig. 5. File content included as /ActualText for a 'fake space', which itself can be tagged as *marked content* linked to an /accesstag structure node.

/MCID number is not needed for this technique to work; these *are* included in the example document[1] which is fully tagged[10].

We refer to these tagged 'fake spaces' as 'Access-tags', since a motivation for their use is to allow Assistive Technology (e.g., a Braille reader) access to the LaTeX source of mathematical content. The spaces are 'fake' in the sense that they are just 1pt in height and have nearly zero width. This makes them hard to select by themselves, but nearly impossible to separate from the mathematical content which they accompany; see Fig. 4c. They act as ordinary spaces when copied, but this is substituted with the /ActualText replacement, if supported. Another aspect of their 'fakeness' is that they take no part in the typesetting, when using pdf-LaTeX (post-2014). Of course the same idea could be implemented in different ways with different PDF-producing software.

One places into the initial 'Access-tag' text of the LaTeX source — with care given to encode special characters[11] — as the value of its /ActualText attribute. The source coding is preceded by the string <latex> and followed by </latex> and <content>, with return-characters (in octal \015) to allow these 'delimiters' to ultimately copy onto lines by themselves. The trailing 'Access-tag' just takes </content>. As a result, the eventual Paste gives text as shown in Fig. 4b.

Assistive Technology (e.g., a Braille reader) works either by (a) emulating Copy/Paste of on-screen portions of the document's window; or (b) by directly accessing the 'Accessible Text' view of the PDF document's contents. In both cases the /ActualText replacements are extracted. (The 'Accessible Text' view can be exported directly using Adobe's 'Acrobat Pro' software, see [12].) For mathematical symbols, where Figures 1b, 3a and 5a show the presence of both /Alt and /ActualText attributes, then the /Alt contributes to the 'Accessible Text', whereas /ActualText supplies what is copied to the Clipboard for Copy/Paste. In either case, a human reader when encountering <latex> on a line by itself can choose whether to read (or listen to) the following lines of LaTeX coding, else use a Find action to skip down to where the next </latex> occurs. This is followed by a line containing <content>. Similarly the human can read/listen or skip down to where </content> occurs.

Acknowledgements. The author wishes to acknowledge James Davenport and Emma Cliffe, at the University of Bath, for valuable discussions regarding the needs of mathematics students having a visual disability. The 'fake space' idea emerged as a result. Thanks also to reviewers for suggesting, among other ideas, inclusion of the PDF language overview as in Sect. 2, allowing later sections to be written more succinctly. Finally, I wish to thank members of ISO TC 171 for encouragement and support regarding 'fake spaces' and other aspects.

"The committee is really happy to have someone actually implementing math accessibility in PDF ..." [15]   — Neil Soiffer, Senior Scientist, Design Science Inc., 12 April 2014.

[10] In PDF 2.0 this will also need an association of /accesstag to /Custom within the /RoleMap dictionary. The /AccessTag *PDF name* can be replaced by /Span.
[11] Octal codes: \134 for backslash, \050 for '(' and \051 for ')', \015 for line-end.

References

1. Adobe Systems Inc.; PDF Reference 1.7 (November 2006), Also available as [5], http://www.adobe.com/devnet/pdf/pdf_reference.html
2. ISO 19005-1:2005; Document Management — Electronic document file format for long term preservation — Part 1: Use of PDF 1.4 (PDF/A-1); Technical Committee ISO/TC 171/SC 2 (September 2005). Revisions via Corrigenda: ISO19005-1:2005/Cor 1:2007 (March 2007); ISO 19005-1:2005/Cor 2:2011 (December 2011), http://www.iso.org/iso/catalogue_detail?csnumber=38920
3. ISO 19005-2:2011; Document Management — Electronic document file format for long term preservation — Part 2: Use of ISO 32000-1 (PDF/A-2); Technical Committee ISO/TC 171/SC 2 (June 2011), http://www.iso.org/iso/catalogue_detail?csnumber=50655
4. ISO 19005-3:2012; Document Management — Electronic document file format for long term preservation — Part 3: Use of ISO 32000-1 with support for embedded files (PDF/A-3); Technical Committee ISO/TC 171/SC 2 (October 2012), http://www.iso.org/iso/catalogue_detail?csnumber=57229
5. ISO 32000-1:2008; Document management — Portable document format (PDF 1.7); Technical Committee ISO/TC 171/SC 2 (July 2008), Also available as [1], http://www.iso.org/iso/catalogue_detail?csnumber=51502
6. ISO 32000-2-20140220; Document management — Portable document format — Part 2: PDF 2.0; Technical Committee ISO/TC 171/SC 2, in draft form (February 2014)
7. Document management applications — Electronic document file format enhancement for accessibility — Part 1: Use of ISO 32000-1 (PDF/UA-1); Technical Committee ISO/TC 171/SC 2 (July 2012), http://www.iso.org/iso/home/store/catalogue_tc/catalogue_detail.htm?csnumber=54564
8. Technical Implementation Guide; AIIM Global Community of Information Professionals, http://www.aiim.org/Research-and-Publications/standards/committees/PDFUA/Technical-Implementation-Guide. Also available as [7]
9. Moore, R.R.: Tagged Mathematics in PDFs for Accessibility and other purposes. In: CICM-WS-WiP 2013, Workshops and Work in Progress at CICM. CEUR Workshops Proceedings (2013), http://ceur-ws.org/Vol-1010/paper-01.pdf
10. Moore, R.R.: Ongoing efforts to generate "tagged PDF" using pdfTEX. In: Sojka, P. (ed.) DML 2009, Towards a Digital Mathematics Library, Proceedings. Muni Press, Masaryk University (2009) ISBN 978-80-20-4781-5, Reprinted as: TUGboat 30(2), 170–175 (2009), http://www.tug.org/TUGboat/tb30-2/tb95moore.pdf
11. Moore, R.R.: serendiPDF, with searchable math-fields in PDF documents. TUGboat 23(1), 65–69 (2002), http://www.tug.org/TUGboat/tb23-1/moore.pdf
12. Moore, R.R.: DMTH237 Assignment 2 Solutions, with tagged PDF; web address for downloading the example document having attachments and 'fake spaces'; also other views of this document's contents exported using 'Acrobat Pro' software, http://rutherglen.science.mq.edu.au/~maths/CICM/
13. Raman, T.V.: An Audio View of (LA) TEXDocuments. TUGboat 13(3), 372–379 (1992), http://tugboat.tug.org/TUGboat/tb13-3/raman.pdf
14. Raman, T.V.: An Audio View of (LA)TEXDocuments–Part II. TUGboat 16(3), 310–314 (1995), http://tugboat.tug.org/TUGboat/tb16-3/tb48rama.pdf
15. Soiffer, N.: Posting to 'PDF-Access' mailing list server, April 12 (2014), http://listserv.aiim.org/scripts/wa.exe?A0=PDF-ACCESS

Which One Is Better: Presentation-Based or Content-Based Math Search?

Minh-Quoc Nghiem[1,4], Giovanni Yoko Kristianto[2],
Goran Topić[3], and Akiko Aizawa[2,3]

[1] Ho Chi Minh City University of Science, Vietnam
[2] The University of Tokyo, Japan
[3] National Institute of Informatics, Japan
[4] The Graduate University for Advanced Studies, Japan
{nqminh,giovanni,goran_topic,aizawa}@nii.ac.jp

Abstract. Mathematical content is a valuable information source and retrieving this content has become an important issue. This paper compares two searching strategies for math expressions: presentation-based and content-based approaches. Presentation-based search uses state-of-the-art math search system while content-based search uses semantic enrichment of math expressions to convert math expressions into their content forms and searching is done using these content-based expressions. By considering the meaning of math expressions, the quality of search system is improved over presentation-based systems.

Keywords: Math Retrieval, Content-based Math Search, MathML.

1 Introduction

The issue of retrieving mathematical content has received considerable critical attention [1]. Mathematical content is a valuable information source for many users and is increasingly available on the Web. Retrieving this content is becoming more and more important.

Conventional search engines, however, do not provide a direct search mechanism for mathematical expressions. Although these search engines are useful to search for mathematical content, these search engines treat mathematical expressions as keywords and fail to recognize the special mathematical symbols and constructs. As such, mathematical content retrieval remains an open issue.

Some recent studies have proposed mathematical retrieval systems based on the structural similarity of mathematical expressions [2–7]. However, in these studies, the semantics of mathematical expressions is still not considered. Because mathematical expressions follow highly abstract and also rewritable representations, structural similarity alone is insufficient as a metric for semantic similarity.

Other studies [8–13] have addressed semantic similarity of mathematical formulae, but this required content-based mathematical formats such as content MathML [14] and OpenMath [15]. Because almost all mathematical content available on the Web is presentation-based, these studies used two freely available toolkits, SnuggleTeX [16] and LaTeXML [17], for semantic enrichment of

S.M. Watt et al. (Eds.): CICM 2014, LNAI 8543, pp. 200–212, 2014.

mathematical expressions. However, much uncertainty remains about the relation between the performance of mathematical search system and the performance of the semantic enrichment component.

Based on the observation that mathematical expressions have meanings hidden in their representation, the primary goal of this paper is making use of mathematical expressions' semantics for mathematical search. To accomplish this problem of retrieving semantically similar mathematical expressions, we use the results of state-of-the-art semantic enrichment methods. This paper seeks the answers to two questions.

- What is the contribution of semantic enrichment of mathematical expressions to content-based mathematical search systems?
- Which one is better: presentation-based or content-based mathematical search?

To implement a *mathematical search system*, various challenges must be overcome. First, in contrast to text which is linear, mathematical expressions are hierarchical: operators have different priorities, and expressions can be nested. The similarity between two mathematical expressions is decided first by their structure and then by the symbols they contain [18, 19]. Therefore, current text retrieval techniques cannot be applied to mathematical expressions because they only consider whether an object includes certain words. Second, mathematical expressions have their own meanings. These meanings can be encoded using special markup languages such as Content MathML or OpenMath. A few existing mathematical search systems also make use of this information. Such markup, however, is rarely used to publish mathematical knowledge related to the Web [18]. As a result, we were only able to use presentation-based markup, such as Presentation MathML or TEX, for mathematical expressions.

This paper presents an approach to a *content-based mathematical search system* that uses the information from *semantic enrichment of mathematical expressions* system. To address the challenges described above, the proposed approach is described below. First, the approach used Presentation MathML markup, a widely used markup for mathematical expressions. This makes our approach more likely to be applicable in practice. Second, a *semantic enrichment of mathematical expressions* system is used to convert mathematical expressions to Content MathML. By getting the underlying semantic meanings of mathematical expressions, a *mathematical search system* is expected to yield better results.

The remainder of this paper is organized as follows. Section 2 provides a brief overview of the background and related work. Section 3 presents our method. Section 4 describes the experimental setup and results. Section 5 concludes the paper and points to avenues for future work.

2   Mathematical Search System

As the demand for mathematical searching increases, several mathematical retrieval systems have come into use [20]. Most systems use the conventional text search techniques to develop a new mathematical search system [2, 3]. Some systems use specific format for mathematical content and queries [4–7, 11]. Based on the markup schema they use, current mathematical search systems are divisible into presentation-based and content-based systems. Presentation-based systems deal with the presentation form whereas content-based systems deal with the meanings of mathematical formulae.

2.1 Presentation-Based Systems

Springer LaTeXSearch. Springer offers a free service, Springer LaTeX Search [3], to search for LaTeX code within scientific publications. It enables users to locate and view equations containing specific LaTeX code, or equations containing LaTeX code that is similar to another LaTeX string. A similar search in Springer LaTeX Search ranks the results by measuring the number of changes between a query and the retrieved formulae. Each result contains the entire LaTeX string, a converted image of the equation, and information about and links to its source.

MathDeX. MathDeX (formerly MathFind [21]) is a math-aware search engine under development by Design Science. This work extends the capabilities of existing text search engines to search mathematical content. The system analyzes expressions in MathML and decomposes the mathematical expression into a sequence of text-encoded math fragments. Queries are also converted to sequences of text and the search is performed as a normal text search.

Digital Library of Mathematical Functions. The Digital Library of Mathematical functions (DLMF) project at NIST is a mathematical database available on the Web [2, 22]. Two approaches are used for searching for mathematical formulae in DLMF. The first approach converts all mathematical content to a standard format. The second approach exploits the ranking and hit-description methods. These approaches enable simultaneous searching for normal text as well as mathematical content.

In the first approach [4], they propose a textual language, Textualization, Serialization and Normalization (TexSN). TeXSN is defined to normalize non-textual content of mathematical content to standard forms. User queries are also converted to the TexSN language before processing. Then, a search is performed to find the mathematical expressions that match the query exactly. As a result, similar mathematical formulae are not retrieved.

In the second approach [23], the search system treats each mathematical expression as a document containing a set of mathematical terms. The cited paper introduces new relevance ranking metrics and hit-description generation techniques. It is reported that the new relevance metrics are far superior to the conventional tf-idf metric. The new hit-descriptions are also more query-relevant and representative of the hit targets than conventional methods.

Other notable math search systems include Math Indexer and Searcher [24], EgoMath [25], and ActiveMath [26].

2.2   Content-Based Systems

Wolfram Function. The Wolfram Functions Site [8] is the world's largest collection of mathematical formulae accessible on the Web. Currently the site has 14 function categories containing more than three hundred thousand mathematical formulae. This site allows users to search for mathematical formulae from its database. The Wolfram Functions Site proposes similarity search methods based on MathML. However, content-based search is only available with a number of predefined constants, operations, and function names.

MathWebSearch. The MathWebSearch system [10, 12] is a content-based search engine for mathematical formulae. It uses a term indexing technique derived from an automated theorem proving to index Content MathML formulae. The system first converts all mathematical formulae to Content MathML markup and uses substitution-tree indexing to build the index. The authors claim that search times are fast and unchanged by the increase in index size.

MathGO! [9] proposed a mathematical search system called the MathGO! Search System. The approach used conventional search systems using regular expressions to generate keywords. For better retrieval, the system clustered mathematical formula content using K-Som, K-Means, and AHC. They did experiments on a collection of 500 mathematical documents and achieved around 70–100 percent precision.

MathDA. Yokoi and Aizawa [11] proposed a similarity search method for mathematical expressions that is adapted specifically to the tree structures expressed by MathML. They introduced a similarity measure based on Subpath Set and proposed a MathML conversion that is apt for it. Their experiment results showed that the proposed scheme can provide a flexible system for searching for mathematical expressions on the Web. However, the similarity calculation is the bottleneck of the search when the database size increases. Another shortcoming of this approach is that the system only recognizes symbols and does not perceive the actual values or strings assigned to them.

System of Nguyen et al. [13] proposed a math-aware search engine that can handle both textual keywords and mathematical expressions. They used Finite State Machine model for feature extraction, and representation framework captures the semantics of mathematical expressions. For ranking, they used the passive–aggressive on-line learning binary classifier. Evaluation was done using 31,288 mathematical questions and answers downloaded from Math Overflow [27]. Experimental results showed that their proposed approach can perform better than baseline methods by 9%.

3   Methods

The framework of our system is shown in Fig. 1. First, the system collects mathematical expressions from the web. Then the mathematical expressions are converted to Content MathML using the *semantic enrichment of mathematical expressions* system of Nghiem et. al [28]. Indexing and ranking the mathematical expressions are done using Apache Solr system [29] following the method described in Topić et. al [30]. When a user submits a query, the system also converts the query to Content MathML. Then the system returns a ranked list of mathematical expressions corresponding to the user's queries.

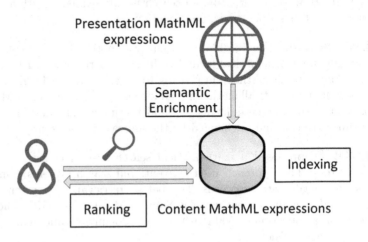

Fig. 1. System Framework

3.1 Data Collection

Performance analysis of a mathematical search system is not an easy task because few standard benchmark datasets exist, unlike other more common information retrieval tasks. Mathematical search systems normally build their own mathematical search dataset for evaluation by crawling and downloading mathematical content from the web. Direct comparison of the proposed approach with other systems is also hard because they are either unavailable or inaccessible.

Recently, simpler and more rapid tests of mathematical search system have been developed. The NTCIR-10 Math Pilot Task [1] was the initial attempt to develop a common workbench for mathematical expressions search. Currently, the NTCIR-10 dataset contains 100,000 papers and 35,000,000 mathematical expressions from ArXiv [31] which includes Content MathML markup. The task was completed as an initial pilot task showing a clear interest in the mathematical search. However, the Content MathML markup expressions are generated automatically using the LaTeXML toolkits. Therefore, this dataset is unsuitable to serve as the gold standard for the research described in the present paper.

As Wolfram Functions Site [8] is the only website that provides high-quality Content MathML markup for every expression, data for the search system was collected from this site. The Wolfram Functions Site data have numerous attractive features, including both Presentation and Content MathML markups, and category for each mathematical expression. In the experiment, the performance of *semantic enrichment of mathematical expressions* component will be compared directly with the system performance obtained using correct Content MathML expressions on Wolfram Functions Site data.

3.2 Semantic Enrichment of Mathematical Expressions

The mathematical expressions were preprocessed according to the procedure described in Nghiem et. al [28]. Given a set of training mathematical expressions in MathML parallel markup, rules of two types are extracted: segmentation rules and translation rules. These rules are then used to convert mathematical expressions from their presentation to their content form. Translation rules are used to translate (sub)trees of Presentation MathML markup to (sub)trees of Content MathML markup. Segmentation rules are used to combine and reorder the (sub)trees to form a complete tree.

After using mathematical expression enrichment system to convert the expressions into content MathML, we use these converted expressions for indexing. The conversion is not a perfect conversion, so there are terms that could not be converted. The queries submitted to the search system are also processed using the same conversion procedure.

3.3 Indexing

The indexing step was prepared by adapting the procedure used by Topić et. al [30]. This procedure used *pq*-gram-like indexing for Presentation MathML expressions. We modified it for use with Content MathML expressions. There are three fields used to encode the structure and contents of a mathematical expression: opaths, upaths, and sisters. Each expression is transformed into a sequence of keywords across several fields. opaths (ordered paths) field gathers the XML expression tree in vertical paths with preserved ordering. upaths (unordered paths) works the same as opaths without the ordering information. sisters lists the sister nodes in each subtree. Figure 2 presents an example of the terms used in the index of the expression $\sin(\frac{\pi}{8})$:$< apply >< sin/ >< apply >< times/ >< pi/ >< apply >< power/ >< cntype =$ "integer" $> 8 < /cn >< cntype =$ "integer" $> -1 < /cn >< /apply >< /apply >< /apply >$.

3.4 Searching

In the mathematical search system, users can input mathematical expressions using presentation MathML as a query. The search system then uses the *semantic enrichment of mathematical expressions* module to convert the input expressions to Content MathML. Figure 3 presents an example of the terms used in the query of the expression $\sin(\frac{\pi}{8})$. Matching is then performed using eDisMax, the default query parser of Apache Solr. Ranking is also done using the default modified TF/IDF scores and length normalization of Apache Solr.

```
opaths:
    1#1#apply 1#1#1#sin 1#1#2#apply 1#1#2#1#times 1#1#2#2#pi
    1#1#2#3#apply 1#1#2#3#1#power 1#1#2#3#2#cn#8 1#1#2#3#3#cn#-1
opaths:
    1#apply 1#1#sin 1#2#apply 1#2#1#times 1#2#2#pi 1#2#3#apply
    1#2#3#1#power 1#2#3#2#cn#8 1#2#3#3#cn#-1
opaths:
    apply 1#sin 2#apply 2#1#times 2#2#pi 2#3#apply 2#3#1#power 2#3#2#cn#8
    2#3#3#cn#-1
opaths: sin
opaths: times
opaths: pi
opaths: apply 1#power 2#cn#8 3#cn#-1
opaths: power
opaths: cn#8
opaths: cn#-1
upaths:
    ##apply ###sin ###apply ####times ####pi ####apply #####power
    #####cn#8 #####cn#-1
upaths:
    #apply ##sin ##apply ###times ###pi ###apply ####power ####cn#8
    ####cn#-1
upaths:
    apply #sin #apply ##times ##pi ##apply ###power ###cn#8 ###cn#-1
upaths: sin
upaths: apply #times #pi #apply ##power ##cn#8 ##cn#-1
upaths: times
upaths: pi
upaths: apply #power #cn#8 #cn#-1
upaths: power
upaths: cn#8
upaths: cn#-1
sisters: power cn#8 cn#-1
sisters: times pi apply
sisters: sin apply
sisters: apply
```

Fig. 2. Index terms of the expression $\sin\left(\frac{\pi}{8}\right)$

4   Experimental Results

4.1   Evaluation Setup

We collected mathematical expressions for evaluation from the Wolfram Function Site. At the time collected, there were more than 300,000 mathematical expressions on this site. After collection, we filtered out long expressions containing more than 20 leaf nodes to speed up the semantic enrichment because the processing time increases exponentially with the length of the expressions.

opaths:

```
1#1#apply 1#1#1#sin 1#1#2#apply 1#1#2#1#times 1#1#2#2#pi
1#1#2#3#apply 1#1#2#3#1#power 1#1#2#3#2#cn#8 1#1#2#3#3#cn#-1
```

upaths:

```
##apply ###sin ###apply ####times ####pi ####apply #####power
#####cn#8 #####cn#-1
```

upaths:

```
#apply ##sin ##apply ###times ###pi ###apply ####power ####cn#8
####cn#-1
```

```
sisters: power cn#8 cn#-1
sisters: times pi apply
sisters: sin apply
sisters: apply
```

Fig. 3. Query terms of the expression $\sin(\frac{\pi}{8})$

The number of mathematical expressions after filtering is approximately 20,000. Presumably, this number is adequate for evaluating the mathematical search system.

Evaluation was done by comparing three systems:

- Presentation-based search with Presentation MathML (PMathML): indexing and searching are based on the Presentation MathML expressions.
- Content-based search with semantic enrichment (SE): indexing and searching are based on the Content MathML expressions. The Content MathML expressions are extracted automatically using semantic enrichment module.
- Content-based search with correct Content MathML (CMathML): indexing and searching are based on the Content MathML expressions. The Content MathML expressions are those from the Wolfram Function Site.

We used the same data to train the semantic enrichment module by 10-fold cross validation method. The data is divided into 10 folds. The semantic enrichment result of each fold was done by using the other 9 folds as training data.

4.2   Evaluation Methodology

We used "Precision at 10" and "normalized Discounted Cumulative Gain" metrics to evaluate the results. In a large-scale search scenario, users are interested in reading the first page or the first three pages of the returned results. "Precision at 10" (P@10) has the advantage of not requiring the full set of relevant mathematical expressions, but its salient disadvantage is that it fails to incorporate consideration of the positions of the relevant expressions among the top k. In a ranked retrieval context, normalized Discounted Cumulative Gain (nDCG) as given by Equation 1 is a preferred metric because it incorporates the order of the retrieved expressions. In Equation 1, Discounted Cumulative Gain (DCG) can be calculated using the Equation 2, where rel_i is the graded relevance of the result at position i. Ideal DCG (IDCG) is calculable using the same equation, but IDCG uses the ideal result list which was sorted by relevance.

$$\text{nDCG}_\text{p} = \frac{DCG_p}{IDCG_p} \tag{1}$$

$$\text{DCG}_\text{p} = rel_1 + \sum_{i=2}^{p} \frac{rel_i}{\log_2(i)} \tag{2}$$

For performance analysis of the mathematical search system, we manually created 15 information needs (queries) and used them as input queries of our mathematical search system. The queries are created based on NTCIR queries with minor modification. Therefore, the search system always gets at least one exact match. Table 1 shows the queries we used. The top 10 results of each query were marked manually as relevant ($rel = 1$), non-relevant ($rel = 0$), or partially relevant ($rel = 0.5$). The system then calculates P@10 and an nDCG value based on the manually marked results.

Table 1. Queries

No.	Query
1	$\int_0^\infty x\,dx$
2	$x^2 + y^2$
3	$\int_0^\infty e^{-x^2}\,dx$
4	$arcsin(x)$
5	k^2
6	$\frac{\cosh ez + \sinh ez}{e}$
7	$\mathcal{R}_z \Psi^\nu(z), \tilde{\infty}$
8	$\int \frac{a^{d+bz}}{z} dz$
9	$\lim_{\nu \to \infty} \frac{L_{\alpha+\nu}}{L_\nu}$
10	$\mathcal{BP}_z \mathfrak{B}_\nu^\mu(z)$
11	$\nu \in \mathbb{N}$
12	$\Psi^\nu(z)$
13	$\log(z+1)$
14	$H_n(z)$
15	$\frac{1}{\pi} \int_0^\pi (\cos tn - z \sin t)dt$

4.3   Experimental Results

Comparisons among the three systems were made using P@10 and nDCG scores. Table 2 and figure 5 show the P@10 and nDCG scores obtained from the search. Figure 4 depicts the top 10 precision of the search system. The x axis shows the k number, which ranges from 1 to 10. The y axis shows the precision score. The precision score decreased, while k increased, which indicates that the higher results are more relevant than lower results.

Table 2. nDCG and P@10 scores of the search systems

Method	nDCG	P@10
PMathML	0.941	0.707
CMathML	0.962	0.747
SE	0.951	0.710

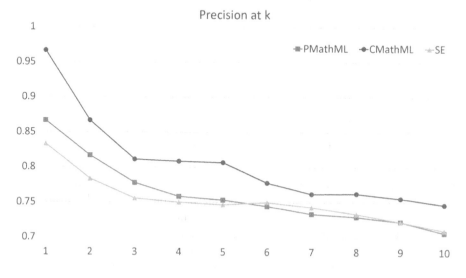

Fig. 4. Top 10 precision of the search system

In the experiment, a strong relation between *semantic enrichment of math-ematical expressions* and *content-based mathematical search system* was found. As shown in Nghiem et. al [28], the error rate of *semantic enrichment of math-ematical expressions* module is around 29 percent. With current performance, using this module for the mathematical search system still improves the search performance. The system gained 1 percent in nDCG score and 0.3 percent in P@10 score compared to the Presentation MathML-based system. Overall, the system using perfect Content MathML yielded the highest results. In direct com-parison using nDCG scores, the system using semantic enrichment is superior to the Presentation MathML-based system, although not by much. Out of 15 queries, the semantic enrichment system showed better results than Presenta-tion MathML-based system in 7 queries, especially when the mathematical sym-bols contain specific meanings, e.g. Poly-Gamma function (query 10), Hermite-H function (query 14). In case the function has specific meaning but there is no ambiguity representing the function, e.g. Legendre-Q function (query 12), both systems give similar results. Presentation MathML system, however, produced better results than semantic enrichment systems in 5 queries when dealing with elementary functions (query 2, 8, 15), logarithm (query 13), and trigonomet-ric functions (query 6) because of its simpler representation using Presentation

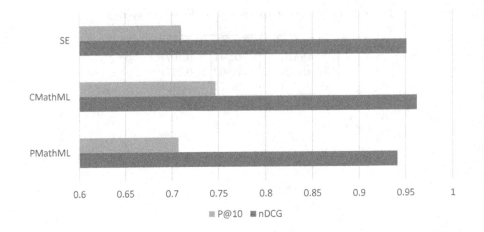

Fig. 5. Comparison of different systems

MathML. One exception is the case of query 4, when there is more than one way to represent an expression with a specific meaning, e.g. sin^{-1} and $arcsin$, Presentation MathML system gives unstable results.

This finding, while preliminary, suggests that we can choose either search strategy depending on the situation. We can use Presentation MathML system for elementary functions or when there is no ambiguity in the Presentation MathML expression. Otherwise, we can use a Content MathML system while dealing with functions that contain specific meanings. Another situation in which we can use a Content MathML system is when there are many ways to present an expression using Presentation MathML markup.

The average time for searching for a mathematical expression is less than one second on our Xeon 32 core 2.1 GHz 32 GB RAM server. The indexing time, however, took around one hour for 20,000 mathematical expressions. Because of the unavailability of standard corpora to evaluate content-based mathematical search systems, the evaluation at this time is quite subjective and limited. Although this study only uses 20,000 mathematical expressions for the evaluation, the preliminary experimentally obtained results indicated that the semantic enrichment approach showed promise for content-based mathematical expression search.

5 Conclusion

By using semantic information obtained from semantic enrichment of mathematical expressions system, the content-based mathematical search system has shown promising results. The experimental results confirm that semantic information is helpful to the mathematical search. Depending on the situation, we can choose to use either presentation-based or content-based strategy for searching. However, this is only a first step; many important issues remain for future studies. Considerably more work will need to be done using a larger collection of queries. In addition, there are many other valuable features that are worth considering besides the semantic markup of an expression, such as the description of the formula and its variables.

Acknowledgments. This work was supported by JSPS KAKENHI Grant Numbers 2430062, 25245084.

References

1. Aizawa, A., Kohlhase, M., Ounis, I.: NTCIR-10 Math pilot task overview. In: National Institute of Informatics Testbeds and Community for Information access Research 10 (NTCIR-10), pp. 654–661 (2013)
2. National Institute of Standards and Technology: Digital library of mathematical functions, http://dlmf.nist.gov (visited on March 01, 2014)
3. Springer: Springer LaTeX Search, http://www.latexsearch.com/ (visited on March 01, 2014)
4. Youssef, A.S.: Information search and retrieval of mathematical contents: Issues and methods. In: The ISCA 14th International Conference on Intelligent and Adaptive Systems and Software Engineering, pp. 100–105 (2005)
5. Altamimi, M.E., Youssef, A.S.: A math query language with an expanded set of wildcards. Mathematics in Computer Science 2, 305–331 (2008)
6. Youssef, A.S., Altamimi, M.E.: An extensive math query language. In: SEDE, pp. 57–63 (2007)
7. Miner, R., Munavalli, R.: An approach to mathematical search through query formulation and data normalization. In: Kauers, M., Kerber, M., Miner, R., Windsteiger, W. (eds.) MKM/CALCULEMUS 2007. LNCS (LNAI), vol. 4573, pp. 342–355. Springer, Heidelberg (2007)
8. Wolfram: The Wolfram Functions Site, http://functions.wolfram.com/ (visited on March 01, 2014)
9. Adeel, M., Cheung, H.S., Khiyal, S.H.: Math go! prototype of a content based mathematical formula search engine. Journal of Theoretical and Applied Information Technology 4(10), 1002–1012 (2008)
10. Kohlhase, M., Sucan, I.: A search engine for mathematical formulae. In: Calmet, J., Ida, T., Wang, D. (eds.) AISC 2006. LNCS (LNAI), vol. 4120, pp. 241–253. Springer, Heidelberg (2006)
11. Yokoi, K., Aizawa, A.: An approach to similarity search for mathematical expressions using mathml. In: 2nd Workshop Towards a Digital Mathematics Library, DML 2009, pp. 27–35 (2009)
12. Kohlhase, M., Prodescu, C.C.: Mathwebsearch at NTCIR-10. In: National Institute of Informatics Testbeds and Community for Information access Research 10 (NTCIR-10), pp. 675–679 (2013)
13. Nguyen, T.T., Chang, K., Hui, S.C.: A math-aware search engine for math question answering system. In: Proceedings of the 21st ACM International Conference on Information and Knowledge Management (CIKM 2012), pp. 724–733 (2012)
14. Ausbrooks, R., Buswell, S., Carlisle, D., Chavchanidze, G., Dalmas, S., Devitt, S., Diaz, A., Dooley, S., Hunter, R., Ion, P., et al.: Mathematical markup language (MathML) version 3.0. W3C recommendation. World Wide Web Consortium (2010)

15. Buswell, S., Caprotti, O., Carlisle, D.P., Dewar, M.C., Gaetano, M., Kohlhase, M.: The openmath standard. Technical report, version 2.0. The Open Math Society (2004)
16. McKain, D.: SnuggleTeX version 1.2.2, http://www2.ph.ed.ac.uk/snuggletex/ (visited on March 01, 2014)
17. Miller, B.R.: LaTeXML a LaTeX to XML converter, http://dlmf.nist.gov/LaTeXML/ (visited on March 01, 2014)
18. Kamali, S., Tompa, F.W.: Improving mathematics retrieval. In: 2nd Workshop Towards a Digital Mathematics Library, pp. 37–48 (2009)
19. Kamali, S., Tompa, F.W.: Structural similarity search for mathematics retrieval. In: Carette, J., Aspinall, D., Lange, C., Sojka, P., Windsteiger, W. (eds.) CICM 2013. LNCS (LNAI), vol. 7961, pp. 246–262. Springer, Heidelberg (2013)
20. Zanibbi, R., Blostein, D.: Recognition and retrieval of mathematical expressions. IJDAR 15, 331–357 (2012)
21. Munavalli, R., Miner, R.: Mathfind: a math-aware search engine. In: Proceedings of the 29th Annual International ACM SIGIR Conference on Research and Development in Information Retrieval, p. 735. ACM (2006)
22. Miller, B.R., Youssef, A.S.: Technical aspects of the digital library of mathematical functions. Annals of Mathematics and Artificial Intelligence 38, 121–136 (2003)
23. Youssef, A.S.: Methods of relevance ranking and hit-content generation in math search. In: Kauers, M., Kerber, M., Miner, R., Windsteiger, W. (eds.) MKM/Calculemus 2007. LNCS (LNAI), vol. 4573, pp. 393–406. Springer, Heidelberg (2007)
24. Sojka, P., Líška, M.: The Art of Mathematics Retrieval. In: Proceedings of the ACM Conference on Document Engineering, DocEng 2011, Mountain View, CA, pp. 57–60. Association of Computing Machinery (2011)
25. Mišutka, J., Galamboš, L.: System description: EgoMath2 as a tool for mathematical searching on wikipedia.org. In: Davenport, J.H., Farmer, W.M., Urban, J., Rabe, F. (eds.) Calculemus/MKM 2011. LNCS, vol. 6824, pp. 307–309. Springer, Heidelberg (2011)
26. Siekmann, J.: Activemath, http://www.activemath.org/eu/ (visited on March 01, 2014)
27. MathOverflow: Math overflow, http://mathoverflow.net/ (visited on March 01, 2014)
28. Nghiem, M.Q., Kristianto, G.Y., Aizawa, A.: Using mathml parallel markup corpora for semantic enrichment of mathematical expressions. Journal of the Institute of Electronics, Information and Communication Engineers E96-D(8), 1707–1715 (2013)
29. Apache: Apache solr, http://lucene.apache.org/solr/ (visited on March 01, 2014)
30. Topic, G., Kristianto, G.Y., Nghiem, M.Q., Aizawa, A.: The MCAT math retrieval system for NTCIR-10 Math track. In: National Institute of Informatics Testbeds and Community for Information access Research 10 (NTCIR-10), pp. 680–685 (2013)
31. Cornell University Library: arxiv, http://arxiv.org/ (visited on March 01, 2014)

POS Tagging and Its Applications for Mathematics

Text Analysis in Mathematics

Ulf Schöneberg and Wolfram Sperber

FIZ Karlsruhe/Zentralblatt MATH, Franklinstr. 11, 10587 Berlin, Germany

Abstract. Content analysis of scientific publications is a nontrivial task, but a useful and important one for scientific information services. In the Gutenberg era it was a domain of human experts; in the digital age many machine-based methods, e.g., graph analysis tools and machine-learning techniques, have been developed for it. Natural Language Processing (NLP) is a powerful machine-learning approach to semiautomatic speech and language processing, which is also applicable to mathematics. The well established methods of NLP have to be adjusted for the special needs of mathematics, in particular for handling mathematical formulae. We demonstrate a mathematics-aware part of speech tagger and give a short overview about our adaptation of NLP methods for mathematical publications. We show the use of the tools developed for key phrase extraction and classification in the database zbMATH.

1 Methods and Tools

1.1 Part of Speech Tagging and Noun Phrases

We describe our approach for Part Of Speech (POS) tagging and Noun Phrases (NPs) extraction of mathematical documents as basic tool for key phrase identification and classification of mathematical publications. NLP methods arising from the field of computer linguistics constitute a statistics- and rule-based machine-learning approach to the processing of speech and language. Natural language analysis and understanding is a central aim of NLP. In western languages, Noun Phrases (NPs) are the most significant parts of sentences. Extraction of the NPs and finding rules for which of them are relevant are the key aspects of automatic key phrase extraction in documents.

An important part of capturing NPs from a text is POS tagging. It presupposes the availability of information about the tokens in a sentence, especially the linguistic types of the tokens. Almost all state-of-the-art POS taggers rely on dictionaries. Some NLP tools are provided as Open Source software. In our project, the Stanford POS tagger [9] is used. We extended the dictionaries of the POS tagger with large amounts of mathematical text data. We put a lot of work into these dictionaries as we already mentioned [4] at CICM 2013. Especially, they contain names of mathematicians, acronyms and special terms that only exist in the domain of mathematics.

The Stanford tagger uses the Penn Treebank POS scheme [3], a classification scheme of linguistic types with 45 tags for tokens and punctuation symbols. This scheme has a relevant drawback for mathematical texts: there is no special tag for mathematical symbols or mathematical formulae (in the following we subsume both to mathematical

S.M. Watt et al. (Eds.): CICM 2014, LNAI 8543, pp. 213–223, 2014.

formulae). We did not change that. Formulae are handled by an auxiliary construct. This simple and straightforward method allows a slim and easy maintenance of the POS tagging software.

In our approach, mathematical formulae (which are available as TeX code) are transformed to unique but random character sequences.

POS tagging has two main problems: new words and the ambiguity of POS tags (many tokens of the corpus can belong to more than one word class) are addressed by determining a suitable POS tag of a token using contextual statistical models and the Viterbi algorithm, a dynamic programming technique. The Viterbi algorithm uses information about the surrounding tokens to predict the probable POS tag of the ambiguous or unknown token, e.g., a formula is mainly tagged as an adjective or a noun.

We illustrate that sequence of substitutions in the following example. It starts with the original LaTeX sentence and ends with the tagged sequence. The original sentence:

```
The classical Peano theorem states that in finite dimensional
spaces the Cauchy problem $x'(t)=f(t,x(t))$, $x(t\sb 0)=x\sb 0$,
has a solution provided $f$ is continuous.
```

The TeX formulae are translated into unique, but randomly generated, character sequences:

```
The classical Peano theorem states that in finite dimensional
spaces the Cauchy problem formula-kqnompjyomsqomppsk,
formula-kqomolugwpjqolugwk, has a solution provided formula-kyk
is continuous.
```

This sentence is fed into the POS tagger. The Stanford tagger assigns an appropriate tag to ervey token.

```
The_DT Classical_JJ Peano_NNP theorem_NN states_VBZ that_IN
in_IN finite_JJ dimensional_JJ spaces_NNS the_DT Cauchy_NNP
problem_NN formula-kqnompjyomsqomppsk_NN,
formula-kqomolugwpjqolugwk_NN ,_, has_VBZ a_DT solution_NN
provided_VBN formula-kyk_NN is_VBZ continuous_JJ ._.
```

The tagged text is transformed back to its TeX representation without touching the tags:

```
The_DT Classical_JJ Peano_NNP theorem_NN states_VBZ that_IN
in_IN finite_JJ dimensional_JJ spaces_NNS the_DT Cauchy_NNP
problem_NN $x'(t)=f(t,x(t))$_NN,
$x(t\sb 0)=x\sb 0$_NN ,_, has_VBZ a_DT solution_NN
provided_VBN $k$_NN is_VBZ continuous_JJ ._.
```

NP extraction is done by chunking with regular expressions for special patterns of POS tags. A very basic form of such a regular expression is:

```
<DT>?<JJ>*<NN>
```

This rule means that an NP chunk should be formed whenever the chunker finds an optional determiner (DT) followed by any number of adjectives (JJ) and then a noun (NN). However, it is easy to find many more complicated examples which this rule will not cover. The actual set of expressions we used is much more complex. Especially, NPs can be combinations of tokens and formulae, e.g., 'the Cauchy problem $x'(t) = f(t, x(t))$, $x(t_0) = x_0$'. After identifying the NPs in a document, we get a collection of NPs in the first step.

If you would like to experiment with our solution we provide a web-based demo at `http://www.zentralblatt-math.org/mathsearch/rs/postagger`

1.2 Noun Phrase and Key Phrase Extraction

A key phrase in our context, information retrieval in the mathematical literature, is a *phrase that captures the essence of the topic of a document* [7].

Mathematical publications, especially journal articles, have a more or less standardized metadata-structure covering important bibliographic data: authors, title, abstract, keywords (key phrases) and sometimes also a classification corresponding to the Mathematical Subject Classification (MSC). Typically, key phrases are short phrases characterizing

- embedding a publication in its general mathematical context as *Diophantine equations* or *optimal control*
- special objects, methods, and results of a publication as *bipartite complex networks, k-centroids clustering*

Often key phrases are descended form the title, the abstract or review or the fulltext, but this is not mandatory. key phrases of a document must not be part of the document.

NPs are natural candidates for key phrases. Key phrase identification via noun phrases is an usual technique. [5]

1.3 Classification with NPs

Classification is also an important task within NLP. The normal approach which uses the full text of a document and favours stemming and Term Frequency/Inverse Document Frequency (TF/IDF) to get rid of redundant words, but we chose a different approach. We use the extracted noun phrases from the texts, and than apply text classification methods. We tested several machine learning techniques. The best results were provided by a Support Vector Machine (SVM). The SVM we used is John Platt's sequential minimal optimization algorithm for support vector classifiers, the kernel is a polynomial kernel, the training data was every item from the database zbMATH from the beginning to the end of 2013. In particular, we used the Sequential Minimal Optimisation (SMO) technique from WEKA.

1.4 The Big Picture

A few words about the Fig. 1 block diagram: We start with an article, it is then tokenised into sentences. For every sentence the formulae have to be preprocessed: if there is an acronym in the sentence, it needs to be expanded. After that, the POS tagger runs and the noun phrases are extracted. The NPs extracted are sent to the classifier. The

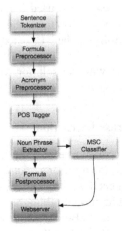

Fig. 1. The tagger with it's related processes

candidates for key phrases and classification codes are laid before to human experts. Their evaluations are used to improve our machine-learning techniques. The POS block in the middle of the diagram is really big in terms of complexity. As said above it, if a new release comes from Stanford, the new block can easily be integrated into our system.

General remark: There are a couple of emerging machine-learning techniques which have also been used for semantic analysis of documents. They work with Deep Belief Networks (deep Neural Networks) and the outcomes of these experiments are more than promising.

2   Reviewing Services in Mathematics

Reviewing journals have a long tradition in mathematics, and nowadays take the form of electronic databases. Today zbMATH and MathSciNet are the most important bibliographic mathematics databases. They are important tools used by the mathematical community in searching for relevant publications. These databases provide the most comprehensive bibliographic information about mathematics enhanced by a deep content analysis of the publications. In the Gutenberg era, all this information was created manually. The digital age has changed the situation dramatically. The digitisation of information allows automation like that we are developing to make the production of databases more efficient and uniformly to improve the quality of the database zbMATH.

The math databases have three different layers which are directly geared towards content analysis of a publication:

- bottom layer: reviews or abstracts
- second layer: key phrases
- top level: classification

Every layer has its own characteristics, but these layers interact. We will show in the following how we have used reviews or abstracts for key phrase extraction and classification.

3 Key Phrase Extraction in zbMATH

The relevance of NPs for key pheases identification is also valid for our data. The Zentralblatt Math, today the database zbMATH, has reacted to the increasing number of key phrases in the mathematical literature and has collected them since the 70s. The field UT *Uncontrolled Terms* was introduced to accentuate single terms or phrases) of a publication, e.g., *marginal function, quasi-differentiable function, directional differentiability, distance function.* This field lists key phrases created by authors and/or reviewers and/or editors of zbMATH. Typically, key phrases of authors will be extended by reviewers and editors within the workflow of zbMATH. The key phrases presented in zbMATH are searchable (by the specification $ut :$ in the search field) and clickable. The key phrases in zbMATH are different in size and quality. The current number of all key phrases in the database zbMATH is greater than (not disjunct) 10,200,000 entities. That means, the average number of key phrase of a publication is not more than 3, which is not sufficient for a description of the content below the MSC level.

The dominant majority of key phrases in zbMATH are noun phrases including formulae as C^*-algebra. Only, a small number of the manually created key phrases are single adjectives as key phrases, e.g., *quasiconvex*. No verbs were used as key phrases. So, we have focused us to identification of noun phrases (with formulae) as the most important candidates for key phrases until now.

For automatic creation of key phrases, typically only the titles and reviews or abstracts are available. This has the advantage that the number of noun phrases which is the list of candidates is small. Moreover, titles and reviews or abstracts are perfectly suited to detecting and extracting key phrases because they are generally understandable and summarise the content of a publication in a highly condensed form.

But of course, the NLP methods using tokenizing, POS tagging, and chunking, have to be adapted to specific requirements of our data.

3.1 Problems

Relevance: The NPs extracted are of different values for content analysis. Such phrases as *in the following paper (chapter etc.)* or *an important theorem* are of marginal value for content analysis. Therefore we allocate a weight to each extracted noun phrase. A noun phrase is given a *very high score* if it is

- a named mathematical entity which is defined in a mathematical vocabulary such as Wikipedia, PlanetMath, Encyclopaedia of Mathematics, etc. The number of named mathematical entities in these vocabularies is limited, and not more than 50,000 entities. Typically, such phrases are important in assigning the publication to its mathematical context.
- identical with a proposed key phrase of the publication. Most mathematical publications have a (limited) number of key phrases created by the author(s).

- identical with an existing key phrase in zbMATH. The existing key phrases describe general or special aspects. (The total number of existing key phrases in the database zbMATH is more than 10,200,000, the number of distinct key phrases is 2,900,000.)

high score if it

- contains names of mathematicians: if a noun phrase contains names of mathematicians, it is an indicator that the noun phrase is a name for a special conjecture, theorem, approach or method.
- is a acronym: Acronyms are artificial words and have a special spelling. Acronyms are used as abbreviations for longer noun phrases. Acronyms are *per se* relevant noun phrases. We compare the extracted candidates for acronyms with our dictionary and resolve them. Generally, the resolution is not unique and depends from the area; some acronyms have up to 20 different meanings.
- is or contains specific mathematical formulae: A special mathematical formula in a term, e.g., H^∞-control, is a relevant noun. At least all formulae which are not one-character mathematical notations, are important.

marginal or negative score

- if it provides no additional information about the content. Then, the extracted noun phrase is removed from the candidate list.

Incomplete Chunking: Sometimes, relevant mathematical key phrases involve a larger number of tokens, e.g., *Browder–Ky Fan and Ky Fan–Glicksberg fixed-point theorems*. Sometimes, the extracted phrases are incomplete. To solve these problems, the rules for chunking have to be adapted permanently.

3.2   Processing of NPs

In the following, the used methods are listed:

- Weighting: The weighting of key phrases is done as described above.
- Redundancy: Very often, some of the extracted NPs are similar. A simple measure for the similarity is the number of different tokens between two phrases. The method used is the LCS (Longest Common Subsequence) algorithm. The NPs are grouped by similarity.
- Filtering: Groups of similar phrases are replaced by a representative. Selecting a representative is done by using the base vocabulary. Existing key phrases and other resources (e.g., the labels of the MSC classes) are used to select the most suitable phrase.
- Evaluation by experts: The resulting list of possible key phrases is shown to human experts, e.g., editors or reviewers who can remove, change or add phrases.

3.3   Results

Number of Key Phrases and Quality: In the average, 3 – 4 key phrases were assigned manually to a publication. The average number of extracted NPs is significantly higher: 10 – 20 NPs. By the methods described above, the number of candidates is reduced to 7 – 10 phrases for a publication.

Up to now, the evaluation of key phrase extraction by human experts has been started only for particular classes because it means additional expense for human involvement. In the first phase, under 40% of the phrases were accepted by the experts. The feedback led to a redesign and essential improvements of the methods. The acceptable proportion of automatically generated key phrases increased to more than 60% by removing irrelevant phrases. It is planned to integrate the machine-based key phrase extraction in a semiautomatic workflow for zbMATH.

Of course, the quality of the proposed key phrases is dependent on the title and review (abstract).

Controlled vocabulary: We applied our tools to the complete zbMATH database. All resulting key phrases and all changes are stored and used for further enrichment and improvement of key phrases. The set of all positive evaluated key phrases is a first controlled vocabulary of mathematics; the irrelevant noun phrases define the bad list. The first version of the prototype of the controlled vocabulary contains 3,500,000 different phrases. The controlled vocabulary can be structured by topic (MSC classification, see below) and weighted by frequency.

3.4   Further Remarks

Key Phrases and Classification: The automatically created key phrases were also used for classification as will be described below in detail. Basing classifiers on the extracted key phrases instead of on reviews has significantly improved the quality of automatic classification.

Structuring key Phrases: Using our method we get only key phrases which are within a text. For a further enhancement of the key phrases and the controlled vocabulary, we have to know additional relations between the phrases, e.g., synonyms, hypernyms, hyponyms, meronyms. Such ontological relations could be used for structuring and improving the extracted key phrases.

Deeper Analysis of Mathematical Formulae: Mathematical symbols and formulae form an important part of mathematical publications but they are more important in the full texts of publications than in reviews.

An analysis of the symbols and formulae found in zbMATH has shown that the reviews, or abstracts, contain over 10,000,000 symbols and formulae. Most of them are simple one-character symbols. Nevertheless, the analysis of symbols and formulae and its combination with text analysis is of great interest, e.g., the correspondence between a text phrase and a formula seems relevant. Formulae were integrated in POS tagging and noun phrase extraction as described above. A deeper analysis of mathematical symbols and formulae is planned in cooperation with the MathSearch project.

4   Classification in zbMATH

Classification is a well-established concept for organising information and knowledge. Although it is a well known method, it is not a trivial task. The reasons for difficulties are numerous. Two main reasons are the classification scheme and the classifying process.

Classes are defined by one or more common properties of the members. Abstracting from individual objects, classification schemes assign the objects to classes. Classification schemes are not given *a priori*, they are intellectually designed and depend on the topic, aims, time, interests and views of the developers of the classification scheme.

The MSC was designed by the American Mathematical Society in the 1970s. The primary goal was to support subject-oriented access to the increasing number of mathematics-relevant publications, e.g., zbMATH lists 35,958 journal articles, books and other publications in mathematics and application areas in 1975. For a sufficiently fine-grained access to these thousands of documents annually, a hierarchical three-level deep classification scheme with more than 5,000 classes was developed. In particular, the top level of the MSC has 63 classes.

Typically, the classes of a classification scheme are not pairwise disjoint. Often, an overlapping of classes is part of a concept. This is also valid for the MSC, e.g., Navier-Stokes equations are listed in two main topics: 35-XX Partial differential equations (this is the mathematical point of view) and 76-XX Fluid mechanics (here, the application aspect is dominant). The MSC shows not only hierarchical relations but also different kinds of similarity. Moreover, an object can possess the properties of different classes. Typically, a mathematical publication cannot be reduced to a unique aspect or property. Publications develop or analyse mathematical models or objects investigated, make quality statements about them, or develop methods or tools to solve problems. This implies that a publication can be a member of more than one class.

A second reason for the difficulties is the classification process. This means that the classification codes (the handles indexing the classes) assigned an object, e.g., the MSC classification codes given a mathematical publication, are subjective (there is a certain range for classification codes). The classifications can be weighted: zbMATH and MathSciNet differentiate between primary and secondary MSC classifications.

What is the true classification of an object? What is the most important class? Are the classification codes given complete? Generally, there are often no objective answers to these questions. The uncertainty of the classification scheme and the subjectivity of the human experts work against the objective value of classification. For the reviewing services in mathematics, authors, editors and reviewers are involved in the classification process. In more detail, the final classification codes in zbMATH are the result of the workflow process. This means, each classification of the author(s) is checked by the reviewer and the editor. This workflow reduces the impact of subjective decision.

To the best of our knowledge, there has been no serious analysis of the quality and reliability of assigned classification codes. Checking classification would cover two steps a.) correctness of a proposed classification code b.) completeness of proposed classification codes (often a publication belongs to more than one class, e.g., look for a control system described by ordinary differential equations and its stability is investigated).

But for the development and evaluation of automatic classifiers we are assuming to start with that the classification in zbMATH is correct.

One aim of our work is to provide tools for automatic classifying, especially SVM methods. In detail, we took 63 different SVM classifiers, one for every top-level of the MSC and trained them on key phrases from the corresponding sections of the zbMATH database.

4.1 Quality

Quality of classification is usually measured by precision & recall and the F1-score which is their harmonic mean. Precision is the proportion of all publications of a class which are assigned to a specific class by the automatic classifier. For recall we look at all publications which are not in a class (in the complement of the class: this means the publication is in some other class) but assigned to this class by the automatic classifier.

4.2 Results

The results of the classification provide a differentiated picture. Roughly speaking; the precision is sufficient (for 26 of the 63 top-level classes the precision is higher than 0.75 and only for 4 classes is it smaller than 0.5), the recall is not. In other words publications which were classified as elements of a particular MSC class i are mostly correctly classified. The classifier is precision weighted: for all MSC classes the precision is higher than recall.

In the following, we discuss the results in more detail. A central idea in discussing the quality is the overlapping of classes. Therefore we have built a matrix, indexed by the top-level classes of the MSC, which lists the numbers of publications according to the following definition: Let a_{ii} be the number of publications which are classified exclusively with the MSC class i and a_{ij} the number of publications with the primary classification i and secondary classification j.

We normalize the elements a_{ij} of the matrix by with a_{ij}/A_i where A_i denotes the number of all publications with primary MSC code i

As a first result, it becomes clear that there is a correlation between the overlapping of classes and the results of the automatic classifier.

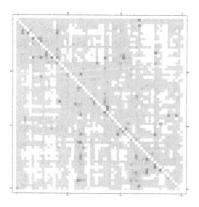

Fig. 2. The axes are the MSC class numbers, and the entries express the degrees of overlap a_{ij}; white indicates no overlap; the darker the cell the more the overlap

Easy Classes: If a_{ii}/A_i is near 1, then both precision and recall are high. The overlap with other classes is small and the vocabulary differs significantly from the vocabulary of other classes. Giving examples of such classes are those publications which have primary classifications from an application areas. This seems to be natural, because each application area has its own specific language and terminology.

Difficult Classes: There are have different types of overlap. Overlap with other classes can be focused on some MSC classes, e.g., for the class 31-XX *Potential theory* and 43 *Harmonic analysis*, or distributed, e.g., MSC 97-XX *Mathematical education*. In the first case, we propose to further cluster some MSC classes which are similar.

In addition, the total number of all publications with the primary classification i is relevant: a small number of documents has a negative impact on the classification quality. It seems that vocabulary and terminology of these classes may not be stable enough. The difficult classes have less than a few thousand documents.

4.3 Remarks

Use of Classification: A high precision means a high reliability that publications of the class i will be also automatically assigned to this class. This is important for preclassification where precision is more important than recall. Until now we have been deploying the classification tool for preclassification of publications. We propose to improve recall by a second step of classification analysis. The key phrases of each publication assigned to the class i by the automatic classifier will be analyzed in more detail.

Controlled Vocabulary and Classification: Classification is – in addition to key phrases – an important piece of metadata in the content analysis of a publication. Each zbMATH item can bear more than one classification code. The database zbMATH does not contain a relation between key phrases and classification codes. It is a $n : m$ relation. Also the hierarchical structure of the MSC is a problem too. To begin we have applied to top classes and assign a MSC class to a key phrase if the MSC classification (at the top level) is unique. This allows creating a first vocabulary for each top MSC class which has a higher precision of the definition of a MSC class as the existing. Moreover, the structure of the MSC scheme can be analyzed e.g. by the studying the intersection between the controlled vocabularies of different MSC classes.

5 Conclusion and Next Steps

It seems that the machine-based methods we have developed for key phrase extraction and classification are already useful in improving the content analysis of mathematical publications and making the workflow at zbMATH more efficient. We note some positive effects:

– Quantity and quality of key phrases is increased by automatic key phrase extraction.
– The integration of formulae into key phrase extraction lays the foundations for including formulae in content analysis. This could essentially improve content analysis of mathematical documents.
– Results of classification can be used to redesign and improve the MSC.
– The use of standardized methods guarantees a balanced and standardized quality of content analysis in zbMATH.

References

1. The database zbMATH, http://www.zentralblatt-math.org/zbmath/
2. Mathematics Subject Classification (MSC 2010), http://www.msc2010.org
3. Santorini, B.: Part-of-Speech-Tagging guidelines for the Penn Treebank Project (3rd Revision, 2nd printing) (June 1990), ftp://ftp.cis.upenn.edu/pub/treebank/doc/tagguide.ps.gz
4. Schöneberg, U., Sperber, W.: The DeLiVerMATH project: Text analysis in mathematics. In: Carette, J., Aspinall, D., Lange, C., Sojka, P., Windsteiger, W. (eds.) CICM 2013. LNCS (LNAI), vol. 7961, pp. 379–382. Springer, Heidelberg (2013), http://arxiv.org/pdf/1306.6944.pdf
5. Nguyen, T.D., Kan, M.-Y.: Keyphrase extraction in scientific publications. In: Goh, D.H.-L., Cao, T.H., Sølvberg, I.T., Rasmussen, E. (eds.) ICADL 2007. LNCS, vol. 4822, pp. 317–326. Springer, Heidelberg (2007)
6. Hall, M., Frank, E., Holmes, G., Pfahringer, B., Reutemann, P., Witten, I.H.: The WEKA Data Mining Software: An Update. SIGKDD Explorations 11(1) (2009)
7. Wikipedia contributors, 'Index term', Wikipedia, The Free Encyclopedia (January 13, 2014), http://en.wikipedia.org/wiki/Index_term
8. Platt, J.C.: Fast Training of Support Vector Machines Using Sequential Minimal Optimization. MIT Press, Cambridge (1999)
9. Samuelsson, C., Voutilainen, A.: Comparing a linguistic and a stochastic tagger. In: Proceedings of the 35th Annual Meeting of the Association for Computational Linguistics, pp. 246–253 (1997)
10. Encyclopedia of Mathematics, http://www.encyclopediaofmath.org/index.php/Main_Page
11. PlanetMath, http://planetmath.org/

Mathoid: Robust, Scalable, Fast and Accessible Math Rendering for Wikipedia

Moritz Schubotz[1] and Gabriel Wicke[2]

[1] Database Systems and Information Management Group,
Technische Universität Berlin, Einsteinufer 17, 10587 Berlin, Germany
schubotz@tu-berlin.de
[2] Wikimedia Foundation, San Francisco, California, U.S.A.
gwicke@wikimedia.org

Abstract. Wikipedia is the first address for scientists who want to recap basic mathematical and physical laws and concepts. Today, formulae in those pages are displayed as Portable Network Graphics images. Those images do not integrate well into the text, can not be edited after copying, are inaccessible to screen readers for people with special needs, do not support line breaks for small screens and do not scale for high resolution devices. Mathoid improves this situation and converts formulae specified by Wikipedia editors in a TeX-like input format to MathML, with Scalable Vector Graphics images as a fallback solution.

1 Introduction: Browsers Are Becoming Smarter

Wikipedia has supported mathematical content since 2003. Formulae are entered in a TeX-like notation and rendered by a program called `texvc`. One of the first versions of `texvc` announced the future of MathML support as follows:

> "As of January 2003, we have TeX markup for mathematical formulas on Wikipedia. It generates either PNG images or simple HTML markup, depending on user prefs and the complexity of the expression. In the future, as more browsers are smarter, it will be able to generate enhanced HTML or even MathML in many cases." [11]

Today, more then 10 years later, less than 20% of people visiting the English Wikipedia, currently use browsers that support MathML (e.g., Firefox) [27]. In addition, `texvc`, the program that controls math rendering, has made little progress in supporting MathML. Even in 2011, the MathML support was "rather pathetic" (see [17]). As a result, users expected MathML support within Wikipedia to be a broken feature. Ultimately, on November 28, 2011, the user preference for displaying MathML was removed [24].

Annoyed by the Portable Network Graphics (PNG) images, in December 2010, user Nageh published a script, `User:Nageh/mathJax.js`, that enables client-side MathJax rendering for individual Wikipedia users [23]. Some effort and technical expertise was required to use the script. The user had to install additional fonts

S.M. Watt et al. (Eds.): CICM 2014, LNAI 8543, pp. 224–235, 2014.

on his system manually, to import the script, into his Wikipedia account settings and to change the Math setting in his Wikipedia user account page to "Leave it as TEX". With client-side MathJax rendering, the visitor was able to choose from the context menu of each equation with the PNG image being replaced by either: (1) a Scalable Vector Graphics (SVG) image, (2) an equivalent HTML + CSS representation, or (3) MathML markup (this requires a MathML capable browser).

MathJax needs a significant amount of time to replace the TEX-code on the page with the above replacements. This amount of time is dependent on the operating system, browser, and hardware configuration. For instance, we measured 133.06 s to load the page *Fourier transform* in client side MathJax mode, as compared to 14.05 s for the page loading without math rendering (and 42.9 s with PNG images) on a typical Windows laptop with Firefox.[1] However, improvements in the layout motivated many users to use that script, and in November 2011, client-side MathJax became an experimental option for registered users [25].

As of today, there are almost 21M registered Wikipedia users, of which 130k have been active in the last 30 days. Of these users, 7.8k use the MathJax rendering option which causes long waiting times for pages with many equations. Also 47.6k users chose the (currently disabled) HTML rendering mode, which if possible, tries to use HTML markup to display formula, and the PNG image otherwise. Furthermore, 10.1k users chose the MathML rendering mode (disabled in 2011). Thus, the latter 57.7k users are temporarily forced to view PNG images, even though they explicitly requested against this. This demonstrates that there is an significant demand for math rendering, other than for the use of PNG images.

Currently, the MediaWiki Math extension is version 1.0. Our efforts have been to make an improvement on that extension. We refer to our update of the extension as version 2.0. Furthermore, we refer to Mathoid as all the ingredients mentioned in this paper which go into developing Math 2.0. One essential ingredient in Mathoid, is what we refer to as Mathoid-server. This is a tool, which we describe in this paper, which converts the TEX-like math input used in MediaWiki to various formats that we describe in this paper. Our paper is organized as follows.

In Section 2, we list the requirements for Math rendering in Wikipedia, explain how one may contribute to these requirements, and elaborate on how one may make math accessible for people with special needs. In this section we introduce the Mathoid-server. In Section 3, we explain how by using Mathoid-server, math can be displayed in browsers that do not support MathML. In Section 4, we discuss how math rendering can be offered as a globally distributed service. In Section 5, we discuss and compare the performance of reviewed rendering tools,

---

[1] The measurement was done on a Lenovo T420 Laptop with the following hardware: 8GB RAM, 500GB HDD, CPU Intel Core i7-2640M, Firefox 24.0 on Windows 8.1, download speed 98.7 MB/s upload speed 9.8 MB/s, ping to en.wikpedia.org was 25($\pm$1) ms.

in regard to layout and speed. Finally, in Section 6, we conclude with results from our comparison and give an overview of future work.

2   Bringing MathML to Wikipedia

For Wikipedia, the following requirements and performance measures are critical.

coverage: The converter must support all commands currently used in Wikipedia.

scalability: The load for the converter may vary significantly, since the number of Wikipedia edits heavily depends on the language. Thus, a converter must be applicable for both small and large Wikipedia instances.

robustness: Bad user input, or a large number of concurrent requests, must not lead to permanent failure for the converter.

speed: Fast conversion is desirable for a good user experience.

maintainability: A new tool for a global site the size of Wikipedia must be able to handle tasks with a large management overhead. Therefore, active development over a long period of time is desirable.

accessibility: Providing accessible content to everyone is one of the key goals of Wikipedia.

There are a large variety of TeX to MathML converters [2]. However, most of them are no longer under active development, or their licenses are not compatible with MediaWiki. In 2009, [22] showed that LaTeXML has the best coverage (but not a very high conversion speed) as compared to the LaTeX converters which were analysed in that paper. Since 2009, a new converter, MathJax [3], has become popular. After negotiations with Peter Krautzberger (of MathJax) and his team, we regard MathJax (executed in a headless browser on a web server), as a useful alternative to LaTeXML. One strong point about MathJax with regard to coverage, is that it is already used by some Wikipedia users on the client-side (as described in Section 1). Therefore the risk of unexpected behavior is limited. For LaTeXML, Bruce Miller has written a set of custom macros for MediaWiki specific TeX commands which supports the MediaWiki TeX markup. A test using these macros based on a Wikipedia dump, has shown very good coverage of the mathematics commands currently used in Wikipedia. LaTeXML is maintained by the United States Department of Commerce Laboratory, the National Institute of Standards and Technology (NIST) and MathJax is maintained by the MathJax project which is a consortium of the American Mathematical Society and the Society for Industrial and Applied Mathematics. We analyze and compare MathJax and LaTeXML in detail, since these are the most promising tools we could discover.

In regard to accessibility, we note that Wikipedia has recently made serious efforts to make the site more accessible. However, images which represent equations are currently not accessible. The only available information from PNG images (which is not very helpful) is the alt-text of the image that contains the TeX code. Based upon recent unpublished work of Volker Sorge [21], we would like to provide meaningful semantic information for the equations. By providing this, more people will be able to understand the (openly accessible) content [5]. One must also consider that there is a large variety of special needs. People with blindness are only a small fraction of the target group which can benefit from increased accessible mathematical content. Between 1 and 33% [6, 7] of the population suffer from dyslexia. Even if we calculate with the lower boundary of 1%, 90,000 people per hour visit the English Wikipedia and some of them could benefit from improvements of the accessibility while reading articles that contain math. However, our the main goal with regard to accessibility is to make Wikipedia accessible for blind people that have no opportunity to read the formulae in Wikipedia today.

Furthermore, the information provided in the tree structure of mathematics by using MathML, helps one to orientate complex mathematical equations, which is useful for general purpose use as well. With regard to accessibility, a screen reader can repeat only part of an equation to provide details that were not understood. PNG images do not give screen readers detailed related mathematical information [4, 14]. In 2007, [10] states that MathML is optimal for screen readers. *The Faculty Room*[2] website, lists four screen readers that can be used in combination with MathPlayer [20] to read Mathematical equations. Thus Mathoid-server and LATEXML server [8] that generate MathML output, contribute towards better accessibility within the English Wikipedia.

3   Making Math Accessible to MathML Disabled Browsers

For people with browsers that do not support MathML, we would like to provide high quality images as a fallback solution. Both LATEXML and MathJax provide options to produce images. LATEXML supports PNG images only, which tend to look rasterized if they are viewed using large screens. MathJax produces scalable SVG images. For high traffic sites like Wikipedia with 9 million visitors per hour, it is crucial to reduce the server load generated by each visitor. Therefore rendered pages should be used for multiple visitors. Rendering of math elements is especially costly. This is related to the nested structure of mathematical expressions. As a result, we have taken care that the output of the MediaWiki Math extension should be browser independent. Since MathJax was designed for client-side rendering, our goal is to develop a new component. We call this new component the Mathoid-server. Mathoid-server, a tool written in JavaScript, uses MathJax to convert math input to SVG images. It is based on `svgtex` [9] which uses `nodejs` and `phantomjs` to run MathJax in a headless browser. It exports SVG images. Mathoid-server improves upon the functionality of `svgtex` while offering a more robust alternative. For instance, it provides a restful interface which supports `json` input and output as well as the support of MathML output. Furthermore, Mathoid-server is shipped as a Debian package for easy installation and maintenance. Many new developments in MediaWiki use JavaScript. This increases the probability of finding volunteers

---

[2] http://www.washington.edu/doit/Faculty/articles?404

to maintain the code and to fix bugs. For general purpose, Mathoid-server can be used as a stand-alone service which can be used in other content management platforms such as Drupal or Wordpress. This implies that Mathoid-server will have a larger interest group for maintenance in the future. The fact that Mathoid-server automatically adapts to the number of available processing cores, and can be installed fully unattended via tools like Puppet, indicates that the administrative overhead for Mathoid-server instances should be independent of the number of machines used. In the latest version, the Mathoid-server supports both LaTeX and MathML input and is capable of producing MathML and SVG output.

To support MathML disabled browsers, we deliver both MathML markup, and a link to the SVG fallback image, to the visitor's browser. In order to be compatible with browsers that do not support SVG images, in addition, we add links to the old PNG images. In the future those browsers will disappear and this feature will be removed.

To prevent users from seeing both rendering outputs, the MathML element is hidden by default, and the image is displayed. For Mozilla based browsers (these support MathML rendering), we invert the visibility by using a custom CSS style, hide the fallback images and display the MathML-markup. This has several advantages. First, no browser detection, neither on the client-side (e.g., via JavaScript) nor on server-side is required. This eliminates a potential source of errors. Our experiments with client-side browser detection showed that the user will observe a change in the Math elements if pages with many equations are loaded. Second, since the MathML element is always available on the client-side, the user can copy equations from the page, and edit it visually with tools such as Mathematica. If the page content is saved to disk, all information is preserved without resolving links to images. If afterwards the saved file is opened with a MathML enabled browser, the equations can be viewed off-line. This feature is less relevant for users with continuous network connections or with high-end hardware and software. However, for people with limited resources and unstable connections (like in some parts of India [1]), they will experience a significant benefit.

The current Firefox mobile version (28.0) passes the MathML Acid-2 test, indicating that there is generally good support for MathML on mobile devices. This allows for customized high quality adaptable rendering for specific device properties. The W3C MathML specification[3] discusses the so called best-fit algorithm for line breaking. According to our experiments, Firefox-mobile (28.0) does not pass the W3C line break tests. However, as soon this issue is fixed, mobile users will benefit from the adjusted rendering for their devices. Note that there is active development in this area by the Mathematics in ebooks project[4] lead by Frédéric Wang.

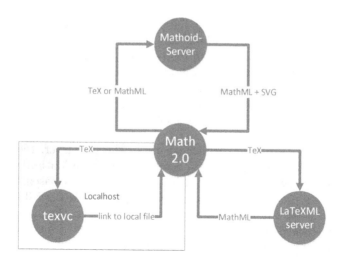

Fig. 1. System overview. The 2.0 release of the MediaWiki Math extension offers new ways to render Math input in MediaWiki. It communicates to LATEXML servers and to instances of our MathJax-based development of the Mathoid-server. Note that we preserve backwards compatibility to the MediaWiki Math extension 1.0.

4   A Global Distributed Service for Math Rendering

To use LATEXML and Mathoid-server for Math rendering within Wikipedia, we have changed the MediaWiki Math extension (see Fig. 1). While preserving backwards compatibility, we pursue our current development [19] by integrating LATEXML and Mathoid-server. System administrators can choose which rendering back-ends are selectable in a MediaWiki instance by logged in users. All rendering modes can be active at the same time, and it is possible to use Mathoid-server to generate SVG images based on the output of the LATEXML server.

For `texvc`, rendering requires one to install a full LATEX distribution (about 1GB) on each web server. This is a huge administrative overhead and the management of files and permissions has caused a lot of problems. These problems were hard to solve and resulted in inconsistent behavior of the website from a user perspective [12, 13, 15, 16]. Furthermore, for MediaWiki instances run by individuals, it is difficult to set up math support. Therefore, one of our major requirements is that future versions of the MediaWiki Math extension should not need to access the file system. Also, one should not need to use shell commands to directly access the server. This has the potential to introduce major security risks. With our separation approach, rendering and displaying of mathematics no longer needs to be done on the same machine. However, for instances with a

---

3 http://www.w3.org/TR/MathML/chapter3.html
4 http://www.ulule.com/mathematics-ebooks

small load, this would still be possible. Furthermore small MediaWiki instances can now enable Math support without any configuration or shell access. By default, public LaTeXML and Mathoid-server instances are used. These will be provided by XSEDE[5]. With this method, no additional security risk is provided for individuals. For Mathoid-server, the security risk for the host is limited as well. This is because the Mathoid process runs on a headless browser on the server without direct access to the file system.

Caching. There are two main caching layers. In the first layer, the database caches the rendering result of the math conversion, i.e., the MathML and SVG output in the database.[6] The second caching layer is browser based. Similar to the SVG rendering process for ordinary SVG images, the MediaWiki Math extension uses cacheable special pages to deliver SVG images. On the server side, those pages are cached by squid servers. In addition, even if images occur on different pages, the browser will only load that image once.

5   Performance Analysis

As a first step towards a visual analysis, we compared our impressions of output from LaTeXML and MathJax using Firefox 24.0. Except for additional `mrow` elements in LaTeXML, the produced presentation MathML is almost identical. The differences that we did notice had no influence on the rendering output. In rare cases, MathJax uses `mi` elements, whereas LaTeXML uses `mo` elements. In contrast to LaTeXML which uses UTF-8 characters to represent special symbols, MathJax uses HTML entities. However, there still remain some minor differences (see Fig. 2).

To illustrate performance differences, we chose a random sample equation, namely

$$\frac{(x-h)^2}{a^2} - \frac{(y-k)^2}{b^2} = 1. \tag{1}$$

With 10 subsequent runs[7], the following conversion times were measured:

- LaTeXML : TeX → MathML (319ms/220ms);
- Mathoid-server : TeX → SVG + MathML (18ms/18ms);
- `texvc` : TeX → PNG (99ms/71ms).

Thus compared to the baseline (`texvc`), Mathoid-server produced a speedup of 5.5 whereas LaTeXML is 3.2 times slower. LaTeXML and PNG seem to benefit from multiple runs, whereas the rendering time with Mathoid-server stays constant.

---

[5] https://www.xsede.org

[6] To keep the lookup time for equations constant, the key for the cache entry is a hash of TeX input.

[7] All measurements were performed using virtual Wikimedia labs instances, with the following hardware specifications: number of CPUs 2 , RAM size 4096 Mb, allocated Storage 40 Gb.

texvc	LaTeXML	LaTeXML-SVG	Mathoid	Mathoid-SVG
`({\begin{array}{cc}a&b\\c&d\end{array}})`				
$\left(\begin{smallmatrix} a & b \\ c & d \end{smallmatrix}\right)$	$\begin{pmatrix} a & b \\ c & d \end{pmatrix}$	$\begin{pmatrix} a & b \\ c & d \end{pmatrix}$	$\left(\begin{smallmatrix} a & b \\ c & d \end{smallmatrix}\right)$	$\left(\begin{smallmatrix} a & b \\ c & d \end{smallmatrix}\right)$
`{\bar {abc}}`				
$\bar{a}bc$	$\bar{a}bc$	$\bar{a}bc$	$\bar{a}bc$	$\bar{a}bc$
`\operatorname {d}\!t`				
dt	dt	dt	dt	dt
`\uplus ,\biguplus`				
$\uplus, \biguplus$	$\uplus, \biguplus$	$\uplus, \biguplus$	$\uplus, \biguplus$	$\uplus, \biguplus$
`{}_{1}^{2}\!\Omega _{3}^{4}`				
${}_1^2\Omega_3^4$	${}_1^2\Omega_3^4$	${}_1^2\Omega_3^4$	${}_1^2\Omega_3^4$	${}_1^2\Omega_3^4$
`{\mathcal {ABC}}`				
$\mathcal{ABC}$	$\mathcal{ABC}$	$\mathcal{ABC}$	ABC	$\mathcal{ABC}$
`a\ b,a\quad b`				
$a\ b, a\quad b$	$a\ b, a\quad b$	$a\ b, a\quad b$	$a\ b, a\quad b$	$a\ b, a\quad b$
`{\begin{array}{|c|c|}a&b\\0&0\\0&1\end{array}}`				
$\begin{array}{\|c\|c\|} a & b \\ 0 & 0 \\ 0 & 1 \end{array}$	$\begin{array}{cc} a & b \\ 0 & 0 \\ 0 & 1 \end{array}$	$\begin{array}{cc} a & b \\ 0 & 0 \\ 0 & 1 \end{array}$	$\begin{array}{\|c\|c\|} a & b \\ 0 & 0 \\ 0 & 1 \end{array}$	$\begin{array}{\|c\|c\|} a & b \\ 0 & 0 \\ 0 & 1 \end{array}$
`{\mathfrak {123}}`				
123	123	123	123	123
`{\big |}{\Big |}{\bigg |}{\Bigg |}\dots {\Bigg |}{\bigg |}{\Big |}{\big |}`				
`{\big /}{\Big /}{\bigg /}{\Bigg /}\dots {\Bigg \backslash }{\bigg \backslash }{\Big \backslash }{\big \backslash }`				
Inline Formulae				
Lorem ipsum sum $\sum_{i=0}^{\infty} 2^{-i}$ con setetur sadipscing	Lorem ipsum sum $\sum_{i=0}^{\infty} 2^{-i}$ con setetur sadipscing	Lorem ipsum sum $\sum_{i=0}^{\infty} 2^{-i}$ con setetur sadipscing	Lorem ipsum sum $\sum_{i=0}^{\infty} 2^{-i}$ con setetur sadipscing	Lorem ipsum sum $\sum_{i=0}^{\infty} 2^{-i}$ con setetur sadipscing

Fig. 2. This figure displays a comparison of possible rendering outputs for the MediaWiki Math extension rendered with Firefox 24.0. Mathoid-server allows one to use either a LaTeXML or MathJax renderer to generate MathML output with SVG fallback. The investigation of the listed corner cases shows that the Mathoid-SVG rendering option, that uses server-side MathJax rendering via `phantomjs`, produces the best results. *Remark: The authors thank Bruce Miller. He improved the LaTeXML implementation based on a preprint version of this paper. This final version of the paper uses the LaTeXML version of 24th of April 2014.*

We also converted all of the English Wikipedia articles and measured the conversion times for each equation therein. The most time consuming equation was the full form of the Ackermann function $A(4,3)$. LaTeXML[8] needs 147 s to answer the HTTP request for $A(4,3)$. According to the self reported LaTeXML-log, approximately 136.18 s was used for parsing. The same request was answered by Mathoid-server in 0.393 s, which is approximately 374 times faster. The old rendering engine needed 1.598 s to produce the image. This does not include the 41 ms it took to load the image from the server.

6 Conclusion, Outlook and Future Work

In the scope of Mathoid, we updated the retrograde MediaWiki Math extension, and developed Mathoid-server which replaces the `texvc` program. This enhanced the security of MediaWiki as proposed in [19]. It is no longer necessary to pass user input via shell access using the command-line. Nor is it necessary to move files on the server via PHP. The MediaWiki Math extension is now capable of using LaTeXML and Mathoid-server to convert TeX-like input to MathML + SVG. Based on the requirements, the user can choose if he prefers to use LaTeXML for the MathML generation (this has the advantage of content MathML output), or he can use Mathoid-server which is much faster (but does not produce content MathML). Mathoid-server takes advantage of LaTeXML since it produces MathML. The MediaWiki math extension, through Mathoid-server, converts MathML to fallback SVG images.[9] For Wikipedia itself, with only a few semantic macros, and no real applications for content MathML produced by LaTeXML, Mathoid-server alone seems to be the best choice.

Table 1. Overview: comparison of different math rendering engines. Values based on articles containing mathematics in the English Wikipedia.

	texvc	LaTeXML	Mathoid
relative speed	1	0.3	5
image output	PNG	PNG	SVG
presentation MathML coverage	low	high	high
content MathML output	no	no	yes
webservice	no	yes	yes
approximate space required on webserver	1GB	0	0
language	OCaml	Perl	JavaScript
maintained by	nobody	NIST	MathJax

We did exhaustive tests to demonstrate the power of Mathoid-server with regard to scalability, performance and enhancement of user experiences. Those test

---

[8] `ltxpsgi` (LaTeXML version 0.7.99; revision ef040f5)

[9] This integral feature of Math 2.0 does not require additional source modifications, and is demonstrated for example at `http://demo.formulasearchengine.com`.

results are summarized in Table 1. Our implementation was finished in October 2013 and is currently being reviewed by the Wikimedia foundation for production use. Our work on the MediaWiki Math extension and Mathoid-server establishes a basis for further math related developments within the Wikimedia platform. Those developments might be realized by individuals or research institutions in the future. For example, we have considered creating an OpenMath content dictionary [18] that is based on Wikidata items [26]. This will augment mathematical formulae with language independent semantic information. Moreover, one can use gathered statistics about formulae currently in use in Wikipedia to enhance the user interface for entering new formulae[10]. This is common for text input associated with mobile devices.

In regard to future potential work, one may note the following. There is a significant amount of hidden math markup in Wikipedia. Many of these have been entered using HTML workarounds (like subscript or superscript) or by using UTF-8 characters. This markup is difficult to find and causes problems for screen readers (since they are not aware that the mode needs to be changed). If desired by the community, by using gathered statistics and edit history, we would be able to repair these damages.

The MediaWiki Math extension currently allows limited semantic macros. For example, one can use `\Reals` to obtain the symbol $\mathbb{R}$. At the moment, semantic macros are seldomly used in the English Wikipedia (only 17 times). One could potentially benefit by increased use of semantic macros by taking advantage of the semantic information associated with these. In the future, one could use this semantic information to take advantage of additional services, such as MathSearch [19], a global search engine which takes advantage of semantic information. Also, the use of semantic macros would provide semantic information, which would provide improved screen reader output.

Acknowledgments. Thanks to Deyan Ginev, Michael Kohlhase, Peter Krautzberger, Bruce Miller and Volker Sorge for fruitful discussions. Thanks also to Frédéric Wang for help with the code review. A special thanks to Howard Cohl for his editorial work on this paper. M. Schubotz would like to express his gratitude to Terry Chay and Matthew Flaschen, for their help in organizing his internship at the Wikimedia Foundation. This work was funded by MediaBotz and the Wikimedia Foundation.

References

[1] Arunachalam, S.: Open access-current developments in India. In: Proceedings Berlin 4 Open Access: From Promise to Practice, Potsdam-Golm, Germany, pp. 1–9 (March 2006), `http://arizona.openrepository.com/arizona/bitstream/10150/105554/1/Berlin4SA.pdf`

[2] Bos, B.: The W3C MathML software list, `http://www.w3.org/Math/Software/` (accessed March 20, 2014)

---

[10] `http://www.formulasearchengine.com/node/189`

[3] Cervone, D.: MathJax: A Platform for Mathematics on the Web. Notices of the American Mathematical Society 59(2), 312–316 (2012)

[4] Chisholm, W., Vanderheiden, G., Jacobs, I.: Web Content Accessibility Guidelines 1.0. Interactions 8(4), 35–54 (2001), doi:10.1145/379537.379550, ISSN: 1072-5520

[5] Cooper, M., Lowe, T., Taylor, M.: Access to Mathematics in Web Resources for People with a Visual Impairment. In: Miesenberger, K., Klaus, J., Zagler, W.L., Karshmer, A.I. (eds.) ICCHP 2008. LNCS, vol. 5105, pp. 926–933. Springer, Heidelberg (2008)

[6] Crystal, D.: The Cambridge Encyclopedia of Language, 2nd edn. Cambridge University Press, Cambridge (1997)

[7] Czepita, D., Lodygowska, E.: Role of the organ of vision in the course of developmental dyslexia. Klinika Oczna 108(1-3), 110–113 (2006) ISSN: 0023-2157

[8] Ginev, D., Stamerjohanns, H., Miller, B.R., Kohlhase, M.: The LaTeXML Daemon: Editable Math on the Collaborative Web. In: Davenport, J.H., Farmer, W.M., Urban, J., Rabe, F. (eds.) Calculemus/MKM 2011. LNCS (LNAI), vol. 6824, pp. 292–294. Springer, Heidelberg (2011)

[9] Grbin, A., Maloney, C.: Svgtex (2013), https://github.com/agrbin/svgtex (accessed March 20, 2014)

[10] Maddox, S.: Mathematical equations in Braille. Maths, Stats and Operations Research (MSOR) Connections 7(2), 45–48 (2007), doi:10.11120/msor.2007.07020045, ISSN: 1473-4869

[11] Mayer, D.: Help: Displaying a formula — Meta, discussion about Wikimedia projects (2003), http://meta.wikimedia.org/w/index.php?title=Help:Displaying_a_formula&oldid=15233 (accessed March 20, 2014)

[12] MediaWiki. Failed to parse (Cannot store math image on filesystem) (2013), https://www.mediawiki.org/w/index.php?oldid=782277 (accessed March 20, 2014)

[13] MediaWiki. Manual: Troubleshooting math display errors — MediaWiki, The Free Wiki Engine (2013), https://www.mediawiki.org/w/index.php?oldid=798098 (accessed March 20, 2014)

[14] Miner, R.: The Importance of MathML to Communication. Notices of the American Mathematical Society 52(5) (2005)

[15] Morris, R.: Bug 54367 - intermittent texvc problems (2013), https://bugzilla.wikimedia.org/show_bug.cgi?id=5436 (accessed March 20, 2014)

[16] Murugan, S.: Bug 54456 - Failed to parse (Cannot store math image on filesystem) (2013), https://bugzilla.wikimedia.org/show_bug.cgi?id=54456 (accessed March 20, 2014)

[17] Netheril96@gmail.com. Option "MathML if possible" doesn't work (2010), https://bugzilla.wikimedia.org/show_bug.cgi?id=25646 (accessed March 20, 2014)

[18] Riem, M.N.: The OpenMath Guide. A practical guide on using Open-Math. Available from the Research Institute for Applications of Computer Algebra (2004)

[19] Schubotz, M.: Making Math Searchable in Wikipedia. In: Conferences on Intelligent Computer Mathematics abs/1304.5475 (2013)

[20] Soiffer, N.: MathPlayer. In: Proceedings of the 7th International Association for Computing Machinery Special Interest Group on Accessible Computing Conference on Computers and Accessibility – ASSETS 2005, p. 204. ACM Press, New York (2005), doi:10.1145/1090785.1090831, ISBN: 1595931597

[21] Sorge, V., et al.: Towards making mathematics a first class citizen in general screen readers. In: 11th Web for All Conference, Seoul, Korea, April 6-9. ACM (2014)

[22] Stamerjohanns, H., et al.: MathML-aware article conversion from LaTeX, A comparison study. In: Sojka, P. (ed.) Towards Digital Mathematics Library, DML 2009 Workshop, pp. 109–120. Masaryk University, Brno (2009)

[23] User: Nageh. User:Nageh/mathJax (2010), `https://en.wikipedia.org/w/index.php?oldid=400482894` (accessed March 20, 2014)

[24] Vibber, B.: Disable the partial HTML and MathML rendering options for Math extension (2011), `https://git.wikimedia.org/commit/mediawiki%2Fextensions%2FMath/09679f2f39e6c6c00e87757292421b26bfa7022a` (accessed March 20, 2014])

[25] Vibber, B.: Experimental option $wgMathUseMathJax to have Extension: Math load things via MathJax (2011), `http://git.wikimedia.org/commit/mediawiki%2Fextensions%2FMath/1042006fd4c2cbe6c62619b860e2e234d04d6d38` (accessed March 20, 2014)

[26] Vrandečić, D.: Wikidata: a new platform for collaborative data collection. In: Proceedings of the 21st International Conference Companion on World Wide Web, pp. 1063–1064. ACM (2012)

[27] Zachte, E.: SquidReportClients@stats.wikimedia.org (2013), `http://stats.wikimedia.org/archive/squid_reports/2013-10/SquidReportClients.htm` (accessed March 20, 2014)

Set Theory or Higher Order Logic to Represent Auction Concepts in Isabelle?[*]

Marco B. Caminati[1], Manfred Kerber[1], Christoph Lange[1,2], and Colin Rowat[3]

[1] Computer Science, University of Birmingham, UK
[2] Fraunhofer IAIS and University of Bonn, Germany
[3] Economics, University of Birmingham, UK
http://cs.bham.ac.uk/research/projects/formare/

Abstract When faced with the question of how to represent properties in a formal proof system any user has to make design decisions. We have proved three of the theorems from Maskin's 2004 survey article on Auction Theory using the Isabelle/HOL system, and we have verified software code that implements combinatorial Vickrey auctions. A fundamental question in this was how to represent some basic concepts: since set theory is available inside Isabelle/HOL, when introducing new definitions there is often the issue of balancing the amount of set-theoretical objects and of objects expressed using entities which are more typical of higher order logic such as functions or lists. Likewise, a user has often to answer the question whether to use a constructive or a non-constructive definition. Such decisions have consequences for the proof development and the usability of the formalization. For instance, sets are usually closer to the representation that economists would use and recognize, while the other objects are closer to the extraction of computational content. We have studied the advantages and disadvantages of these approaches, and their relationship, in the concrete application setting of auction theory. In addition, we present the corresponding Isabelle library of definitions and theorems, most prominently those dealing with relations and quotients.

1 Introduction

When representing mathematics in formal proof systems, alternative foundations can be used, with two important examples being set theory (e.g., Mizar takes this approach) and higher order logic (e.g., as in Isabelle/HOL). Another dimension in the representation is the difference between classical and constructive approaches. Again, there are systems which are predominantly classical (as most first order automated theorem provers) and constructive (e.g., Coq). Isabelle/HOL is flexible enough to enable the user to take these different approaches in the same system (e.g., although it is built on higher order logic, it contains a library for set theory, Set.thy). For instance, participants in an auction, i.e.

---

[*] This work has been supported by EPSRC grant EP/J007498/1 and an LMS Computer Science Small Grant.

S.M. Watt et al. (Eds.): CICM 2014, LNAI 8543, pp. 236–251, 2014.

bidders, can be represented by a predicate `Bidder x` or alternatively as a list of bidders `[b1, b2, b3]`.

The difference between a classical and a constructive definition can be demonstrated by the example of the argument of the maximum for a function (which we need for determining the winner of an auction). Classically we can define it, e.g., as `arg_max f A = {x ∈ A. f x = Max(f'A)}`. This definition is easy to understand but, unlike a constructive one, it does not tell how to compute `arg_max`. The constructive definition is more complicated (and has to consider different cases). It corresponds to a recursive function recurring on the elements of the domain. From a programming perspective, the two kinds of definitions just illustrated (classical versus constructive) can be seen respectively as *specification* versus *implementation*.

The approaches coexist in Isabelle/HOL. For example, for a set `X` one can apply the higher order function `f` by the construction `f ' X` to yield the image of the set. As a result, an author does not have to make a global decision of whether to use sets or higher order functions, but has fine grained control on what to use for a new mathematical object (e.g., sets vs. lists or lambda functions). Such a choice will typically depend on many factors. One factor is the task at hand (e.g., whether one needs to prove a theorem about an object or needs to compute its value). Another factor is naturalness of the constructions. This will typically depend on the authors and their expected audiences. Pragmatically, users also have to consider which of the possible approaches is more viable given the support provided by existing libraries for the system being used.

The ForMaRE project [8] applies formal methods to economics. One of the branches of economics the project focuses on is *auction theory*, which deals with the problem of allocating a set of resources among a set of participants while maximizing one or more parameters (e.g., revenue, or social welfare) in the process. ForMaRE has produced the Auction Theory Toolbox (ATT), containing Isabelle code for a range of auctions and theorems about them.[1] Therefore, there is a good opportunity to practically test the feasibility of the different approaches, as introduced above, in this concrete setting. We adopted a pragmatic attitude: we typically took a set theoretic approach, since firstly we felt most familiar with it and secondly we knew from our ongoing interaction with economists that it would look more natural to them. However, when we needed to generate code, this generally excluded set-theoretical constructions such as the set comprehension notation. As a consequence most of our work was done in the set-theoretical realm, but some done constructively which allowed us to produce the code we wanted. The disadvantage of this approach is that we had to provide supplementary 'compatibility' proofs to show the equivalence of the set theoretical and the computable definitions when needed. This means that, as a byproduct of our efforts, we also generated a good amount of generic set-theoretical mater-

---

[1] See `https://github.com/formare/auctions/`; the state at the time of this writing is archived at `https://github.com/formare/auctions/tree/1f1e7035da2543a0645b9c44a5276229a0aeb478`.

ial which was neither provided by the Isabelle library nor Isabelle's Archive of
Formal Proofs (AFP)[6].

In this paper, rather than discussing our progress in the application domain
(auction theory), we illustrate this material and its relationship to the mathematical objects we needed. We will discuss this by the following three concrete
examples that occurred while we were working on the ATT:

1. We model a function indirectly through the set-theoretical notion of the
   graph of a relation, rather than directly using the HOL primitive notion of
   a lambda-abstracted function. This allowed us to concisely define auction-
   related notions through two natural set-theoretical constructions (extension
   and restriction of a function). Moreover, this in turn allowed us to generalize
   some theorems from functions to relations. We also discuss how this choice
   does not necessarily mean giving up the advantage of computability (section
   2). Incidentally, this will also give insight into the main differences between
   set theory as implemented in Isabelle/HOL and standard set theory.
2. We developed a stand-alone formalization of the definition of functions over
   equivalence classes (section 3). This is a common construction when, e.g.,
   defining the canonically induced operation on the quotient of a group by
   one of its normal subgroups. To define such an operation, we proved the
   invariance (or well-definedness) under an equivalence relation.
3. We applied a set-theoretical, non-constructive definition and a constructive
   one to the same object (the set of all possible partitions of a set). The two
   approaches were used for different purposes (theorem proving and computation, respectively) and proved formally their equivalence by a 'compatibility'
   theorem (section 4). The same approach is adopted for the set of all possible
   injective functions between sets.

In existing proof assistants based on set theory (e.g., Mizar, Metamath), the
desirable quality of being able to compute values of functions (rather than only
the truth value of predicates involving them), is typically lost[2].

We will present examples which show that we can retain this Isabelle/HOL-
induced advantage in our developments.

Overview of the Paper

Section 2 introduces and motivates the set-theoretical encoding of functions, as
an alternative to a lambda representation; this is done by commenting on the
relevant Isabelle definitions, and by illustrating a particular theorem regarding
auctions, which employs those definitions. Section 3 brings this approach to a
more abstract level, by showing how it can handle, given an initial function, the

---

[2] We do not know if this is due to how the existing proof assistants are implemented, or to some fundamental limitation of set theory. Reading the recent
"Computational set theory" thread (http://www.cs.nyu.edu/pipermail/fom/2014
-February/017841.html), and its related threads, on the "Foundations of Mathematics" mailing list, we feel the question is currently open.

definition of a second function on the corresponding equivalence classes. Since this was already done in Isabelle's `Equiv_Relations.thy`, we also illustrate the differences with it.

In some cases, defining a mathematical object in purely set-theoretical terms does not preserve computability of that object: section 4 introduces a technique for such cases. As mentioned in point 3 above, this technique introduces a parallel, computable definition, and then formally proves the equivalence of the two definitions. In the same section, this technique is presented through two examples from our ATT; we then discuss how we took advantage from having preserved both the definitions. Finally, section 5 explains how we applied the general machinery from section 3 to the ATT.

2   Set-Theoretical Definition of Functions in Isabelle/HOL

In higher order logic, functions, function abstraction, and function application are primitives [3]; in set theory, the primitive notion is that of a set, and a function f is represented by its graph, i.e., the *set* of all the ordered pairs $(x, f(x))$. We chose to work mostly with the set-theoretical representation of the functions even though we are using a tool based on higher order logic. In the following we give reasons why.

1. A first reason is that a set theoretical representation more easily allows to enumerate all the (injective) functions from a finite set to another finite set ([4]). In contrast, this seems to be more complicated to do directly in higher order logic, where all the functions are assumed be total.
2. A second reason is that the set-theoretical, graph-based representation works even for generic relations, thus often allowing us to extend the results we proved to the more general situations involving relations, rather than functions (see the next subsection).
3. A third reason is that the operation of function restriction is naturally expressed in terms of two elementary set operations: cartesian product and intersection. Indeed, this allows to extend this operation to relations, immediately giving an instance of what we argued in point (2) above. Restriction is a fundamental operation for representing the concept of weakly dominant strategy, a key concept in auction theory, see e.g., [10, proposition 2]. The definition of function restriction is arguably more complicated in standard higher order logic since functions are always total. That is, restricting them to a set requires carrying the restricted domain set with the function, e.g. by forming a pair $(R, f|_R)$, whereas the set-theoretical representation naturally includes this set.
4. Finally, specific partial, finite set theoretical functions can be very concisely and quickly defined; this made it possible to promptly test Isabelle code while we were working on it. For example, functions can be written in the form of a set as `{(0,10), (1,11), (2,12)}` and fed to the Isabelle definitions very easily in order to test the correctness of related computations empirically, whereas a lambda expression would be more complex to define.

2.1 Two Basic Operators on Relations: 'outside' and Pasting

In this section, we discuss two general mathematical operations we encountered specifically during formalization of auctions. Assume a set of bidders N and a function $b : N \to \mathbb{R}$ that determines the corresponding bids. The first operation removes one bidder i from the domain N of the bid function. The second operation alters the bid function in one point i of its domain, $b +* (i, b_i)$, which is equal to b except for argument i, where the value is changed to b_i. In set theory, a function (or a relation) inherently specifies its domain and range. A generalization of the first operation is thus obtained by writing (in Isabelle):

```
definition Outside ::
"('a  ×    'b) set ⇒ 'a set ⇒ ('a × 'b) set"
(infix "outside" 75) where
"R outside X = R - (X × Range R)",
```

where 75 denotes the binding strength of the infix operator. The following specialization to singletons ...

```
abbreviation singleoutside (infix "--" 75) where
"b -- i ≡ b outside {i}",
```

... turned out handy for our purposes.

Another circumstance making the set theoretical approach convenient is that now the second operation can be obtained in simple terms of the first as follows:

```
definition paste (infix "+*" 75)
where "P +* Q = (P outside Domain Q) ∪ Q",
```

which can be specialized to the important case of Q being a singleton function:

```
abbreviation singlepaste where
"singlepaste F f ≡ F +* {(fst f, snd f)}"
notation singlepaste (infix "+<" 75).
```

While we often applied outside and +* to functions rather than relations (i.e., assuming right uniqueness, see section 2.2), many of its properties were proved for relations in general. For example, the associativity theorem

```
lemma 1153: "(P +* Q) +* R = P +* (Q +* R)"
```

has no hypotheses in its statement and holds for general relations.

2.2 Specializing Relations to Functions: Right-Uniqueness and Evaluation

The operators outside and +* are building blocks on which the statements of many theorems we proved for auctions are based. In turn, a number of preparatory lemmas have been proven about those objects in the generic case of relations; however, others only hold when considering relations which actually are functions. Hence, we need the following predicate for *right-uniqueness* of a relation, which we define in terms of the *triviality* of a set (i.e. being empty or singleton):

```
definition trivial where "trivial x = (x ⊆ {the_elem x})",
```

where `the_elem` extracts an element from a set, being undefined when this cannot be done in a unique way. The predicate for right-uniqueness is called `runiq`, and it uses the operator `R' 'X`, which yields the image of the set `X` through the relation `R`:

```
definition runiq :: "('a × 'b) set ⇒ bool" where
"runiq R = (∀ X . trivial X → trivial (R '' X))".
```

We note that, contrary to other proof assistants based on set-theory (e.g., Mizar), which in general cannot directly compute values of functions (at most the truth or falsity of predicates involving those values), we are able to preserve from Isabelle/HOL the ability of actually computing the evaluation of these set-theoretical flavoured functions, when they are right-unique, through the following operator:

```
fun eval_rel :: "('a × 'b) set ⇒ 'a ⇒ 'b" (infix ",," 75)
where "R ,, a = the_elem (R '' {a})"
```

Now, indeed, set theoretical functions can be evaluated via `,,`. This can be tested through the Isabelle command `value`: for example we can write

```
value "{(0::nat,10),(1,11),(1,12::nat)} ,, 0"
```

and obtain 10 as an answer. This holds also when combining `eval_rel` with the operators as from the beginning of this section; for example

```
value "({(0::nat,10),(1,11),(1,12)} +< (1,13::nat)) ,, 1"
```

yields the answer 13.

A right-unique relation and a standard higher order logic, lambda abstracted function represent the same mathematical object, hence it should be possible to pass from one representation to another. `graph` from `Function_Order.thy`, defined as

```
definition graph where "graph X f = {(x, f x) | x. x ∈ X}"
```

does exactly that. The opposite conversion can be achieved easily as follows:

```
definition toFunction (* inverts graph *)
where "toFunction R = (λ x . (R ,, x))"
```

However, the degree of computability of set-theoretical functions is less than with original HOL functions; for example we cannot evaluate

```
value "(graph {x::nat. x<3} (λx. (10::nat))),,(1::nat)",
```

while the following works as expected

```
value "(graph {0,1,2} (λx. (10::nat))),,(1::nat)".
```

We also note that, since Isabelle formalizes set theory inside higher order logic, types still impose some rigidity, compared to stand-alone set theory: see section 6. For example, the following alternative Isabelle definition would be exactly equivalent to `eval_rel` in a standard (untyped) set theory:

```
abbreviation "eval_rel2 (R::('ax('b set)) set) (x::'a)
≡ ⋃ (R''{x})" notation eval_rel2 (infix ",,," 75),
```

It is, however, actually defined (and equivalent to it) *only* for set-yielding relations:

```
lemma 11182: assumes "runiq (f::(('a × ('b set)) set))"
"(x::'a) ∈ Domain f" shows "f,,x = f,,,x"
```

However, when it is applicable, `eval_rel2` has the desirable qualities of evaluating to the empty set outside the domain of `f`, and in general to something defined when right-uniqueness does not hold (in which case `eval_rel` is undefined). This allowed us to give more concise proofs in such cases.

`runiq` is a central definition in our formalization; its many possible equivalent formulations have turned out to be useful in different steps when proving various lemmas. Here we present the possible alternative definitions we have proven to be equivalent in the ATT:

```
lemma 11133: "runiq P=inj_on fst P"

lemma runiq_alt: "runiq R ↔ (∀ x . trivial (R '' {x}))"

lemma runiq_basic:
"runiq R ⟷ (∀ x y y' . (x, y) ∈ R ∧ (x, y') ∈ R ⟶ y = y')"

lemma runiq_wrt_eval_rel:
"runiq R ⟷ (∀x . R '' {x} ⊆ {R ,, x})"

lemma runiq_wrt_eval_rel':
"runiq R ⟷ (∀x ∈ Domain R . R '' {x} = {R ,, x})"

lemma runiq_wrt_ex1:
"runiq R ⟷ (∀ a ∈ Domain R . ∃! b . (a, b) ∈ R)"

lemma runiq_wrt_THE:
"runiq R ⟷ (∀ a b . (a, b) ∈ R ⟶ b = (THE b . (a, b) ∈ R))"
```

In general, we found that, especially for basic and ubiquitous concepts such as `runiq`, the more equivalent definitions we have, the better. One reason is that this improves the understandability of the formalization: different readers will find different definitions easier to grasp. Another reason is that automated theorem proving tools, such as Sledgehammer[3], will be more likely to find automated justification in single steps of subsequent proofs: by picking the appropriate equivalent definition, sledgehammer can find a justification, while, upon removing that definition, it is no longer able to do that. We actually experienced this phenomenon with proofs involving `runiq`: the form of `runiq` given in `11133` allowed Sledgehammer to find the proof of this technical lemma:

```
lemma 11134: assumes "runiq P" shows "card (Domain P) = card P".
```

`lemma 11134` above was in turn used to formalize proposition 3 from [10].

---

[3] Sledgehammer is an Isabelle tool that applies automatic theorem provers (ATPs) and satisfiability-modulo-theories (SMT) solvers to automatically produce proofs [2].

2.3   Application to Auctions

Next, we give one example of the roles of the operations introduced in this section in our practical setting of auctions, for the simple case of a single-good auction. In this case, the input data for the auction are given through a function b associating to each bidder the amount she bids for the good. Given a fixed bidder i, the outcome of the auction is determined by two functions, a and p. Both take b as an argument: the first yields whether that bidder won the item (a,,b = 1) or not (a,,b = 0); the second, p,,b, yields how much she has to pay. We take the Isabelle formalization of the second proposition in [10], which is theorem th10 in file Maskin2.thy. This result proves, given some general requirements, the logical equivalence of two properties, each binding a and p:

- The first property we called genvick (for generalized Vickrey auction, see below); it states that the payment imposed to i is the sum of a 'fee' term t(b--i), to be paid irrespectively of the outcome, and of a proper price term (a,,b - a1)*w(b--i), which is to be added only in case she obtains the good. Moreover, the first term does not depend on how much i herself bids, but only on others' bids. The proper price is determined by the auxiliary function w, which also does not depend on i's bid (since Vickrey auctions are second price auctions). Hence, i's bid can influence only whether i pays or not the proper price, but not its amount.
- The second property, called dom4, states that i can never be worse off if she changes her bid to her real valuation v of the good.

Hence, genvick assumes the following form:

```
abbreviation genvick where "genvick a p i w ≡
(∃ (a1::allocation) t. (∀ b ∈ Domain a ∩ (Domain p).
p,,b  =  (a,,b - a1)*w(b--i) + t(b--i)  ))",
```

while dom4 is the following inequality:

```
definition dom4 where "dom4 i a p = (∀ b::bid. ∀ v.(
{b,b+<(i,v)}⊆(Domain a ∩ (Domain p)) ∧ i∈Domain b)⟶
v*(a,,b)-(p,,b) ≤ v*(a,,(b<(i,v)))-(p,,(b+<(i,v))))".
```

The operators -- and +< are central in expressing those two conditions; their respective general properties (collected in RelationProperties.thy) permitted to streamline the proof of the theorem, whose thesis reads:

$$\text{genvick a p i w = dom4 i a p.} \tag{1}$$

3   Quotients between Relations

We built a library of basic facts centred around our new constructs of right-uniqueness (runiq), evaluation (,,), pasting (+*), and considering a function outside some subset of its domain. The library also contains more advanced results. In particular, we describe here our approach to building quotients; then, we derive functions on equivalence classes of points from functions defined on single points. These methods are common in many areas of mathematics, especially algebra and topology; they are used when a given property holds on classes of objects, and one wants to abstract away from the specific representative of a class, and rather define the given property on the whole class. Such classes are typically the equivalence classes induced by an equivalence relation over its domain, which form the *quotient* of the original set.

For example, in group theory, the operation of a group G is canonically transported to the set of the cosets yielded by a normal subgroup N. This is what makes the quotient group G/N a group, and is only possible if the group operation is *class-invariant*[4]: the product of two representative elements of two cosets must be in the same coset, irrespective of how those representative elements are selected. This is ensured by the definition of normal subgroup.

In our case, to construct t appearing in the definition of genvick towards the end of the previous section, we need to define some function taking as an argument a bid vector with the i-th component removed (where i is a bidder). The formal way to do that was to aggregate all possible bid vectors differing only on their i-th component and to define that function on the classes obtained this way; hence, using quotients naturally emerged as one elegant approach. More details on this particular application of quotients are in section 5.

As illustrated in [12], existing Isabelle's theory Equiv_Relations.thy already introduces tools for these general mathematical techniques. There are, however, the following problems:

1. The operation of passing from a pointwise function to the 'abstracted' version defined on equivalence classes is done using type-theoretical Abs_ and Rep_. In paper-based mathematics, on the other hand, everything is done using set theory; hence, a mathematician would probably not know how to use this implementation without first getting some knowledge of the underlying type-theoretical foundations.

2. This operation must be performed 'manually' in each separate case: there is no generic definition to do that given a pair (f, R), where f is the function to be abstracted and R is an equivalence relation on its domain. In contrast, in our treatment we introduce a function called quotient doing exactly this.

3. f and R are typed as a lambda function and a set-theoretical relation, respectively, while there is no reason to preemptively preventing f from being a generic relation (i.e., not necessarily being right unique).

For these reasons, we coded a purely set-theoretical Isabelle implementation of this machinery, via three simple definitions and one theorem establishing the right-uniqueness of the abstracted function given basic requirements on the pointwise function.

The first definition gives a map to pass from a point of the domain of a relation R to the corresponding equivalence class:

```
definition projector where
"projector R = {(x,R''{x}) | x. x ∈ Domain R }"
```

---

[4] In this case we can also say that the group operation is *well-defined*, or that it *respects* the corresponding equivalence relation, or even that it is *compatible* with it.

The second definition builds, given a pointwise relation R and two equivalence relations P, Q (working on its domain and codomain, respectively) the corresponding, abstracted relation on the resulting equivalence classes:

```
definition quotient where "quotient R P Q =
{(p,q)| p q. q ∈ (Range (projector Q)) ∧
p ∈ Range (projector P) ∧ p × q ∩ R ≠ {} }".
```

While this definition is typically given for a function R and equivalence relations P and Q, it still makes sense if these additional conditions are not satisfied. This allows us to lift these requirements in some preparatory lemmas before assuming them in the following main result:

```
lemma 123: assumes "compatible f P Q" "runiq f" "trans P" "sym P"
"equiv (Domain Q) Q" shows "runiq (quotient f P Q)",
```

where the predicates equiv X P, trans P, sym P exist in the Isabelle library, and state, respectively, that P is an equivalence relation over the set X; that P is a transitive relation; and that P is a symmetric relation.

Note that in Isabelle there is a definition for a quotient of a relation R written as quotient R as the set of all equivalence classes associated with R. Here, however, we assume a function (or relation) f with respect to relations P and Q and define the quotient of f with respect to R as a function with the domain quotient R.

The notion of compatibility above asks that f respects P and Q:

```
definition compatible where
"compatible R P Q =
(∀ x . (R''(P''{x}) ⊆ Q''(R''{x})))"
```

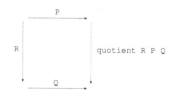

Note that the definition of compatible is a generalization of the usual commutativity of the application of functions as displayed to the right.

It should be noted that this condition is not required at the time of defining a quotient, which can be defined for any triple of relations; rather, it is only required when asking that the quotient behaves in the expected way, as stated by lemma 123 above. This allows us to freely use the construction quotient in advance, to show a number of intermediate results not requiring compatibility themselves. For example, the following lemma holds for general relations:

```
lemma quotientFactors: assumes "equiv (Domain p) p" "equiv (Domain q) q"
shows "quotient r p q = (projector p)⁻ O r O (projector q)".
```

where R⁻ is the converse of the relation R and O stands for relation composition.

We also note that, for similar reasons, we devised a definition of compatibility taking any triple of relations as arguments: as in the definition of quotient, one can use compatible R P Q even before showing that R is right-unique and that P, Q are equivalence relations. In the latter case, however, the definition of compatibility reduces to asking that any two Q-equivalent points have P-equivalent images through R.

4   Injective Functions and Partitions

In section 2.3, we introduced the mathematical description of a single-good auction. An important family of more complex schemes is given by *combinatorial* auctions. In these, there are several objects at stake (a set of goods G), and each bidder (of a set N) can bid for each possible combination of them. The outcome of the auction is still described by a pair of maps (a, p), yielding respectively what a bidder gets and how much she has to pay. However, now the "what a bidder gets" part must be represented by a mathematical object more articulate than a $\{0, 1\}$-valued function. It is represented by a partition of G, and by an injective function (or injection) from that partition to N. Treating the latter function inside set-theory gives an advantage: the pair (partition, injection) is conveniently represented by the injection alone, because the partition of G will be simply the domain of the former. This allowed us to type the relevant objects plainly as follows:

```
type_synonym bidder = "nat"
type_synonym goods = "nat set"
type_synonym allocation_rel = "(goods × bidder) set".
```

Since we wanted to extract code from our Isabelle formalization, we had to implement recursive definitions for both the set of all possible partitions of a finite set and for the set of all possible injections from a finite set to another finite set. In both cases, those recursive definitions turned out to be inconvenient when it came to prove mathematical facts involving them. Hence a separate, more natural definition was needed that is equivalent but not computable.

We illustrate this dual approach in the case of injections.

Table 1. Constructive definition of injection

```
fun injections_alg ::
"'a list ⇒ 'b::linorder set ⇒ ('a × 'b) set list" where
"injections_alg [] Y = [{}]" |
"injections_alg (x # xs) Y = concat
[[R+*{(x,y)}. y←sorted_list_of_set (Y-Range R)].
                    R ← injections_alg xs Y]"
```

In Table 1, we find a definition of all injective functions between two finite sets X and Y, which recurs on X, while in Table 2 the set of all injective functions is defined axiomatically.

Table 2. Axiomatic definition of injection

```
definition injections :: "'a set ⇒ 'b set ⇒ ('a × 'b) set set"
where "injections X Y =
{R. Domain R=X ∧ Range R⊆Y ∧ runiq R ∧ runiq(R⁻)}"
```

We use the constructive definition in computations and the axiomatic for proofs and mathematical manipulations. In order to do that we have to prove their equivalence in the theorem stated in Table 3.

Table 3. Compatibility of constructive and axiomatic definitions of injection

```
theorem injections_equiv:
fixes xs::"'a list"
and Y::"'b::linorder set"
assumes non_empty: "card Y > 0"
shows "distinct xs ⟹
(set(injections_alg xs Y)::('a×'b)set set)=injections (set xs) Y"
```

Similarly, we have a constructive and an axiomatic definition for partitions (see Tables 4 and 5, respectively).

Table 4. Constructive definition of partition

```
definition insert_into_member_list
:: "'a ⇒ 'a set list ⇒ 'a set ⇒ 'a set list"
where "insert_into_member_list new_el Sets S =
    (S ∪ { new_el }) # (remove1 S Sets)"

definition coarser_partitions_with_list
::"'a ⇒ 'a set list ⇒ 'a set list list" where
"coarser_partitions_with_list new_el P = ({ new_el } # P)
   #
   (map ((insert_into_member_list new_el P)) P)"

definition all_coarser_partitions_with_list
::"'a ⇒ 'a set list list ⇒ 'a set list list" where
"all_coarser_partitions_with_list elem Ps =
    concat (map (coarser_partitions_with_list elem) Ps)"

fun all_partitions_list :: "'a list ⇒ 'a set list list" where
"all_partitions_list [] = [[]]" |
"all_partitions_list (e # X) =
    all_coarser_partitions_with_list e (all_partitions_list X)"
```

The constructive definition above represents a partition of a finite set as a list of (disjoint) subsets of it; it works by induction on the cardinality of the set to be partitioned, as follows. Given a set of cardinality $n + 1$, we write it as a disjoint union $X \cup \{x\}$, and, given a partition P of X, we insert x into each set belonging to P (this is done by the operator `insert_into_member_list` above). We thus obtain $|P|$ many partitions of $X \cup \{x\}$, to which we add the distinct partition $\{\{x\}\} \cup P$. Making P range over all possible partitions of X, we obtain in this way all possible partitions of $X \cup \{x\}$.

When implementing this idea into the code above, however, we chose to represent the finite set X as a list (rather than a set) of elements of it, and a partition of it as a list (rather than a set) of subsets of it. This is because the algorithm just described is iterated in two ways: first, x is inserted into each set of a partition; secondly, the construction is iterated for each possible partition of X. Since iterations are easier with lists than with finite sets [11], we adopted that choice.

Table 5. Axiomatic definition of partition

```
definition is_partition where
"is_partition P = (∀ X∈P . ∀ Y∈ P . (X ∩ Y ≠ {} ↔ X = Y))"

definition is_partition_of (infix "partitions" 75)
where "is_partition_of P A = (⋃ P = A ∧ is_partition P)"

definition all_partitions where
"all_partitions A = {P . P partitions A}."
```

It should be noted that passing from a list to a finite set (via the operator `set`) is easier than the converse, hence the equivalence theorem for partitions in Table 6 is stated taking as input data a list `xs`:

Table 6. Compatibility of constructive and axiomatic definitions of partition

```
theorem all_partitions_paper_equiv_alg:
    fixes xs::"'a list"
    shows "distinct xs ⟹
    set(map set (all_partitions_list xs)) = all_partitions(set xs)"
```

Thanks to the constructive definitions above, we were able to extract, from Isabelle code [5], executable code running Vickrey combinatorial auctions [4]. Furthermore, we were able to prove fundamental theoretical properties about those auctions by using the non-constructive versions of those definitions; for example, that the price for any bidder is non-negative. The compatibility theorems in Tables 3 and 6 allow to certify that such theoretical properties hold also for the extracted code.

We note that, while non-constructive versions hold in the general case, constructive versions are obviously limited to the finite case (e.g., calculating all the partitions of a finite set). Therefore, the two compatibility theorems must restrict to the finite case: this is reflected by the fact that the argument of injections and all_partitions, appearing in Tables 3 and 6, respectively, is set xs. This means that such an argument is automatically a finite set, being the result of converting a list (xs) to a set using the Isabelle function set. We stress the fact that our approach allows to prove most theorems (e.g., thesis (1) at the end of section 2) without restricting to the finite case, and to add the additional hypothesis of finiteness only and exactly for the theorems needing it.

5   Application of Quotients to Auctions

We now give enough formalization details to illustrate the exact point in our proofs of auction theory in which we needed to employ projector and quotient, introduced in section 3.

To prove the thesis (1) at the end of section 2, we needed to explicitly build the t appearing in the definition of genvick. This function is uniquely determined by a and p; however, the latter takes b as an argument, while t takes b--i (which represents a bid vector with bidder i's bid removed, and is called a *reduced bid*). The algebraic way to pass from b to b--i is to consider an equivalence relation associating any two b b' differing at most in the point i. Correspondingly, to obtain from p a function on arguments of the form b--i, we form the quotient of p according to that equivalence relation. The equivalence relation we need is exactly the kernel[5] of the following function:

```
definition reducedbid:: "bidder ⟹ (bid × allocation) set ⟹
    (bid × bidder set × bid × allocation) set"
    where "reducedbid i a =
    {(b, (Domain b, b outside {i}, a ,, b))| b. b ∈ Domain a}".
```

---

[5] We recall that the kernel of a function f is the equivalence relation $\circ_f$ given by $x_1 \circ_f x_2 \iff f(x_1) = f(x_2)$. See [1, Definition 1.18]. The kernel notion was missing in the Isabelle library, and we also provided it in ours.

So that the explicit construction of t, used inside the proof for (1), ends up as

```
"λx. reducedprice p i a ,, ({i} ∪ Domain x, x, Min (Range a))",
```

where

```
definition reducedprice:: "(bid × price) set ⇒ bidder ⇒
(bid × allocation ) set ⇒
((bidder set × bid × allocation) × price) set"
where "reducedprice p i a =
(projector ((reducedbid i a)⁻)) O
(quotient p (Kernel (reducedbid i a)) Id) O
((projector Id)⁻)",
```

Id being the identity function. With this definition in place, an important part of the theorem consists in showing that reducedprice is right-unique.

This reduces to showing that quotient p (Kernel(reducedbid i a)) Id appearing in the definition above is right-unique. Thanks to lm23 (see section 3), this in turn means proving the compatibility between p and the equivalence relation just introduced, i.e., Kernel (reducedbid i a), which is provided by

```
lemma 124b: assumes "functional (Domain a)" "Domain a ⊆ Domain p"
"dom4 i a p" "runiq p" shows
"compatible p (Kernel (reducedbid i a)) Id".
```

6   Discussion and Related Work

The work presented here is based on the specific set theory implemented in Isabelle/HOL, sometimes called simply-typed set theory [14, Section 6.1], [13, Section 1]. It differs from standard set theories, as Zermelo-Fraenkel (ZF), in that the primitives of the latter are encoded using primitives of higher order logic, as follows: a given set p is actually a term p of type $\tau \Rightarrow$ bool, and the writing $x \in p$ is actually the application $p\ x$. This means that in Isabelle/HOL one cannot write a set like $\{x, \{x\}\}$ (which is not well-typed): the standard hierarchy of ZF is no longer constructable. As long as the objective is to formalize ordinary (as defined in [15, Section I.1]) mathematics, this usually causes no problem: while this difference prevents the encoding of relevant mathematical objects (e.g., natural numbers, integers, reals, cartesian products) as usually done in ZF, those objects can be directly represented in higher order logic. This means that, for many mathematical branches, what is lost with respect to ZF is limited to its more technical side-effects, as $\pi \cap \mathbb{Q} \neq \emptyset$, or $1 \cap 2 \subseteq 3$. Since the latter are often regarded as strange or meaningless writings [9], [13, Section 1], this could even be a desirable consequence.

On the other hand, when trying to exploit technical ZF 'hacks', as we tried to do with eval_rel2, problems can arise: we discussed that in section 2.2. Moreover, ZF is a more appropriate tool for studying the remaining branches of mathematics, starting with set theory itself.

The Z pattern catalogue [16, Section IV] allows to represent functions as sets of pairs, as done here. However, Z is a specification language, while we are interested in both specification and implementation. Mizar is based on untyped set theory, thus modelling functions and relations as done here, and also provides many relevant existing theorems in its library; however, it is not possible to extract code or to do computations in general. The Ssreflect extension library for the Coq proof assistant is extensive and provides a lot of operations; on the other hand, most of this material seems to apply to functions as represented by lists, thereby inherently limiting to the finite case. More details on how Isabelle/HOL generally compares with other systems are in the comparative study [7].

7   Conclusions

We have built an extensive library of results about functions, represented as right-unique relations as an alternative to the lambda abstractions that have so far been typical of Isabelle/HOL. In this paper we explained how we employed this alternative technique, and the concepts in our library, in the application domain of auction theory. Our library ranges from simpler constructions such as pasting a relation onto another, to more sophisticated ones such as quotients.

We described how set-theoretical constructs can concisely express constructions that frequently occur in formalization, giving concrete examples from the application domain of auction theory. In the particular case of representing functions, we discussed the advantages and disadvantages of the set-theoretical representations with respect to the alternative, more natural (given the foundations of Isabelle/HOL) lambda representation. We also showed that computability, a typical feature of lambda functions, can be achieved for set-theoretical functions in Isabelle/HOL. Moreover, even in those cases in which it cannot be achieved, we showed how we were still able to employ a dual approach, giving non-constructive, more expressive definitions along with constructive (and thus computable) ones, finally proving their equivalence through compatibility theorems. This allows us to obtain the best of both worlds, that is, expressiveness and proof-friendly definitions together with computability, at the cost of having to prove the additional compatibility theorems.

Finally, we took our application of this approach to a more advanced, abstract level by describing in a simple, close-to-paper set-theoretical style notions such as *quotients, compatibility* and *kernel*. We presented a concrete application of this material in a hands-on case encountered in formalizing auction theory within the ForMaRE project. We think that our library is rich and generic enough to be of possible use to other Isabelle users, and we hope that this paper can serve as a first guiding example to show how it can be employed.

References

1. Bergman, C.: Universal Algebra: Fundamentals and Selected Topics. Chapman & Hall Pure and Applied Mathematics. Taylor & Francis (2011)
2. Blanchette, J.C., Paulson, L.C.: Hammering Away. A User's Guide to Sledgehammer for Isabelle/HOL (December 5, 2013),
http://isabelle.in.tum.de/dist/Isabelle2013-2/doc/sledgehammer.pdf

3. Bowen, J., Gordon, M.: Z and HOL. In: Z User Workshop, Cambridge 1994, pp. 141–167. Springer (1994)

4. Caminati, M.B., et al.: Proving soundness of combinatorial Vickrey auctions and generating verified executable code, arXiv:1308.1779 [cs.GT] (2013)

5. Haftmann, F., Nipkow, T.: Code generation via higher-order rewrite systems. In: Blume, M., Kobayashi, N., Vidal, G. (eds.) FLOPS 2010. LNCS, vol. 6009, pp. 103–117. Springer, Heidelberg (2010)

6. Klein, G., et al. (eds.): Archive of Formal Proofs (2014), http://afp.sf.net/ (visited on March 14, 2014)

7. Lange, C., Caminati, M.B., Kerber, M., Mossakowski, T., Rowat, C., Wenzel, M., Windsteiger, W.: A qualitative comparison of the suitability of four theorem provers for basic auction theory. In: Carette, J., Aspinall, D., Lange, C., Sojka, P., Windsteiger, W. (eds.) CICM 2013. LNCS (LNAI), vol. 7961, pp. 200–215. Springer, Heidelberg (2013)

8. Lange, C., Rowat, C., Kerber, M.: The ForMaRE Project – Formal Mathematical Reasoning in Economics. In: Carette, J., Aspinall, D., Lange, C., Sojka, P., Windsteiger, W. (eds.) CICM 2013. LNCS (LNAI), vol. 7961, pp. 330–334. Springer, Heidelberg (2013), arXiv:1303.4194[cs.CE]

9. Leinster, T.: Rethinking set theory. arXiv preprint arXiv:1212.6543 (2012)

10. Maskin, E.: The unity of auction theory: Milgrom's master class. Journal of Economic Literature 42(4), 1102–1115 (2004), http://scholar.harvard.edu/files/maskin/files/unity_of_auction_theory.pdf

11. Nipkow, T., Paulson, L.C.: Proof pearl: Defining functions over finite sets. In: Hurd, J., Melham, T. (eds.) TPHOLs 2005. LNCS, vol. 3603, pp. 385–396. Springer, Heidelberg (2005)

12. Paulson, L.C.: Defining functions on equivalence classes. ACM Transactions on Computational Logic (TOCL) 7(4), 658–675 (2006)

13. Paulson, L.C.: Set Theory for Verification: I. From Foundations to Functions. Journal of Automated Reasoning 11, 353–389 (1993)

14. Paulson, L.C., et al.: Isabelle/HOL. A Proof Assistant for Higher-Order Logic (2013)

15. Simpson, S.G.: Subsystems of second order arithmetic, vol. 1. Cambridge University Press (2009)

16. Valentine, S.H., et al.: AZ Patterns Catalogue II-definitions and laws, v0.1 (2004)

Realms: A Structure for Consolidating Knowledge about Mathematical Theories

Jacques Carette[1], William M. Farmer[1], and Michael Kohlhase[2]

[1] Computing and Software, McMaster University, Canada
http://www.cas.mcmaster.ca/~carette,
http://imps.mcmaster.ca/wmfarmer
[2] Computer Science, Jacobs University Bremen, Germany
http://kwarc.info/kohlhase

Abstract. Since there are different ways of axiomatizing and developing a mathematical theory, knowledge about a such a theory may reside in many places and in many forms within a library of formalized mathematics. We introduce the notion of a *realm* as a structure for consolidating knowledge about a mathematical theory. A realm contains several axiomatizations of a theory that are separately developed. Views interconnect these developments and establish that the axiomatizations are equivalent in the sense of being mutually interpretable. A realm also contains an external interface that is convenient for users of the library who want to apply the concepts and facts of the theory without delving into the details of how the concepts and facts were developed. We illustrate the utility of realms through a series of examples. We also give an outline of the mechanisms that are needed to create and maintain realms.

1 Introduction

In [Far11] the second author calls for the establishment of a "universal digital library of mathematics" (UDLM). In our joint work it is understood that the UDLM will be organized as a theory graph, i.e., a set of theories (collections of symbol declarations, definitions, assertions, and their proofs) interconnected by meaning-preserving views (morphisms). The "little/tiny theories approach" first put forward in [FGT92] has been very fruitful for formal developments of mathematical knowledge, but it has not found its way into mainstream mathematics. One reason may be that there is a mismatch with the way mathematicians — the supposed users of the UDLM — think about and work with theories. [CF08] argues for the development of "high-level theories" that better mesh with these expectations. We will re-examine the issues involved and propose a solution.

In the mathematical community the term *"theory"* is used to describe multiple ideas, from the axiomatic theory of the algebraic structure of a group to "Group Theory" as an entire discipline, and various gradations in between. Looked at more closely, this implies a multi-scale organizational structure to the basic components of mathematics, ranging from individual concepts (e.g., a group) to whole subareas of mathematics (e.g., Group Theory). Here our interest

S.M. Watt et al. (Eds.): CICM 2014, LNAI 8543, pp. 252–266, 2014.
© Springer International Publishing Switzerland 2014

in this structure is purely pragmatic: how can it be leveraged to build a better mechanized mathematics systems and, ultimately, a better UDLM.

We will consider this in a bottom-up manner: what is the most natural structure on *theories* that allows us to abstract away from irrelevant details, yet still allow us to get some practical work done? One such structure is that of *mutual interpretability* between theories. Basically this is the case when we have two equivalent theories T_1 and T_2 (in a sense to be made precise later) with presentations that can be markedly different.

But why should *theory presentations* matter at all? Studies of "theories" in mathematics (e.g., Lawvere theories [Law04; LR11]) or in logic focus on entities that are *complete* in some sense. But such completeness generally also implies that the object at hand is effectively infinite, and thus cannot be directly represented in software. Hence we are immediately forced to work with *finite representations* of these infinite objects. Furthermore, by Gödel's incompleteness theorem, most of the interesting theories will be fundamentally incomplete, in that no finite representation will be able to adequately represent the complete whole. The relevance here is that we are forced to deal with (syntactic) representations, which will generally be *incomplete*. As this is forced on us, we need to gracefully adapt to dealing with theory presentations in place of the theories they represent.

2 The Setting: Theory Graphs

We will now present an abstract notion of a theory graph that is sufficient to introduce the notion of a realm without committing ourselves to a particular approach such as [CO12] or [RK13].

Let a **theory** be a presentation of an axiomatic theory consisting of a finite sequence of symbol and formula declarations. The symbols denote concepts and the formulas denote facts about these concepts. There are three kinds of formula declarations: **axioms, definitions**, and **theorems**.[1] We further assume that for a theory $T = [A_0, A_1, \ldots, A_n]$, for all i with $0 \leq i < n$, A_{i+1} is well formed in the context of $[A_0, A_1, \ldots, A_i]$. A theory thus represents an axiomatization of a mathematical topic. If T is a theory and A is a symbol or formula declaration, $T \vdash A$ **wf**, means that A is well formed in the context of theory T. When $T \vdash A$ **wf**, we define $T \ltimes A$ to mean $[A_0, A_1, \ldots, A_n, A]$; we also extend $\ltimes$ to apply to sequences of declarations (on the right). If T is a theory and φ is a formula, then $T \models \varphi$ means φ is a logical consequence of T. A theory is **primitive** if it contains only symbol declarations and axioms. A primitive theory represents a set of concepts and facts without a development structure. The **empty theory** is the empty sequence. When $T_1 = T \ltimes A$ and $T_2 = T \ltimes B$, we also

---

[1] If we make use of the Curry-Howard isomorphism — as we do in [RK13], then we can get by with typed symbol declarations (with optional definitions) only. In the propositions-as-types paradigm, axioms are typed constant declarations, and theorems are typed definitions — a proof corresponds to the respective definiens of a symbol which is of the respective type.

define $T_1 \oplus T_2 := (T \ltimes A) \ltimes B$ whenever A and B are disjoint (i.e., they declare symbols or formulas with different names). By also extending $\oplus$ to sequences of declarations, we get a **join** operation on theory presentations.

A **theory graph** is a directed graph in which a node is a theory and we have edges from a theory T to a theory $T \ltimes A$. If T and T' are theories in a theory graph G, an edge from T to T' is designated as $T \xrightarrow{G} T'$. A theory graph is a modular representation of a formalized body of mathematical knowledge.

An **axiomatic development** of a theory T to a theory T' in G is a subgraph G' of G in which T is the only source of G' and T' is the only sink of G'. In this case, T and T' are called the **bottom theory** and the **top theory**, respectively. An axiomatic development is thus a lattice of theories in which the top theory is the join of the members of the lattice.

Let T_1 and T_2 be theories in G. A **view of T_1 in T_2** is a homomorphic mapping Φ of the language of T_1 to the language of T_2 such that, for each formula φ of T_1, $T_1 \models \varphi$ implies $T_2 \models \Phi(\varphi)$. We denote a view by $\Phi : T_1 \rightsquigarrow T_2$. A view is thus a meaning preserving mapping that shows how T_1 can be embedded in T_2. It also provides a mapping from the models of T_2 to the models of T_1 (note the reversal of order). T_1 and T_2 are **equivalent** if there is a view in both directions. A view $\Phi : T_1 \rightsquigarrow T_2$ is **faithful** if for each formula φ of T_1, $T_2 \models \Phi(\varphi)$ implies $T_1 \models \varphi$. Views give a second (oriented, multi) graph structure on G, making it into a bigraph. It is important to note that the base theory graph (with edges but not views) is always acyclic.

Let T and T' be theories. T' is an **extension** of T, and T is a **subtheory** of T', if there exists a sequence S such that $T' = T \ltimes S$. In this case, there is a view $\Phi : T \rightsquigarrow T'$ such that Φ is the identity function. We call Φ the **inclusion** of T into T' and denote it by $T \hookrightarrow T'$. This corresponds to the extensions of [CO12], identity structures in [RK13], as well as the display maps of categorical type theory.

An **interface** for T is a view $\Phi : T' \rightsquigarrow T$ such that Φ is injective. T' and T are called, respectively, the **front** and **back** of the interface. Each subtheory of T can be a front of an interface for T. An interface is intended to be a convenient means for accessing (parts of) T. The front of a good interface includes a carefully selected set of symbols and formulas that denote orthogonal concepts and facts that can be easily combined to express the other concepts and facts of T. See section 5 for some concrete examples.

An extension T' of T is **conservative** if there is a view $\Phi :$ $T' \rightsquigarrow T$ such that Φ is the identity function when restricted to T (i.e., $\Phi \circ \iota = \mathrm{Id}_T$ where ι is the inclusion of T in T'). If T' is a conservative extension of T, then for each formula φ of T, $T' \models \varphi$ implies $T \models \varphi$. Common examples of conservative extensions are extensions by symbol declarations, definitions,

or theorems (with proofs). Obviously, if T' is a conservative extension of T, then T and T' are equivalent. We abbreviate the two arrows in a conservative extension with a double inclusion arrow in theory graphs. A subgraph G' of a theory graph G is conservative if T' is a conservative extension of T for each edge $T \xrightarrow{G} T'$ in G'. A **conservative development** is an axiomatic development that is conservative. Note that all the theories in a conservative development are equivalent to each other. We will write a conservative development with bottom theory S and top theory T as $S \Longleftrightarrow T$.

$\Phi : T_1 \rightsquigarrow T_2$ is **expansive** if there is a $\Psi : T_2 \rightsquigarrow T_2$ such that (1) the range of Ψ is the image of Φ and (2) Ψ is the identity on the image of Φ. That is, a view of T_1 in T_2 is expansive if, roughly speaking, T_2 is a conservative extension of the view's image. A view of T is **conservative** if it is faithful and expansive. A conservative view of T is a generalization of a conservative extension of T. If there is a conservative view of T_1 in T_2, then T_1 and T_2 are mutually viewable.

3 Motivation: Developers, Students, and Practitioners

The user of a UDLM can play three different roles. As a *developer*, the user creates new representations of mathematical knowledge or modifies existing ones in the library. As a *student*, the user studies the mathematical knowledge represented in the library. And, as a *practitioner*, the user applies the mathematical knowledge in the library to problems, both theoretical and practical. A user may play different roles at different times and may even sometimes combine roles.

A theory graph does not fully support all three of the user's roles. In fact, it lacks the structure that is necessary to satisfy the following requirements:

R1 There can be many *equivalent theories* in a theory graph that represent different axiomatizations of the *same mathematical topic*. As a result, concepts and facts about this mathematical topic, possibly expressed in different languages, may be widely distributed across a theory graph. The developer and the student would naturally want to have these *different axiomatic developments* and the *set of concepts and facts* that are produced by them in *one convenient place and in one convenient language*.

R2 Developers prefer developments that start with *minimal bottom theories* and are built as much as possible using *conservative extensions*. This approach minimizes the chance of *introducing inconsistencies* (which would render the developments pointless) and maximizes the *opportunities for reusing the development in other contexts*. While these two benefits may not be of primary concern for the student and the practitioner, such a careful development is usually easier to understand and produces concepts and facts that can be more reliably applied.

R3 The developer would like to *create a view from one theory to another in a convenient manner* by starting with a view of a minimal axiomatization of the theory and then building up the view as needed using conservative extensions. Also, there is a desire to use *the most convenient axiomatization* amongst equivalent presentations.

R4 The *application of a mathematical fact* usually does not require an understanding of how concepts and facts were derived from first principles. Hence the parts of the theory graph which were needed by the developer may not be useful to the practitioner, and may well get in the way of the practitioner's work. The practitioner would naturally want the concept or fact to be *lifted out of this tangled development bramble*.

R5 Languages are introduced in a theory graph for the purpose of theory development. They may employ vocabulary that is inconvenient for particular applications. The practitioner would naturally like to have *vocabulary chosen for applications instead of development*.

In summary, the developer, the student, and the practitioner have different concerns that are not addressed by the structure of a theory graph. These different concerns lead us to propose putting some additional structure on a theory graph, a notion we call a "realm", to meet these five requirements.

4   Realms

In a nutshell, a realm identifies a subgraph of a development graph, equips it with a carefully chosen interface theory that abstracts from the development, and supplies the practitioner with the symbols and formulas she needs.

Definition 1. A **realm** R is a tuple $(G, F, \mathcal{C}, \mathcal{V}, \mathcal{I})$ where:

1. G is a theory graph.
2. F is a primitive theory in G called the **face** of the realm R.
3. $\mathcal{C}$ is a set $\{C_1, C_2, \ldots, C_n\}$ of conservative developments in G.
4. $\mathcal{V}$ is a set of views that establish that the bottom theories $\bot_1, \bot_2, \ldots, \bot_n$ of $C_1, C_2, \ldots, C_n$, respectively, are pairwise equivalent.
5. $\mathcal{I}$ is a set $\{I_1, I_2, \ldots, I_n\}$ of conservative interfaces such that F is the front of I_i and the top theory $\top_i$ of C_i is the back of I_i for each i with $1 \le i \le n$.

For each i we call $(\bot_i, C_i, \top_i)$ the i-**th pillar** of R and I_i its **interface**. Note that every subset of pillars of a realm R forms a realm together with its interface and the face of R. We call realms with just one pillar **simple** and realms with more than one pillar **proper**.

Figure 1 shows the general situation, we depict realms by double dashed boxes and faces by dashed ones. All the theories in the realm R are equivalent to each other since *i*) all the bottom theories are equivalent by the views in $\mathcal{V}$, *ii*) all the members of a conservative development are equivalent, and *iii*) the front and back of a conservative interface are equivalent.

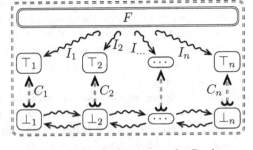

Fig. 1. The Architecture of a Realm

To ensure that realms have a pleasant categorical structure (that of a contractible groupoid), we assume that $\mathcal{V}$ always contains identity views which show self-equivalence.

A realm consolidates a body of formalized mathematics pertaining to one topic. Each bottom theory $\bot_i$ is a different (ideally minimal) axiomatization and each conservative development C_i is a family of extensions of $\bot_i$. F is an (ideally convenient) presentation of the topic without any development structure and without any *scaffolding*, i.e., the concepts and facts that are needed only for development purposes. Finally, each interface I_i establishes that F is indeed a presentation of the topic and how it embeds into each $\top_i$.

The realm $R = (G, F, \mathcal{C}, \mathcal{V}, \mathcal{I})$ minus F and $\mathcal{I}$ records the development structure of the topic; we call $\overline{R} := (G, \mathcal{C}, \mathcal{V})$ the **body** of R. It can be used to study the development structure of the topic or as a basis for extensions. The face F exposes the most important and useful concepts and facts pertaining to that realm. It is also meant to be used as a module for constructing larger bodies of formalized mathematics. In other words, it can be seen as an export facility that only exports carefully selected symbols and formulas from the realm, without duplication or redundancy. Note that, in practice, we will choose for F the "usual symbols" traditionally used for that theory; these will also often correspond to the "original symbols" used in (some of) the bottom theories.

In particular, realms offer the infrastructure to satisfy the five requirements for users of a UDLM described in section 3. **R1** is addressed by the set of theories in the realm R and by the concepts and facts in F. **R2** is addressed by the conservative developments in $\mathcal{C}$. **R3** is addressed by the views in $\mathcal{V}$ and the conservative developments in $\mathcal{C}$. **R4** and **R5** are addressed by F being primitive and the fact that F is the front of an interface to each top theory.

Example 1 (Trivial realm). Any theory S in G induces a simple realm $R = (G, S, \{G_S\}, \{\mathtt{Id}_S\}, \{\mathtt{Id}_S\})$ where G_S is the subgraph G consisting of S alone and $\mathtt{Id}_S$ is the identity view on S. Thus S serves as the top and bottom theories of the trivial conservative development of S, as well as the face of R.

Example 2 (Initial realm). For any theory T in G we can extend G with a copy F_T of T and a conservative interface $F_T \overset{\iota}{\hookrightarrow} T$, where ι maps any symbol to its copy to obtain a theory graph G'. We call $R_G^T := (G', F_T, \{G_T\}, \{\mathtt{Id}_S\}, \{\iota\})$, where G_T is the subgraph of G consisting of T alone, the **initial realm** for T in G.

We can project any realm R to its face F, forgetting all developmental structure. Note that we can lift views between theories to **realm morphisms** (theory morphisms between their faces): given two realms R_1 and R_2 with two interfaces I_i, fronts F_i, and top theories $\top_i$, a view

$\top_1 \overset{v}{\leadsto} \top_1$ induces a partial view $F_1 \overset{\tilde{v}}{\leadsto} F_2$ on the faces, where $\tilde{v} = I_2^{-1} \circ v \circ I_1$ (see Figure 2). In practice, the lifted views will almost always be total, since we prefer to use (in $\top_i$ and F_i) the original symbols from the bottom theories.

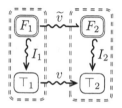

Fig. 2. Lifting

5 Examples

As the development of the last few sections is fairly abstract, we will attempt to give the reader a better feel for realms through a selection of examples. We develop the first one in some detail and then give a more intuitive (and thus shorter) description of the remaining examples.

5.1 Groups

It is well known that groups can alternatively be described in two ways:

> *Definition 1* [KM79]. A **group$_1$** is a set G together with an associative binary operation $\circ \colon G \times G \to G$, such that there is a unit element e for $\circ$ in G, and all elements have inverses.
>
> *Definition 2* [Hal59]. A **group$_2$** is a set G, together with a (not necessarily associative) binary operation $/ \colon G \times G \to G$, such that $a/a = b/b$, $a/(b/b) = a$, $(a/a)/(b/c) = c/b$, and $(a/c)/(b/c) = a/b$ for all $a, b, c \in G$.

For any group$_1$ $(G, \circ)$, we can define a binary operation $/_\circ$ by $a /_\circ b := a \circ b^{-1}$ that shows that $(G, /_\circ)$ is a group$_2$, and vice versa — using $a \circ_/ b := a / b_/^{-1}$ with $b_/^{-1} := (b/b)/b$. Practitioners want to use both group multiplication and division but are usually indifferent to how and where they are introduced.

In Figure 3, we have assembled this situation into a two-pillar realm with face group. The two bottom theories group$_1$ and group$_2$ are equivalent via the views v_1 and v_2 and the back views of c_1 and c_1, respectively.[2] Note that the views $v_1 = \circ \mapsto \circ_/, e \mapsto e_/, i \mapsto i_/$ and $v_2 = / \mapsto /_\circ$ carry proof obligations that show that the newly defined extensions slash$_1$ and circ-i$_2$ behave as expected by the group$_{3-i}$. The face group contains "new" symbols, for which we use underlined symbols, to distinguish them from the ones in the pillars of the realm. The interface views I_i pick out the respective "original operators" $/$, $\circ$, and i, together with the corresponding axioms (and any theorems that may have been proven along the way). Here we have

$$I_1 = \underline{\circ} \mapsto \circ, \underline{e} \mapsto e, \underline{i} \mapsto i, \underline{/} \mapsto /_\circ \quad \text{and} \quad I_2 = \underline{\circ} \mapsto \circ_/, \underline{e} \mapsto e_/, \underline{i} \mapsto i_/, \underline{/} \mapsto /$$

In particular, it is very natural to require that the interfaces of a realm are conservative since they have access to all symbols in the body of the realm.

5.2 Natural Number Arithmetic

A realm $\mathbb{N}$ of natural number arithmetic would naturally contain conservative developments of several different axiomatizations of the natural numbers with

---

[2] This is a very common situation; the base theories differ mainly in which symbols are considered primitive, and the conservative developments mainly introduce definitions for the remaining ones.

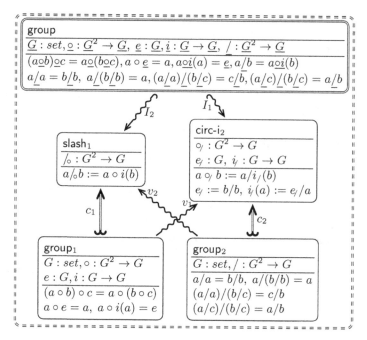

Fig. 3. A Realm of Groups with Face group

the usual arithmetic operations. One conservative development would certainly start with Peano's axiomatization of the natural numbers [Pea89]. The base theory would contain the symbols 0 and S (the successor function) and the (second-order) Peano axioms. The conservative development would include recursive definitions of + (addition) and * (multiplication). This development is particularly useful as it makes the proofs of many properties of the natural numbers simple.

Another kind of conservative development would start with a construction of the natural numbers using machinery available in the underlying logic. There are many such constructions. Some examples are finite von Neumann ordinals constructed from sets, Church numerals constructed from lambda expressions, strings of bits, and various bijective numeration schemes. These constructions define representations of the natural numbers that are semantically equivalent but far from equivalent with respect to computational complexity. It is worth singling out the *sequences of machine-sized words* representation, which tends to be the most efficient.

The face $F_{\mathbb{N}}$ of $\mathbb{N}$ would be restricted to the most basic concepts and facts about natural number arithmetic. These would naturally include symbols for all the natural numbers (i.e., natural number numerals) and all the true equations of the form $n_1 + n_2 = n_3$ and $n_1 * n_2 = n_3$. Thus $F_{\mathbb{N}}$ would contain an infinite number of symbols and facts. An implementation of $F_{\mathbb{N}}$ would require an efficient means to represent and compute with natural number numerals. *Biform theories* [Far07] would be best suited for such a task.

This realm $\mathbb{N}$ that we have described is a multi-pillar presentation of the mathematical topic of natural number arithmetic. It can be used by developers as a module with which to build more complex theories and by students who are interested in understanding what are the basic concepts and facts of this topic and how they are derived from first principles. A realm like $\mathbb{N}$ that contains several pillars and a face of basic concepts and facts is called a **foundational realm**. Used for building new theories and for study, a foundational realm would not be expected to change much over time.

Since the mathematical theory of natural number arithmetic is exceedingly rich, there are a great many concepts and facts about natural numbers that could be of use to practitioners. For developers and students, the usefulness of $\mathbb{N}$ would be greatly reduced if there was an attempt to include all of these concepts and facts in $F_{\mathbb{N}}$. It would be much better for practitioners — who are primarily interested in applications — to create another single-pillar realm $\mathbb{N}'$ of natural number arithmetic whose face would contain all the useful concepts and facts about natural numbers that have been derived someplace in the theory graph. A realm of this type is called a **high-level realm**. Used for applications, a high-level realm would be continuously updated as new concepts and facts are discovered. It would be an implementation of the idea of a *high-level theory* discussed in [CF08].

5.3   Real Numbers

The theory of the real numbers covers the algebraic and topological structure of the real numbers. Rich in concepts and facts, it is one of the most important theories in all of mathematics. It is important to developers since real numbers are needed in most mathematical developments. It is important to students since it includes many of the most important ideas of mathematics. It is important to practitioners since most mathematical problems involve the real numbers in some way.

A realm $\mathbb{R}$ of the real numbers could be used to consolidate and organize all the knowledge about the real numbers that resides in a UDLM. $\mathbb{R}$ would have a structure similar to the realm of natural number arithmetic. It would contain two kinds of conservative developments. The first kind are axiomatizations of a complete ordered field – all complete ordered fields are isomorphic. The second kind are constructions of the real numbers, of which there are many. Some examples are Dedekind cuts in the field of rational numbers, Cauchy sequences of rational numbers, infinite decimal expansions, the quotient of the finite hyperrationals by the infinitesimal hyperrationals, and as a substructure of the surreal numbers. It is worth remarking that most of these constructions leverage $\mathbb{N}$, so that constructing realms is also a modular process.

The realm $\mathbb{R}$ would be a foundational realm like $\mathbb{N}$ for developers and students. There should also be a high-level realm $\mathbb{R}'$ like $\mathbb{N}'$ for practitioners. The face of $\mathbb{R}$ would only contain the basic concepts and facts about the real numbers, while the face of $\mathbb{R}'$ would contain all the useful concepts and facts about the real numbers that have been derived someplace in the theory graph. The prominent role of the real numbers would mean that $\mathbb{R}$ or $\mathbb{R}'$ would be the basis of many of the more sophisticated theories in a UDLM.

5.4 Monads

Category theorists and (advanced) Haskell programmers are familiar with the expressive power of monads. Most know that there are in fact two equivalent presentations of the theory of monads, one using a multiplication operation μ (called `join` in Haskell) and unit η (`return`), the other using *Kleisli triples* with a lifting operation $-^*$ (called `bind` or `>>=` in Haskell). From there, one can define a large list of generic combinators that work for any monad.

These two presentations are equivalent, and are again similar in flavor to the previous ones: one is more convenient for proofs, the other for computational purposes. Again, these basic theories tend to be followed by a substantial tower of *conservative extensions*. In other words, Haskell's `Control.Monad` should really be seen as the *face* of a realm of monads.

5.5 Modal Logic S4

The modal logic **S4** has a large number of equivalent presentations — John Halleck [Hal] lists 28 of them. This gives developers significant flexibility when using views (aka requirement **R3**) to establish that a structure can interpret **S4**. And, of course, **S4** supports rather significant conservative extensions and applications of it are found in a variety of places.

5.6 Models of Computation

The *Chomsky hierarchy* of regular, context-free, context-sensitive and recursively enumerable languages offer names for (the face of) four more, nested, realms. As is well known, each of the above languages contains many different formalisms which are nevertheless equivalent.

Inside the recursively enumerable languages (for example), we would have the pillars of Turing machines, Register Machines, the Lambda Calculus, certain automata, etc, as alternatives. It is difficult to design a suitable face theory for this realm, as the syntax of any high-level programming language could serve; given the heated debates around what language is "best", this is one realm whose face may not settle for a long time.

6 The Realm Idea

The examples of the previous section show the advantages of realms as consolidated structures: a realm hides cumbersome details, while still allowing access to the details for those (such as developers) who must deal with them. Realms thus deal with two structural tensions in the design of theory graphs that formalize a mathematical domain:

Foundational realms can in many ways be understood as the formalization of the ideas of *information hiding* and *modules* coming from software engineering. The face of a realm corresponds to an interface; its secrets, i.e., what it hides, is the actual conservative development of the theory; and its representation details correspond to an axiomatization. Of course, to get substitutivity, we need to ensure equivalence. In an ad hoc manner, Haskell's type classes, ML's modules and functors, Scala's traits, Isabelle's locales (etc) all capture certain aspects of realms. However, the lack of good support for *views* really hampers the use of these proto-realms as a modular development mechanism.

High-level realms give practitioners high-level collections of useful symbols and formulae that function like a tool-chest for applications based on the tiny theories developers use as a fine-grained model of dependencies, symbol visibilities, and consistency. For them theories should be static over time, depicting a completed axiomatic development of a mathematical topic. This gives a persistent base (and rigid designators) to develop against. But this means that conservative extensions (like definitions and theorems) need new theories, leading to a severe pollution of the theory namespace. Practitioners, on the other hand, would naturally prefer dynamic theories that continuously grow as new concepts and facts are introduced (another kind of rigid designator).

The contribution of realms is an overlay structure that can implement information hiding, and mediates between dynamic high-level theories and an underlying, static theory graph. So users can have their cake and eat it.

7   Representing and Growing Realms in a UDLM

Our work on OMDoc/MMT [MMT; KRZ10] and the MathScheme [CFO11] systems have given us a decent intuition (or so we feel) regarding the services that a theory-graph based system should provide. We now extent this to realms.

Marking Up Realms. If we look at the definition of a realm, we see that the body components are already present in the theory graph given by the existing axiomatic developments. Thus, given a theory graph G, we can add a realm R by just tagging a subgraph of G and adding a set of interfaces with their common front (the face of R); all of these are regular components of theory graphs, so we only need to extend the theory graph data structures (and representation languages) by a "realm tagging" functionality. This also shows us that the concept of realms is conservative over theory graphs.

One can easily envision two methods of syntactically identifying realms: globally via a theory-level "realm declaration" which specifies the five components from Definition 1, or locally by extending theory and view declarations with a field that specifies the realm (or realms) it participates in. Given the little theory approach, we tend to use theory extensions and view declarations when the local context is clear (for example, within a single "file"), and the more global approach when drawing from a wider context. This appears to be a good syntactic compromise.

An implementation will have to check the internal constraints from Definition 1, in particular, that interfaces are total. But the idea of simply "discovering" realms that occur in the wild is a bit optimistic. From our case studies, we expect that realms have to be engineered purposefully: they are grown from a seed, and grow over time by coordinated (and system-supported) additions of theories and views.

7.1   Supporting the Life Cycle of Realms

We postulate that three realm-level operations will be needed in practice: *i)* realms are initialized by designating chosen theories as initial realms, which *ii)* can be extended by adding conservative extensions, and *iii)* proper realms are created by merging existing realms. These three operations were sufficient to explain the complex realms in our case studies. We will now discuss them in more detail.

Initializing Realms. Given a theory graph G, we add any realm (e.g., the initial realm R_G^T for a theory T in G; see Example 2) as a starting point of development.

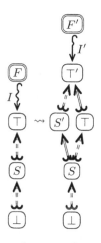

Extending Realms by (internal) conservative extensions. Given a realm $R := (G, F, C, V, \mathcal{I})$, a top theory $\top$ of some $C \in \mathcal{C}$, and an interface I for $\top$, then we can extend R by:

i) adding a conservative extension $S \Leftrightarrow S'$ by declaration c and a commensurate extension $\top \Leftrightarrow \top'$ to C and

ii) (optionally) adding a declaration $\underline{c}$ to F, giving a new face F', and extending I so that $I' := I, \underline{c} \mapsto c$. If we do – e.g., for a high-level view – we have to apply i) to each of the pillars of R, so that all their interfaces are total; the diagram shows the situation for a simple realm.

realm extension

In particular, an implementation of high-level realms must provide a registration functionality for conservative extensions in C that keeps the interface(s) consistent by ensuring new names appear in the face of the realm. Note that this extension operation does not change the number of pillars of a realm, in particular, if realms are started by initial realms, they will only be extended to simple realms. It is predominantly used for high-level realms.

The next operation merges two realms if they are mutually interpretable. This operation is mainly used to build up foundational views, the construction makes sure that all symbols in the face are interpreted in all the pillars.

Merging Realms along Views. Given two realms $R_1 := (G, F_1, C_1, V_1, \mathcal{I}_1)$ and $R_2 := (G, F_2, C_2, V_2, \mathcal{I}_2)$, and views $\bot_1 \overset{v}{\rightsquigarrow} \bot_2$ and $\bot_2' \overset{w}{\rightsquigarrow} \bot_1'$, where $\bot_i$ and

$\perp_i'$ are (arbitrary) bottom theories in $\mathcal{C}_i$, then we can define the union realm $R_1 \cup_w^v R_2$ along v and w as $(G, F_1 \cup F_2, \mathcal{C}_1^{+w} \cup \mathcal{C}_2^{+v}, \mathcal{V}_1 \cup \mathcal{V}_2 \cup \{v,w\}, \mathcal{I}_1^{+w} \cup \mathcal{I}_2^{+v})$. Figure 4 shows the situation for two simple realms. Generally, we define that:

i) $\mathcal{C}_1^{+w}$ is the set of conservative developments $\{C^{+w} \mid C \in \mathcal{C}_1\}$, where C^{+w} is C extended by a copy[3] of the development of $\perp_2'$ to T_2' along w, itself extended to $\mathsf{T} \cup w(\mathsf{T}_2')$. $\mathcal{C}_2^{+v}$ is defined analogously. In Figure 4, $\mathcal{C}_2^{+v}$ and $\mathcal{C}_1^{+w}$ are the two diamonds on the left and right.

ii) $F_1 \cup F_2 \xrightarrow{\mathcal{I}_1^{+w}} \mathsf{T}_1 \cup w(\mathsf{T}_2)$ is $I_1 \cup_w \circ I_2$ and $F_1 \cup F_2 \xrightarrow{\mathcal{I}_2^{+v}} \mathsf{T}_2 \cup v(\mathsf{T}_1)$ is $I_2 \cup v \circ I_1$. An implementation of this construction would take great care to merge corresponding symbols in the two faces to minimize the union. Moreover, the copying operation can be optimized to only copy over those conservative extensions that are mentioned in the interface extension.

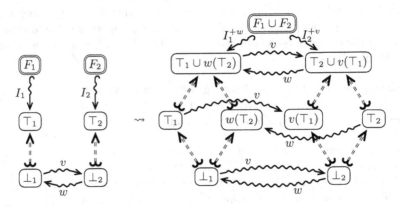

Fig. 4. Union Realm

7.2 Modular Realms

Note that the extension and merging operations highlight an internal invariant of realms that may not have been obvious until now: All pillars of a realm must interpret the full vocabulary of the face to admit total interfaces. This duplication can become quite tedious in practice. Therefore it is good practice to modularize realms in the spirit of a "little realms approach". For instance, the groups realm from Figure 3 could be extended by the usual group theorems via conservative extensions in both pillars. But we can also build a simple realm with base theory group and extend that conservatively (once per theorem). Unless there are proofs that directly profit from the particulars of the concrete formulations in the pillars below, the modular approach is more efficient representationally and thus more manageable.

---

[3] A copy of a development (sub)graph H along a view v is an isomorphic graph H', where for any theory S in H, S' in H' consists of the declarations $c : v(\tau) = v(\delta)$, for all $c : \tau = \delta$ in S. This construction gives us a view $S \xrightarrow{v} S'$.

7.3   Interface Matters

As the realms are the main interaction points for mathematicians with the UDLM, realms must be discoverable and provide a range of convenient information retrieval methods (after all, realms will get very large in practice). These can range from community tools like peer reviewed periodicals (aka. academic journals) to technical means like intra- and cross-realm search engines (as realms are built upon theory graphs, specialization of the ♭search engine [KI12] will be a good starting point.)

It will be very important to provide a set of interactions for the interface of a realm that users can understand. It will be important to look up the definienda and proofs of interface items, even though this will usually mean that we need to descend into (conservative extensions of) one of the fronts of the interface, which employ different languages. This needs to be transparent enough to be understandable to users/mathematicians.

Similarly, the equivalence relation of the (tiny) theories that make up the realm should be made transparent and easy to browse for the user.

8   Conclusion

We have presented an extension of the theory graph approach to representing mathematical knowledge. Realms address the mismatch between the successful practice of the little/tiny theory approach natural for developing theory graphs and the high-level theories most useful for practitioners utilizing such mathematical knowledge representations. We have proposed a formal definition for realms that is conservative over theory graphs and shown its adequacy by applying it to examples from various areas of mathematics and computation.

As a step towards an implementation we have investigated a set of realm-level operations that can serve as a basis for system support of realm management. The next step in our investigation will be to realize and test such support in the OMDoc/MMT [MMT; KRZ10] and the MathScheme [CFO11] systems, fully develop the examples sketched in this paper, and test the interactions on developers, students, and practitioners (see section 3).

References

[CF08]   Carette, J., Farmer, W.M.: High-Level Theories. In: Autexier, S., Campbell, J., Rubio, J., Sorge, V., Suzuki, M., Wiedijk, F. (eds.) AISC/Calculemus/MKM 2008. LNCS (LNAI), vol. 5144, pp. 232–245. Springer, Heidelberg (2008)

[CFO11]   Carette, J., Farmer, W.M., O'Connor, R.: MathScheme: Project description. In: Davenport, J.H., Farmer, W.M., Urban, J., Rabe, F. (eds.) Calculemus/MKM 2011. LNCS (LNAI), vol. 6824, pp. 287–288. Springer, Heidelberg (2011)

[CO12]   Carette, J., O'Connor, R.: Theory Presentation Combinators. In: Jeuring, J., Campbell, J.A., Carette, J., Dos Reis, G., Sojka, P., Wenzel, M., Sorge, V. (eds.) CICM 2012. LNCS (LNAI), vol. 7362, pp. 202–215. Springer, Heidelberg (2012)

[Far07]   Farmer, W.M.: Biform Theories in Chiron. In: Kauers, M., Kerber, M., Miner, R., Windsteiger, W. (eds.) MKM/Calculemus 2007. LNCS (LNAI), vol. 4573, pp. 66–79. Springer, Heidelberg (2007)

[Far11]   Farmer, W.M.: Mathematical Knowledge Management. In: Schwartz, D., Te'eni, D. (eds.) Encyclopedia of Knowledge Management, 2nd edn., pp. 1082–1089. Idea Group Reference (2011)

[FGT92]   Farmer, W.M., Guttman, J.D., Thayer, F.J.: Little Theories. In: Kapur, D. (ed.) CADE 1992. LNCS, vol. 607, pp. 567–581. Springer, Heidelberg (1992)

[Hal]   Halleck, J.: http://home.utah.edu/ nahaj/logic/structures/systems/s4 .html (accessed: March 14, 2014)

[Hal59]   Hall, M.: The Theory of Groups. The Macmillan Company, New York (1959)

[KI12]   Kohlhase, M., Iancu, M.: Searching the Space of Mathematical Knowledge. In: Sojka, P., Kohlhase, M. (eds.) DML and MIR 2012. Masaryk University, Brno (2012) (in press) ISBN: 978-80-210-5542-1

[KM79]   Kargapolov, M.I., Merzljakov, J.I.: Fundamentals of the Theory of Groups. Graduate Texts in Mathematics. Springer (1979)

[KRZ10]   Kohlhase, M., Rabe, F., Zholudev, V.: Towards MKM in the Large: Modular Representation and Scalable Software Architecture. In: Autexier, S., Calmet, J., Delahaye, D., Ion, P.D.F., Rideau, L., Rioboo, R., Sexton, A.P. (eds.) AISC/Calculemus/MKM 2010. LNCS (LNAI), vol. 6167, pp. 370–384. Springer, Heidelberg (2010), arXiv:1005.5232v2[cs.OH]

[Law04]   Lawvere, W.F.: Functorial Semantics of Algebraic Theories. Reprints in Theory and Applications of Categories 4, 1–121 (2004)

[LR11]   Lack, S., Rosický, J.: Notions of Lawvere Theory. English. Applied Categorical Structures 19(1), 363–391 (2011), doi:10.1007/s10485-009-9215-2, ISSN: 0927-2852

[MMT]   Rabe, F.: The MMT Language and System, https://svn.kwarc.info/repos/MMT (visited on November 10, 2011)

[Pea89]   Peano, G.: Arithmetices principia nova methodo exposita. Bocca, Turin (1889)

[RK13]   Rabe, F., Kohlhase, M.: A Scalable Module System. Information & Computation (230), 1–54 (2013)

Matching Concepts across HOL Libraries

Thibault Gauthier and Cezary Kaliszyk

University of Innsbruck, Austria
{thibault.gauthier,cezary.kaliszyk}@uibk.ac.at

Abstract. Many proof assistant libraries contain formalizations of the same mathematical concepts. The concepts are often introduced (defined) in different ways, but the properties that they have, and are in turn formalized, are the same. For the basic concepts, like natural numbers, matching them between libraries is often straightforward, because of mathematical naming conventions. However, for more advanced concepts, finding similar formalizations in different libraries is a non-trivial task even for an expert.

In this paper we investigate automatic discovery of similar concepts across libraries of proof assistants. We propose an approach for normalizing properties of concepts in formal libraries and a number of similarity measures. We evaluate the approach on HOL based proof assistants HOL4, HOL Light and Isabelle/HOL, discovering 398 pairs of isomorphic constants and types.

1 Introduction

Large parts of mathematical knowledge formalized in various theorem provers correspond to the same informal concepts. Basic structures, like integers, are often formalized not only in different systems, but sometimes also multiple times in the same system. There are many possible reasons for this: the user may for example want to investigate special features available only for certain representations (like code extraction [4]), or simply check if the formal proofs can be done in a more straightforward manner with the help of alternate definitions. With multiple proof assistants, even the definitions of basic concepts may be significantly different: in Isabelle/HOL [21] the integers are defined as a quotient of pairs of naturals, while in HOL Light [6] they are a subset of the real numbers. Typically the proofs concerning a mathematical concept formalized in one system are not directly usable in the other, so a re-formalization is necessary.

The idea of exchanging formal developments between systems has been investigated both theoretically and practically many times [10, 14, 16]. Typically when a concept from the source systems is translated to a target system, and the same concept exists in the target system already, a new isomorphic structure is created and the relation between the two is lost. The properties that the two admit are the same and it is likely that the user formalized many similar ones.

In this work we investigate automatic discovery of such isomorphic structures mostly in the context of higher order logic. Specifically the contributions of this work are:

S.M. Watt et al. (Eds.): CICM 2014, LNAI 8543, pp. 267–281, 2014.

- We define patterns and properties of concepts in a formal library and export the data about constants and types from HOL Light, HOL4, and Isabelle/HOL together with the patterns.
- We investigate various scoring functions for automatic discovery of the same concepts in a library and across formal libraries and evaluate their performance.
- We find 398 maps between types and constants of the three libraries and show statistics about the same theorems in the libraries, together with normalization of the shape of theorems.

There exists a number of translations between formal libraries. The first translation of proofs that introduced maps between concepts was the one of Obua and Skalberg [16]. There, two commands for mapping constructs were introduced: type-maps and const-maps that let a user map HOL Light and HOL4 concepts to corresponding ones in Isabelle/HOL. Given a type (or constant) in the maps, during the import of a theorem all occurrences of this type in the source system are replaced by the given type of the target system. In order for this construction to work, the basic properties of the concepts must already exist in the target system, and their translation must be avoided. Due to the complexity of finding such existing concepts and specifying the theorems which do not need to be translated, Obua and Skalberg were able to map only small number of concepts like booleans and natural numbers, leaving integers or real numbers as future work.

The first translation that mapped concepts of significantly different systems was the one of Keller and Werner [14]. The translation from HOL Light to Coq proceeds in two phases. First, the HOL proofs are imported as a defined structures. Second, thanks to the *reflection* mechanism, native Coq properties are built. It is the second phase that allows mapping the HOL concepts like natural numbers to the Coq standard library type N.

The translation that maps so far the biggest number of concepts has been done by the second author [10]. The translation process consists of three phases, an exporting phase, offline processing and an import phase. The offline processing provides a verification of the (manually defined) set of maps and checks that all the needed theorems will be either skipped or mapped. This allows to quickly add mappings without the expensive step of performing the actual proof translation, and in turn allows for mapping 70 HOL Light concepts to their corresponding Isabelle/HOL counterparts. All such maps had to be provided manually.

Bortin et al. [1] implemented the AWE framework which allows the reuse of Isabelle/HOL formalization recorded as a proof trace multiple times for different concepts. Theory morphisms and parametrization are added to a theorem prover creating objects with similar properties. The use of theory morphisms together with concept mappings is one of the basic features of the MMT framework [17]. This allows for mapping concepts and theorems between theories also in different logics. So far all the mappings have been done completely manually.

Hurd's OpenTheory [9] aims to share specifications and proofs between different HOL systems by defining small theory packages. In order to write and read such theory packages by theorem prover implementations a fixed set of concepts is defined that each prover can map to. This provides highest quality standard among the HOL systems, however since the procedure requires manual modifications to the sources and inspection of the libraries in order to find the mappings, so far only a small number of constants and types could be mapped. Similar aims are shared by semi-formal standardizations of mathematics, for example in the OpenMath content dictionaries. For a translation between semi-formal mathematical representation again concept lookup tables are constructed manually [2, 19].

The proof advice systems for interactive theorem proving have studied similar concepts using various similarity measures. The methods have so far been mostly restricted to similarity of theorems and definitions. They have also been limited to single prover libraries. Heras and Komendantskaya in the proof pattern work [8] try to find similar Coq/SSReflect definitions using machine learning. Hashing of definitions in order to discover constants with same definitions in Flyspeck has been done in [12]. Using subsumption in order to find duplicate lemmas has been explored in the MoMM system [20] and applied to HOL Light lemmas in [11].

The rest of this paper is organized as follows: in Section 2 we describe the process of exporting the concepts like types and constants from three provers. In Section 3 we discuss the classification of patterns together with the normalization of theorems, while in Section 4 we define the scoring functions and an iterative matching algorithm.We present the results of the experiments in Section 5 and in Section 6 we conclude and present an outlook on the future work.

2   The Theorem and Constant Data

In this section we shortly describe the data that we will perform our experiments on and the way the theorems and constants are normalized and exported. We chose three proof assistants based on higher-order logic: HOL4 [18], HOL Light [6] and Isabelle/HOL [21]. The sizes of the core libraries of the three are significantly different, so in order to get more meaningful results we export library parts of the same order of magnitude. This amounts to all the theories included with the standard distribution of HOL4. In case of HOL Light we include multivariate analysis [7], HOL in HOL [5] and the 67 files that include the proofs of the 100 theorems [22] compatible with the two. For Isabelle we export the theory Main.

The way to access all the theorems and constants in HOL Light has been described in detail in [13] and for HOL4 and Isabelle/HOL accessing values of theories can be performed using the modules provided by the provers (DB.thms and @{theory} object respectively). We first perform a minimal normalization of the forms of theorems (a further normalization will be performed on the common representation in Section 3) and export the data. We will focus on HOL4, the procedures in the other two are similar.

The hypotheses of the theorems are discharged and all free variables are generalized. In order to avoid patterns arising from known equal constants, all theorems of the form $\vdash c_1 = c_2$ (in HOL4 four of them are found by calling DB.match) are used to substitute c_1 by c_2 in all theorems.

The named theorems and constants are prefixed with theory names and explicit category classifiers (c for constants, t for theorems) to avoid ambiguities. Similarly, variables are explicitly numbered with their position of the binding λ (this is equivalent to the de Bruijn notation, but possible within the data structure used by each of the three implementations). We decided to include the type information only at the constant level, and to skip it inside the formulas.

Example 1. $\forall x : int.\ x = x \longrightarrow cHOL4.bool.\forall\ (\lambda V.((cHOL4.min. = V)\ V))$

Analogously, for all the constants their most general types are exported. Type variables are normalized using numbers that describe their position and type constructors are prefixed using theory identifiers and an explicit type constructor classifier.

Example 2. $(num, a) \longrightarrow tHOL4.pair.prod(tHOL4.num.num, Aa)$

The numbers of exported theorems and constants are presented in Table 1.

Table 1. Number of theorems and constants after the exporting phase

	HOL Light	HOL4	Isabelle/HOL
Number of theorems	11501	10847	18914
Number of constants	871	1962	2214

3   Patterns and Classification

In this section we will look at the concept of *patterns* created from theorems, which is crucial in our classification of concepts and the algorithms for deriving patterns and matching them. In the following we will call the constants and types already mapped to concepts as *defined*.

Definition 1 (pattern). *Let f be a formula with no free variables and C the set of its constants. Let $D = \{d_1, \ldots, d_n\}$ be a set of defined constants and $A = C \setminus D = \{a_1, \ldots, a_m\}$ a set of undefined constants. Its pattern is defined by:*

$$P(f[a_1, \ldots, a_m, d_1, \ldots, d_n]) := \lambda a_1 \ldots a_n.f[a_1, \ldots, a_n, d_1, \ldots, d_n]$$

Example 3. The pattern of $\forall x\ y.\ x * y = y * x$ is:
- with $D = \{\forall, =\}$, $\lambda a_1.\ \forall x\ y.\ a_1\ x\ y = a_1\ y\ x.$
- with $D = \{\forall\}$, $\lambda a_1 a_2.\ \forall x\ y.\ a_1(a_2\ x\ y)(a_2\ y\ x).$
- with $D = \varnothing$, $\lambda a_1 a_2 a_3.\ a_1\ \lambda x\ y.\ a_2(a_3\ x\ y)(a_3\ y\ x).$

Patterns are equal when they are α-equivalent. In practice, we order the variables and constants by the order in which they appear when traversing the formula from top to bottom. This means that checking if two formulas are α-equivalent amounts to verifying the equality of their patterns with no constants abstracted.

The formulas exported from all proof assistant libraries are parsed to a standard representation (λ-terms). The basic logical operators of the different provers are mapped to the set of defined constants and the theorems are rewritten using these mappings before further normalization. Finally, the patterns of the normalized formulas are extracted according to the specified defined constants.

We define three ways in which patterns are derived from the formula, each corresponding to a certain level of normalization:

$norm_0$: Given $D = \varnothing$ we can define a pattern corresponding to the theorem without any abstraction (identity).

$norm_1$: With $D = \{\forall, \exists, \wedge, \vee, \Rightarrow, \neg, =\}$ ($\Leftrightarrow$ is considered as $=$). The procedure is similar to the normalization done by first order provers (to the conjunctive normal form) with the omission of transformations on existential quantifiers, as we do not want do perform skolemization. We additionally normalize associative and commutative operations. The procedure performs the following steps at every formula level:

- remove implication,
- move negation in,
- move universal quantifiers out (existential quantifiers are not moved out to maximize the number of disjunctions in the last step),
- distribute disjunction over conjunctions,
- rewrite based on the associativity of $\forall, \exists, \wedge$ and $\vee$,
- rewrite based on the commutativity of $\forall, \exists, \wedge, \vee$ and $=$,
- separate disjunctions at the top formula level (example below).

Example 4. $\forall x\ y.\ (x \geq 0 \wedge x \leq y) \longrightarrow (\forall x.\ x \geq 0) \wedge (\forall x\ y.\ x \leq y)$

$norm_2$: Aside from all the normalizations performed by $norm_1$, we additionally consider a given list of associative and commutative constants (see Table 2 in Section 5) that is used to further normalize the formula. The set of defined constants stays the same as $norm_1$, which in particular means that the associative - commutative (AC) constants stay undefined and can be abstracted over.

Given the normalized theorems we will look at patterns relative to constants. In the following, we will assume that the constants are partitioned in ones that have been defined (mapped to a constant) and undefined.

Definition 2 (pattern relative to a constant). *Let a_{i-1} be an undefined constant appearing in a formula f in the ith position. The pattern of f relative to a_{i-1} is defined by:*

$$P_{a_{i-1}}(f) := (P(f), i - 1)$$

Example 5. Suppose $D = \varnothing$. Then the only two patterns that the reflexivity principle induces are:

$$P_\forall(\forall x.\ x = x) = (\lambda a_0 a_1.\ a_0\ (\lambda v_0.\ a_1\ v_0\ v_0), 0)$$
$$P_=(\forall x.\ x = x) = (\lambda a_0 a_1.\ a_0\ (\lambda v_0.\ a_1\ v_0\ v_0), 1)$$

Typically, we will be interested in patterns where D includes the predicate logic constants, so the reflexivity principle will not produce any patterns. The patterns will be properties of operations like commutativity or associativity. In order to find all such properties we define:

Definition 3. *The set of patterns associated with a constant c in a library lib is defined by:*

$$P^{set}(lib, c) = \bigcup_{f \in lib} P_c(f)$$

Let (abs, i) be a relative pattern. Its associated set of constants, in library lib, is:

$$C^{set}(lib, (abs, i)) := \{c \in lib, \exists f \in lib,\ P_c(f) = (abs, i))\}$$

We can now define one of the basic measures we will use for comparing similarity of constants:

Definition 4. *The set of common relative patterns shared by a constant c_1 in lib_1, and a constant c_2 in lib_2 is:*

$$P^{set}(lib_1, c_1) \cap P^{set}(lib_2, c_2)$$

In the remaining part of this paper, we will not always specify if a pattern is relative or not.

We proceed with forming type patterns. Type patterns are defined in a similar way to formula patterns. Types are partitioned into already defined types (initially the type of booleans – propositions) and undefined types. Type variables are also considered as undefined to enable their instantiation, and the list of leaf and node types involved is saved to allow matching.

Example 6. Let $D^{type} = \{fun\}$ and a be a type variable. Then:

$$P^{type}((a \to a, int \to int)) = P^{type}((pair(fun(a, a), fun(int, int))))$$
$$= (\lambda a_0 a_1 a_2.\ (a_0(fun(a_1, a_1), fun(a_2, a_2))), [pair, a, int])$$

Suppose we are given two types with respective patterns $(abs_1, [t_1 \ldots t_n])$ and $(abs_2, [u_1 \ldots u_m])$. They match if abs_1 is α-equivalent to abs_2. The list of their derived type matches is $[(t_1, u_1), \ldots, (t_n, u_n)]$, from which the pairs containing at least one type variable are removed.

4   Matching Concepts across Libraries

In this section, we will investigate measures of similarity in order to find the same types and constants between libraries. First, we will define a similarity score for each pair of constants. Then, we will suppose that the best match is correct and use it to update the similarity scores of the other pairs iteratively.

4.1   Similarity Score

The easiest way to tell if two constants are related is to look at the number of patterns they share. However, the more a pattern has associated constants, the less relevant it is. To test each of these possibilities, two weighting functions are defined:

$$w_0(lib, p) = 1, \quad w_1(lib, p) = \frac{1}{card(C^{set}(lib, p))}$$

where p is a pattern in library lib. The weighting functions presented here do not consider the size of the pattern, nor the numbers of defined and undefined constants. Considering more complicated weighting functions may be necessary for formal libraries with significantly different logics.

Based on the weighting functions two scoring functions are defined. Let c_1 be a constant from library lib_1 and c_2 a constant from library lib_2. Let $P = \{p_1, \ldots, p_k\}$ be the set of patterns c_1 and c_2 have in common. Then:

$$score_0(c1, c2) = \sum_{i=1}^{k} w_0(lib_1, p_i) * w_0(lib_2, p_i)$$

$$score_1(c1, c2) = \sum_{i=1}^{k} w_1(lib_1, p_i) * w_1(lib_2, p_i)$$

In order to account for the fact that constants with a high number of associated patterns are more likely to have common patterns with unrelated constants, we further modify $score_1$. Let n_1 be the number of patterns associated to c_1 and n_2 be the number of patterns associated to c_2. We define a third similarity scoring function by:

$$score_2(c1, c2) = \frac{\sum_{i=1}^{k} w_1(lib_1, p_i) * w_1(lib_2, p_i)}{log(2 + n_1 * n_2)}$$

4.2   Iterative Approach

In our initial experiments, a direct computation of the $score_i$ functions for all constants in two libraries after an initial number of correct pairs would find incorrect pairs (false positive matches). Such pairs can be quickly eliminated if the information coming from the first successful matches is propagated further.

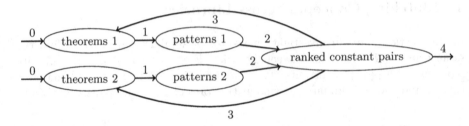

Fig. 1. Graphical representation of the iterative procedure

In order to do this, we propose an iterative approach (presented schematically in Fig. 1):

The iterative approach returns a sorted list of pairs of constants and a sorted list of pair of types from two libraries by following this steps:

0. Export theorems from a library as well as constants with their types and parse them.
1. Normalize theorems and create theorem patterns, constant patterns and type patterns according to the current defined constants and types.
2. Score every pair of constants.
3. Take the highest ranked pair of constants (c_1, c_2). Check if their type matches, if not take the next one and so on. When their type matches, rewrite all the theorems inside lib_1 with the substitution $c_1 \to d$ and all the theorems inside lib_2 with the substitution $c_2 \to d$, where d is a fresh defined constant. Then, get the derived pairs of types from the pair of constant and substitute every pair member with the same fresh defined type as for the other member.
4. Return the pairs of constants and the pairs of types, in the order they were created, when you have reached the number of iteration desired.

The single-pass approach is defined by doing only one iteration, where the list of pairs of constants are returned ranked by their score. A type check performed after a single-pass can discard a number of wrong matches efficiently.

In the presented approach, we assume that the constants and types inside one library are all different, which we tried to ensure by the initial normalization. Thus, we will not match constant from the same library. Furthermore, if a constant is matched, then it can no longer be matched again and the same reasoning applies for types. This first statement will turn out not to be true for a few constants in Section 5.

The complexity of the iterative approach is obviously larger than that of the single-pass approach. On an IntelM 2.66GHz CPU, the single-pass approach between HOL4 and HOL Light with $score_2$ and $norm_2$ takes 6 minutes to complete. The main reason is that it has to compare the patterns of all possible pairs of constants (about two million). Thus, the bottleneck is the time taken by the comparison function which intersects the set of patterns associated with each constant and scores the resulting set. However, the iterative method can use

Table 2. Most frequent properties of one constant

HOL Light			HOL4			Isabelle/HOL		
Pattern	Consts	Thms	Pattern	Consts	Thms	Pattern	Consts	Thms
Inj	37	37	Inj	54	68	Inj	83	137
Asso	32	36	Asso	50	65	App	17	18
Comm	25	44	Comm	40	48	Inj1	16	16
Refl	22	22	Trans	32	33	Comm	14	51
Lcomm	19	20	Refl	23	23	Inj2	12	35
Idempo	12	12	Idempo	20	15	App2	11	12

the first pass to remove pairs of constants that have no common patterns. This reduces the number of possible matches to ten thousand. As a consequence, it takes only 3 minutes more to do 100 iterations.

5 Experiments

In order to verify the correctness of our approach we first investigate the most common patterns and shapes of theorems in each of the three formal libraries and then we look at the results of the matching constants across libraries. The data given by these experiments is available at http://cl-informatik.uibk.ac.at/users/tgauthier/matching/.

5.1 Single Library Results

Tables 2 and 3 show the most common properties when applying the standard normalization $norm_1$ of a single constant and of two constants respectively in the three considered proof assistant libraries. The tables are sorted with respect to the total number of different constants in the theorems from which the patterns are derived. In Table 2, Inj stands for injectivity, $Asso$ for associativity and $Comm$ for commutativity. In Table 3, the pattern $Class$ and Inv are defined by $Class\ (c_0, c_1) = c_0\ c_1$, $Inv(c_0, c_1) = \forall x_0.\ c_0\ (c_1\ x_0) = x_0$.

As expected, HOL Light and HOL4 show the most similar results and injectivity is the most frequent property. Commutativity and associativity are also very common, and their associated constants are used to apply $norm_2$ as stated in Section 3.

The common patterns immediately show constants defined to be equivalent to the defined equality in each of the libraries, through an extensional definition. There is one such constant in HOL4, one in HOL Light and three in Isabelle/HOL. In order to avoid missing or duplicate patterns we mapped all these constants to equality manually.

Table 3. Most frequent properties of two constants

HOL Light			HOL4			Isabelle/HOL		
Pattern	Consts	Thms	Pattern	Consts	Thms	Pattern	Consts	Thms
Class	71	87	Inv	131	89	Class	188	642
Inv	64	34	Neutr	64	55	Inv	114	75
Imp	52	76	Class	63	70	Equal	58	40

Furthermore, in Table 3, the third row of the Isabelle/HOL column shows 40 equalities between two different constants that were created during the normalization. We have also found 10 such equalities in HOL4 and 1 in HOL Light. Often a constant with a less general type can be replaced by the other, but without type-class information in Isabelle/HOL we decided not to do such replacements in general.

5.2   Cross-Library Results

The way we analyze the quality of the matching, is by looking at the number of correct matches of types and constants between the libraries, in particular we consider the occurrence of the first incorrect match, also called *false positive* below. It is very hard to spot same concepts in two large libraries, therefore a manual evaluation of the false negatives (pairs that could be mapped but are not) is a very hard task and requires the knowledge of the whole libraries.

In the first three experiments, we test how much normalization, scoring, iteration and types contribute to better matches. This will be used to choose the best parameters for matching constants and types between each pair of provers.

The first experiment (Fig. 2) evaluates the similarity of the libraries. We match the provers using the (a-priori) strongest normalization ($norm_2$) with a single-pass approach with no types. In this setting, the constant with the most similar properties is 0 between HOL Light and HOL4, and between HOL4 and Isabelle/HOL. And it is ∅ between HOL Light and Isabelle/HOL. Form this perspective, the most similar pairs of provers are in decreasing order (HOL Light,HOL4), (HOL4,Isabelle/HOL) and (HOL Light-Isabelle/HOL). We test the four other parameters relative to the pair of provers (HOL Light, HOL4) as we should have most common patterns to work with.

The second experiment (Fig. 3) is meant to evaluate the efficiency of normalization on the number of patterns. It is also run as a single-pass with no types. We observe an increase in number of patterns from $norm_0$ and $norm_1$ which is mostly due to the splitting of disjunctions. Moreover, the difference between $norm_2$ and $norm_1$ is negligible, which means that associative and commutative constants are used in almost the same way across the two libraries. In

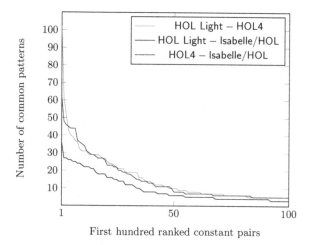

Fig. 2. Number of patterns by constant pairs in different provers

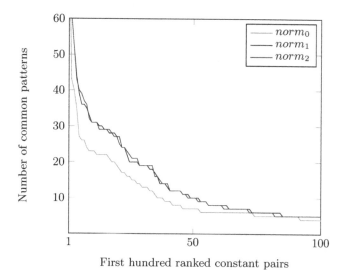

Fig. 3. The normalization effect

the following experiments we will only use $norm_2$ assuming it is the strongest normalization also in the other scenarios.

We next evaluate the scoring functions, the contribution of iterations, and of the use of type information. Table 4 shows the effect of iterative method and scoring function on the occurrence of the first wrong match (false positive). It has been inspected manually. Fig. 4, shows the positive effects of the iterative effect on the $score_1$ and $score_2$ curves. Some patterns are ranked higher after an iteration, as they become more scarce. The iterative method also has an opposite effect that is not directly visible in the figure: the score of pairs of constants diminishes by removing false pattern matches. Table 5 shows how type information contributes to matches. Types do help, but become less important with better scoring functions combined with the iterative approach.

Table 4. Rank of the first wrong match for (HOL Light, HOL4)

	$score_0$	$score_1$	$score_2$
Single-pass	39	69	88
Iterative	49	68	113

Table 5. Number of pairs of constants discarded, due to type matching

	$score_0$	$score_1$	$score_2$
Single-pass	31	19	21
Iterative	224	18	6

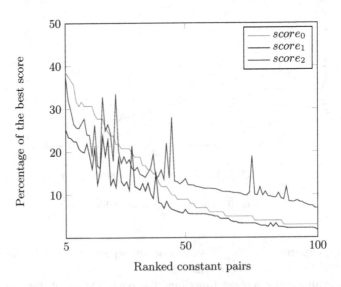

Fig. 4. Effect of different scoring functions on the iterative approach

The last experiment is run with the best parameters found by the previous experiments, namely $norm_2$, $score_2$ and the iterative approach with types. Three numbers are presented in each cell of Tables 6 and 7. The first one is the number of correct matches obtained before the first error. The second one is number of correct matches we have found. In the case of constants, the third one is the number of matches we have manually checked. We stop at a point where a previously found error propagates. In the case of types, the third number is the rank of the last correct match. As seen previously, the best results come from comparing the HOL4 and HOL Light libraries, where we have verified 177 constant matches and 16 type matches.

Table 6. Number of constants accurately matched

HOL Light-HOL4		HOL4-Isabelle/HOL		HOL Light-Isabelle/HOL	
112	177/203	65	109/131	55	78/98

Table 7. Number of types accurately matched

HOL Light-HOL4		HOL4-Isabelle/HOL		HOL Light-Isabelle/HOL	
11	16/22	8	11/17	6	7/13

6   Conclusion

We have investigated the formal mathematical libraries of HOL Light, HOL4 and Isabelle/HOL searching for common types and constants. We defined a concept of patterns that capture abstract properties of constants and types and normalization on theorems that allow for efficient computation of such patterns. The practical evaluation of the approach on the libraries let us find hundreds of pairs of common patterns, with a high accuracy.

Formal mathematical libraries contain many instances of the same algebraic structures. Such instances have many same properties therefore their matching is non-trivial. Our proposed approach can match such instances correctly, because of patterns that link such concepts to other concepts. For example integers and matrices are instances of the algebraic structure ring. However each of the libraries we analyzed contains a theorem that states that each integers is equal to a natural number or its negation. A pattern derived from this fact, together with many other patterns that are unique to integers match them across libraries correctly.

The work gives many correct matches between concepts that can be directly used in translations between proof assistants. In particular HOL/IMPORT would immediately benefit from mapping the HOL Light types and constants to their Isabelle/HOL counterparts allowing for further merging of the results.

The approach has been tested on three provers based on higher-order logic. In principle the properties of the standard mathematical concepts defined in many other logics are the same, however it remains to be seen how smoothly does the approach extend to provers based on different logics.

In order to further decrease the number of false positive matches, more weighting and scoring functions could be considered. One could imagine functions that take into account the length of formulas, and numbers of mapped concepts per pattern. Similarly, the scoring functions could penalize pairs of constants with only one pattern in common (these have been the first false positives in all our experiments). Further, the equalities between constants created during normalization could be used for further rewriting of theorems into normal forms. Other ideas include normalizing relatively to distributive constants and trying weaker kind of matching for example on atoms or subterms.

Building a set of basic mathematical concepts together with their foundational properties has been on the MKM wish-list for a long time. It remains to be seen if a set of common concepts across proof assistant libraries can be extended by minimal required properties to automatically build such "interface theories", and if automatically found larger sets of theories can complement the high-quality interface theories built in the MKM community.

Acknowledgments. We would like to thank Josef Urban for his comments on the previous version of this paper.

This work has been supported by the Austrian Science Fund (FWF): P26201.

References

1. Bortin, M., Johnsen, E.B., Lüth, C.: Structured formal development in Isabelle. Nordic Journal of Computing 13, 1–20 (2006)
2. Carlisle, D., Davenport, J., Dewar, M., Hur, N., Naylor, W.: Conversion between MathML and OpenMath. Technical Report 24.969. The OpenMath Society (2001)
3. Furbach, U., Shankar, N. (eds.): IJCAR 2006. LNCS (LNAI), vol. 4130. Springer, Heidelberg (2006)
4. Haftmann, F., Krauss, A., Kunčar, O., Nipkow, T.: Data refinement in isabelle/HOL. In: Blazy, S., Paulin-Mohring, C., Pichardie, D. (eds.) ITP 2013. LNCS, vol. 7998, pp. 100–115. Springer, Heidelberg (2013)
5. Harrison, J.: Towards self-verification of HOL Light. In: Furbach, Shankar (eds.) [3], pp. 177–191
6. Harrison, J.: HOL Light: An overview. In: Berghofer, S., Nipkow, T., Urban, C., Wenzel, M. (eds.) TPHOLs 2009. LNCS, vol. 5674, pp. 60–66. Springer, Heidelberg (2009)
7. Harrison, J.: The HOL Light theory of euclidean space. J. Autom. Reasoning 50(2), 173–190 (2013)
8. Heras, J., Komendantskaya, E.: Proof pattern search in Coq/SSReflect. arXiv preprint, CoRR, abs/1402.0081 (2014)
9. Hurd, J.: The OpenTheory standard theory library. In: Bobaru, M., Havelund, K., Holzmann, G.J., Joshi, R. (eds.) NFM 2011. LNCS, vol. 6617, pp. 177–191. Springer, Heidelberg (2011)
10. Kaliszyk, C., Krauss, A.: Scalable LCF-style proof translation. In: Blazy, S., Paulin-Mohring, C., Pichardie, D. (eds.) ITP 2013. LNCS, vol. 7998, pp. 51–66. Springer, Heidelberg (2013)

11. Kaliszyk, C., Urban, J.: Lemma mining over HOL Light. In: McMillan, K., Middel-dorp, A., Voronkov, A. (eds.) LPAR-19. LNCS, vol. 8312, pp. 503–517. Springer, Heidelberg (2013)
12. Kaliszyk, C., Urban, J.: HOL(y)Hammer: Online ATP service for HOL Light. arXiv preprint abs/1309.4962, accepted for publication in Mathematics in Computer Science (2014)
13. Kaliszyk, C., Urban, J.: Learning-assisted automated reasoning with Flyspeck. arXiv preprint abs/1211.7012, accepted for publication in Journal of Automated Reasoning (2014)
14. Keller, C., Werner, B.: Importing HOL Light into Coq. In: Kaufmann, M., Paulson, L.C. (eds.) ITP 2010. LNCS, vol. 6172, pp. 307–322. Springer, Heidelberg (2010)
15. Mohamed, O.A., Muñoz, C., Tahar, S. (eds.): TPHOLs 2008. LNCS, vol. 5170. Springer, Heidelberg (2008)
16. Obua, S., Skalberg, S.: Importing HOL into Isabelle/HOL. In: Furbach, Shankar (eds.) [3], pp. 298–302
17. Rabe, F.: The MMT API: A generic MKM system. In: Carette, J., Aspinall, D., Lange, C., Sojka, P., Windsteiger, W. (eds.) CICM 2013. LNCS (LNAI), vol. 7961, pp. 339–343. Springer, Heidelberg (2013)
18. Slind, K., Norrish, M.: A brief overview of HOL4. In: Mohamed, et al. (eds.) [15], pp. 28–32
19. So, C.M., Watt, S.M.: On the conversion between content MathML and OpenMath. In: Proc. of the Conference on the Communicating Mathematics in the Digital Era (CMDE 2006), pp. 169–182 (2006)
20. Urban, J.: MoMM - fast interreduction and retrieval in large libraries of formalized mathematics. Int. J. on Artificial Intelligence Tools 15(1), 109–130 (2006)
21. Wenzel, M., Paulson, L.C., Nipkow, T.: The Isabelle framework. In: Mohamed, et al. (eds.) [15], pp. 33–38
22. Wiedijk, F. (ed.): The Seventeen Provers of the World. LNCS (LNAI), vol. 3600. Springer, Heidelberg (2006)

Mining State-Based Models from Proof Corpora

Thomas Gransden, Neil Walkinshaw, and Rajeev Raman

Department of Computer Science, University of Leicester, Leicester, UK
tg75@student.le.ac.uk, n.walkinshaw@mcs.le.ac.uk, r.raman@le.ac.uk

Abstract. Interactive theorem provers have been used extensively to reason about various software/hardware systems and mathematical theorems. The key challenge when using an interactive prover is finding a suitable sequence of proof steps that will lead to a successful proof requires a significant amount of human intervention. This paper presents an automated technique that takes as input examples of successful proofs and infers an Extended Finite State Machine as output. This can in turn be used to generate proofs of new conjectures. Our preliminary experiments show that the inferred models are generally accurate (contain few false-positive sequences) and that representing existing proofs in such a way can be very useful when guiding new ones.

Keywords: Interactive Theorem Proving, Model Inference, Extended State Machines.

1 Introduction

Interactive theorem provers (ITPs) provide a semi-automatic environment in which a user can reason about the correctness of hardware and software systems and verify the proofs of significant mathematical theorems. Given a desired property expressed in a formal logic, provers such as Coq [26] and Isabelle [28] provide a framework by which to construct higher-order logic proofs in a stepwise manner, drawing upon libraries of existing proven theorems. In the context of computer mathematics, ITPs have successfully been used in the verification of the Four Color Theorem [9], the Kepler Conjecture [13] and the Feit-Thompson Theorem [10].

ITPs rely on the ability of an expert to choose suitable proof steps to apply. Clearly this requires not only the selection of the correct proof steps, but also knowledge about how to sequence these proof steps in order to arrive at a successful proof. To complicate matters further, the user must select suitable parameters for these proof steps. In a significant development, the overall proof effort can contain tens of thousands of lines. For example, Gonthier's machine checked proof of the Feit-Thompson theorem amounted to 170,000 lines of code. This shows that a lot of human effort was needed to complete the proof.

Over the past decade, several semi-automatic tools have been developed to simplify the verification process [6, 7, 11, 21]. These tools adopt data mining and heuristic search strategies to identify proof patterns and to conjecture new

S.M. Watt et al. (Eds.): CICM 2014, LNAI 8543, pp. 282–297, 2014.

proofs. One outstanding challenge, recently highlighted by Grov *et al.* [12], is the need to identify *proof strategies*. There is a desire not only to recognize common syntactic patterns (as achieved by current techniques), but to take this one step further and to capture the rules that govern the possible ordering of the proof steps required to yield a successful proof. This is what motivates the work in this paper.

Accordingly, we present a technique to derive sequential models (in the form of Extended Finite State Machines (EFSM) [31]) from existing corpora of proofs. These models can be interpreted as an instance of the proof strategies referred to by Grov *et al.* The models capture the reasoning patterns that bind groups of proofs together, and in doing so capture the possible sequences of proof steps that have led to successful proofs. Corpora that contain tens or hundreds of proofs can be collapsed into (relatively) compact, graphical models. We show how these models can be used to the benefit of interactive theorem prover users. The specific contributions of this paper are:

- A technique to automatically derive EFSM models from libraries of interactive proofs (Section 3).
- An evaluation that indicates that the models are broadly precise and can be used as an aid to yield proofs of new propositions, and to shorten existing ones (Section 4).

All of the example data used in this paper, along with links to EFSM inference tool can be found online.[1].

2   Background and Related Work

This section discusses the problem that ITPs demand a significant amount of time and effort to complete the proof process. After reviewing some previous work focussed on aiding with this problem, we introduce Extended Finite State Machines as a possible mechanism to improve proof development by representing existing proofs by means of a descriptive, sequential model.

2.1   Interactive Theorem Provers

The expressiveness of interactive theorem provers has led to an abundance of formal proofs becoming available in proof libraries that are distributed with each system. These proof libraries can then be used during the development of new proofs. As with conventional programming languages, developers can build up and exchange their own libraries of proofs to suit their particular domain. Nevertheless, most non-trivial proofs still require an extensive manual effort - Wiedijk states that it takes as long as one week to formalize one page from an undergraduate mathematics textbook [34].

The emergence of tools such as Sledgehammer [29] has reduced this effort to an extent, by enabling Isabelle to call on powerful external Automated Theorem

---

[1] http://www.cs.le.ac.uk/people/tg75/efsmdata/

Provers (ATPs) that attempt to solve the goal automatically. Although such tools have been proven to have great value, they require extensive research into the translation between different logics as ATPs utilize different logics to the higher order varieties typically used in ITPs [27]. An empirical study of Sledgehammer [3] indicated a success rate of 45% at proving goals from 7 Isabelle theories (known collectively as the Judgement Day benchmark).

Another method of reducing human intervention is called *proof planning* [4]. Proof planning allows the encoding of reusable strategies that are used to guide proof search - for example many inductive proofs follow a similar pattern which can be encoded as a proof plan. A typical proof plan contains preconditions to state when then plan is applicable, a postcondition stating the effects of executing the proof plan, and the relevant proof steps to apply. Proof planning has been implemented for Isabelle by a tool called IsaPlanner [6].

An interesting strand of research is the use of machine learning techniques to improve theorem proving by guiding the proof search or suggesting hints to users. One area that has benefitted greatly from machine learning is the *premise selection problem* [1, 18]. Informally, this is the problem of selecting useful premises (from a large collection) to automatically solve a new proposition. By utilizing machine learning techniques, the performance of ATPs on large theory reasoning significantly improved on the state of the art [19]. Recently, machine learning capabilities have been added to Sledgehammer [22]. By using the same empirical study (Judgement Day [3]) that was used to evaluate the original Sledgehammer, it was shown that using machine learning can improve the percentage of completed proofs to 70%.

Recently, there has been the emergence of a tool called ML4PG [21] that uses statistical machine learning techniques to identify commonalities between Coq proofs. Given a proposition that a user is trying to prove, ML4PG can automatically identify clusters of existing lemmas that follow a common proof strategy. The user can then interpret the results and formulate the proof themselves by analogy, using the suggestions provided. ML4PG has been shown to work in a variety of areas such as computer algebra [15] and industrial proofs [16].

2.2 Motivating Scenario

To motivate our work, let us consider the following scenario. A novice user is trying to prove (in this case using Coq) the following *app_nil_l* proposition stating that an empty list appended to a list l should result in l:

Lemma app_nil_l: forall l:list A, [] ++ l = l.

We assume that the user will be aware of the possible proof methods available in Coq. However it may be unclear how one would sequence these proof methods to arrive at a successful proof. One approach that a user could try might be to manually scour existing proofs to find a sequence of proof steps that will prove *app_nil_l*. However, keeping track of the relevant proofs and identifying useful reasoning patterns is time-consuming.

Even for this relatively simple example, finding a proof requires some careful manual processing of the relevant proof libraries. In large scale developments, the task of manually searching through proof corpora to identify the correct steps is generally not practical. We present an automated method based on state machine inference techniques. By providing examples of successful proofs we can generate a model capturing all of the reasoning patterns that occur within the chosen corpora of proofs. This model can then be used to drive the proof search by presenting options to the user about which proof steps to try.

2.3   State Machine Inference

State machine inference techniques can address the challenge of identifying the rules that govern a particular sequencing of events. The problem of deriving a model from sequences of events was introduced by Gold in 1967 [8]. Since then it has become a well-established problem, spawning several families of algorithms for different types of models, learning settings and problem domains. The archetypal model for sequences of events is the Finite State Machine (FSM).

Definition 1. Finite State Machine *A Finite State Machine (FSM) is defined as a tuple (S, s_0, F, L, T). S is a set of states, $s_0 \in S$ is the initial state, and $F \subseteq S$ is the set of final states. L is as defined as the set of labels. T is the set of transitions, where each transition takes the form (a, l, b) where $a, b \in S$ and $l \in L$. When referring to FSMs, this paper assumes that they are deterministic.*

In the past 40 years numerous algorithms have been developed to infer FSMs (equivalently regular grammars) from observed sequences of events [2, 23, 33]. These sequences are referred to as *traces*, and are recorded from the system under analysis. The challenge is to derive from the set of traces a FSM that accurately captures the set of all valid sequences of events, even if they do not belong to the initial set of traces.

Such techniques have previously been applied to proof planning. Jamnik *et al.* [17] used an Inductive Logic Programming technique to infer what are ultimately regular expressions from well chosen sets of proof methods. For example, if we have the following two proofs (where a-d are proof methods): [a, a, c, d] and [a, b, d] they may be generalized as the following: [a*, [b|c], d].

The value of even such a basic model is intuitive. Jamnik *et al.* demonstrated that the models were useful for the automatic generation of new proofs in the ΩMEGA prover. However, the proof steps that were learned in the examples do not contain any parameters, they are simply method names meaning that this kind of model is too basic to be applied to provers such as Isabelle and Coq. A proof in either of these provers not only relies on the sequencing of the proof steps, but also the values of the parameters provided to these steps.

To combat this problem, this paper explores the use of the Extended Finite State Machines [5] as a means of modelling examples of successful proofs. EFSMs extend the traditional FSM. Transitions are labelled with guards on an underlying data store (although the update functions on the store are not explicitly modelled).

Definition 2. Extended Finite State Machine *An Extended Finite State Machine (EFSM) M is a tuple* $(S, s_0, F, L, V, \Delta, T)$, *where* S, s_0, F *and* L *are defined as in a conventional FSM. V is a store represented by a set of variables, and v represents a set of variable values. Δ is the set of* data guards, *where each guard* δ *takes the form* (l, v), *where* $l \in L$, $v \in V$ *is the set of possible data variable configurations specified by the guard. The set of transitions T is an extension of the conventional FSM version, where transitions take the form* (a, l, δ, b), *where* $a, b \in S$, $l \in L$, *and* $\delta \in \Delta$.

Definition 3. Traces *A trace* $T = \langle e_0, \ldots, e_n \rangle$ *is a sequence of n trace elements. Each element e maps to a tuple* (l, v), *where l is a label representing the names of function calls or input / output events, and v is a set of corresponding variable values (this may be empty).*

In recent years, algorithms have been developed to infer EFSMs from traces of events [25,31], where events are paired with a selection of variable values. In this work we choose the EFSMInfer tool by Walkinshaw *et al.* [31], which has been shown to be reasonably accurate when applied to the task of reverse-engineering models of software modules. We provide a brief overview of the essential steps of the approach below.

Given a set of traces (see Definition 3), the approach first infers the guard conditions. For each symbol $l \in L$ the trace is scanned, identifying every instance where l is applied, the variable values v at that instance, and the label of the subsequent step in the trace. This is used to construct a training set where, with the use of standard machine learning algorithms (e.g. decision tree learners like [30]), it is possible to construct a model that predicts from a given pair label and data configuration what the subsequent label will be. In terms of EFSMs, this gives us L, V, Δ, and implies some constraints on the order in which particular configurations of labels and variables can occur.

The subsequent task is to derive an underlying state transition model that obeys and incorporates these data guards. To achieve this EFSMInfer applies an augmented version of the standard FSM state merging algorithm (Lang's Blue-Fringe algorithm [23]). The set of traces is first arranged as a prefix tree [33], where traces with the same prefix also share the same path from the root. Subsequently, states in the tree are merged according to the likelihood that they represent the same state, based on the similarity of their outgoing paths.

Since this model incorporates data, the merging process includes a step to ensure that the model remains consistent with the data guards. Each transition in the tree is mapped to its corresponding variable configurations. Pairs of states are only merged if the resulting model completely obeys the data classifiers (guard conditions) that were obtained in the previous step. If the inferred data model predicts that the data value for a given transition is followed by a label l, any merge involving the target state can only occur if the resulting state machine contains an outgoing transition that is labelled by l. After each merge, the resulting state machine is further post-processed to ensure that each transition is deterministic [31].

EFSMInfer has several optional parameters. The most important parameter is the choice of data classifier algorithm, which is used to infer the guards on the transitions. For this, EFSMInfer incorporates several standard algorithms that were implemented as part of the Weka [14] toolset. In our experiments, we will adopt the default parameters in EFSMInfer.

3   Inferring EFSMs from Proof Corpora

This paper shows how the EFSMInfer tool can be used to derive models from proofs that not only describe the possible sequences of proof steps that have been used in existing proofs, but also the necessary parameter values associated with these proof steps. Although previous work on EFSMs has focussed on program execution traces, they also appear to be well suited to the domain of interactive proofs where we want to capture the interplay between control (proof steps) and data (parameters).

In this section, we describe the process of inferring EFSMs from proofs, and provide a description about how such a model can be interpreted. We begin by showing how existing proofs can be converted into traces, before demonstrating how the model is inferred from these proof traces. The example model shown in this section is for a set of proofs called ListNat, that contains proofs about simple properties of lists and natural numbers.

3.1   Turning Existing Proofs into Proof Traces

A typical tactical proof script[2] contains many examples of propositions that have been proven, along with the sequence of proof steps that the expert user entered to complete the proof. Each proof step has the structure: $proof\_method \ p_1 \ldots p_n$ where $proof\_method$ refers to a Coq command (e.g. rewrite, apply, intros) and $p_1 \ldots p_n$ constitutes the parameters provided to the Coq command. The parameters refer to many different entities such as existing lemmas, rewrite rules or may be related to variables in the goal.

As shown in Table 1, the encoding of Coq proofs is a straightforward translation into the trace format shown in Definition 3. With respect to the tuple of labels and variables (l, v), the $proof\_method$ would correspond to l whilst the parameters $p_1 \ldots p_n$ correspond to v. If a proof method doesn't have any parameters provided to it, we indicate this by appending 0 to the end of the proof method (i.e. in Table 1 we see $intros_0$). Also, if proof steps are part of a combination, which is denoted by the presence of a semicolon separating individual proof steps, we encode this information as part of the trace. If two proof steps are put in combination, it means that the first proof step is applied, and then the next one applied to *all* subgoals generated. Including this information in the model is useful so that we know when applying proof steps whether they should be combined.

---

[2] Although this work concentrates on Coq, the method in principle can be applied to other ITPs.

Table 1. Original proof and proof trace for an example lemma

(a) Proof Script

(b) Trace

```
Lemma ex : (n*m = 0)->(n=0)\/(m=0).
  intros.
  induction n.
  tauto.
  simpl in H.
  right.
  assert (m <= 0);
  try omega.
  rewrite <- H.
  auto with arith.
Qed.
```

Event e	Label l	Values v
e_0	$intros_0$	$\langle\rangle$
e_1	induction	$\langle p_1 = \text{"}n\text{"}\rangle$
e_2	$tauto_0$	$\langle\rangle$
e_3	simpl	$\langle p_1 = \text{"}in\ H\text{"}\rangle$
e_4	$right_0$	$\langle\rangle$
e_5	assert	$\langle p_1 = \text{"}m \leq 0\text{"}, p_2 = \text{"};\text{"}\rangle$
e_6	try	$\langle p_1 = \text{"}omega\text{"}\rangle$
e_7	rewrite	$\langle p_1 = \text{"} \leftarrow H\text{"}\rangle$
e_8	auto	$\langle p_1 = \text{"}with\ arith\text{"}\rangle$

3.2 Inferring the Model

After converting each proof into its corresponding trace, it becomes possible to infer a model from a collection of these traces. We choose the standard configuration for the EFSMInfer tool and, for the sake of illustration select the J48 decision tree learner (a Weka implementation of the C4.5 algorithm [30]). Having chosen the classifier we can run the EFSMInfer tool and generate an EFSM.

To begin with data classifiers are inferred that, for each *proof_method*, produce a function that uses the parameters to predict the subsequent transition in the model. An example data classifier can be seen in Figure 1(a) for the induction proof method. The data classifier is interpreted as follows: if the parameter p_1 is equal to n, a or l, then the subsequent proof method should be simpl. If p_1 is equal to m then the following proof method should be trivial. Although not the case here, the C4.5 algorithm can produce more complex trees of if-then-else rules governing the possible value ranges for parameters if necessary.

Once the data classifiers have been inferred, the state merging can commence. Initially, the set of proof traces is arranged as a prefix tree. The tree for our example is shown in Figure 1(b). The labels are unreadable, but the purpose is merely to give an intuition of what the tree might look like, and to illustrate the ensuing state merging challenge. Each transition in the tree is associated with a label (which is linked to one of the inferred data classifiers), along with the variable values that correspond to that transition. The inference challenge for the merging algorithm is to select compatible pairs of states to be merged. These states should have similar outgoing paths, should not entail the merging of states that are incompatible (e.g. accepting and rejecting), and should not raise contradictions with the inferred data classifiers (as discussed in Section 2).

The final EFSM is shown in Figure 1(c). The constraints on the transitions detail the parameter configurations that are associated with each transition. The model is deterministic; for any state there is never more than one outgoing transition for a given combination of label and variable configuration.

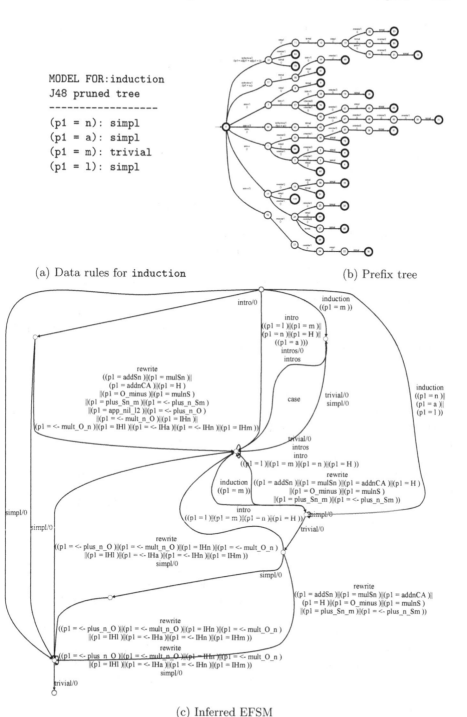

MODEL FOR:induction
J48 pruned tree

(p1 = n): simpl
(p1 = a): simpl
(p1 = m): trivial
(p1 = l): simpl

(a) Data rules for `induction`

(b) Prefix tree

(c) Inferred EFSM

Fig. 1. PTA and inferred EFSM for `ListNat` traces

4   Using EFSMs in Interactive Theorem Proving

This section seeks to determine the potential value of inferred EFSMs as a guidance mechanism for users of interactive theorem provers. To assess the EFSMs when applied to proofs, we conduct an automated experiment to measure the accuracy of the models, whilst also producing a more informal, qualitative case study to show how the models can be used to manually derive new proofs. Our notion of accuracy revolves around the inferred EFSM's ability to distinguish whether a proof should be accepted by the model or not. We conclude the section by highlighting some future work to improve our current technique.

4.1   Assessing the Accuracy of Inferred EFSMs

Measuring accuracy. Measuring the accuracy of inferred models is challenging, especially in the absence of "gold-standard" models that could be used as a reference. In machine learning this problem is common. One of the most popular evaluation techniques that can be used in such a situation is known as *k-folds cross validation* [20]. The dataset is randomly partitioned into k non-overlapping sets (also known as folds). Over k iterations all bar one of the folds are used to infer a model, and the remaining fold is used to evaluate the model according to some metric (discussed below). For each iteration a different fold is used for the evaluation. The final accuracy score is the average of the k accuracy scores.

Of course, given the probability that the given set of proofs is not "rich enough", the accuracy score cannot be interpreted as an *absolute* score of the accuracy of the model. However, if we accept that the test set captures a representative set of proofs for a given domain, then the resulting scores can be interpreted as being at least indicative of the actual accuracy score.

To assess 'accuracy' there are many metrics that we can choose as a measurement such as precision, sensitivity and specificity. All are computed from the set of true-positives (TP), true-negatives (TN), false-positives (FP) and false-negatives (FN). In this experiment we choose sensitivity (TP/(TP+FN)), and specificity (TN/(TN+FP)).

Negative examples. As indicated above, fully assessing sensitivity and specificity implies the existence of "negative" proof traces - traces that do not correspond to valid proofs (and therefore should not lead to accepting states in an inferred model). For the purposes of this evaluation we have selected some of the positive examples and mutated the sequences of proof steps by randomizing them. In addition to this, we provide sequences of proof steps from theories that are different from the ones we have inferred a model from. In practice, these negative examples could be captured from proof attempts that have failed to prove a proposition. In each experiment, we use approximately 30 negative examples.

Evaluation process. To get an idea how accurate our inferred models are we use k-folds cross validation each of our datasets. We set the number of folds $k = 5$ to ensure that we have an adequate sized evaluation set at each iteration. At

Table 2. Proof libraries, and the accuracy of the inferred models

Data Set	Proofs	Lines	Sensitivity	Specificity
ListNat	70	660	0.84	0.81
Bool	100	809	0.95	0.55
Coqlib	100	1326	0.22	0.96
Values	85	1188	0.24	0.98

each iteration of the k-folds an EFSM is inferred from the traces contained in the training set. We then run the traces from the evaluation set and the negative traces through the model, logging whether they are accepted or not. From this we can compute sensitivity and specificity. Since the inference of the guards in the EFSM depends on the selection of a suitable data classifier algorithm, we repeat the experiment for five data classifiers provided in EFSMInfer.

Data sets. We chose four sets of Coq proofs, which are listed (along with the number of proofs and lines of code) in Table 2. ListNat contains proofs regarding the basic properties of lists and natural numbers. Bool contains proofs about boolean values. To complement these datasets, we also chose two theories contained in CompCert [24], which is a formally verified C compiler. The Coqlib theory contains proofs about functions used throughout CompCert, whilst Values focuses on proofs related to run-time values. All of our datasets are composed of hand curated proofs so that the models don't simply contain calls to automated tactics that may solve the goal instantly.

Results. For each proof set, the choice of data classifier algorithm made a negligible difference to the results. The five classifiers (all part of the Weka distribution [14]) were J48, NaiveBayes, NNGE, AdaBoostDiscrete and JRIP. Our results in Table 2 show the values obtained from using the J48 classifier. For all systems apart from Bool, the specificity measures are all 85% and above. In these cases there were very few false-positives (meaning that a low proportion of negative examples were falsely accepted by the model).

The sensitivity values vary substantially depending on the dataset used. The ListNat and Bool datasets have reasonably high sensitivity values (both over 80%), indicating that they were good at predicting new proofs that did not belong to the training data. Coqlib and Values had low sensitivity scores, meaning that the inferred models failed to predict a large proportion of proofs that were not in the training set.

In the cases of Coqlib and Values, the low sensitivity scores are not particularly surprising and is to an extent inevitable. Whereas the proofs in ListNat and Bool are relatively homogeneous because they are concerned with specific, simple data structures, the proofs in Coqlib and Values are highly diverse and have less common reasoning patterns than the other libraries. Coqlib provides a general library of proofs that are intended for use in almost *any* context. Values provides proofs that apply to the values of variables in a compiler and, given that CompCert is entirely concerned with a compiler verification, plays a central role in the diverse range of contexts within CompCert.

In such cases, the EFSMInfer tool is inevitably only provided with a small fraction of the proofs that are required to constitute a truly 'representative' training set. Accordingly, the tool is bound to under-generalize, resulting in models that are too conservative; the proofs that they predict are largely valid, but they invariably miss out many other proofs that are in fact valid.

4.2 Case Studies

Although the results from the previous section provide us with a qualitative assessment of the accuracy of the inferred models, they only provide a limited insight into the practical value of the models from a user's perspective. We conclude this section with a detailed walk through the process of how a user can derive a proof using an EFSM as guidance. We show two case studies that demonstrate the process of using an EFSM during the proof process.

In the subsequent examples, we model the following scenario. Let us assume that we have a collection of existing proofs available; ListNat contains proofs about lists and natural numbers and is used in case study 1. The Bool dataset contains proofs about boolean values, which we use in case study 2. We then suppose that we task a Coq user to prove one of the lemmas in the dataset (and allow them to use the remaining proofs to infer a model from). We demonstrate that in each case we can be led to a proof using the model as guidance. We then compare the EFSM based proof with the original proof from each dataset. In both cases, we see interesting results when we compare.

Example 1. Let us refer back to the motivating example from Section 2, where the user is tasked with proving the *app_nil_l* proposition, which is part of the ListNat dataset. As an exercise, let us assume that the user has been given the remainder of the ListNat proofs (minus the proof for *app_nil_l*). Figure 1(c) shows the EFSM associated with this example and was inferred from every proof in ListNat minus *app_nil_l*. The process that one might follow to derive a proof from the model is as follows:

- Our main choices to start the proof are induction or intros. We know that typically proofs containing lists begin with induction, and the model also suggests parameter $p_1 = 1$, so we select the first step of the proof as induction 1.
- The first subgoal that needs proving is the base case showing that appending 2 empty lists together results in the empty list. The options that the model suggests are the following - trivial, simpl or rewrite. This particular subgoal is a simple equality, so it suffices to choose trivial as the next proof step.
- We can now move on to the inductive step. The model then presents us with 3 more options - intro, simpl or rewrite. None of the parameters suggested for rewrite seem to be applicable, they are more suited to natural number

proofs. There is nothing we can introduce, so we choose to simplify using `simpl`.

- There is only one possible step that can follow, which is `rewrite`. Besides a couple of existing lemmas regarding natural numbers, the model seems to be suggesting rewriting the inductive hypothesis. By analogy with the model we choose the parameter $p_1 = $ `<- IH1`.
- Finally, we can complete the proof (and arrive at an accept state) by using `trivial`.

We have shown by using an EFSM that one way of solving *app_nil_l* would be to use the following sequence of proof steps:

```
induction 1. trivial. simpl. rewrite <- IH1. trivial.
```

So how does this proof compare against the original proof for the same proposition in ListNat? The existing proof was the following:

```
intro 1. case 1. simpl. trivial. intros a0 10. simpl. trivial.
```

Interestingly, the proof found by using the EFSM was two steps shorter, and also required less effort in identifying the parameters required for the proof steps. Additionally, the sequence found from traversing the EFSM was (at least not in its entirety) part of the training data, and was only found as a result of inferring an EFSM.

Example 2. In our second example, we try to prove the following proposition: `negb(b1 || b2) = negb b1 && negb b2`, which states that (for two boolean values b1 and b2) if b1 or b2 is false, then b1 is false and b2 is false. We infer a model from all of the other proofs available in the Bool dataset. The corresponding model for this example can be found on the authors webpage[1]. The process of using the model to arrive at a proof is the following:

- To begin the proof, the model suggests either `destruct` or `intros`. We try the `intros` path first as there are quantified variables that we can introduce, but we are then led to a state where nothing is applicable. So we must use `destruct` instead. There are numerous parameters that are suggested, but we see that we have b1 and b2 in our goal, so it makes sense to choose parameters that include one of these. We decide to set $p_1 = $ b1 and $p_2 = $; to make the proof step `destruct b1 ;`
- We are presented with a number of options, most of which we can rule out due to not being applicable e.g. `rewrite, case`. We do have a boolean b2 in our goal, so we follow the suggested step - `destruct b2 ;`
- At the next node, there are 2 possible paths. One involves the `rewrite,case` steps that we still cannot apply. We take the path that uses the simplifying method `simpl` and the suggested parameters which are $p_1 = $ in `|- *` and $p_2 = $;
- Finally, the model suggests `trivial` or `reflexivity` to complete the proof. Either of these lead to the proof, but we choose the `trivial` method for the purposes of this example.

We have again been led to a proof by following the guidance provided by the EFSM. The proof that corresponds to the sequence above is the following:

```
destruct b1; destruct b2; simpl in |- *; trivial.
```

The original proof from the Bool dataset corresponds to the following proof steps:

```
intros; destruct_all bool; simpl in |- *;trivial;try discriminate
```

Although only 1 step shorter this time, we have again shown that using an EFSM to complement the proving process can yield useful results. In this particular example, we have shown that there is a smaller number of distinct proof methods used in the newly found EFSM based proof than in the original one.

4.3   Threats to Validity

It is important to bear in mind that these results are primarily intended to be indicative, and as such there are several elements in the design that could potentially invalidate the results. Firstly, we have only chosen four data sets for our experiments. Clearly there may be other collections of proofs that could lead to much better or worse results than the ones described here. Nevertheless, we chose these data sets to ensure a highly diverse selection that covered a wide broad variety of examples. Another potential threat is the generation of the negative examples that factor into the calculation of sensitivity and specificity. By manually inspecting the generated examples we tried to not select negative traces that were too easy to identify as such.

4.4   Improvements and Future Work

Although we have shown that inferring models can be useful in the proof process, we haven't yet discussed the current limitations of the approach. We have identified the following areas where our EFSM-based approach can be improved, and in doing this can lead us towards our overall aim, which is to automatically complete proofs using EFSM-based approaches.

The way we choose to represent parameters in the EFSMs can be improved. Currently everything is treated entirely textually, so an interesting avenue for future work would be to abstract away from the actual variable names and investigate the inclusion of the *types* of the variables instead. This would help to simplify the models, whilst also making them applicable to a larger range of propositions.

Another limitation is being able to identify the relevant paths through the model for any given proof. From a user's perspective, when presented with a small model such as the one shown in Section 3) they can simply evaluate the options and each step and make an informed choice. We are ultimately interested in a system that can execute the EFSM automatically to derive proofs. This could be done in a number of ways, for example by using a Breadth-First Search of the EFSM to check the applicability of proof steps, or by using evolutionary algorithms.

The negative information that we used in the experiments is not entirely accurate, in the sense that a more robust selection of negative examples could be actual failed proof attempts. In addition to improving the quality of the negative examples, we are also interested in the incorporation of this negative information within the model [32]. By including this information within the model, we may be able to infer much more accurate models of proofs.

A final consideration is the selection of proofs that we infer EFSMs from. The approach we use in this paper is to select *similar* proofs in the sense that proofs are grouped together because they all deal with a similar data structure, or are contained within the same theory file. An interesting addition to our tool would be to make use of proof filtering tools such as ML4PG [21]. By using ML4PG, we could inspect the proof obligation that we are trying to prove, before being presented with the most relevant proofs (as suggested by ML4PG). We can then use these suggestions as input to EFSMInfer, instead of a collection of manually selected proofs.

5   Conclusion

We have shown how EFSMs can be derived from existing proof corpora. These state machines have proven to be useful as they can reduce large, complex proof files into a more manageable, concise representation. In our evaluation, we have demonstrated that the models are reasonably accurate and that they can be used to derive new proofs. We have also shown that in comparison to existing proofs, the EFSM based ones can be shorter and less complex that the original. The models not only show a user the possible sequencing of proof methods (which is valuable enough information on its own), but also help to suggest the parameters that may be useful in completing a proof. Finally, we have highlighted some areas for improving our technique in the future.

References

1. Alama, J., Heskes, T., Kühlwein, D., Tsivtsivadze, E., Urban, J.: Premise Selection for Mathematics by Corpus Analysis and Kernel Methods. Journal of Automated Reasoning 52(2), 191–213 (2014)
2. Biermann, A., Feldman, J.A.: On the Synthesis of Finite-State Machines from Samples of Their Behavior. IEEE Transactions on Computers C-21(6), 592–597 (1972)
3. Böhme, S., Nipkow, T.: Sledgehammer: Judgement Day. In: Giesl, J., Hähnle, R. (eds.) IJCAR 2010. LNCS (LNAI), vol. 6173, pp. 107–121. Springer, Heidelberg (2010)
4. Bundy, A.: The Use of Explicit Plans to Guide Inductive Proofs. In: Lusk, E., Overbeek, R. (eds.) CADE 1988. LNCS, vol. 310, pp. 111–120. Springer, Heidelberg (1988)

5. Cheng, K.T., Krishnakumar, A.S.: Automatic Functional Test Generation Using the Extended Finite State Machine Model. In: Proceedings of the 30th International Design Automation Conference, DAC 1993, pp. 86–91. ACM, New York (1993)
6. Dixon, L., Fleuriot, J.: IsaPlanner: A Prototype Proof Planner in Isabelle. In: Baader, F. (ed.) CADE-19. LNCS (LNAI), vol. 2741, pp. 279–283. Springer, Heidelberg (2003)
7. Duncan, H.: The Use of Data Mining for the Automatic Formation of Tactics. Ph.D. thesis, University of Edinburgh (2007)
8. Gold, E.M.: Language Identification in the Limit. Information and Control 10(5), 447–474 (1967)
9. Gonthier, G.: Formal Proof - The Four-Color Theorem. Notices of the American Mathematical Society 55(11), 1382–1393 (2008)
10. Gonthier, G., et al.: A Machine-Checked Proof of the Odd Order Theorem. In: Blazy, S., Paulin-Mohring, C., Pichardie, D. (eds.) ITP 2013. LNCS, vol. 7998, pp. 163–179. Springer, Heidelberg (2013)
11. Grov, G., Kissinger, A., Lin, Y.: A Graphical Language for Proof Strategies. In: McMillan, K., Middeldorp, A., Voronkov, A. (eds.) LPAR-19. LNCS, vol. 8312, pp. 324–339. Springer, Heidelberg (2013)
12. Grov, G., Komendantskaya, E., Bundy, A.: A Statistical Relational Learning Challenge – Extracting Proof Strategies from Exemplar Proofs. In: ICML 2012 Workshop on Statistical Relational Learning (2012)
13. Hales, T.C.: Introduction to the Flyspeck Project. In: Coquand, T., Lombardi, H., Roy, M.F. (eds.) Mathematics, Algorithms, Proofs. Dagstuhl Seminar Proceedings, vol. 05021. Internationales Begegnungs- und Forschungszentrum für Informatik (IBFI), Schloss Dagstuhl, Germany (2006)
14. Hall, M., et al.: The WEKA Data Mining Software: An Update. SIGKDD Explorations 11(1), 10–18 (2009)
15. Heras, J., Komendantskaya, E.: ML4PG in Computer Algebra Verification. In: Carette, J., Aspinall, D., Lange, C., Sojka, P., Windsteiger, W. (eds.) CICM 2013. LNCS (LNAI), vol. 7961, pp. 354–358. Springer, Heidelberg (2013)
16. Heras, J., Komendantskaya, E., Johansson, M., Maclean, E.: Proof-Pattern Recognition and Lemma Discovery in ACL2. In: McMillan, K., Middeldorp, A., Voronkov, A. (eds.) LPAR-19. LNCS, vol. 8312, pp. 389–406. Springer, Heidelberg (2013)
17. Jamnik, M., Kerber, M., Pollet, M., Benzmüller, C.: Automatic Learning of Proof Methods in Proof Planning. Logic Journal of the IGPL 11(6), 647–673 (2003)
18. Kaliszyk, C., Urban, J.: Learning-assisted Automated Reasoning with Flyspeck. CoRR abs/1211.7012 (2012)
19. Kaliszyk, C., Urban, J.: MizAR 40 for Mizar 40. CoRR abs/1310.2805 (2013)
20. Kohavi, R.: A Study of Cross-Validation and Bootstrap for Accuracy Estimation and Model Selection. In: IJCAI, pp. 1137–1145. Morgan Kaufmann (1995)
21. Komendantskaya, E., Heras, J., Grov, G.: Machine Learning in Proof General: Interfacing Interfaces. In: UITP. EPTCS, vol. 118, pp. 15–41 (2013)
22. Kühlwein, D., Blanchette, J.C., Kaliszyk, C., Urban, J.: MaSh: Machine Learning for Sledgehammer. In: Blazy, S., Paulin-Mohring, C., Pichardie, D. (eds.) ITP 2013. LNCS, vol. 7998, pp. 35–50. Springer, Heidelberg (2013)
23. Lang, K.J., Pearlmutter, B.A., Price, R.A.: Results of the Abbadingo One DFA Learning Competition and a New Evidence-Driven State Merging Algorithm. In: Honavar, V.G., Slutzki, G. (eds.) ICGI 1998. LNCS (LNAI), vol. 1433, pp. 1–12. Springer, Heidelberg (1998)

24. Leroy, X.: Formal Verification of a Realistic Compiler. Communications of the ACM 52(7), 107–115 (2009)
25. Lorenzoli, D., Mariani, L., Pezze, M.: Automatic Generation of Software Behavioral Models. In: ACM/IEEE 30th International Conference on Software Engineering, ICSE 2008, pp. 501–510 (May 2008)
26. The Coq Development Team: The Coq Proof Assistant Reference Manual. Version 8.4. LogiCal Project, `http://coq.inria.fr/refman`
27. Meng, J., Paulson, L.C.: Translating Higher-Order Clauses to First-Order Clauses. Journal of Automated Reasoning 40(1), 35–60 (2008)
28. Nipkow, T., Paulson, L.C., Wenzel, M.T.: Isabelle/HOL. LNCS, vol. 2283. Springer, Heidelberg (2002)
29. Paulson, L.C., Susanto, K.W.: Source-Level Proof Reconstruction for Interactive Theorem Proving. In: Schneider, K., Brandt, J. (eds.) TPHOLs 2007. LNCS, vol. 4732, pp. 232–245. Springer, Heidelberg (2007)
30. Quinlan, J.R.: C4.5: Programs for Machine Learning. Morgan Kaufmann (1993)
31. Walkinshaw, N., Taylor, R., Derrick, J.: Inferring Extended Finite State Machine models from software executions. In: 20th Working Conference on Reverse Engineering (WCRE), pp. 301–310 (October 2013)
32. Walkinshaw, N., Derrick, J., Guo, Q.: Iterative Refinement of Reverse-Engineered Models by Model-Based Testing. In: Cavalcanti, A., Dams, D. (eds.) FM 2009. LNCS, vol. 5850, pp. 305–320. Springer, Heidelberg (2009)
33. Walkinshaw, N., Lambeau, B., Damas, C., Bogdanov, K., Dupont, P.: STAMINA: a competition to encourage the development and assessment of software model inference techniques. Empirical Software Engineering 18(4), 791–824 (2013)
34. Wiedijk, F.: Formal Proof – Getting Started. Notices of the American Mathematical Society 55(11), 1408–1414 (2008)

Querying Geometric Figures Using a Controlled Language, Ontological Graphs and Dependency Lattices

Yannis Haralambous[1] and Pedro Quaresma[2]

[1] Institut Mines-Télécom, Télécom Bretagne Computer Science Department
UMR CNRS 6285 Lab-STICC Technopôle Brest Iroise CS 83818,
29238 Brest Cedex 3, France
[2] CISUC/Departament of Mathematics, University of Coimbra
P-3001-454 Coimbra, Portugal

Abstract. Dynamic geometry systems (DGS) have become basic tools in many areas of geometry as, for example, in education. Geometry Automated Theorem Provers (GATP) are an active area of research and are considered as being basic tools in future enhanced educational software as well as in a next generation of mechanized mathematics assistants. Recently emerged Web repositories of geometric knowledge, like TGTP and Intergeo, are an attempt to make the already vast data set of geometric knowledge widely available. Considering the large amount of geometric information already available, we face the need of a query mechanism for descriptions of geometric constructions.

In this paper we discuss two approaches for describing geometric figures (declarative and procedural), and present algorithms for querying geometric figures in declaratively and procedurally described corpora, by using a DGS or a dedicated controlled natural language for queries.

Introduction

Dynamic geometry systems (DGS) distinguish themselves from drawing programs in two major ways. The first is their knowledge of geometry: from a initial set of objects drawn freely in the Cartesian plane (or maybe, on another model of geometry), one can specify/construct a given geometric figure using relations between the objects, e.g., the intersection of two non-parallel lines, a line perpendicular to a given line and containing a given point, etc. Another major feature of a DGS is its capability to introduce dynamics to a given geometric construction moving a (free) basic object always preserving the geometric properties of the construction [18].

That is, one uses a DGS by constructing a geometric figure with geometric objects and geometric relations between them, and not by placing points on specific Cartesian coordinates. Most (if not all) DGS possess a formal language for the specification of geometric constructions. In some systems this formal language is explicit, in others it is hidden from the user by the graphical interface. The Intergeo project designed a common format, called I2G, for this

S.M. Watt et al. (Eds.): CICM 2014, LNAI 8543, pp. 298–311, 2014.

formal language which is already accepted by some DGS [16,17]. Geometry automated theorem provers (GATP), being formal systems, need a formal language to describe geometric conjectures. GATPs are nowadays mature tools capable of proving hundreds of geometric conjectures [2,8]. The I2GATP formal language is an extension of the formal language used by the DGS. The I2GATP project goal is to define a common language, an extension of the I2G language, to the DGS/GATP tools [12].

The design of common languages, and the emergence of Web repositories of geometric knowledge is an attempt to make the already vast data set of geometric knowledge widely available. The Intergeo project [9], GeoThms [7] and TGTP [11] systems already meet some of these goals, having provided a large data set of geometric information widely available. In these systems the question of querying the geometric construction is not solved, that is, it is not yet possible to query the data set for a construction similar to some other construction, or to query for all constructions having some common geometric properties. The goal of our research is to develop a search mechanism for geometric constructions (done by a DGS or a GATP) using the different ways of geometric construction descriptions.

1 What You See and How to Get It: Declarative vs. Procedural vs. Analytic Figure Description

On Fig. 1 the reader can see (a visual representation of) the centroid theorem, a simple geometric figure taken from the TGTP corpus [11, Fig. 13].

There are many approaches for describing this figure:

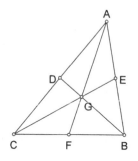

Fig. 1. Construction 13 of the corpus

- the *declarative* one: "we have points A, E, B on the same line such that $AE = EB$, points A, D, C on the same line such that $AD = DC$, points C, F, B on the same line such that $CF = FB$, and displayed are AC, BC, AB, AF, BD, CE. The intersection of BD, CE, AF is called G";
- the *procedural* one: "draw segments AB, AC, BC, take their midpoints E, D, F, draw AF, BD and CE, take the intersection of BD and CE and call it G";
- the *analytic* one: "points A, B, C have coordinates $(35, 40)$, $(10, 10)$, $(40, 10)$; points D, E, F, have coordinates $(37.5, 25)$, $(22.5, 25)$, $(25, 10)$; segments have coordinates $((35, 40), (40, 10))$, $((35, 40), (10, 10))$)", etc.

In this paper we will concentrate on the *procedural* and *declarative* descriptions of a figure.

The declarative description is about "what the parts of the figure are and how they relate to each other," while the procedural description is about "how to construct the figure." In the former we can supply arbitrary (and potentially redundant) information about the figure; in the latter we provide only instructions that result into the given figure.

The first problem we encountered when querying figures was the fact that a given figure can often be constructed (and hence, procedurally described) in several ways. For example, Fig. 1 can be procedurally obtained in (at least) the following two ways (cf. [4] for the second):

start with points A, B, C	start with points D, E, F
draw midpoints of AB, AC, BC	draw a line at F parallel to DE
call them D, E, F	draw a line at E parallel to DF
draw the segments	draw a line at D parallel to EF
	call the intersections A, B and C
	remove the lines
	draw the segments

That is, we can start with the triangle and find the midpoints, or we can start with the midpoints and find the triangle.

Both the DGS to be used (GeoGebra [6]) and the controlled language we will define (§ 3) are *procedural*, and hence describe a figure by its construction. As there are many constructions resulting in the same figure, we concluded that our search system should better use a *declarative* approach. For this (see also [13]), we convert procedural descriptions into declarative ones and represent the information they contain by the use of *ontological graphs*. This operation is done both for the search corpus and for the queries, so that a figure query becomes the search of a graph pattern inside a corpus of ontological graphs.

The second problem we encountered is that procedural descriptions are sometimes lacunary, provided the correct visual result is obtained. For example, in the procedural description of Fig. 1, as it is included in our corpus, the creator of the figure has defined G as being the intersection of BD and CE, without going any further. Since the goal was to obtain the correct visual representation of the figure, it was not necessary to state that G is also the intersection of AF and BD as well as of AF and CE. This means that, after conversion into the declarative representation, the information provided in it will lack these facts.

Inference can fill some of the gaps and make a declarative description more complete. For example it can detect parallelisms or orthogonality relations that are not explicitly stated.[1]

The way we propose to solve this problem is by going the "other way around": instead of making the corpus richer, we can weaken the query. This method is called *query reduction* and is useful when the query contains too much information and cannot be found in the corpus.

The problem then is, how do we reduce the query? Indeed, when in front of an ontological graph query where all ingredients of the query figure have become

---

[1] In a future development we plan to use the deductive database method to find all the fix-points for a given construction, finding in this way the missing facts [3].

nodes, and their relations have become edges, how do we choose the most suitable nodes or edges to remove?

It is the procedural description of the query that provides us with an answer[2] to this question. From the procedural data, we build a *dependency lattice* of the query figure. The lattice structure provides us with the nodes to remove, and the order in which to remove them.

For these reasons, we have developed, and will discuss in this paper, both procedural and declarative descriptions of geometric figures. Thanks to their complementarity we obtain en efficient geometric figure search system.

2   Ontological Graphs

2.1   Describing a Geometric Figure by an Ontological Graph

In the following we will use an ontology specific to geometric figures on the plane. This ontology contains concepts:

- *point*: a point of the plane;
- *segment*: a segment, defined by two points. It has an attribute "length" which induces a relation of "ratio" among segment instances;
- *line*: a line, defined by two points or in some other way (for example, by a point and a property like perpendicularity);
- *conic*: a conic defined in various ways, and, in particular, a circle, defined by its center point and another point;
- *angle*: the angle of two segments/lines, it has an attribute *value* which can have a numeric value or the modal value "straight".

The relations between instances will be[3]:

- *belongs_to*: a relation whose domains are both points (belonging to segments, lines, circles and angles), and segments (belonging to lines);
- *has_ratio*: can be used for lengths and angle values. It is a 3-ary reified relation, the members of which are the nominator, denominator and ratio value;
- *is_center_of*: connects a point with the circle of which it is the center;
- *is_parallel_to*: connects two parallel lines (using inference, we will find all parallel lines);
- *is_perpendicular_to*: connects two perpendicular lines or segments (using inference, we will find all perpendicular lines or segments, knowing that the perpendicular of a perpendicular is a parallel);
- *is_radius_of*: connects a segment with the circle of which it is the radius.

---

[2] Well understood, the answer is not unique since it strongly depends on the way the figure has been constructed, which is not unique.

[3] The list is not exhaustive.

These concepts and relations have been inspired by the element types of DTD GeoCons.dtd [10] (the ontology does not cover XML elements towards, translation, rotation which are useful for drawing but do not affect the ontological graph of the figure) and of GeoGebra XML schema ggb.xsd [6].

Every figure becomes a graph of instances of concepts and of relations. Not only this approach is independent of the way the figure has been constructed, but it is also independent of instance names and allows to focus on the network of relations between the ingredients of the figure.

Our choice of concepts and relations makes some graphical constructions obtainable by a single relation, for example: "$AB \perp BC$"

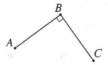

is represented by instances A, B, C (points), AB and BC (segments) and $\angle ABC$ (angle), and the following graph of relations:

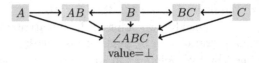

where solid arrows denote the *belongs_to* relation, and $\perp$ is the "right angle" value of the value attribute.

Other constructions, although trivial, are more difficult to encode. For example: "AB is tangent at circle c at point B"

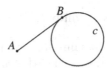

cannot be encoded by a single relation. We need to use the radius BO and say that $B \in c \wedge AB \perp BO$

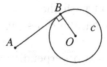

and the graph of relations will be

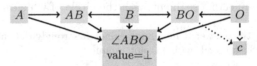

where the dotted arrow represents the relation *is_center_of* and the dashed one, the relation *is_radius_of*.

The ontological graph of a geometrical figure can rapidly increase in size. Its generation is done in a two-step process:

1. every XML element of the figure description is converted into a set of concepts and relations;
2. then, inference is applied to generate additional relations:
    (a) we calculate the transitive closure of parallelism and orthogonality relations $(a \mathbin{/\!/} b \wedge b \mathbin{/\!/} c \vdash a \mathbin{/\!/} c$ and $a \perp b \wedge b \perp c \vdash a \mathbin{/\!/} c)$;
    (b) nodes having equal lengths or equal angle values by construction[4] obtain a *has_ratio* relation with value attribute equal to 1;
    (c) if necessary, angles are instantiated for every pair of segments with a common point.

Our corpus of 134 figures, encoded as an XML file of 3,137 elements resulted into graphs of a total of 5,282 concept instances and 10,211 relation instances.

2.2 Example

Take Fig. 1 representing Figure 13 of the corpus (illustrating the fact that medians intersect at the barycenter of the triangle). The construction, as given in the XML file, takes arbitrary points A, B, C, defines D (resp. E, F) as the midpoint of AC (resp. AB, BC), and G as the intersection of BD and CE. Furthermore, the segment AF is drawn.

The ontological graph will contain concepts for points A, B, C, D, E, F, G, and segments $AB, AC, AD, AE, AF, BC, BD, BE, BF, CD, CE, CF$. The relations will all be of type *belongs_to*, except for some 3-ary *has_ratio* relations representing equal lengths. In Fig. 2, unlabelled arrows denote the *belongs_to* relation.

2.3 Querying Ontological Graphs

To be able to search in a corpus, we convert all figures of the corpus into ontological graphs and we store them in a graph database (we use a neo4j database [15]). The user query is a figure drawn by using a DGS, or a query using the controlled query language (§3). This figure or CQL statement is converted into an ontological graph on-the-fly, and then into a Cypher query (Cypher is the query language of the neo4j graph database system). The query is send to the database and returns graph instances containing the query as sub-graph.

At this step, ranking is performed to present the results to the user in a pertinent way. Our ranking criterion (which we will try to improve in the future) is the ratio between number of nodes and relations of the query and the number of nodes and relations of the matched graphs. Using this criterion we obtain first

---

[4] We emphasize the fact that equality is explicitly given in the construction and is not the result of measurements between objects of the figure.

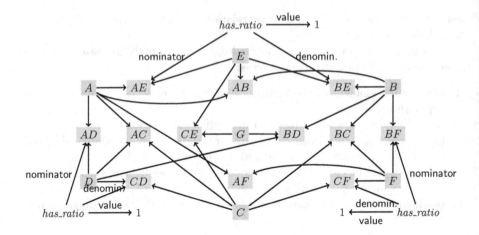

Fig. 2. The ontological graph of Fig. 1

the smallest figure possible figure containing the query subgraph. We intend to use graphical mechanisms to highlight the matched pattern in the resulting graphs by using, for example, a different color.

3   The Controlled Query Language

In some cases the user may not wish to use the DGS to build the query, either because it is cumbersome to use or because it does not provide the necessary abstractions. We propose, as an alternative to the DGS, a *controlled query language* that allows the (procedural) description of a figure in a way that is simple and close to natural language.

3.1   Description of the Controlled Query Language

Here is the grammar of the controlled query language:

```
S -> query
query -> sents drawvp PERIOD
query -> sents PERIOD
drawvp -> DRAW ents
sents -> sent SEMICOLON sents
sents -> sent
sent -> nps vrb pps
sent -> nps vrb
ent -> INST LABEL
ent -> LABEL
vrb -> VERB ADJE
vrb -> VERB NOUN
```

```
vrb -> VERB
pps -> ents pents
pps -> pents
pps -> ents
pents -> pent pents
pents -> pent
pent -> PREP ents
ents -> ent AND ents
ents -> ent
nps -> ents
NOUN -> /(midpoint|foot|mediatrix|intersection|bisector)[s]?/
INST -> /(points|segments|lines|angles|circles|centers|point|
          segment|line|angle|circle|center)/
VERB -> /(is|are|intersect[s]?|connect[s]?)/
ADJE -> /(perpendicular|parallel|defined|right)/
SEMICOLON -> /;/
AND -> /(,|and)/
DRAW -> /draw/
PREP -> /(at|of|by|to|on)/
PERIOD -> /\./
LABEL -> /[A-Z]([_]?[0-9]+)?(-[A-Z]([_]?[0-9]+)?)*/
```

and here are its rules[5]:

1. Every query is of the form:
   [list of sentences separated by ;] draw [list of instances separated by ,].
2. An *instance* consists of a type and a name (or just a name, if there is no
   ambiguity). It is written in the form "type [name]".
3. A *primitive instance* is an instance of a point, a line or a circle.
4. The name of a primitive instance matches the regular expression
   [A-Z]([_]?[0-9]+)?.
5. Names of non-primitive instances are composite: they are written by joining
   names of points using the hyphen character (for example: segment A-B).
6. An instance type can be followed by more than one instance, in that case
   it is written in plural form and the instances are separated by commas (for
   example: points A, B, C).
7. The first part of a query defines (and draws) new instances, the second draws
   already known instances.
8. The following sentences can be used[6]:
   (a) line ? intersects line ? at point ?
   (b) point ? is the midpoint of segment ?
   (c) line ? is perpendicular to line ? at point ?
   (d) point ? is the foot of point ? on line ?
   (e) line ? is the mediatrix of segment ?

---

[5] We will use sans serif font for illustrating the controlled language.
[6] All type names are optional except for center in (h).

(f) line ? is parallel to line ? at point ?
(g) line ? connects points ?, ?
(h) circle ? is defined by center ? and point ?
(i) points ?, ? are the intersections of circles ?, ?
(j) points ?, ? are the intersections of circle ? and line ?
(k) line ? is the bisector of angle ?
(l) angle ? is right

9. All sentences have plural forms where arguments are distributed at all positions and separated by commas (for example: lines L_1, L_2, L_3 connect points A_1, B_1, A_2, B_2, A_3, B_3, which means that $\{A_i, B_i\} \subset L_i$).

10. In all sentences except (8e), the terms segment and line are synonymous, with the syntactic difference that segment must be followed by a composite name (for example A-A_1), while line must be followed by a primitive name, since "line" is a primitive instance.

11. Some variation is allowed, for example and is a synonym of the comma, articles the in front of nouns are optional.

12. Queries end by a period '.'.

Here is, for example, a description of Fig. 1 in the controlled query language:

D, E, F are midpoints of A-C, A-B, B-C ; C-E intersects B-D at G ; draw A-C, A-F, A-B, B-C, B-D, C-E.

The query language is compiled, producing a Cypher query, which is then submitted to the graph database exactly as when using the DGS. The compiler has been developed using the Python PLY package [1].

3.2   Future Plans for the Controlled Language

In future versions of the controlled language, we plan to introduce the possibility of extending the query ontology by introducing new concepts and/or new relations. For example, it may be interesting to define a type square as

Points A, B, C, D form a square A-B-C-D when we draw equal segments A-B, B-C, C-D, D-A where angles A-B-C, B-C-D, C-D-A, D-A-B are right.

This would allow queries of the form (which will draw the notorious figure of the Pythagorean theorem)

Angle A-B-C is right ; A-C-C_1-A_2, A-A_1-B_1-B, C-B-B_2-C_2 are squares.

4   Reduced Queries

The algorithms we describe in this paper can be quite successful in finding exact matches of queries in the corpus. But what happens when the figures in the corpus match only partially the query?

4.1   Ontological Graphs

For example, let us consider Fig. 2 anew. The ontological graph of the figure has been build solely using the XML data of Figure 13 of the corpus (cf. Fig. 1). What is not visible on Fig. 1 is the fact that G has not been defined as lying on AF, and hence the *belongs_to* edge between G and AF is missing in the ontological graph.

This is also reflected in the CQL query example we gave in § 3.1, where we request that C-E intersects B-D at G but not that A-F intersects B-D at G, probably because this could be inferred from the previous one, if we had the external Euclidean Geometry knowledge of the fact that the three medians of a triangle have a common intersection.

Nevertheless, the user seeking Fig. 1 is not necessarily aware of this subtlety, and will search for "a triangle with three medians," which will result in an ontological graph similar to the one of Fig. 2 but containing an additional edge $G \rightarrow AF$, and this graph, of course, will not match Figure 13 of the corpus, since it is not a sub-graph of it.

To solve this problem, as long as a query does not return any results, we retry with *reduced queries*, in the sense of the same query graph with one or more instances (or relations) removed.

But how do we decide which nodes and edges to remove from a query, and in what order? The answer to this question is provided by *dependency lattices*, described in the next section.

4.2   Dependency Lattices

Let us return to the *procedural* approach of describing geometric figures. How do we describe a figure using the operations that led to its construction?

Strictly speaking, such a description would require a Berge-acyclic hypergraph [5, §3], where each operation would be a hyper-edge, connecting the input (the set of known nodes) and the output (the set of new nodes), for example, in the case of the *midpoint* operation on segment AC, the hyper-edge would connect $\{A, C\}$ (input) and $\{B\}$ (output).

But there is a simpler way. In fact, it suffices for our needs to represent *dependencies* as edges of a directed graph. For example, in the midpoint example, B is dependent of A and C, since the latter have been used to calculate the former:

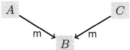

By adding a "global source node" (located above all source nodes) and a "global sink node" (underneath all "final results"), this graph becomes a *lattice*, the partial order of which is the dependency relation.

On Fig. 3, the user can see the dependency lattice of Fig. 1. We have used only nodes that are used in calculations, so that, for example, segments AB, AE, etc. do not appear in the lattice. S and T are the global source and global sink

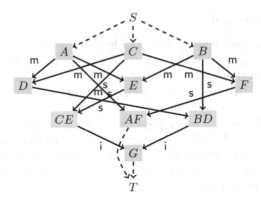

Fig. 3. The dependency lattice of Fig. 1, as it is procedurally described in the corpus

nodes, they are connected to nodes of the lattice by dashed arrows. Full arrows represent operations and are labelled by their initial letters (m = midpoint, s = segment drawing, i = intersection).

4.3 Using Dependency Lattices for Reduced Queries

Let us now see how the dependency lattice would be affected if the XML description of Fig. 1 had an additional instruction, saying that G is (also) the intersection of BD and AF. On Fig. 4 one can compare the two graphs, on the right side one can see the one with the additional instruction.

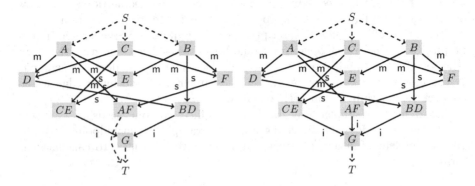

Fig. 4. The dependency lattice of Figure 13 of the corpus (left) and the dependency lattice of Figure 13 plus an additional instruction segment A-F intersects segment B-D at point G (right)

Indeed, the new dependency lattice has one additional edge $AF \to G$. On the other hand, the edge $AF \to T$ disappears since there is a path from F to T, and AF is not a sink anymore.

Data: A corpus of declaratively described figures, a query
Result: One or more figures matching the query
if *using controlled query language* **then**
   | write the query in controlled query language;
else
   | draw the query in a DGS;
end
convert query into ontological graph;
apply inference to ontological graph;
convert ontological graph to Cypher;
submit to neo4j database;
if *no results returned* **then**
   convert query into dependency lattice;
   while *no results returned* **do**
      extract node or relation from bottom of dependency lattice;
      remove that node or that relation from the ontological graph;
      convert ontological graph to Cypher;
      submit to neo4j database;
   end
end

Algorithm 1. The Query Algorithm for a Declaratively Described Corpus

As dependencies have to be respected, if we remove a node from the upper part of the lattice, we will have to remove all descendants of it. For this reason, the only reasonable query reduction strategy would be to remove nodes or edges from the *lower* part of the lattice.

If we remove, for example, the node G (and hence the edges $CE \rightarrow G$, $AF \rightarrow G$ and $BD \rightarrow G$), then we obtain a triangle with three medians but where the barycenter has not been explicitly drawn. (Interestingly, we still obtain a figure that is visually identical to Fig. 1.)

If we go further and remove one of the relations among $CE \rightarrow G$, $AF \rightarrow G$ and $BD \rightarrow G$ (for symmetry reasons it doesn't matter which relation we remove), then the query will succeed, while the visual representation of the figure still has not changed.

Algorithm 1 is a synthesis of the geometric figure query algorithm we propose.

5   Evaluation

We plan to evaluate the algorithms described in this paper, in the following ways:

Querying a sub-figure in a corpus of declaratively described figures gives a binary result: either the figure is matching the sub-figure or it is not, so evaluation is simply counting the number of successes.

An interesting parameter to observe is the number and nature of query reductions that were necessary to obtain results, correlated with the number of results obtained.

We will proceed as follows: after visually inspecting the corpus (and hence with no knowledge about the procedural and declarative descriptions of the figures) we will formulate 20 queries and manually annotate the figures we expect to find.

After using the algorithm, we will count the number of successes and study the number of results vs. the parameters of query reduction.

6   Future Work

As future work, besides extending the controlled natural language (§ 3.2), we plan to integrate this search mechanism in repositories such as TGTP and Intergeo, and in learning environments like the Web Geometry Laboratory [14]. In a more generic approach, we should use a common format and develop an application programming interface that will allow to integrate the searching mechanism in any geometric system in need of it.

7   Conclusion

In this paper we have presented algorithms for querying geometric figures in either declaratively or analytically described corpora, by using either a DGS or a dedicated controlled query language.

At the time of submission of the article, evaluation was not completed, hence it is presented as a plan.

References

1. Beazley, D.: Python Lex-Yacc, http://www.dabeaz.com/ply/
2. Chou, S.C., Gao, X.S.: Automated reasoning in geometry. In: Handbook of Automated Reasoning, pp. 707–749. Elsevier Science Publishers (2001)
3. Chou, S.C., Gao, X.S., Zhang, J.Z.: A deductive database approach to automated geometry theorem proving and discovering. Journal of Automated Reasoning 25, 219–246 (2000)
4. DiffeoR: Answer to "Is it possible to reconstruct a triangle from the midpoints of its sides?" (February 20, 2014), http://math.stackexchange.com/a/683496/122762
5. Fagin, R.: Degrees of acyclicity for hypergraphs and relational database schemes. Journal of the ACM (JACM) 30(3), 514–550 (1983)
6. Hohenwarter, M., Preiner, J.: Dynamic mathematics with GeoGebra. The Journal of Online Mathematics and Its Applications 7, ID 1448 (2007)
7. Janičić, P., Quaresma, P.: System description: GCLCprover + GeoThms. In: Furbach, U., Shankar, N. (eds.) IJCAR 2006. LNCS (LNAI), vol. 4130, pp. 145–150. Springer, Heidelberg (2006)
8. Jiang, J., Zhang, J.: A review and prospect of readable machine proofs for geometry theorems. Journal of Systems Science and Complexity 25, 802–820 (2012)
9. Kortenkamp, U., Dohrmann, C., Kreis, Y., Dording, C., Libbrecht, P., Mercat, C.: Using the Intergeo platform for teaching and research. In: Proceedings of the 9th International Conference on Technology in Mathematics Teaching, ICTMT-9 (2009)

10. Quaresma, P., Janičić, P., Tomašević, J., Vujošević-Janičić, M., Tošić, D.: XML-Bases Format for Descriptions of Geometric Constructions and Proofs. In: Communicating Mathematics in The Digital Era, pp. 183–197. A. K. Peters, Ltd. (2008)
11. Quaresma, P.: Thousands of Geometric Problems for Geometric Theorem Provers (TGTP). In: Schreck, P., Narboux, J., Richter-Gebert, J. (eds.) ADG 2010. LNCS (LNAI), vol. 6877, pp. 169–181. Springer, Heidelberg (2011)
12. Quaresma, P.: An XML-format for conjectures in geometry, Aachen. CEUR Workshop Proceedings, vol. 921, pp. 54–65 (2012), http://ceur-ws.org/Vol-921/
13. Quaresma, P., Haralambous, Y.: Geometry construction recognition by the use of semantic graphs. In: RECPAD 2012: 18th Portuguese Conference on Pattern Recognition, Coimbra, October 26, pp. 47–48 (2012)
14. Quaresma, P., Santos, V., Bouallegue, S.: The web geometry laboratory project. In: Carette, J., Aspinall, D., Lange, C., Sojka, P., Windsteiger, W. (eds.) CICM 2013. LNCS (LNAI), vol. 7961, pp. 364–368. Springer, Heidelberg (2013)
15. Robinson, I., Webber, J., Eifrem, E.: Graph Databases. O'Reilly (2013)
16. Santiago, E., Hendriks, M., Kreis, Y., Kortenkamp, U., Marquès, D.: I2G Common File Format Final Version. Tech. Rep. D3.10, The Intergeo Consortium (2010), http://i2geo.net/xwiki/bin/view/I2GFormat/
17. The Intergeo Consortium: Intergeo implementation table (2012), http://i2geo.net/xwiki/bin/view/I2GFormat/ImplementationsTable
18. Wikipedia: List of interactive geometry software (February 2014), http://en.wikipedia.org/wiki/List_of_interactive_geometry_software

Flexary Operators for Formalized Mathematics

Fulya Horozal, Florian Rabe, and Michael Kohlhase

Computer Science, Jacobs University Bremen, Germany
http://kwarc.info

Abstract. We study representation formats that allow formally defining what we call *flexary* operators: functions that take arbitrarily many arguments, like $\sum_{k=1}^{n} a_k$ or binders that bind arbitrarily many variables, like $\forall x_1, \ldots x_n. F$. Concretely, we define a flexary logical framework based on LF, and use it as a meta-language to define flexary first-order logic and flexary simple type theory. We use these to formalize several flexary mathematical concepts including arithmetical and logical operators, matrices, and polynomials.

1 Introduction and Related Work

Ellipses (...) such as in $a_1, \ldots, a_n$ are commonly and indispensably used in mathematical texts. However, representation formats for formalized mathematics typically do not provide a structural analog for ellipses. This is problematic because many common operators are naturally defined using a primitive ellipsis operator, and formalizations have to work around the missing language infrastructure. We will now lay out the problem, survey and discuss the most commonly used workarounds and then present a solution which introduces sequences as a language feature at the meta-level.

1.1 Flexary Operators and Ellipses

We say that an operator is **of flexible arity** or **flexary** if it can take arbitrarily many arguments. Common examples are set-construction $\{a_1, \ldots, a_n\}$ or addition $a_1 + \ldots + a_n$. We speak of **fixary** operators if the number of arguments is fixed.

Ellipses. Flexary operators are closely related to the ellipsis operator .... For presentation-oriented formats, ellipsis are no challenge. For example, LaTeX offers \ldots, and presentation MATHML [ABC+03] marks up the corresponding Unicode character as an operator via the mo element. However, the content-oriented formats that we need for formalized mathematics have devoted much less attention to ellipses. This is surprising considering how ubiquitous they are in mathematical practice.

We can distinguish 4 kinds of ellipses. We speak of a **sequence** ellipsis if we give a sequence of arguments to a flexary operators as in $a_1 + \ldots + a_n$.

The **nesting** ellipsis uses a characteristic double-... pattern to compose a sequence of functions as in $f_1(\ldots f_n(x) \ldots)$. It corresponds to folding the function

S.M. Watt et al. (Eds.): CICM 2014, LNAI 8543, pp. 312–327, 2014.

f over the list of arguments. In the presence of a flexary operator for function composition, we can recover the nesting ellipsis as a special case of the sequence ellipsis via $f_1(\ldots f_n(x)\ldots) := (f_1 \circ \ldots \circ f_n)(x)$.

When working with matrices, we use **2-dimensional** ellipses as in the matrix on the right. If we define vectors using a flexary constructor $(a_1, \ldots, a_n)$ and matrices as vectors of vectors, we can recover the 2-dimensional ellipses by combining two sequence ellipses. Of course, that would still leave the problem of presenting vectors with ellipses.

$$E_n = \begin{pmatrix} 1 & 0 & \cdots & 0 \\ \vdots & \ddots & & \vdots \\ 0 & \cdots & 0 & 1 \end{pmatrix}$$

Finally, we have the **infinite** ellipsis used mainly for infinite series as in $a_1 + a_2 + \ldots$.

Flexary Binders We can generalize the above concepts to binding operators. We speak of a **flexary binder** if it can bind an arbitrary number of variables as in $\forall x_1, \ldots, x_n.F$. Most unary binders such as quantifiers and λ are usually assumed to be flexary in this sense.

1.2   Flexary Notations

A common approach is to use representation languages that are fixary at the content level but flexary at the presentation level. The connection between the two is performed by **flexary notations**. Typically, these use associativity constraints on binary infix operators.

For example, we can define flexary addition $+(a_1, a_2, a_3)$ as identical to $a_1 + a_2 + a_3$, which in turn is an abbreviation for $(a_1 + a_2) + a_3$. Here we use the logical property of associativity to justify a left-associative notation.

We can also use associative notations for logically non-associative operators. For example, we can define flexary implication $\Rightarrow (a_1, a_2, a_3)$ as $a_1 \Rightarrow (a_2 \Rightarrow a_3)$, i.e., by using a right-associative notation for binary implication.

While there is no established terminology, we can apply similar notations to binders. We call a flexary binder **associative** if $Qx_1.Qx_2.F = Qx_1, x_2.F$. In that case, we can define the flexary version of the binder from the unary version. This is very common because the important binders of universal and existential quantifier are associative and (up to currying) so are λ, integral (e.g. $\int dx\, dy\, dz$), sum (e.g. $\sum_{i,j\in\mathbb{N}}$), and product (e.g. $\prod_{n+m<k}$). A notable exception is the quantifier of unique existence.

Using flexary notations has the advantage that the content level remains simpler: Flexary operators are always implicitly reduced to fixary ones. This is important for language analysis and perfectly sufficient in informal mathematics. However, in formalized mathematics, the implicit conversions must be explicitly implemented. Usually, this is achieved by using the notation declarations to direct the parser and printer to convert between the seemingly-flexary human-facing syntax and the fixary official syntax.

This is unsatisfactory for several reasons. Firstly, for logically associative operators, there is no canonical choice between using a left- or a right-associative notation. In a flexary content representation both $(a_1+a_2)+a_3$ and $a_1+(a_2+a_3)$ would normalize to the canonical $+(a_1, a_2, a_3)$.

Secondly, this trick only works well if the domains and codomain of the operator are equal. Consider the flexary set construction operator $\{a_1, a_2, a_3\}$. We can only approximate it using a left-associative notation for an adjoin operator $a\&b := a \cup \{b\}$ and then write $\varnothing\&a_1\&a_2\&a_3$.

Thirdly, the flexary representation is often the more natural one for implementation, e.g. for the left-associative application $f@t$ in simple type theory. In the usual fixary type theory, the n-ary functions $f(x, y)$ are represented in the curried form $f@x@y$, which internally expands to $@(@(f, x), y)$. Thus, the head f of the term is not available at the root of the syntax tree and has to be looked up by traversing the tree. In practice, this traversal is so awkward that many implementations of type theory, e.g., Twelf [PS99], internally use a flexary application operator after all so that $f(x, y)$ can be represented as $@(f, x, y)$. This leads to the strange situation that both the user and the developer effectively use flexary operators, and only the official language definition uses fixary ones.

Finally, this approach only works in general for the case where the number of arguments is constant: We cannot use it to represent a sequence ellipsis like $+(a_1, \ldots, a_n)$, where the number of arguments is a variable. For the special case of conjunction and disjunction, notations for ellipses were realized in Mizar [Kor12]. For example, a special binary connective $\&\ldots\&$ is used for the conjunction of a sequence ellipsis, and the parser expands $F(m)\&\ldots\&F(n)$ into $\forall i.\, m \leq i \leq n \Rightarrow F(i)$.

1.3   Flexary Representation Languages

Instead of simulating flexary operators through notations, we can use a representation language that supports flexary operators at the content level. There are several content features that can be used.

Lists We can represent flexary operators as unary operators that take a list as an argument. In that case, we represent $a_1 + a_2 + a_3$ as $+(List(a_1, a_2, a_3))$. Actually, this tacitly assumes that we have at least a flexary list constructor. In a pure fixary language, we would have to represent it as $+(cons(a_1, cons(a_2, cons(a_3, nil))))$, which is quite different from the informal mathematical object.

This approach permits using variables that quantify over sequences, and – using map and fold – it is easy to represent ellipses. This is widely used in programming languages, where lists are an accepted foundational data type.

In mathematics however, it is artificial to use lists since any formal mathematical theory for flexary operators would depend on the theory of lists, which itself is rarely used in informal mathematics. Another drawback is that all arguments must have the same domain. To permit different argument domains, we must allow lists whose elements have different types (or use sufficiently imprecise types).

Sequences. Sequence types use a monadic type constructor $Seq : type \to type$ like lists and enjoy the same advantages. The difference is that sequences are always flattened, i.e., the canonical functions $Seq(Seq(A)) \to SeqA$ and $A \to Seq(A)$ are inclusions. For example, $Seq(a, b, Seq(c, d), e, f) = Seq(a, b, c, d, e, f)$. This makes sequence types closer to informal mathematics because they need less representational artifacts. Variants of sequence types occur in some programming languages but are rare in typed languages for formalized mathematics.

Sequences are more common in untyped languages. In the absence of a type system and in the presence of flattening, there is no need to write $f(Seq(a, b, c))$ at all. Instead, we can simply write $f(a, b, c)$ (even if one of the arguments is another sequence).

This approach is used in Common Logic (CL [Com07]), an untyped flexary variant of first-order logic. There, every non-logical symbol is flexary and variables may quantify over sequences. This substantially complicates the semantics because models must interpret every function symbol as a function that takes an arbitrary sequence of arguments; incidentally a proof theoretical semantics is not defined in CL.

[KB04] defines a flexary first-order logic and studies its semantics. The signature defines the arity of each non-logical symbol, and the arity can either be fixed or flexible. Similarly, variables are divided into individual and sequence variables.

Mathematica [Wol12] also uses untyped sequences, including sequence variables. Functions are fixary, but flexary functions can be defined by matching arguments against sequence patterns. Because Mathematica focuses on computation rather than logic, this is less problematic than for CL.

The untyped approaches to sequences usually cannot represent ellipses well because they tend to lack higher-order functions.

Indexed types. Mixed-type lists can be represented concisely in Martin-Löf type theory [ML74], calculus of constructions [CH88] and related languages. Example implementations are Agda [Nor05] and Coq [Coq14]. If we write $[n]$ for the type containing $0, \ldots, n - 1$, we call objects of type $T : [n] \to type$ indexed types. Mixed-type lists can be defined as indexed terms $l : \Pi i : [n]. T(i)$. Then flexary functions can be declared concisely as binary functions that take a natural number n and term indexed by $[n]$.

Ellipses can be represented very elegantly now, e.g., $a_1, \ldots, a_n$ is simply $\lambda i. a_i$. Moreover, contrary to all of the above, the length of a sequence is statically known, which permits static index-within-bounds checking when accessing an element of a sequence. Quantification only affects sequences of a certain length, e.g., $\forall x : [n] \to A. F$; to quantify over all sequences, we can use $\forall n. \forall x : [n] \to A. F$.

A disadvantage is the substantial commitment at the language level, which goes far beyond simply adding sequences: The language must be able to express the types $[n]$ and $[n] \to type$ (e.g. via inductive constructions and a universe hierarchy in Coq).

Type sequences. We introduce a novel approach: we use term sequences $a_1, \ldots, a_n$ that are typed component-wise by a type sequence $A_1, \ldots, A_n$. Importantly, type sequences $A_1, \ldots, A_n$ are not types themselves – they are simply sequences of types.

Like sequences and contrary to indexed types and list types, this has the advantage that we do not change the underlying type theory. No representational artifacts are needed to flatten sequences or to apply a function to a sequence of arguments. And like for indexed types, the length of a sequence is statically known.

1.4   Flexary Meta-Languages

For content representations of flexary operators, the previous section discussed which representational primitives to use. An orthogonal question is at which language level they should be introduced.

Consider the first-order theory of monoids in which we want to define the flexary version of the composition operator in the obvious way. This should also include the power $a^n = a \circ \ldots \circ a$ for a natural number n as a special case. We might do that by importing the theory of sequences and then using some kind of induction. But this is awkward because the theory of monoids would become much more complex. We might even say that it becomes polluted by the imported operations.

We might try to move the definition to a special enriched theory, which includes both sequences and monoids. But that would contradict mathematical practice, where the definition of the flexary composition is likely to be found in the same paragraph where the binary one is.

Thus, we should add sequences to the logic as a fixed interpreted sort. However, now a similar argument applies: The logic is complicated. Moreover, it does not account for the fact that we would like to use sequences in any logic. Therefore, our goal is to add sequences at the level of the logical framework. This corresponds most closely to informal mathematics where sequences and ellipses are assumed to be given at the informal meta-level and not explicitly defined at the logical or set theoretical level.

The approach of type sequences is most suitable in this respect because it is already orthogonal to the type theoretical and logical foundations. Thus, it can be added easily as a framework feature. Once we go down that road, it also becomes very easy and natural to add constructors for sequence and nesting ellipses at the framework level.

1.5   Overview

Following the above analysis, we develop a logical framework with type sequences. We choose the logical framework LF [HHP93] as an example logical framework since it has been used to represent a large variety of formalisms. But our approach can be transfered to other frameworks (e.g. Isabelle [Pau94]) as well, because it is orthogonal to the underlying type theory.

We briefly summarize LF in Sect. 2 and then extend it to LFS (LF with sequences) in Sect. 3. Notably, our extension is minimally invasive, keeping the essence of the LF type theory unchanged (and reusing the existing rules). We use LFS in Sect. 4 to define flexary versions of first-order logic and simple type theory. In both cases, we declare flexary versions of all operators (where reasonable) and show that LFS can formally define the flexary versions in terms of the fixary ones. Finally, we use our two flexary logics to formalize a collection of common mathematical examples in Sect. 5.

2 The Edinburgh Logical Framework

In this section, we briefly revisit LF [HHP93], a dependently-typed λ-calculus that can be used well as a logical framework [Pfe01]. We give the LF grammar in Fig. 1.

Kinds	K	$::=$	$\mathbf{type} \mid \Pi x : U.K$
Type Families	$U, V \ldots$	$::=$	$a \mid \Pi x : U.V \mid \lambda x : U.V \mid U\,S$
Terms	$S, T \ldots$	$::=$	$x \mid \lambda x : U.S \mid S\,T$
Signatures	Σ	$::=$	$\cdot \mid \Sigma, x : U[= S] \mid \Sigma, a : K[= U]$

Fig. 1. LF Grammar

LF expressions are grouped into kinds K, kinded type-families $U : K$, and typed terms $S : U$. The kinds are the base kind $\mathbf{type}$ and the dependent function kinds $\Pi x : U.K$. The type families are the symbols a, the dependent function type $\Pi x : U.V$, abstractions $\lambda x : U.V$, applications $U\,S$; type families of kind $\mathbf{type}$ are called types. The terms are symbols x, abstractions $\lambda x : U.S$, and applications $S\,T$. Signatures Σ consist of typed or kinded symbols $x : U[= S]$ or $a : K[= U]$ with optional definiens.

As usual, we write $U \to E$ instead of $\Pi x : U.E$ if x does not occur free in E, and we omit the types of bound variables if they can be inferred. Free variables are implicitly bound on the outside of the expression (implicit arguments). Substitution of T for x in E is written $[T/x]E$.

We use the signatures given in Fig. 2 as running examples throughout this paper. The signature SFOL on the left defines the syntax and proof rules of sorted first-order logic. The signature STT on the right defines the syntax and β-conversion rule of simple type theory.

In order to emphasize the similarity between LF and LFS, we also give the judgments for well-formed LF expressions in Fig. 3 and the inference rules in Fig. 4. For brevity, we omit the equality judgement, whose rules consist of equivalence, congruence, and $\alpha\beta\eta$-conversion.

318   F. Horozal, F. Rabe, and M. Kohlhase

$$
\begin{array}{ll}
sort & : \textbf{type} \\
tm & : sort \rightarrow \textbf{type} \\
form & : \textbf{type} \\
ded & : form \rightarrow \textbf{type} \\[6pt]
true & : form \\
\wedge & : form \rightarrow form \rightarrow form \\
\forall & : \Pi S : sort.\,(tm\,S \rightarrow form) \rightarrow form
\end{array}
$$

$$
\begin{array}{ll}
tp & : \textbf{type} \\
tm & : tp \rightarrow \textbf{type} \\
\Longrightarrow & : tp \rightarrow tp \rightarrow tp \\
= & : tm\,A \rightarrow tm\,A \rightarrow \textbf{type} \\
lam & : tm\,A \rightarrow tm\,B \rightarrow tm\,(A \Longrightarrow B) \\
app & : tm\,(A \Longrightarrow B) \rightarrow tm\,A \rightarrow tm\,B \\
beta & : app\,(lam(\lambda x : tm\,A.T))\,X = (T\,X)
\end{array}
$$

$$
\begin{array}{ll}
true_I & : ded\ true \\
\wedge_I & : ded\,F \rightarrow ded\,G \rightarrow ded\,F \wedge G \\
\wedge_{El} & : ded\,F \wedge G \rightarrow ded\,F \\
\wedge_{Er} & : ded\,F \wedge G \rightarrow ded\,G \\
\forall_I & : (\Pi x : tm\,S.\,ded\,F\,x) \rightarrow ded\,\forall\,F \\
\forall_E & : ded\,\forall\,F \rightarrow \Pi x : tm\,S.\,ded\,F\,x
\end{array}
$$

Fig. 2. LF Signatures for SFOL (left) and STT (right)

Judgment	Meaning
$\Sigma \vdash S : U$	S is a well-formed term of type U over Σ
$\Sigma \vdash U : K$	U is a well-formed type family of kind K over Σ
$\Sigma \vdash K\ Kind$	K is a well-formed kind over Σ.

Fig. 3. LF Judgments

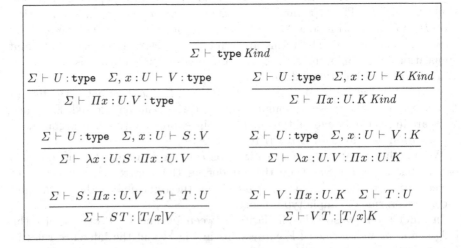

Fig. 4. LF Inference Rules

3   A Flexary Logical Framework

3.1   Natural Numbers

In this section we extend LF to form our logical framework LFS (LF with Sequences). LFS adds type sequences and ellipsis constructors.

Therefore, we also need natural numbers as indices to access elements of sequences and to form ellipses. We do this by assuming that the LF signature on the right is always present. Moreover, we assume declarations that formalize the usual computation rules to normalize expressions of type $\mathtt{nat}$.

$$
\begin{aligned}
&\mathtt{nat} : \mathtt{type}\\
&\leq\ : \mathtt{nat} \to \mathtt{nat} \to \mathtt{type}\\
&=\ : \mathtt{nat} \to \mathtt{nat} \to \mathtt{type}\\
&0\ \ : \mathtt{nat}\\
&1\ \ : \mathtt{nat}\\
&+\ : \mathtt{nat} \to \mathtt{nat} \to \mathtt{nat}\\
&-\ : \Pi n : \mathtt{nat}.\,\Pi m : \mathtt{nat}.\\
&\qquad\quad m \leq n \to \mathtt{nat}
\end{aligned}
$$

Note that $-$ is a partial subtraction operator: It takes as a third argument a proof that $n - m$ is defined. We will omit that argument whenever the needed proof is straightforward. This restriction guarantees that we work with natural numbers but not with negative integers. We additionally assume proof irrelevance, i.e., an axiom of type $-(m, n, P) = -(m, n, Q)$. Moreover, we will use the symbols $\leq$, $=$, $+$, and $-$ in infix notation.

Here, for simplicity, we do not formally restrict the use of natural numbers within LFS. However, we consider their status to be similar to that of types. In particular, we assume that all terms of type $\mathtt{nat}$ refer only to the above signature and free variables introduced in the toplevel context.

3.2   Syntax

We can now give the grammar of LFS in Fig. 5 by extending the grammar of LF. All productions for LF are retained but generalized to sequences. All our extensions are underlined: The singly underlined productions add sequences, the doubly underlined ones add ellipses.

Kinds	K	$::=$	$\mathtt{type}^{\underline{S}} \mid \Pi x : U.\,K$
Type Seq. Families	U, V	$::=$	$\underline{\cdot} \mid \underline{U, V} \mid \underline{U_S} \mid \underline{[U]_{x=1}^{S}} \mid a \mid \Pi x : U.\,V \mid \lambda x : U.\,V \mid U\,S$
Term Sequences	S, T	$::=$	$\underline{\cdot} \mid \underline{S, T} \mid \underline{S_T} \mid \underline{\underline{[S]_{x=1}^{T}}} \mid \underline{\underline{;\,S}} \mid x \mid \lambda x : U.\,S \mid S\,T$

Fig. 5. LFS Grammar

The **term sequences** S, T are formed by the **empty term sequence** $\cdot$ and **concatenation** S, T. If $n : \mathtt{nat}$, then S_n accesses the n-**th element** of a sequence S. **Type sequences** are formed in the same way. We write E^n for the sequence $[E]_{x=1}^{n}$ if x does not occur free in E.

If a type family sequence has a length other than 1, all its elements are types. Consequently, the only **kind sequences** we need are $\mathtt{type}, \ldots, \mathtt{type}$, which we write as $\mathtt{type}^S$ for $S : \mathtt{nat}$. We recover $\mathtt{type}$ as an abbreviation for $\mathtt{type}^1$.

The term **sequence ellipses** constructor is $[S(x)]_{x=1}^n$. It takes an argument n : nat and binds the symbol x : nat in S. Its intended meaning is that it reduces to the term sequence $S(1), \ldots, S(n')$ whenever n reduces to a natural number n'. We use an analogous constructor $[U(x)]_{x=1}^n$ for type sequence ellipses.

The constructor for **nesting ellipses** is more complicated. After several experiments, we opted for a **flexary function composition** operator as a primitive concept. The intended meaning of $; S$ is that it takes a sequence of functions and returns their composition. Thus, $; (f_1, \ldots, f_n)\, s$ reduces to $f_n (\ldots (f_1\, s) \ldots)$. Notably, this is more general than a fold operator because the type of each f_i may depend on i.

Finally, we have to clarify the intuitions of the now-flexary primitives of LF. **Flexary variable bindings** $x : U$ formalize variable sequences, i.e., binding $x : (U_1, \ldots, U_n)$ corresponds to binding $x_1 : U_1, \ldots, x_n : U_n$. Thus, the type sequence ellipses immediately induces a corresponding ellipses constructor for variable bindings. Accordingly, a **flexary application** $f\, T$ applies a function f to an argument sequence T.

Much of the intuition behind our grammar becomes clear from the function $|E|$ for the **length of a sequence** defined in Fig. 6. It

$\|x\|$	$= \|U\|$ if $x : U$ in Σ
$\|a\|$	$= \|K\|$ if $a : K$ in Σ
$\|\text{type}^n\|$	$= n$
$\|\Pi x : E. F\|$	$= 1$
$\|\lambda x : E. F\|$	$= 1$
$\|E\, F\|$	$= 1$
$\|\cdot\|$	$= 0$
$\|E, F\|$	$= \|E\| + \|F\|$
$\|E_n\|$	$= 1$
$\|[E]_{x=1}^n\|$	$= \|n\|$
$\|; S\|$	$= 1$

Fig. 6. Length of a Sequence

maps LFS expressions to expressions of type nat (where E and F range over any expression allowed by the grammar). We already mention that the type system given below will respect length, i.e., $S : U : \text{type}^n$ only if $|S|$, $|U|$, and n are provably equal.

Functions f and applications $f\, T$ always have length 1 and so have the bodies of the binders. This forbids computations that returns sequences. This restriction could be lifted, but we find it is more reasonable to introduce such computations in object languages defined within LFS.

Before giving the type system, we fortify our intuitions by defining a few useful abbreviations that we will use later on. The **reversal** of a sequence is defined by:

$$\text{revert } E = [E_{|E|+1-i}]_{i=1}^{|E|}.$$

The **generalized sequence ellipses** $a_n, \ldots, a_1$ and $a_m, \ldots, a_n$ (if $m \le n$ for natural numbers m, n) are defined by

$$[E(x)]_1^{x=n} = \text{revert } [E(x)]_{x=1}^n \qquad [E(x)]_{x=m}^n = [E(m + x - 1)]_{x=1}^{n-m+1}$$

We obtain the usual fold operator in terms of flexary composition:

$$\text{foldl } f\, S\, a = (; [\lambda x : A. f\, x\, S_i]_{i=1}^n)\, a$$

Thus, $\text{foldl } f\, S\, a$ reduces to $(f \ldots (f\, (f\, a\, S_1)\, S_2) \ldots S_n)$ for a folding function $f : A \to B \to A$, a start element $a : A$, and a sequence $S : B^n$. foldr is defined analogously.

3.3   Type System

LFS uses the same judgments as LF, i.e. the ones from Fig. 3. However, we write the typing judgment $\Sigma \vdash S : U$ (where $U : type$ is implied in LF) as $\Sigma \vdash S : U : \text{type}^n$ to keep track of the length of S and U. This is redundant because the length is statically known, but it makes the notations much simpler.

Most importantly, term sequences are typed by type sequences of the same length, and type sequences are kinded by $type^n$, where n is their length.

$$\frac{\Sigma \vdash n : \textbf{nat} : \text{type}}{\Sigma \vdash \textbf{type}^n \; Kind} \; kindSeq$$

$$\frac{\vdash \Sigma \; Sig}{\Sigma \vdash \cdot : \textbf{type}^0} \; emptyType \qquad \frac{\Sigma \vdash U : \textbf{type}^m \quad \Sigma \vdash V : \textbf{type}^n}{\Sigma \vdash U,V \; : \textbf{type}^{m+n}} \; typeSeq$$

$$\frac{\vdash \Sigma \; Sig}{\Sigma \vdash \cdot : \cdot : \text{type}^0} \; emptyTerm \qquad \frac{\Sigma \vdash S : U : \text{type}^m \quad \Sigma \vdash T : V : \text{type}^n}{\Sigma \vdash S,T \; : \; U,V \; : \text{type}^{m+n}} \; termSeq$$

$$\frac{\Sigma \vdash S : U : \text{type}^n \quad \Sigma \vdash x^* : 1 \leq x : \text{type} \quad \Sigma \vdash x_* : x \leq |S| : \text{type}}{\Sigma \vdash S_x : U_x : \text{type}} \; termIndex$$

$$\frac{\Sigma \vdash U : \textbf{type}^n \quad \Sigma \vdash x^* : 1 \leq x : \text{type} \quad \Sigma \vdash x_* : x \leq n : \text{type}}{\Sigma \vdash U_x : \textbf{type}} \; typeIndex$$

$$\frac{\Sigma \vdash n : \textbf{nat} : \text{type} \quad \Sigma, x : \textbf{nat}, x^* : 1 \leq x, x_* : x \leq n \vdash S : U : \text{type}}{\Sigma \vdash [S]_{x=1}^n : [U]_{x=1}^n : \text{type}^n} \; termSeqEll$$

$$\frac{\Sigma \vdash n : \textbf{nat} : \text{type} \quad \Sigma, x : \textbf{nat}, x^* : 1 \leq x, x_* : x \leq n \vdash U : \textbf{type}}{\Sigma \vdash [U]_{x=1}^n : \textbf{type}^n} \; typeSeqEll$$

$$\frac{\Sigma \vdash U : \textbf{type}^{n+1} \quad \Sigma \vdash S : [U_i \to U_{i+1}]_{i=1}^n}{\Sigma \vdash \; ; S : U_1 \to U_{n+1}} \; nestEll$$

Fig. 7. Inferece Rules for Sequences and Ellipses

The inference rules essentially reuse the rules of LF from Fig. 4. We only make two minor changes to the four rules for binders. Firstly, the four occurrences of type are replaced with type^n for $\Sigma \vdash n : \textbf{nat}$. This permits binders to bind variables sequences $x : U_1, \ldots, U_n$. The LF rules are recovered as the special case $n = 1$. Secondly, we add a premise to each of the four rules that ensure that the body of a binder always has length 1.

Then we add the rules of Fig. 7 for sequences and ellipses. *kindSeq* makes $type^n$ a valid kind. The rules for the empty sequences and the concatenation of sequences are obvious. The rules *termIndex* and *typeIndex* for taking an element of a sequence implement the index-within-bounds check: E_i is only valid if $1 \leq i \leq |E|$.

The rules *termSeqEll* and *typeSeqEll* handle the sequence ellipsis. Note that $[E]_{x=1}^n$ actually binds three variables in E: The index x and two assumptions x_* and x^* that guarantee that x is within 1 and n. These assumptions can be used later on to discharge the proof obligations posited by the rules *termIndex* and *typeIndex* and by the subtraction of natural numbers.

Finally, *nestEll* handles the nesting ellipsis $; S$ by checking that the function in S are actually composable. This is easiest if we restrict attention to the composition of simple functions.

We only sketch the **conversion rules** that we add to the ones of LF. Binders distribute over sequences:

$$\lambda x : U, V. E = \lambda x_1 : U. \lambda x_2 : V. [x_1, x_2/x]T \qquad \lambda x : \cdot . T = [\cdot/x]T$$

and similarly for Π. Sequence elements can be projected if the sequence is normal

$$(E^1, \ldots, E^n)_x = E_x \qquad \text{if } |E^1| = 1, \ldots, |E^n| = 1$$

Ellipses are expanded if enough information is available:

$$[E]_{x=1}^n = [1/x]E, \ldots, [n/x]E \qquad \text{if } n = 1 + \ldots + 1$$

$$;\cdot = \lambda x. x \qquad ;(f, g) = \lambda x. (; g)((; f) x) \qquad ; f = f \text{ if } |f| = 1$$

These conversions have the effect that LFS is conservative over LF in the following sense: If $\Sigma \vdash S : U : type^n$ and all terms of type **nat** reduce to $1 + \ldots + 1$, then n reduces to m, and S and U reduce to $S^1, \ldots S^m$ and $U^1, \ldots, U^m$, and $S^i : U^i$ for $i = 1, \ldots, m$ in the LF type theory. This means that if the involved natural number expressions normalize, then LFS-expressions reduce to sequences of LF-expressions, and LFS-judgments reduce to sequences of LF-judgments. Under this condition, the canonical LFS expressions are sequences of canonical LF expressions. Consequently, an adequate encoding of objects as LF-expressions, yields an adequate encoding of sequences of objects as LFS-expressions. Additionally, reducing full LFS to LF would require a formalization of sequences in LF already, which is doable, but also very costly.

4   Flexary Logics

Now we use LFS to define the flexary analogues of two languages commonly used for formalized mathematics.

We define **flexary sorted first-order logic** SFOL* by extending the syntax of SFOL from Fig. 2 with flexary symbols $\wedge^*$, $\Rightarrow^*$ and $\forall^*$ along with their proof rules. All are defined in terms of their fixary counterparts.

$$\wedge^* : form^n \to form$$
$$= \lambda F : form^n.\,\texttt{foldl} \wedge F\ true$$

$$\wedge_I^* : \Pi F : form^n.\,[ded\ F_x]_{x=1}^n \to ded \wedge^* F$$
$$= \lambda n.\,\lambda F.\,\lambda D : [ded\ F_x]_{x=1}^n.$$
$$;[\lambda x : ded\ (\wedge^*\ [F_j]_{j=1}^{i-1}).\ \wedge_I\ \underline{(\wedge^*\ [F_j]_{j=1}^{i-1})}\ \underline{F_i}\ x\ D_i]_{i=1}^n\ true_I$$

$$\wedge_E^* : \Pi F : form^n.\,\Pi i : \textbf{nat}.\,\Pi i_* : 1 \le i.\,\Pi i^* : i \le n.\,ded \wedge^* F \to ded\ F_i$$
$$= \lambda F.\lambda i.\,\lambda i_*.\lambda i^*.\lambda D.\ \wedge_{Er}\ \underline{(\wedge^*\ [F_k]_{k=1}^{i-1})}\ \underline{F_i}$$
$$;[\lambda x.\ \wedge_{El}\ \underline{(\wedge^*\ [F_k]_{k=1}^{n-j-1})}\ \underline{F_{n-j}}\ x]_{j=1}^{n-i}\ D$$

$$\text{abbreviation: } S_a^b := [tm\ S_j]_{j=a}^b$$

$$\forall^* : \Pi S : sort^n.\,(S_1^n \to form) \to form$$
$$= \lambda S.\lambda F.\ ;[\lambda f : S_1^i \to form.\,\lambda y : S_1^{i-1}.\,\forall \lambda x : tm\ S_i.\ f\,(y,x)]_1^{i=n}\ F$$

$$\forall_I^* : (\Pi x : S_1^n.\,ded\ F\ x) \to ded \forall^* F$$
$$= \lambda D.\ ;[\lambda d : \big(\Pi y : S_1^i.\,ded \forall^* \lambda x : S_{i+1}^n.\,D\,(y,x)\big).$$
$$\lambda y : S_1^{i-1}.\,\forall_I\ \lambda x : tm\ S_i.\ d\,(y,x)$$
$$]_1^{i=n}\ D$$

$$\forall_E^* : ded \forall^* F \to \Pi x : S_1^n.\,ded\ F\ x$$
$$= \lambda D.\lambda x.\ ;[\lambda d : ded \forall^* \lambda y : S_i^n.\,F([x_j]_{j=1}^{i-1}, y).\,\forall_E\ d\ x_i]_{i=1}^n\ D$$

The flexary conjunction $\wedge^*$ takes a natural number n and then a sequence of n conjuncts. We have $\wedge^* F_1 \ldots, F_n = (\ldots(true \wedge F_1)\ldots) \wedge F_n$ and $\wedge^* \cdot = \ true$. The introduction rule uses essentially the same folding: Without implicit arguments, it would simply read $\wedge_I^* D_1 \ldots, D_n = \wedge_I\,(\ldots(\wedge_I\ true_I\ D_1)\ldots)\,D_n$ and $\wedge_I^* \cdot = true_I$. However, to demonstrate that it is in fact definable, we also give the implicit arguments in detail: They are underlined. That complicates the definition because each occurrence of $\wedge_I$ uses different implicit arguments, which themselves need ellipses to write down. The elimination rule proceeds along the same lines except that we have to take guards i_* and i^* to make sure the indices F_i are within bounds.

For disjunction, we would use $\vee^* F = \texttt{foldl} \vee F\ false$ accordingly. For implication, which is not associative, we define $\Rightarrow^*: form^n \to form \to form$ and $\Rightarrow^* F G = \texttt{foldr} \Rightarrow F G$.

The definition of flexary quantifiers is more involved. Intuitively, we want $\forall^* S F = \forall \lambda x^1 : S_1.\ldots.\forall \lambda x^n : S_n.\,F(x^1, \ldots, x^n)$. Let $[;G(i)]_1^{i=n}$ be the ellipsis in the definiens of $\forall^*$. Then the type of $G(i)$ is $(S_1^i \to form) \to (S_1^{i-1} \to form)$, and when constructing $G(n)\,(\ldots(G(1)\ F)\ldots)$, each $G(i)$ introduces $\forall x^i$. Note that all variables are called x, the names x^i are introduced when LFS α-renames x during capture-avoiding substitution.

The definitions of the proof rules are conceptually straightforward but equally subtle. The flexary existential quantifier can be defined in the same way.

Next, we define **flexary simple type theory** STT* by extending the syntax of STT from Fig. 2. We define a flexary function type constructor, flexary λ-abstraction, and flexary application in terms of the fixary ones. We omit the proof of the flexary β-reduction.

$$\Longrightarrow^* : \ tp^n \to tp \to tp$$
$$= \lambda A : tp^n. \lambda B : tp. \texttt{foldr} \Longrightarrow A B$$
$$lam^* : \ ([tm\,A_i]_{i=1}^n \to tm\,B) \to tm\,(A \Longrightarrow^* B)$$
$$= \lambda F. \, ; [\lambda f : [tm\,A_j]_{j=1}^i \to tm\,([A_j]_{j=i+1}^n \Longrightarrow^* B).$$
$$\lambda y : [tm\,A_j]_{j=1}^{i-1}.\ lam\,\lambda x : tm\,A_i.\ f\,(y,x)$$
$$]_1^{i=n}\,F$$
$$app^* : \ tm\,(A \Longrightarrow^* B) \to [tm\,A_i]_{i=1}^n \to tm\,B$$
$$= \lambda F. \lambda a. \, ; [\lambda f : A_i^n \Longrightarrow B.\ app\,f\,x_i]_{i=1}^n\,F$$
$$beta^* : \ app^*\,(lam^*(\lambda x : [tm\,A_i]_{i=1}^n.T))\,X = (T\,X) = \ldots$$

5   Flexary Mathematics

Now we use SFOL* and STT* to formalize a collection of common mathematical examples.

Monoid Operations Like we did for conjunction, we can define the flexary version of any associative binary operator. Consider a monoid with carrier α : type, binary operation $\circ : \alpha \to \alpha \to \alpha$ and unit element $e : \alpha$. Then we define the flexary operation $\circ^*$ as follows:

$$\circ^* : \ \Pi n : \texttt{nat}. \alpha^n \to \alpha \ = \ \lambda n. \lambda x : \alpha^n. \texttt{foldl} \circ x\,e$$

This immediately yields the power operator:

$$power : \ \alpha \to \texttt{nat} \to \alpha \ = \ \lambda x. \lambda n. \circ^*\,x^n$$

By specializing to the monoid of endofunctions on a type, we obtain the iteration of functions as follows:

$$iter : \ (\alpha \to \alpha) \to \texttt{nat} \to (\alpha \to \alpha) \ = \ \lambda f. \lambda n. \, ; f^n$$

Multi-relations Multi-relations like $a \in b \subseteq c$ are routinely used in informal mathematics but cannot be defined as single operators within a fixary logic. They can only be defined as notations within an implementation of the logic. Using SFOL*, we can define it as a flexary operator elegantly:

$$multirel : \ \Pi A : sort^n.\,(tm\,A)^{n+1} \to (tm\,A \to tm\,A \to form)^n \to form$$
$$= \lambda x : (tm\,A)^{n+1}. \lambda r : (tm\,A \to tm\,A \to form)^n. \wedge^*\,[r_i\,x_i\,x_{i+1}]_{i=1}^n$$

For example, we can now write the above multi-relation as $multirel\,(a,b,c)\,(\in, \subseteq)$. More generally, we can define multi-relations for relations between different types:

$$multirel' : \ \Pi A : sort^n.\,[tm\,A_i]_{i=1}^{n+1} \to [tm\,A_i \to tm\,A_{i+1} \to form]_{i=1}^n \to form$$
$$= \lambda A. \lambda x. \lambda r. \wedge^*\,[r_i\,x_i\,x_{i+1}]_{i=1}^n$$

Vectors and Matrices Now we formalize vectors in STT*. a' : tp
For simplicity, we assume a fixed base type a with the a : $\mathbf{type} = tm\,a'$
structure of a ring as given on the right 0_a : a
Then we axiomatize a type constructor of fixed-length 1_a : a
vectors as follows, where n is an implicit argument every- $+$: $a \to a \to a$
where: $\times$: $a \to a \to a$

$$
\begin{aligned}
&Vec' && : \mathbf{nat} \to tp \\
&Vec && : \mathbf{nat} \to \mathbf{type} = \lambda n.\, tm\,(Vec'\,n) \\
&vec && : a^n \to Vec\,n \\
&index && : \Pi m : \mathbf{nat}.\, \Pi m_* : 1 \le m.\, \Pi m^* : m \le n.\, Vec\,n \to a \\
&compute && : \Pi m : \mathbf{nat}.\, \Pi m_* : 1 \le m.\, \Pi m^* : m \le n.\, \Pi x : a^n. \\
&&& \quad index\,m\,p\,q\,(vec\,x) = x_m \\
&complete && : \Pi v : Vec\,n.\, v \; = \; vec\,[index\,i\,i^*\,i_*\,v]_{i=1}^n
\end{aligned}
$$

It may appear strange that we call $vec : \mathbf{nat} \to tp$ an example in *simple* type theory. However, recall that $\mathbf{nat}$ is not a type of the logic STT* but a feature of the framework. Thus, there is no substantial structural difference between our vec and type operators like $list : tp \to tp$. Indeed, our treatment of implicit arguments of type $\mathbf{nat}$ is very similar to the treatment of free type variables in higher-order logics.

We can now define the addition and scalar multiplication elegantly:

$$
\begin{aligned}
\vec{+} : \; & Vec\,n \to Vec\,n \to Vec\,n \\
= \; & \lambda v.\, \lambda w.\, vec\,[(index\,i\,i_*\,i^*\,v) + (index\,i\,i_*\,i^*\,w)]_{i=1}^n \\
\vec{\times} : \; & a \to Vec\,n \to Vec\,n \\
= \; & \lambda x.\, \lambda v.\, vec\,[x \times (index\,i\,i_*\,i^*\,v)]_{i=1}^n
\end{aligned}
$$

In implementations, these definitions would resemble the ones in informal mathematics even more: We would use appropriate notations for vec and $index$, and it is straightforward for a theorem prover to find the guard arguments of $index$ automatically.

More generally, we can formalize the type of vectors $Vec : tp \to \mathbf{nat} \to tp$ over an arbitrary base type and define matrices as vectors of vectors. Then matrix addition and multiplication can be defined accordingly.

Polynomials Finally, we axiomatize a type of polynomials (over the same fixed base type $a : \mathbf{type}$ as above) as follows:

$$
\begin{aligned}
&Poly' && : tp \\
&Poly && : \mathbf{type} = tm\,Poly' \\
&poly && : a^{n+1} \to Poly \\
&deg && : Poly \to \mathbf{nat} \\
&coeff && : Poly \to \mathbf{nat} \to a \\
&comp\_deg && : \Pi c : a^{n+1}.\, deg\,(poly\,c) = n \\
&comp\_coeff_1 && : \Pi c : a^{n+1}.\, \Pi m : \mathbf{nat}. \\
&&& \quad 1 \le m \to m \le n+1 \to coeff\,(poly\,A)\,m = A_m \\
&comp\_coeff_2 && : \Pi c : a^{n+1}.\, \Pi m : \mathbf{nat}.\, n+2 \le m \to coeff\,(poly\,A)\,m = 0_a \\
&complete && : \Pi p : Poly.\, p = poly\,[coeff\,p\,i]_{i=1}^{1+deg\,p}
\end{aligned}
$$

Here *poly* constructs a polynomial from coefficients, *deg* returns an upper bound on the degree, and *coeff* returns a coeffiecient. *comp_deg*, *comp_coeff*$_1$, and *comp_coeff*$_2$ compute the degree and coefficients of an explicitly given polynomial. And *complete* makes every polynomial equal to the one formed from its coefficients.

We can now define the evaluation of a polynomial for a given x concisely:

$$eval : poly \to a \to a = \lambda p. \lambda x. +^* \left[(coeff\, p\, i) \times (\times^* x^{i-1}) \right]_{i=1}^{1+\deg p}$$

where $+^*$ and $\times^*$ are the flexary versions of $+$ and $\times$, respectively.

The ring operations on polynomials can be defined accordingly.

6   Conclusion and Future Work

It is almost impossible to write about mathematical objects without using sequence ellipsis ($\ldots$). This observation is independent of the language used, or the formal system employed: if we eliminate ellipses, then expressions get more complicated and communication can become quite awkward. This universality strongly suggests that sequences and sequence ellipsis are a meta-level feature of mathematical languages.

Guided by this realization, we present LFS, an extension of LF with a novel feature of type sequences. Using the extended framework for logic development enables us to specify flexary logics with flexary operators and calculi that deal with them in proofs. We exemplify this ability with flexary sorted first-order logic and flexary simple type theory. As they use LFS as their meta-language, both can define flexary operators in terms of their fixary counterparts. A central theme is that the type of a flexary operator depends on a natural number argument, which instantiates the flexible arity: We call this **arity polymorphism** because it is very similar to the well-known type polymorphism where the type of an operator depends on a type argument.

Numerous examples from everyday mathematics show that the flexary languages allow more adequate formalizations of complex objects like vectors, matrices, and polynomials.

In the future, we want to extend/complete MKM support for LFS. In particular, we want to *i*) look at additional examples, e.g. the complex matrix representations in [SS06], *ii*) investigate type reconstruction of LFS and implement it based on MMT [RK13, Rab13], which will in particular infer omitted natural number arguments, *iii*) extend the support for sequences and elisions to flexiformal mathematics markup systems like LaTeX and sTeX, *iv*) integrate native support for argument sequences into OpenMath and content MathML completing the work started in [HKR11], and finally *v*) develop semantics reconstruction techniques that transform $1, 2, \ldots, n$ into $[i]_{i=1}^{n}$ or $1, 2, 4, 8, \ldots$ into $[2^i]_{i=1}^{\infty}$.

Acknowledgements. Work on the concepts presented here has been partially supported by the German Research Foundation (DFG) under grant KO 2428/13-1.

References

[ABC+03] Ausbrooks, R., Buswell, S., Carlisle, D., Dalmas, S., Devitt, S., Diaz, A., Froumentin, M., Hunter, R., Ion, P., Kohlhase, M., Miner, R., Poppelier, N., Smith, B., Soiffer, N., Sutor, R., Watt, S.: Mathematical Markup Language (MathML) Version 2.0, 2nd edn. (2003), http://www.w3.org/TR/MathML2

[CH88] Coquand, T., Huet, G.: The Calculus of Constructions. Information and Computation 76(2/3), 95–120 (1988)

[Com07] Information technology — Common Logic (CL): a framework for a family of logic-based languages. Technical Report 24707:2007, ISO/IEC (2007)

[Coq14] The Coq Development Team. The Coq Proof Assistant: Reference Manual. Technical report, INRIA (2014)

[HHP93] Harper, R., Honsell, F., Plotkin, G.: A framework for defining logics. Journal of the Association for Computing Machinery 40(1), 143–184 (1993)

[HKR11] Horozal, F., Kohlhase, M., Rabe, F.: Extending OpenMath with Sequences. In: Asperti, A., Davenport, J., Farmer, W., Rabe, F., Urban, J. (eds.) Intelligent Computer Mathematics, Work-in-Progress Proceedings, pp. 58–72. University of Bologna (2011)

[KB04] Kutsia, T., Buchberger, B.: Predicate logic with sequence variables and sequence function symbols. In: Asperti, A., Bancerek, G., Trybulec, A. (eds.) MKM 2004. LNCS (LNAI), vol. 3119, pp. 205–219. Springer, Heidelberg (2004)

[Kor12] Korniłowicz, A.: Tentative experiments with ellipsis in Mizar. In: Jeuring, J., Campbell, J.A., Carette, J., Dos Reis, G., Sojka, P., Wenzel, M., Sorge, V. (eds.) CICM 2012. LNCS (LNAI), vol. 7362, pp. 453–457. Springer, Heidelberg (2012)

[ML74] Martin-Löf, P.: An Intuitionistic Theory of Types: Predicative Part. In: Proceedings of the 1973 Logic Colloquium, pp. 73–118. North-Holland (1974)

[Nor05] Norell, U.: The Agda WiKi (2005), http://wiki.portal.chalmers.se/agda

[Pau94] Paulson, L.C.: Isabelle: A Generic Theorem Prover. LNCS, vol. 828. Springer, Heidelberg (1994)

[Pfe01] Pfenning, F.: Logical frameworks. In: Handbook of Automated Reasoning, pp. 1063–1147. Elsevier (2001)

[PS99] Pfenning, F., Schürmann, C.: System description: Twelf - A meta-logical framework for deductive systems. In: Ganzinger, H. (ed.) CADE 1999. LNCS (LNAI), vol. 1632, pp. 202–206. Springer, Heidelberg (1999)

[Rab13] Rabe, F.: The MMT API: A Generic MKM System. In: Carette, J., Aspinall, D., Lange, C., Sojka, P., Windsteiger, W. (eds.) CICM 2013. LNCS (LNAI), vol. 7961, pp. 339–343. Springer, Heidelberg (2013)

[RK13] Rabe, F., Kohlhase, M.: A Scalable Module System. Information and Computation 230(1), 1–54 (2013)

[SS06] Sexton, A., Sorge, V.: Processing textbook-style matrices. In: Kohlhase, M. (ed.) MKM 2005. LNCS (LNAI), vol. 3863, pp. 111–125. Springer, Heidelberg (2006)

[Wol12] Wolfram Research, Inc. Mathematica 9.0 (2012)

Interactive Simplifier Tracing and Debugging in Isabelle

Lars Hupel

Technische Universität München, Germany
lars.hupel@tum.de

Abstract. The Isabelle proof assistant comes equipped with a very powerful tactic for term simplification. While tremendously useful, the results of simplifying a term do not always match the user's expectation: sometimes, the resulting term is not in the form the user expected, or the simplifier fails to apply a rule. We describe a new, interactive tracing facility which offers insight into the hierarchical structure of the simplification with user-defined filtering, memoization and search. The new simplifier trace is integrated into the Isabelle/jEdit Prover IDE.

Keywords: Isabelle, simplifier, term rewriting, tracing, debugging.

1 Introduction

Isabelle is a generic proof assistant [10]. It comes with some very powerful tactics which are able to discharge large classes of proof goals automatically. This work is concerned with the rewriting tactic, often called the *simplifier*. It can be used to rewrite subterms according to a user-definable set of equations, which generally means simplifying a term to a normal form. These equations can have conditions which are recursively solved by the simplifier itself. Hence, there can be quite a huge number of steps between the original term and its normal form. Because of that complex work in the background, it is not obvious to the user how the input gives rise to certain terms in the output.

By default, this process is completely opaque: the only observable effect is—given that the simplification succeeded—the (hopefully) simpler term it produced. If it failed, only an error message without any indication of the reasons is printed, or it might not even terminate at all. Currently, there is a tracing facility which can be enabled by the user. It collects data about the steps the simplifier executed, and prints each of them without any high-level structure. The resulting trace can easily contain many screenfuls of items which the user has to laboriously search for interesting pieces of information. This could also easily lead to sluggish GUI behaviour in the Isabelle/jEdit Prover IDE because of the huge amount of content which has to be rendered on the screen. There are various techniques to alleviate that problem, notably limiting the recursion depth.

However, this does not solve the fundamental problem of showing only the interesting parts of the trace to the user. This paper describes the design of a new

S.M. Watt et al. (Eds.): CICM 2014, LNAI 8543, pp. 328–343, 2014.

tracing and debugging mechanism which can be configured to filter trace messages and render a semantically meaningful representation of the simplification trace (Section 2). For example, rewriting the term $a < b \implies 0 < b \implies 0 < c \implies 0 < (c+c)/(b-a)$ already produces over 300 trace messages with the old tracing. Navigating the linear presentation of the trace is non-trivial. The new tracing shows a hierarchical view of the trace where the user can hide the traces of uninteresting subexpressions. The user can simply filter the trace for various keywords, e.g. "ac" (which usually denotes an associativity/commutativity rule if its name contains this string). For our example expression, this uncovers that for solving the goal—the term is rewritten to `True`—the simplifier used commutativity of multiplication (it rewrote $2 \cdot c$ to $c \cdot 2$).

The new tracing also offers interactive features like breakpoints and stepping into or over recursive calls (Section 3), which are well-known from debuggers for imperative programming languages such as `gdb` [12]. In combination with the filtering, this is very effective in narrowing down causes for failure.

Furthermore, this paper contributes a mechanism for incremental debugging, i.e. upon changes in the source code the system continues execution where it left off instead of starting from scratch again (Section 3.2), which is especially useful during exploratory proof development where assumptions and rewrite rules change frequently. Suppose that in our example, the precondition $a < b$ were absent. Upon stepping through the simplification, the user realizes that the precondition is missing, and adds it to the goal. The simplifier runs again, but the user does not have to step through identical subtraces again; instead, only new parts are shown.

The tracing has been integrated into the Isabelle/jEdit Prover IDE, which is implemented in Scala [15]. The interplay between the ML process and the JVM process hosting the prover and the IDE, respectively, works via an underlying protocol, of which only a reasonably high-level interface is exposed [14, §3]. The new simplifier trace works on both sides: Trace messages are produced and preprocessed in Isabelle/ML and formatted and renderered in Isabelle/Scala. In order to reduce the amount of data sent between the processes, there is a multistaged and user-configurable message filter. The implementation also supports parallel simplifier runs. Although the new tracing produces more metadata than previously and incurs an overhead for the communication between the prover and the IDE, the performance is acceptable for user interaction (Section 6).

We compare our work with existing approaches in SWI-Prolog [6,8,11], Maude [4,5] and other proof assistants (Section 5). Both SWI and Maude offer tracing mechanisms that solve similar problems, e.g. conditional rules and failing rewrite (or deduction) steps. Finally, we evaluate the performance of the system and give prospects for further research (Section 6).

Terminology A *(trace) message* is a piece of structured information about the current state of the simplifier. Messages requiring a response (possibly by the user) are called *interactive*. A *rewrite rule* is a theorem stating an equality. If a rewrite rule is conditional, its preconditions are (in most situations) solved recursively by the simplifier itself.

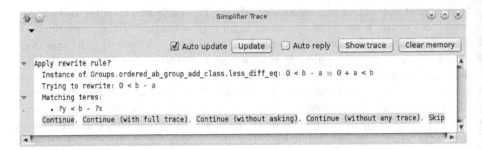

Fig. 1. New trace panel in Isabelle/jEdit

2   Design Principles

The fundamental idea for tracing and debugging is to instrument the simplifier to collect data and influence the flow. The previous tracing only does the former: at certain points in the program flow, it prints messages indicating the current state. For example, when the simplifier is invoked, the function `trace_term` is called, which takes a `term` and a `string`. The system then prints the fixed string "Simplifier invoked..." and the term to the regular output channel, where it appears alongside other status information, e.g. the goal state.

In the new implementation, the simplifier has been extended with the possibility to install hook functions. When the simplifier is invoked, the hook function `invoke` is called which takes a `term`. The precise formatting of the message, or if it gets printed at all, is left as an implementation detail to the new module. The internal details are not observable for the simplifier.

This has several advantages: First, it reduces code clutter, because trace formatting and output is not mixed with the program logic. The semantics of the program flow is not lost, since it is clear that a call to the `invoke` hook denotes a (possibly recursive) invocation of the simplifier. Lastly, influencing the simplifier becomes easy. For example, the `apply` hook which denotes an application of a conditional rewrite rule takes a *continuation*[1] as an argument, which allows the tracing to replay the program flow if necessary. This means that instead of just observing the current state of the simplifier, the user is actually able to interactively manipulate it. Because of pervasive immutability in Isabelle's internal data structures, the simplifier does not have to track any tracing state explicitly for that case.

In the user interface, this is realized by presenting questions to the user at certain points in the program flow (Fig. 1). A question contains the instantiated rewrite rule, the redex, and matching term breakpoints (see Section 2.2). The simplifier is blocked until the user answers the question, and upon answering,

---

[1] Continuation passing is a progamming technique in which a function does not return a value, but rather takes a function as argument and invokes this function with the value it computed. In essence, a function `'a -> 'b` is transformed into a function `'a -> ('b -> unit) -> unit`.

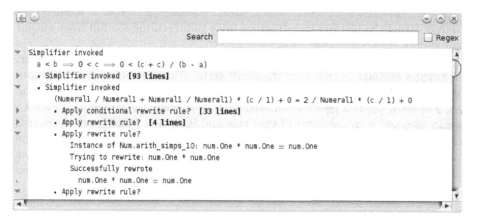

Fig. 2. New trace window in Isabelle/jEdit

the program flow might deviate from the usual one, e.g. certain steps may be skipped. At any point in time, the user can open a trace window which shows the accumulated trace output (Fig. 2). There, the trace can be filtered and sub-parts can be folded for easier navigation.

The new tracing facility is highly configurable and goes to great lengths to keep the number of unwanted messages low. A related design goal is efficiency, so that the user does not experience long delays. Together, configurability and efficiency should guarantee that the user interface stays responsive.

2.1 Hooks and Message Types

On a high level, these are the types of trace messages the system can send:

invoke tells the tracing system that the main entry point of the simplifier has been called. This happens on initial invocation of the tactic and on recursive calls when the conditions of rewrite rules are being solved.

apply guards the application of a (potentially conditional) rewrite rule by the simplifier. It is invoked *before* the rule is applied. It depends on the user input how the simplification proceeds after that.   *(interactive)*

hint indicates whether a rewrite step failed or succeeded. If it failed, the user is given a chance to inspect the failure, and can decide if the failing step should be tried again (with different settings).   *(possibly interactive)*

ignore marks a specific part of the trace as obsolete. This message is produced when the user wants to retry a failing step. It is generated by the tracing system and thus cannot be sent explicitly by the simplifier.

log emits a log message (with arbitrary content) which will not be further interpreted by the system.

Messages may have children. For example, **apply** messages are naturally associated with the **invoke** message emitted in the simplifier invocation. Hence, each message carries a pointer to the parent message. The IDE is in turn able to reconstruct the hierarchical structure prior to presenting the full trace output to the user.

Recall the example from the introduction. When the user calls the simplifier, an `invoke` message is issued for the whole term. During simplification, the sub-term $0 < c \cdot 2/(b - a)$ arises. To simplify it, a rewrite rule with the precondition $0 < x$ is used, where x gets instantiated with $b - a$. Here, the system first issues a `apply` message with the name of that rule and its instantiation, and afterwards `invoke` with the term $0 < b - a$. This can be solved by the simplifier (after some more rule applications), which yields a `hint(success)` message, indicating that the original subterm was rewritten to $0 \cdot (b - a) < c \cdot 2$. The trace window then groups these messages and presents them in one subtree.

2.2   Settings

It is important to determine which questions are relevant. Showing a question for each step is not feasible, because it might take thousands of steps until a term is rewritten into normal form. Hence, by default, the system shows no questions at all. The user is able to specify that behaviour via multiple configuration axes: Most importantly, *verbosity modes* (which can be changed during a simplification run) and *breakpoints*.

Modes   The three regular modes of operation are:

`disabled` do not produce any trace messages at all
`normal` produce messages, but display them only if their contents are triggering a breakpoint
`full` produce messages and display all of them

Each of these cases can be combined with a flag which denotes whether the output should just be logged, or presented as a question to the users. The user can only interactively influence the simplification process in the latter case.

In most cases, it is reasonable to avoid the `full` mode: even for seemingly small terms, the potential amount of applied rewrite rules can get quite high. While the system has no problem producing and transmitting these messages, displaying them takes a while.

Breakpoints   The user can specify the set of interesting rewrite steps by defining breakpoints. If a step triggers such a breakpoint, the simplifier is intercepted and the system displays a question.

In debuggers for imperative languages, the concept of breakpoints is well-known. Usually, breakpoints are set to lines in the source code, and when the sequential execution of the program hits a marked line, the execution is halted. Furthermore, many debuggers support conditional breakpoints, where users can specify a condition, and the breakpoint only triggers if that condition is met.

Here, the implementation obviously has to differ from traditional debuggers, because it does not follow a strict sequential execution model. The principle is easy, though: each rewrite step in the simplifier consists of a term (the redex) and a theorem (the rewrite rule). Breakpoints can be set for either of them. Term breakpoints usually contain patterns and can refer to locally fixed variables. For example, the breakpoint _ > 0 matches when the term $y + z > 0$ is to be rewritten, where y and z can be any fixed or free variables. A theorem breakpoint is triggered when the corresponding rewrite rule is used.

Users can declare breakpoints with the usual Isabelle *attribute* mechanism, i.e. by adding the string declare theorem_name[break_thm] into the theory sources before the point where the simplifier is invoked (and similarly for term breakpoints). For example, the panel in Fig. 1 shows a step which matches a breakpoint declared with break_term "?y < b - ?x" (where ? indicates a pattern variable).

3   User Interaction

User interaction is fundamental to the new tracing. The system might present the user some questions about the current progress which it deems to be relevant. The user can then decide how to progress. In this section, the handling of interactive messages is described.

3.1   Interactive Messages

As seen earlier, there are two types of interactive messages which allow the user to influence the outcome of the simplifier: apply before a simplification step is attempted, and hint for when a simplification step failed.

Message apply When a step is attempted, the message contains the instantiated rewrite rule, the redex, and a number of different possible replies. The user can choose to continue the computation, which instructs the simplifier to apply the specified rule (which requires solving the rule's preconditions first) and thus does not influence the result of the simplifier. The other option is to skip the current step, even if it would have succeeded.

As a result, the outcome of a simplification run is potentially different from when tracing would be disabled. Hence, skipping should be used sparingly: the most common use case would be to find overlapping rewrite rules, i.e. multiple rules which match on the same term.

Message hint(failed) Often, the user wants feedback immediately when the simplification failed. Prior to this work, in case of failure the simplifier just does not produce any result, or produces an unwanted result.

On the other hand, with this message type, the new tracing provides insight into the simplification process: It indicates that the simplifier tried to solve the preconditions of a rewrite rule, but failed. There are a number of different reasons for that, including that the preconditions do not hold or that a wrong theorem has been produced in the recursive call, which would indicate a programming error in the tactic. Regardless of the reason, it is possible to *redo* a failed step if (and only if) the original step triggered a user interaction previously.

Consider an example: The term $f\ t_1$ is to be rewritten. The rewrite rule $P_1\ x \Longrightarrow f\ x \equiv g\ x$ is applicable and gets instantiated to $P_1\ t_1 \Longrightarrow f\ t_1 \equiv g\ t_1$. Assume that there is a breakpoint on that particular rule, hence the user is presented a question whether the rule should be applied. The user chooses to continue, and the simplifier recursively tries to solve the precondition $P_1\ t_1$. Now assume that this entails application of another conditional rule which does not trigger a breakpoint (hence no question), but this step fails. In turn, the rewriting of $f\ t_1$ fails. The system now displays the "step failed" message to the user for the outermost failing step. No messages are displayed for the inner failing steps which caused the outermost one to fail. This is by design for two reasons:

- Often, the simplifier has to try multiple rules to prove a precondition. This is the case when there are multiple, overlapping rules for a predicate. Were the panel to notify the user for each of those steps, this would quickly become very confusing because of a flood of unrelated, and in the end, unimportant failures.
- If the innermost failure is several layers of recursions away from the original, interesting step, it becomes difficult for the user to establish a causal relationship between the previously answered `apply` and the subsequent `hint(failed)` message.

Should the user choose to redo the computation, the simplifier state will be rolled back to *before* the last question. In the above example, the system would ask for the application of the rewrite rule $P_1\ t_1 \Longrightarrow f\ t_1 \equiv g\ t_1$ again. Of course, answering that question the same way a second time would not change anything. But it is possible to change the settings and obtain more detailed information. This requires the simplifier to run anew, which is a consequence of the message filtering (see Section 4).

Suppose the user invoked the simplifier with a set of rewrite rules and realizes during a simplification run that a rule is missing. The user then adds the rule to the simplifier and expects not to be asked the same questions they already answered again. This is similar to debugging an imperative program: While stepping through the execution, an error is found and the source code is changed accordingly. After restarting the process, the user reasonably expects to continue at the same point (or a bit earlier) where the debugging session has been exited previously.[2] This is not always possible, but as long as editing the source code

---

[2] In fact, the Java debugger of the *NetBeans* IDE offers a similar feature. Code changes while debugging can be applied to the running program, which requires reloading the affected classes in the JVM instance. This completely avoids the problem of reconstructing the original state after restarting the process, because the process is not even being terminated.

preserves certain invariants (e.g. the arrangement of global variables), resuming at the old execution point is safe.

However, this contradicts the stateless nature of Isabelle's document model: any changes in the proof text causes the system to discard and reinitialize all active command states.

In the new tracing, a *memoization* system helps in mitigating that issue by trying to reconstruct the original tracing state. Each time the user answers an `apply` question, that answer is recorded in a (global) storage. When the same question appears again later, be it in the same simplification run or in another one, it is automatically answered identically. All of this happens in the JVM process to avoid cluttering the tracing logic with mutable state.

Redoing a simplification step creates an interesting feature overlap with memoization. Since fundamentally, redoing a step makes the system display the original question again, a naively implemented cache would auto-answer that question. As a consequence, the (unchanged) computation would fail again, which would obviously diminish the utility of this question type. Hence, care has been taken in the implementation to partially clear the cache.

Recall that every time a user chooses to redo a failed computation, the system generates an `ignore` message. The memoization mechanism in the IDE picks up the parent of such a message from its store, and deletes its answers and the answers of its children from the memory. The user is also able to clear (or even disable) the memory completely in case that behaviour is unwanted.

Note that despite what the name "memory" might suggest, no fuzzy matching of any sort is done. At the moment, questions are compared using simple textual identity of their contents. If the text of a message is slightly different, it will not be considered (this includes renaming of free variables). This is essentially a trade-off: the notion of "fuzziness" is extremely context-dependent. For example, for some predicates, a simple change of a constant is meaningless, whereas for other predicates, a conditional rule depends on it. Designing a reasonable fuzzy matcher is outside of scope here, but an interesting starting point for future work.

4   Message Filtering

Based on the settings described in the previous sections, messages get *filtered*. The process for `apply` messages is depicted in Fig. 3 and consists of these steps:

1. Using *normal* verbosity, messages which have not been triggered by a breakpoint are discarded right at the beginning. This happens immediately after the simplifier creates them. Unless tracing is disabled completely, accepted interactive messages are then transferred to the IDE, where they will be treated as potential questions for the user.

2. If the tracing operates without user intervention (e.g. if the user explicitly disabled it earlier), messages are merely logged and answered with a default reply. The default reply is chosen so that it does not influence the simplifier in any way, i.e. it proceeds as if tracing would be disabled.

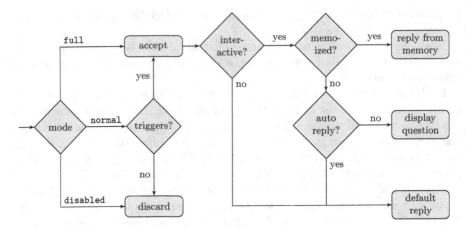

Fig. 3. Message filtering for `apply` messages

3. Some questions are eligible for memoization. The memory is queried to check for a match.
4. If *auto reply* is enabled, all remaining questions are also automatically answered with a default reply. Otherwise, they are finally being displayed. This is scoped to the current focus, i.e. only applies to the active questions of the selected command in the proof text. A use case for this facility arises when interactive tracing is globally enabled, but the user wants to discharge active questions of some selected commands without having to modify the proof text.

At a first glance, this pipeline might seem a little convoluted. However, these steps are necessary to match the user's expectation to only get asked if desired, which (ideally) should happen rarely. Filtering keeps the number of unwanted messages at a minimum.

5 Related Work

In this section, we will compare our contributions with the *SWI* implementation of Prolog and the *Maude* rewriting language. Both systems offer tracing and debugging facilities where the user is able to step through the computation. The theorem provers Coq, HOL4, HOL Light and PVS have only rudimentary, non-interactive trace facilities, which is why we omit a thorough comparison.

5.1 Debugging and Tracing in SWI-Prolog

Prolog is a logic programming language [6, 11]. A program consists of a set of *clauses*, namely *rules* and *facts*. Rules are (usually) Horn clauses, with a head and a body. Facts are merely rules with an empty body. Heads may contain

```
                              [trace]  ?- descendant(a, c).
                                 Call: (6) descendant(a, c) ? creep
                                 Call: (7) child(a, _G1949) ? creep
                                 Exit: (7) child(a, b) ? creep
                                 Call: (7) descendant(b, c) ? creep
                                 Call: (8) child(b, _G1949) ? creep
                                 Exit: (8) child(b, c) ? creep
                                 Call: (8) descendant(c, c) ? creep
                                 Exit: (8) descendant(c, c) ? creep
                                 Exit: (7) descendant(b, c) ? creep
                                 Exit: (6) descendant(a, c) ? creep
                              true ;
                                 Redo: (8) descendant(c, c) ? creep
child(a, b).                     Call: (9) child(c, _G1949) ? creep
child(b, c).                     Fail: (9) child(c, _G1949) ? creep
                                 Fail: (8) descendant(c, c) ? creep
descendant(X, X).                Fail: (7) descendant(b, c) ? creep
descendant(X, Z) :-              Fail: (6) descendant(a, c) ? creep
  child(X, Y), descendant(Y, Z). false.
```

(a) Input database   (b) Query with tracing enabled

Fig. 4. Prolog example

variables, and bodies may contain variables not occurring in the head. Variable names must begin with an upper-case letter or an underscore, whereas *atoms* must begin with a lower-case latter.

The example in Fig. 4a defines a program with two predicates, child and descendant. A query is basically a predicate with possibly uninstantiated variables, and Prolog tries to instantiate those. In Prolog terminology, such an expression is a *goal*, and the interpreter attempts to *prove* it.

When proving a goal, Prolog tries to unify the current goal with any of the available clause heads, and proceeds recursively with each item in the body as new subgoals. This is similar to how the simplifier works in Isabelle: The left-hand side of a rewrite rule is matched to the current term, and if matches, it tries to solve the preconditions of the rule recursively.

The Prolog implementation *SWI-Prolog* provides a tracing facility for queries [8, §§2.9, 4.38]. An example for the tracing output can be seen in Fig. 4b (the term creep denotes continuing the normal process).[3]

Apart from continuing the process, SWI offers some additional commands. The commands relevant for this work are:

---

[3] A discussion of tracing in Prolog can be found in [6, §8], and further analyses in [7]. SWI uses a slightly extended variant thereof.

abort exits the whole proof attempt

fail the current goal is explicitly marked as failure, regardless whether it could have been proved

ignore the current goal is explicitly marked as success

retry discards the proof of a subgoal and backtracks to the original state of the parent goal

In contrast to SWI, marking a goal as success is not supported in our work. The simplifier—just like any other tactic—has to justify its steps against Isabelle's proof kernel, which would not accept such a user-declared success. A possible workaround would be to introduce such theorems "by cheating", i.e. by explicitly declaring an "oracle" which adds an axiom to the proof. It is an interesting discussion whether or not it is sensible to grant the simplifier the "privilege" to generate invalid theorems in tracing mode.

Finally, it is possible to declare breakpoints on predicates. SWI allows to refine breakpoints with the specific event (referred to as *port*). For example, the user can specify that they are interested only in `fail` messages, but not `call` messages. However, as soon as such a breakpoint is set, the tracing ceases to be interactive and switches to a log-only mode.[4] In our work, the filtering concept is more sophisticated and allows a fine-grained control over what is being asked. In SWI, it is not possible to set a breakpoint on terms.

In summary, SWI's features are quite similar to what we have implemented, but differ in a few conceptual points. First and foremost, the "execution" model of Prolog and Isabelle theories differ significantly. Evaluation of Prolog queries happens sequentially, and any changes in the underlying source code must be explicitly reloaded. That also means that running queries need to be aborted. On the other hand, using Isabelle/jEdit, changes in a theory get reloaded immediately and affects pending computations directly. We can conclude that a memoization mechanism as implemented in our work is not necessary in Prolog.

5.2 Debugging and Tracing in Maude

Maude is a logic language based on term rewriting [4, 5]. A program consists of data type declarations and equations. Then, the user can issue a `reduce` command, which successively applies rewrite rules to an input term. A short example modelling natural numbers can be seen in Fig. 5a.

This snippet declares a data type for natural numbers, along with its two constructors `zero` and `s`. Additionally, it defines an addition function, a predicate to check whether a number is non-zero and two equalities for that predicate. We have left out the equations for addition because they are not needed for the example.

Similar to Isabelle's simplifier, rewrite rules can be *conditional*. In the trace, it becomes obvious that those are handled exactly like in Isabelle. A potentially

---

[4] Switching to log-only mode is possible in our work, too. When in non-interactive mode, trace is only produced for steps matching breakpoints, but no questions are presented to the user.

```
                              Maude> reduce nonzero (s zero + s s zero) .
                              *********** trial #1
                              ceq nonzero (N + M) = true
                                 if nonzero N = true /\ nonzero M = true .
                              N --> s zero
                              M --> s s zero
                              *********** solving condition fragment
                              nonzero N = true
                              *********** equation
fmod SIMPLE-NAT is            eq nonzero s N = true .
  sort Nat .                  N --> zero
  op zero : -> Nat .          nonzero s zero
  op s_ : Nat -> Nat .        --->
  op _+_ : Nat Nat -> Nat .   true
  op nonzero_ : Nat -> Bool . (...)
                              *********** success #1
  vars N M : Nat .            *********** equation
  ceq nonzero (N + M) = true  (...)
     if nonzero N /\ nonzero M . nonzero (s zero + s s zero)
  eq nonzero (s N) = true .   --->
endfm                         true
```

(a) A simple program   (b) Reducing a term with trace enabled

Fig. 5. Maude example (based on [4, §2.2])

applicable rule gets instantiated with the concrete term, and the preconditions are solved recursively.

The Maude tracing is purely sequential and offers little to no insight into the hierarchical structure when conditional rules are involved. The trace can be tuned in various ways [4, §§14.1, 18.6]: for example, Maude allows filtering for named rules and operations (albeit only the outermost operation in the redex is considered). There is also a wealth of settings which control verbosity of the trace, e.g. whether solving preconditions or the definition of rules should be included in the trace.

Apart from controlling the textual output, it is also possible to enable output colouring, similarly to the highlighting in Isabelle. The major difference here is that Maude distinguishes between *constructor* and *nonconstructor* symbols, whereas such a distinction is not made in Isabelle. An indicator for problems in the set of rewrite rules of a Maude program is when a term does not get fully rewritten, which is defined as nonconstructor symbols still occur after reducing [4, §14.1.2]. Hence, colouring symbols differently greatly improves debugging experience in Maude, because it also gives hints into when exactly a problem has been introduced.

In addition to tracing, there is also a debugging facility. It can be configured with breakpoints in the same way as the tracing. When a breakpoint is hit, the user can resume or abort the whole computation, but also (on request) observe the stack of recursive invocations. The latter also includes a textual explanation, e.g. that the current term is being rewritten to solve a precondition.

Table 1. Execution times to simplify $10^x \cdot 10^y$ (in seconds, without rendering)

x	y	disabled	old tracing	normal	full
10	10	0.02	0.56	0.39	0.92
20	10	0.04	1.61	1.21	2.98
20	20	0.08	3.15	2.44	6.00

A distinguishing feature of the debugger is that it allows to execute a new `reduce` command when a debugging session is active. This allows the user to quickly check related terms and hence better understand an issue with the original term.

Maude has been an active target for research for refining the trace even further, providing insights into when and how a particular term emerged during reducing (e.g. Alpuente et.al. [1]). Term provenance could certainly be an interesting extension for our work on the simplifier trace, but would require significantly more instrumentation of the simplifier.

6   Evaluation and Future Work

In this section, we will briefly discuss the performance of the new tracing and the practical usability. Furthermore, possibilities for future work are explored.

6.1   Performance

Logging the simplification process obviously incurs a measurable overhead. For example, consider the expression scheme $10^x \cdot 10^y$ (which will be evaluated to a literal number by the simplifier given concrete values of x and y). The test machine is an Intel Core i7-3540M with a peak frequency of about 3.7 GHz, and 8 GiB of memory. Execution times have been collected using the `Timing` panel in Isabelle/jEdit without having the IDE render the trace data. The results can be seen in Table 1.

As can be seen in the table, the simplifier itself is pretty fast, but enabling the trace slows down the process significantly. Note that for measuring in `normal` mode, no breakpoints have been set, hence these numbers show just the overhead of the tracing: In every single rewrite step, the tracing hooks have to be called and various lookups are performed. The most interesting comparison is between the old tracing and `full` mode, where the ratio is roughly 1 : 2. This can be explained by the fact that the full mode collects more information and processes it more thoroughly than the old tracing.

The slowest component overall is the GUI though (times not shown in the table), which requires about 5 s to display the full trace for $x = y = 10$. The old tracing needs just 1 s. The difference here can again be explained by the amount of information processed; in particular, the old tracing does not have to reconstruct the hierarchical structure of the trace messages. For $x = 20, y = 10$, the GUI already needs 37 s to render the full trace, resulting in roughly 200 000 lines. However—once rendered—scrolling, collapsing and expanding nodes is instantaneous. Hence, it is generally advisable to enable the simplifier trace only if necessary when dealing with huge traces. For smaller traces which are in the order of hundreds of messages, e.g. the example from the introduction, all GUI actions are instantaneous.

6.2 Future Work

There are multiple dimensions in which this work can be extended in the future.

Integration into the Isabelle system would benefit by adapting more tactics to use the new tracing mechanisms, since many of them can be modelled in a hierarchical manner, and more message types could be introduced. For example, the simplifier supports custom simplification procedures ("ML functions that prove rewrite rules on demand", [16, §9.3.5]). In the new tracing, invocations of those "simprocs" are ignored. A future extension could be to provide hook functions to those simprocs. Besides rewriting, proof reconstruction tactics [3] fit naturally into the recursive model; for example the `metis` resolution prover to replay proofs from external ATPs (automated theorem provers).

The user experience could be improved by asking even less questions (introduce fuzzy matching in the memoization) or providing more information per question (e.g. the whole term context of the redex). Interesting context data includes for example term provenance, i.e. tracking how a particular subterm in the result emerged during the rewriting process. Also, there still some oddities arising from the asynchronous document model which leads to undesired delays in the user interface under certain circumstances. Resolving this would most likely require nontrivial changes in the message passing mechanisms in Isabelle/Scala.

Furthermore, the current implementation would hang indefinitely if the simplifier fails to terminate. The old tracing in recent versions of Isabelle pauses the simplification process after a certain number of messages have been emitted and waits for user confirmation to continue. (Previously, the IDE could easily be rendered unresponsive when the memory pressure became too high.) In the new tracing, this problem occurs less often, because—if running in `normal` mode— most messages are discarded, so users will most likely see no messages. This is not a big problem, since non-termination may happen in other commands in Isabelle as well, and the user can transparently cancel such invocations just by deleting the corresponding text in the theory source. To allow for better debugging, we could additionally perform some basic termination checks, along with a new message type informing the user of the problem.

Support for other Isabelle IDEs is currently not supported. IDEs based on the document model (e.g. Isabelle/Eclipse[5]) would require reimplementing the pure GUI parts of the tracing. As for Proof General [2], all Scala parts of this work would need to be completely reimplemented, possibly in a different programming language.

---

[5] Developed by Andrius Velykis, `http://andrius.velykis.lt/research/`

6.3   Case Study: A Parallelized Simplifier

As indicated earlier, because of the loose coupling between simplifier and tracing, a parallelized simplifier can easily be supported. Since Isabelle's simplifier is quite complex, we implemented an extremely stripped-down but parallel version in order to test this capability. The "proof of concept" simplifier is not nearly as powerful as the "real" one—given that it consists only of roughly 100 lines of code—but it splits off some of the work into parallel tasks: Preconditions of a conditinal rewrite rule are resolved simultaneously, and if multiple rewrite rules are applicable, they are tried in parallel.

The implementation overhead of parallelization is low, because Isabelle/ML offers library support for both concurrent and parallel programming. The primary abstraction are *future* values [9, 13]: the type constructor `'a future` denotes a value of type `'a` becoming available at a later time. Parallelizing existing purely (i.e. not side-effecting) functional code is simple, because it only requires to wrap subexpressions into the `fork` constructor of futures, which automatically evaluates them on a pool of worker threads. There are combinators for combining parallel computations and to wait on the completion of the result.

A little bit more intervention is required when parallelizing side-effecting code. However, since the Isabelle document model is asynchronous by its nature, it already offers message-oriented ways for communication between the prover and the IDE process. Consequently, the system already deals with common concurrency issues; in this case the only remaining task was to ensure that no deadlocks occur.

For the simple examples we tried, a parallelized simplifier did not yield a significant speed-up compared to the sequential simplifier. However, the key insight lies in the user interaction. When multiple preconditions are solved simultaneously, the user might see more than one active question at the same time which could potentially be confusing. Furthermore, race conditions could lead to nondeterministic traces, a problem well-known when debugging parallel imperative programs. Hence, the question of a reasonable user interface for parallel simplification remains open.

7   Conclusion

We presented a generic tracing facility for Isabelle tactics which replaces the old simplifier trace. That new facility is interactive, highly configurable, and intuitive to operate. The performance and implementation impact on the rest of the system turned out to be rather small. Nonetheless, it became possible to provide more insights for the user into the simplification process.

The design goal that the amount of interaction with the user should be kept low has been achieved. Various sophisticated filtering and memoization techniques help maintaining a good trade-off between flexibility and opacity of the system.

The new tracing mechanism is available in the main Isabelle development repository since early February 2014.

Acknowledgements. I thank Tobias Nipkow and Lars Noschinski for encouraging me to implement a new simplifier trace. I am grateful to Makarius Wenzel who commented multiple times on various stages of the code and patiently answered my questions about internals of the Isabelle system. Dmitriy Traytel and Cornelius Diekmann commented on early drafts of this paper.

References

1. Alpuente, M., Ballis, D., Frechina, F., Sapiña, J.: Slicing-Based Trace Analysis of Rewriting Logic Specifications with iJULIENNE. In: Felleisen, M., Gardner, P. (eds.) ESOP 2013. LNCS, vol. 7792, pp. 121–124. Springer, Heidelberg (2013)
2. Aspinall, D.: Proof General: A generic tool for proof development. In: Graf, S. (ed.) TACAS 2000. LNCS, vol. 1785, pp. 38–43. Springer, Heidelberg (2000)
3. Blanchette, J.C., Bulwahn, L., Nipkow, T.: Automatic Proof and Disproof in Isabelle/HOL. In: Tinelli, C., Sofronie-Stokkermans, V. (eds.) FroCoS 2011. LNCS (LNAI), vol. 6989, pp. 12–27. Springer, Heidelberg (2011)
4. Clavel, M., Durán, F., Eker, S., Lincoln, P., Martí-Oliet, N., Meseguer, J., Talcott, C.: Maude manual (version 2.6)
5. Clavel, M., Eker, S., Lincoln, P., Meseguer, J.: Principles of Maude. In: Meseguer, J. (ed.) Electronic Notes in Theoretical Computer Science, vol. 4. Elsevier Science Publishers (1996)
6. Clocksin, W.F., Mellish, C.S.: Programming in Prolog: Using the ISO standard. Springer (2003)
7. Ducassé, M., Noyé, J.: Logic programming environments: Dynamic program analysis and debugging. The Journal of Logic Programming 19-20(suppl. 1), 351–384 (1994)
8. Fruehwirth, T., Wielemaker, J., De Koninck, L.: SWI Prolog Reference Manual 6.2.2. Books on Demand (2012)
9. Matthews, D.C., Wenzel, M.: Efficient parallel programming in Poly/ML and Isabelle/ML. In: Proceedings of the 5th ACM SIGPLAN Workshop on Declarative Aspects of Multicore Programming, pp. 53–62. ACM (2010)
10. Nipkow, T., Paulson, L.C., Wenzel, M.T.: Isabelle/HOL. LNCS, vol. 2283. Springer, Heidelberg (2002)
11. Sterling, L., Shapiro, E.Y.: The Art of Prolog: Advanced Programming Techniques. MIT Press, Cambridge (1994)
12. The GNU Project: GDB: The GNU Project Debugger, https://www.gnu.org/software/gdb/
13. Wenzel, M.: Parallel proof checking in Isabelle/Isar. In: Proceedings of the ACM SIGSAM 2009 International Workshop on Programming Languages for Mechanized Mathematics Systems, pp. 13–29. ACM (2009)
14. Wenzel, M.: Asynchronous proof processing with Isabelle/Scala and Isabelle/jEdit. Electronic Notes in Theoretical Computer Science 285, 101–114 (2012)
15. Wenzel, M.: Isabelle/jEdit – A Prover IDE within the PIDE framework. In: Jeuring, J., Campbell, J.A., Carette, J., Dos Reis, G., Sojka, P., Wenzel, M., Sorge, V. (eds.) CICM 2012. LNCS (LNAI), vol. 7362, pp. 468–471. Springer, Heidelberg (2012)
16. Wenzel, M.: The Isabelle/Isar Reference Manual (2013)

Towards an Interaction-based Integration of MKM Services into End-User Applications

Constantin Jucovschi

Jacobs University Bremen

Abstract. The Semantic Alliance (SAlly) Framework, first presented at MKM 2012, allows integration of Mathematical Knowledge Management services into typical applications and end-user workflows. From an architecture allowing invasion of spreadsheet programs, it grew into a middle-ware connecting spreadsheet, CAD, text and image processing environments with MKM services. The architecture presented in the original paper proved to be quite resilient as it is still used today with only minor changes.

This paper explores extensibility challenges we have encountered in the process of developing new services and maintaining the plugins invading end-user applications. After an analysis of the underlying problems, I present an augmented version of the SAlly architecture that addresses these issues and opens new opportunities for document type agnostic MKM services.

1 Introduction

A major part of digital mathematical content today is created in Computer-Aided Design (CAD) systems, spreadsheet documents, wiki pages, slide presentations, program source code. Taking advantage of Mathematical Knowledge Management (MKM) representations and services, to better manage mathematical content in these documents, is still a complex and time consuming task, often, because there is no adequate tool support. Imagine a complex CAD model composed of hundreds of components. Readers of the CAD model would benefit a lot if functional information (e.g. specifying role) could be attached to components so that: *i)* semantic services generate descriptive information about the role of the object in the model; *ii)* safety requirements matching that role could be fetched from a "safety-requirements.tex" document etc. Adding functional information (e.g. encoded as a URI to an ontology concept) to CAD components can be achieved in most systems by adding a custom property to the CAD component. The services generating descriptive information and fetching safety requirements could be implemented as web services which, given the concept URI, would fetch required information. In this scenario, the engineer would rightfully consider that *the MKM services are not adequately supported because he has to manually get the functional information associated to an object, change application context to a web browser and supply the functional information into the MKM service.* The feeling of tool support inadequacy, is not caused by the

S.M. Watt et al. (Eds.): CICM 2014, LNAI 8543, pp. 344–356, 2014.

MKM services themselves, but rather by the steps (workflow) that the engineer had to perform in order to be able to consume the MKM services. I call this workflow *User-Service Interaction* (USI).

The importance of developing adequate tool support, especially for authoring, is a reoccurring topic at MKM [KK04] and also in the wider semantic technology community [SL04; Joo11; SS10]. The arguments come from different directions. [SS10] argues that some tasks, required for semantic content creation, can be performed *only* by humans; making development of computer support for tasks that allow semi-automation ever more important. [SL04; Joo11] base their intuitions on the experiences and lessons learned from deploying semantic technologies in real-world scenarios. They stress the importance of *"user-friendly modeling tools and procedures"*[Joo11] and share lessons learned such as *"KM systems should not be introduced as explicit stand-alone applications that user intentionally must interact with in addition to their other job responsibilities"*[SL04]. [KK04] uses Prisoner's dilemma to explain why the long-term benefits of having semantic content fail at motivating semantic content creation. Further on, authors suggest that, improving authoring support and letting content authors reap the results of semantic content creation early on, would increase motivation for semantic content creation.

The Semantic Alliance (SAlly) Framework [Dav+12], set the goal of supporting the process of building adequate tool support for Mathematical Knowledge Management services by integrating them into typical applications and end-user workflows. The framework allowed MKM services to be simultaneously integrated in several end-user applications that share the same media-type (e.g. spreadsheet documents). In this way, once an MKM service was integrated, through the SAlly Framework, with Microsoft Excel, it would work "out of the box" in Open/Libre Office Calc. Later work, added support for CAD [Koh+13], text and image [Bre+14] editors as well as allowed creation of cross-application MKM services, e.g. allowing seeing costs of a CAD component in the pricing spreadsheet document, leading to the Multi-Application Semantic Alliance Framework (MA-SAlly) [Koh+13].

This paper explores extensibility challenges we have encountered in the process of integrating new MKM services into end-user applications. These challenges, described in section 3, did not depend so much on the MKM service functionality, but rather, on the type of User-Service Interaction it required. Section 4, describes how USI requirements can be decomposed into modular interfaces, that are easier to standardize but also extend. Afterwards, I present an augmented version of the Semantic Alliance Architecture that incorporates the USI modular interfaces and thus solve the extensibility problem motivating this research. The paper ends with implementation results and a conclusion.

2 Integration Analysis of the Semantic Alliance Framework

In this section, I want to analyze the integration strategies used in the SAlly framework [Dav+12] to integrate MKM services into end-user applications. I will shortly introduce each identified strategy and assess its impact by using the following cost model:

> $n \geq 2$ applications must be integrated with $m \geq 2$ MKM services. There is a cost function C that computes the cost of implementing any part of the integration such that the cost of implementing any part is equal to the sum of the costs implementing its subparts.

The properties of the cost function C clearly oversimplify the software development process. Yet, the extent to which these properties are used in this paper should not significantly change the validity of the computations. Also, the goal of integrating n applications with m services might seem unrealistic. Indeed, there are MKM services for which integration into certain applications makes little or no sense at all. On the other hand, from the experience gained with the MA-SAlly framework, MKM services such as definition lookup and semantic navigation were integrated in all invaded applications.

Figure 1a, shows the Model-View-Adapter (MVA) architectural pattern that, along with Model-View-Controller (MVC), are widely used in the development of applications with heavy user interaction. The MVA pattern structures application components into three categories: model, adapter and view. The components in the view category are responsible for the visual interface; components in the model category are responsible for application logic and components in the adapter category, mediate the interaction between view and model components. The MVA architectural pattern is also suitable for representing the architecture of MKM services where the model implements the MKM representation, service logic

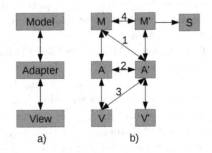

Fig. 1. a) the Model-View-Adapter pattern b) direct integration of a MKM service into an application. M, M' are the models, V, V' are the views and A, A' are the adapter of the application and MKM service respectively. S is an optional semantic (web-) service.

or just sends requests to an external MKM service. The view of an MKM service implements any custom dialogs, toolbars and service configuration pages. The MKM service adapter wires everything together.

MKM services that are directly integrated into an application, i.e. implemented as a plugin for that application and share the same memory space, enjoy several practical benefits. Namely, they can directly access all relevant resources the host application can provide. Figure 1b, shows typical resource access patterns among end-user application components (M, A, V) and MKM service components (M', A', V'). The MKM service adapter (A'), often influences how application events are handled, how application view and model get updated (hence the edges 1, 2 and 3). The MKM model (box M') may both listen to changes and modify application model M (edge 4). When MKM services are directly integrated into an application, implementing edges 1, 2, 3 and 4 is equivalent to a performing simple API calls and hence very straightforward. A disadvantage is that directly integrating m such services into n applications will result in a huge cost

$$\sum_{a \in App} \sum_{s \in Serv} C(M_a'^s) + C(A_a'^s) + C(V_a'^s) \tag{1}$$

where $M_a'^s$, $A_a'^s$ and $V_a'^s$ are the model, adapter and view that needs to be implemented to integrate MKM service s into an application a.

A natural way of reducing the cost in equation 1, is to refactor MKM service models and adapters into standalone web services (Figure 2a). The MKM adapter and model (A', M') are no longer part of the application but still need to communicate with it. This is achieved through an end-user application plugin A'', that allows A' and M' to access the same resources as before, i.e., edges 1, 2, 3 and 4, in Figure 1b by communicating through a network channel COM. Additionally, A'' also needs to allow A' to communicate with V' (edge 5 in Figure 2a). The cost of this integration strategy is:

$$\sum_{a \in App} \sum_{s \in Serv} C(A_a''^s) + C(V_a'^s) + \sum_{s \in Serv} C(A_*'^s) + C(M_*'^s) \tag{2}$$

where $A_*'^s$ and $M_*'^s$ are the adapter and model of MKM service s implemented as a standalone service; $A_a''^s$ is the application plugin allowing standalone service s to communicate with application a. At this point, $A_*'^s$ and $M_*'^s$ can be running on a different server, can be implemented in any language and optionally even be part of the service S from Figure 1b. On the other hand, implementation costs of the application plugin $A_a''^s$ can become very high as they depend on the complexity of access patterns between MKM service and application resources (i.e. edges 1, 2, 3, 4 and 5).

To further reduce costs, SAlly refactors V' into the same standalone service as M' and A' (Figure 2b). Generally, this may be achieved in two ways: *i)* using the graphical toolkit of the language in which V' is implemented — this requires V' to run on the same computer as the end-user application. *ii)* using an external interaction interpreter (e.g. web browser) running on the same computer as the end-user application. The SAlly architecture uses the second approach where a general purpose screen manager, called Theo (Figure 4), runs on the same computer as the application. The cost is further reduced to

$$\sum_{a \in App} \sum_{s \in Serv} C(A_a''^s) + \sum_{s \in Serv} C(A_*'^s) + C(M_*'^s) + C(V_*'^s) \tag{3}$$

Further optimizations come from standardizing and reusing plugins $A_a''^s$ for multiple services. The SAlly architecture choses to standardize $A_a''^s$ by application type (e.g. spreadsheet documents). Namely, for application type T and a set of services $S(T)$ for that application type, one could define an API $\mathcal{A}(T)$ that can serve all services in $S(T)$ i.e.

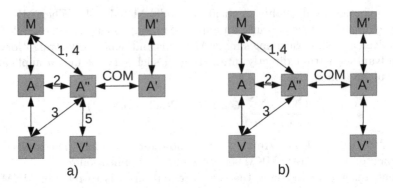

Fig. 2. a) refactoring MKM service model and adapter into a Web-Service b) refactoring MKM service view into the Web-Service

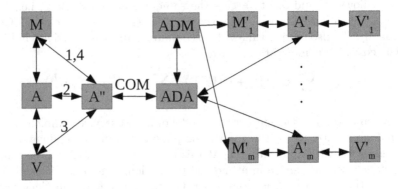

Fig. 3. Architecture of the Semantic Alliance Framework from an integration perspective

$$\forall a \in App(T).\forall s \in S(T).A_a''^{s} \subset \mathcal{A}(T) \tag{4}$$

Hence, the cost equation can be updated to

$$\sum_{T \in AppType} \left(\sum_{a \in App(T)} C(\mathcal{A}_a(T)) + \sum_{s \in S(T)} C(A_*'^{s}) + C(M_*'^{s}) + C(V_*'^{s}) \right) \tag{5}$$

where $\mathcal{A}_a(T)$ is the implementation of API $\mathcal{A}(T)$ in application a and corresponds to the application dependent invaders called Alexes in [Dav+12] (Figure 4).

Considering that the SAlly Framework, was specifically designed for integrating semantic services into applications, most MKM service models $M_*'^{s}$ require persistent storage for semantic information in the document. For example, most MKM services for spreadsheet documents, require functionality allowing semantic information to be attached to sheets and cell ranges. Instead of letting each MKM service model implement its own persistence strategy, an Abstract Document Model $ADM(T)$ (application-type dependent), was introduced (Figure 3), that served as a common semantic layer that all MKM services could reuse. Similarly, an Abstract Document Adapter (ADA) was implemented to ease communication and coordinate how multiple MKM services process events. The ADM, ADA, A_*^{ls}, M_*^{ls}, V_*^{ls} components in Figure 3 are part of the box "Sally" in Figure 4.

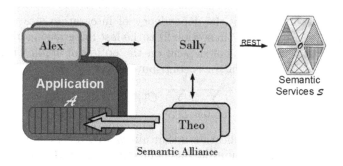

Fig. 4. SAlly Architecture from [Dav+12]

3   Problem Description

The first problem of the SAlly architecture comes from equation 4, namely, $\mathcal{A}(T)$ is defined as an API that can serve all services $S(T)$. Defining an $\mathcal{A}(T)$ for a finite set $S(T)$ is not very hard but generally, $S(T)$ is infinite. In practice, one defines an $\mathcal{A}(T)$ to support currently available MKM services. However, when a new MKM service needs to be integrated, which requires support to e.g. create a new sheet and the current $\mathcal{A}(spreadsheet)$ does not support this operation, S cannot be implemented unless $\mathcal{A}(spreadsheet)$ is extended. Considering that multiple applications implement API $\mathcal{A}(T)$ (through plugin $\mathcal{A}_a(T)$), constantly changing $\mathcal{A}(T)$ becomes a bottleneck and a source of errors. Conversely, if an application can only support 95% of $\mathcal{A}(T)$, it cannot be (safely) integrated even with services that do not need the rest 5% of the $\mathcal{A}(T)$.

Defining the abstract document model $ADM(T)$ for an application type, shares similar issues as defining $\mathcal{A}(T)$. This became very apparent when we tried to define an abstract document model for CAD systems [Koh+13]. Most CAD systems share the concept of an assembly which defines the position in space of CAD parts or nested CAD assemblies. At the assembly level, one could define an $ADM(T)$ capturing various relationships among assembly components that could be applied to most CAD systems. The interesting geometrical properties, however, are only represented in the CAD parts, which, may vastly differ even within the same CAD system (especially if CAD parts are imported from other systems). Defining an $ADM(T)$ capturing parts data is so time-consuming that it is not worth the integration benefits it brings.

Defining an application type dependent $\mathcal{A}(T)$ and $ADM(T)$ makes a lot of sense when the invaded applications are similar in most respects as LibreOffice Calc and Microsoft Excel (used in the original SAlly paper) are. But the list of such examples is not very long. Most applications are, to some extent, unique but still share a great deal of concepts with other applications.

4   Method

This paper aims at reducing the dependency of integration strategies on the application type. Equation 3 (in section 2) is the last integration strategy in the SAlly architecture that does not depend on the notion of application-type. At that point, the biggest cost factor is contributed by

$$\sum_{a \in App} \sum_{s \in Serv} C(A_a''^s) \tag{6}$$

where $A_a''^s$ is a plugin in application a that allows MKM service s to e.g. insert new data into the active document, highlight some part of the document or get notified when another object is selected and so on. $A_a''^s$ *is the critical component that enables User-Service Interaction between MKM service s and application a.* An important aspect is that, while $A_a''^s$ enables service s to interact with application a, it does not implement the interaction itself. The implementation of the User-Service Interaction is part of the MKM service.

One of the main purposes of MKM User-Service Interactions is to support the process of aligning document content with concepts in some ontology. Typical tasks include: annotating content, making relationships explicit, validating content and managing changes. The features and the types of interactions offered by USIs strongly depend on the semantic format/ontology and often dependent only slightly on the format of the document that is semantically annotated. Consider a high-level description of the OMDoc document ontology:

> An OMDoc document organizes content into theories that relate to each other either by import or view relations. Theories may contain symbols that are transported from one theory to another using the relations among theories.

This description clearly specifies requirements regarding the types of annotations (theories and symbols) and relationships among them (theories contain symbols and relate to other theories) that OMDoc authoring services need to support. The requirements on the host document format are rather implicit: relations need to be persistent and react in sensible ways to document changes. Furthermore, validation services that make sure that imports are valid, that there are no cyclic dependencies, that all symbols are defined — require only information regarding OMDoc relations are hence completely independent of host document format.

From the experience gained through the Semantic Alliance Framework, I compiled a list of common and application-type independent functionality that $A_a''^s$ typically need to implement. Namely:

content selection — is by far the most common functionality USIs require. It allows the user to point MKM services to document objects, and reversely, MKM services to point the user to document objects. It is heavily used for enriching and modifying semantic content.

semantic storage — stores and retrieves semantic information associated to document objects. One can differentiate between document and object level semantic information.

context menus — provide a natural way of accessing object specific interactions.

application focus — used in the multi-application context and allows changing focus to a certain application. It is often used in combination with content selection.

content marking — used to visually mark (e.g. highlight) document content in a non-persistent way. Very useful for projecting multiple types of semantic content to document objects as well as selecting/deselecting multiple objects without the danger of clicking the wrong button and loosing the whole selection.

The intuition behind this paper, is that major parts of the User-Service Interactions that MKM services require, can be implemented on top of document-type independent interactions such as the ones above. So an annotation service for OMDoc format could be defined as follows: when the user requests the context menu for some selected object X, and there is no semantic annotation about X in the semantic storage, add the menu item "Annotate as OMDoc module" to the context menu, and so on. Depending on document type, selected object X can be a text fragment, a cell range, a CAD assembly. In the same time, MKM services might require very specific, application dependent information which also need to be supported e.g. position of geometrical assemblies.

The problem of reducing the cost in equation 6, is essentially a problem of defining reusable interfaces in a distributed setting. Any interface X, that can be reused by $k \geq 2$ MKM services, reduces the cost in equation 6 by $n(k-1)C(X)$ i.e. each application must implement X once to then reuse it for k services. One can iterate this process and define set of reusable, modular interfaces $\mathcal{M} = \{M_1, M_2, ...\}$ such that

$$A_a''^s = M_a^{s_1} \circ M_a^{s_2} \circ ... \circ M_a^{s_k}, \tag{7}$$

i.e. one can represent $A_a''^s$ as composition of several implementations of modular interfaces $(M^{s_1}, ... M^{s_k} \in \mathcal{M})$ for application a. Substituting this representation in equation 6 and removing duplicate module implementations for the same application, results in

$$\sum_{a \in App} \sum_{m \in \mathcal{M}(S)} C(M_a^m) \tag{8}$$

where $\mathcal{M}(S)$ is the set of all modules in $\mathcal{M}$ that are required to modularize all $A_a''^s$ in $s \in Serv$.

The total cost in equation 8, depends on the ability of representing the functionality, a MKM User-Service Interaction requires, in a reusable modular manner. Setting $\mathcal{M} = \{A''^s | s \in Serv\}$, i.e. no reuse possible, makes cost in equation 8 equal to the one in equation 6.

5 Augmented Semantic Alliance Architecture

The Augmented Semantic Alliance Architecture assumes, that a set $\mathcal{M} = \{M_1, M_2, ...\}$ of reusable modular interfaces are defined and publicly available to the MKM community. It also assumes that the User-Service Interaction of any MKM service s, can be achieved by combining some set of modular interfaces that I define as $\mathcal{M}(s) \subseteq \mathcal{M}$. If for some service s this is not possible, the set $\mathcal{M}$ is extended with the necessary functionality.

Each application a, can implement a subset of the modular interfaces denoted with $\mathcal{M}(a) \subseteq \mathcal{M}$. A MKM service s can be integrated into application a if $\mathcal{M}(s) \subseteq \mathcal{M}(a)$ i.e. application a implements all modules required by MKM service s. To allow the possibility of having the equivalent of Abstract Document Models from the SAlly architecture, modular interfaces may depend on other modular interfaces (typical restrictions on circular dependencies apply).

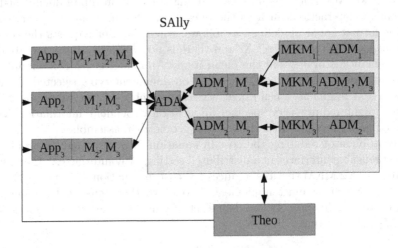

Fig. 5. Integration of three MKM services with three applications using the Augmented Semantic Alliance Architecture

Figure 5, shows an example how three applications (App_1, App_2, App_3) can be integrated with three services (MKM_1, MKM_2, MKM_3). Applications App_1, App_2, App_3 implement modular interfaces $\{M_1, M_2, M_3\}$, $\{M_1, M_3\}$ and $\{M_2, M_3\}$ respectively. Service MKM_1, requires abstract document model ADM_1 which, in turn, requires module M_1 hence $\mathcal{M}(MKM_1) = \{M_1\}$. This means that one can integrate service MKM_1 into App_1 and App_2. Similarly, $\mathcal{M}(MKM_2) = \{M_1, M_3\}$ and so can be integrated only in App_1 and $\mathcal{M}(MKM_3) = \{M_2\}$ and can be integrated into App_1 and App_3.

From this example, one can see that the reuse strategy the Augmented Semantic Alliance Framework has, is more flexible than the one presented in [Dav+12]. Namely, App_1, App_2 and App_3 clearly share common concepts, but assigning them an application type that guides reuse strategy, reduces reuse opportunities. Also, the augmented Sally architecture solves the extensibility problems described in section 3 as reuse is no longer associated with application type. Additionally, the augmented architecture also allows abstract document models to be implemented in the end-user applications themselves, if that makes integration easier.

6   Implementation

To test the interaction based method of integrating MKM services into applications, I chose three simple MKM services that cover four, out of five types of application independent interactions presented in section 4. Namely: content selection, semantic storage, context menu and application focus. These services were integrated into five applications that are part of the LibreOffice suite: Writer (rich text processor), Calc (spreadsheet application), Impress (slide presentations), Draw (graphic documents) and Base (database manager). In this section, I want to shortly introduce the MKM services that I integrated into the LibreOffice suite. These services are quite different and yet have almost identical User-Service Interaction requirements. The section will end in a discussion about the effort it took to perform this integration.

The **Concept Linker** service, allows linking document contents to ontology concepts stored in the `MathHub.info`[Ian+] portal. The service adds a new "Link to concept" item in the context menu of the application, if currently selected object(s) can store semantic information. After clicking on the "Link to concept" context item, a window is generated by the Theo screen manager (for more information see [Dav+12]) where one can choose the concept to which selected objects should be linked to. Figure 6 depicts the concept linker service in a LibreOffice Writer document. One can see that text fragments, formulas, drawings (segment BE) and text boxes can be linked to ontology concepts. Also, one can see the Theo window allowing the user to link text box "O" to a concept from the `MathHub.info` portal.

Definition lookup service retrieves the definition of the ontology concept associated to the currently selected object. The service adds a new "Get definition" item in the context menu of the application, if currently selected object has a concept linked to it. The definition associated to the concept is displayed in a window generated by the Theo screen manager.

Semantic Navigation presents the user with a graphical representation of ontology relations associated to the concept selected in the document. Just like the concept linker and definition lookup services, it adds a context menu item "Semantic navigation" which triggers creation of a new window presenting the ontology relations (Figure 7). Additionally, if the user right-clicks on a node of a related concept e.g. vertice, and there is an object V in the document linked to

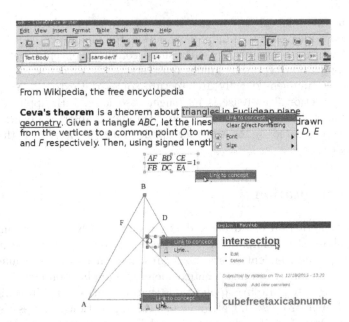

Fig. 6. Mash-up of several screenshots demonstrating the types of LibreOffice Writer objects that can be connected to ontology concepts by the Concept Linker

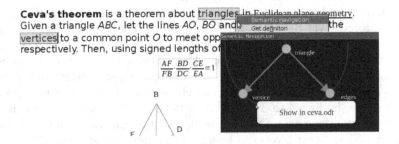

Fig. 7. Mash-up of several screenshots demonstrating Semantic Navigation service

the concept of vertice, the user is given the possibility to change the document focus to object V.

The following table summarizes the types of objects that can be annotated and support definition lookup and semantic navigation services.

Application	Types of document object
All applications	images, math formulas, text boxes and shapes
Writer	text fragments
Calc	cell ranges, charts
Impress	slides, text fragments
Draw	text fragments
Base	tables, forms, queries and reports

Even though the document models of the applications in the LibreOffice suite are very different, the LibreOffice API provides several document type independent mechanisms to access/modify document's content and meta-data. The implementation of the content selection, context menu, semantic storage and application focus interactions required by the three MKM services, were implemented using the document model independent API provided by LibreOffice. This meant that only one LibreOffice plugin had to be implemented which required a similar amount of effort as creating one Alex invader in the SAlly architecture ($\approx$ 1 week). Integrating the same services into the SAlly architecture from [Dav+12], would have required a 5 weeks investment into the Alex (invader) plugins.

7   Conclusion

This paper tackles the problem of MKM services lacking adequate tool support when integrated into end-user applications. The Semantic Alliance framework [Dav+12], developed for reducing MKM service integration costs without sacrificing usability, was successfully used for integrating MKM services into spreadsheet[Dav+12], CAD[Koh+13], text and image [Bre+14] processing software. In the process of developing these integrations, several extensibility problems of the framework became very apparent and yet could not be predicted from the original architecture presented in [Dav+12].

First contribution of this paper is the in-depth integration analysis that helped identifying the reasons for the extensibility challenges in the SAlly framework. In particular, this analysis captures and categorizes the hidden costs associated with decoupling a MKM service into a standalone entity (cost of implementing $A_a''^s$ in section 2). While the integration analysis is strictly used for the SAlly framework, it can be reused for other integration efforts in MKM such as integration of theorem provers into authoring solutions.

The second and main contribution of this paper is the Augmented Semantic Alliance architecture that solves the extensibility problems presented in section 3 and enables more flexible reuse strategies. First experiments with the new architecture show that end-user applications, sometimes provide major reuse opportunities and that the new architecture can take full advantage of them.

References

[Bre+14]   Breitsprecher, T., et al.: Semantic support for engineering design process. In: DESIGN 2014 (to appear, 2014)

[Dav+12]   David, C., Jucovschi, C., Kohlhase, A., Kohlhase, M.: Semantic Alliance: A Framework for Semantic Allies. In: Jeuring, J., Campbell, J.A., Carette, J., Dos Reis, G., Sojka, P., Wenzel, M., Sorge, V. (eds.) CICM 2012. LNCS (LNAI), vol. 7362, pp. 49–64. Springer, Heidelberg (2012)

[Ian+]   Iancu, M., Jucovschi, C., Kohlhase, M., Wiesing, T.: System Description: MathHub.info. In: Watt, S.M., Davenport, J.H., Sexton, A.P., Sojka, P., Urban, J. (eds.) CICM 2014. LNCS (LNAI), vol. 8543, pp. 436–439. Springer, Heidelberg (2014), http://kwarc.info/kohlhase/submit/cicm14-mathhub.pdf

[Joo11]   Joo, J.: Adoption of Semantic Web from the perspective of technology innovation: A grounded theory approach. International Journal of Human-Computer Studies 69(3), 139–154 (2011)

[KK04]   Kohlhase, A., Kohlhase, M.: CPOINT: Dissolving the Author's Dilemma. In: Asperti, A., Bancerek, G., Trybulec, A. (eds.) MKM 2004. LNCS, vol. 3119, pp. 175–189. Springer, Heidelberg (2004), http://dx.doi.org/10.1007/978-3-540-27818-4_13

[Koh+13]   Kohlhase, A., Kohlhase, M., Jucovschi, C., Toader, A.: Full Semantic Transparency: Overcoming Boundaries of Applications. In: Kotzé, P., Marsden, G., Lindgaard, G., Wesson, J., Winckler, M. (eds.) INTERACT 2013, Part III. LNCS, vol. 8119, pp. 406–423. Springer, Heidelberg (2013), http://kwarc.info/kohlhase/papers/Interact2013_FST.pdf

[SL04]   Stenmark, D., Lindgren, R.: Integrating knowledge management systems with everyday work: Design principles leveraging user practice. In: Proceedings of the 37th Annual Hawaii International Conference on System Sciences, 9 p. IEEE (2004)

[SS10]   Siorpaes, K., Simperl, E.: Human intelligence in the process of semantic content creation. World Wide Web 13(1-2), 33–59 (2010)

Towards Knowledge Management for HOL Light

Cezary Kaliszyk[1] and Florian Rabe[2]

[1] University of Innsbruck, Austria
[2] Jacobs University, Bremen, Germany

Abstract. The libraries of deduction systems are growing constantly, so much that knowledge management concerns are becoming increasingly urgent to address. However, due to time constraints and legacy design choices, there is barely any deduction system that can keep up with the MKM state of the art. HOL Light in particular was designed as a lightweight deduction system that systematically relegates most MKM aspects to external solutions — not even the list of theorems is stored by the HOL Light kernel.

We make the first and hardest step towards knowledge management for HOL Light: We provide a representation of the HOL Light library in a standard MKM format that preserves the logical semantics and notations but is independent of the system itself. This provides an interface layer at which independent MKM applications can be developed. Moreover, we develop two such applications as examples. We employ the MMT system and its interactive web browser to view and navigate the library. And we use the MathWebSearch system to obtain a search engine for it.

1 Introduction and Related Work

Deduction systems such as type checkers, proof assistants, or theorem provers have initially focused on soundness and efficiency. Consequently, many follow the LCF approach which amounts to a kernel that implements the logic, a theorem prover that produces input for the kernel, and a read-eval-print loop for user interaction. But over time formalization projects like [Hal05,GAA+13] have reached larger scales that call for more sophisticated *knowledge management* (KM) support.

However, it has proved non-trivial to retrofit KM support to an existing system. Moreover, even where it is possible, developers' resources are usually stretched already by improving and maintaining the kernel, the proof assistant, and performing actual formalizations. Therefore, a long-standing goal of MKM has been to provide generic KM support for deduction systems in a parametric way. This can create a valuable separation of concerns between deduction and KM systems.

We follow up on this by combining one of each: the HOL LIGHT [Har96] proof assistant and the MMT KM system [Rab13].

HOL LIGHT implements the HOL logic [Pit93], which is a variant of Church's simple type theory extended by shallow polymorphism. Polymorphism is achieved

S.M. Watt et al. (Eds.): CICM 2014, LNAI 8543, pp. 357–372, 2014.

by adding arbitrary type operators (to allow the construction of compound types), and (implicitly universally quantified) type variables. HOL LIGHT's implementation follows the LCF tradition: the kernel defines a private ML type for each of the HOL constructs. This means that the only way a user can construct HOL objects is with the help of exported kernel functions, which only allow the building of valid types, correctly-typed terms and provable theorems. The proofs of the theorems are *ephemeral*: they are only known at theorem construction time. To this end, the HOL LIGHT kernel maintains lists (ML references) of defined types, constants and axioms.

HOL LIGHT does not use its own syntax for the structure of proofs and libraries, instead inheriting that of OCAML (only changing OCAML's capitalization with the help of a camlp5 module). The only parser that HOL LIGHT defines is for the object level (inner syntax). The only library management feature is a special function used to obtain the list of all theorems from the OCAML toplevel [HZ], which is also used to offer an internal search mechanism based on term matching. This means that the HOL LIGHT library is hard to browse and search, especially for inexperienced users.

On the other hand, this lightness makes it relatively easy to export the basic structures manipulated internally by HOL LIGHT [Wie09], which can be combined with library information gathered by patching its internal functions and processing developments multiple times [KK13].

MMT [RK13] is a generic type theory and knowledge management system. It combines the general syntax of OPENMATH [BCC+04] and OMDOC [Koh06] with the formal semantics of logical frameworks like LF [HHP93] and ISABELLE [Pau94]. The MMT system [Rab13] implements both logical algorithms such as module system and type reconstruction as well as KM support such as IDE and change management.

MMT systematically avoids commitment to specific type systems or logics and focuses on using open interfaces and extensible algorithms. Thus, MMT-based implementations are maximally reusable, and MMT is a prime candidate for providing external KM support at extremely low cost.

Following our experiences with the representation of the Mizar library in MMT [IKRU13], we know that the export of a library in an KM-friendly interchange format is the biggest bottleneck: it is so much work to get a complete and hack-free export, that in practice none exist. Therefore a lot of interesting KM research is blocked and cannot be applied. Our export proceeds in two steps. Firstly, we define HOL LIGHT as a theory H of LF-theory, which in turn is implemented within the MMT framework. Secondly, we implement an exporter as a part of the HOL LIGHT system that writes the HOL LIGHT library as a set of MMT theories that extend H.

Our export includes type and constant definitions, theorems, and notations. For proofs, we use a simplified export, which stores only dependencies of a theorem but not the structure of the proof. This is reasonable because most of the structure has already been eliminated at the OCAML level anyway before the kernel is involved.

The exported theories are ready-to-use for KM systems. Apart from the fact that we skipped the proofs, they formally type-check within LF. And the OMDOC-based concrete syntax as well as the high level API of MMT make it easy to access them.

We exemplify and exploit this infrastructure by obtaining two major KM services for HOL LIGHT. Firstly, we use a native MMT service: an interactive library browser based on HTML+presentation MathML. Secondly, we apply a third-party service: the MathWebSearch [KŞ06] search engine.

Our work flow and infrastructure is not only interesting per se but also serves as a first test case for future multi- and cross-library knowledge management services. We discuss applying such ideas on the KM level in Section 5.

Related Work. Several previous exports of the HOL LIGHT library have focused on translations to other proof assistants. All such approaches use a patched kernel, that records the applications of all kernel theorem creation steps. [OS06] and [KW10] automatically patch the sources, replacing the applications of **prove** by recording of the theorem and its name. This has been further augmented in [KK13], which obtains the list of top-level theorems from the OCAML toplevel. [OAA13] has additionally focused on exporting the proof structure of HOL LIGHT proofs (to the extent it can be recovered), by patching tactics.

These approaches are orthogonal to ours: We focus on exporting the knowledge present in the library, keeping the proof structure transparent. In fact, this is the reason why we do not export the low-level proof structure at all (except for the calls to extension principles and the dependency information). We believe that future work can combine our export with the high level but more brittle exports à la [OAA13].

The OpenTheory project [Hur09] similarly patches the HOL LIGHT kernel in order to export theorems with proofs in the OpenTheory format. By manually annotating the library files, users can group theorems into theories. This refactoring is crucial for reuse across systems because it permits making dependencies between theories explicit so that they can be abstracted.

Several knowledge management tools have been developed for HOL LIGHT. [TKUG13a] export a HOL LIGHT development into the AGORA Wiki system. It heuristically HTMLizes/Wikifies the OCAML sources with anchors and references. The focus of the work is to present the informal Flyspeck text alongside with the formalization. Deconstructing compound tactics can be performed with the help of Tactician [AA12], and viewing recorded intermediate steps can be done with Proviola [TGMW10]. Online editing of HOL LIGHT proof scripts has been added [TKUG13b]. While the browsing capability is similar to the one presented here, AGORA does not look into the inner syntax of HOL LIGHT and displays almost unmodified original OCAML files (only folding of proofs is added). The heuristic HTMLization is able to pick up a number of HOL constants, but even some of the basic ones, like conjunction, are not linked to their definitions.

None of the above systems focus on connecting proof assistants to arbitrary MKM systems. Moreover, none of them used a logical framework to formally define and thus document the HOL LIGHT logic, instead focusing on an import that would match it with the intended target system.

2   Exporting the HOL Light Library

2.1   Defining the HOL Light Logic

MMT itself does not force the choice of a particular logical framework: Individual frameworks are defined as MMT plugins that provide the typing rules. We define the logic of HOL LIGHT declaratively in the logical framework LF. LF is a dependently-typed λ-calculus, and we will use the notations $[x:A]t$: $\{x:A\}B$ for $\lambda x : A.t : \Pi x : A.B$.

By using a logical framework, we can formalize and reason about the relation between different logics. This is a crucial step towards obtaining verified logic translations like in [NSM01], where the method of translation guarantees the correctness of the translated theorems. It also lets us reuse generic algorithms that depend on awareness of the logical primitives of HOL LIGHT. These include parsing, which we will use for search queries, and type inference, which we use as an interactive feature in the browser.

HOL LIGHT prides itself in a very small and simple to understand kernel, and its embedding in LF documents this nicely. Our definition is conceptually straightforward is mainly notable for systematically following HOL LIGHT down to the choice of identifier names.

The three basic OCAML datatypes to represent types, terms and theorems correspond to three LF types. The OCAML type `holtype` becomes an LF type, and LF variables of this type naturally represent HOL type variables. Initially two primitive type constructors are present – `bool` and `fun`:

```
holtype : type
bool : holtype
fun   : holtype → holtype → holtype                    "1 ⇒ 2"
```

Here, we also use MMT to define appropriate notations for all operators, e.g., 1 ⇒ 2 introduces the usual infix notation.

Next the type of HOL terms of a given type together with application and abstraction operators (denoted ' and λ respectively) are introduced. LF variables of the term type represent bound variables, or implicitly universally quantified free variables. The only primitive constant of HOL LIGHT (equality) is also specified here:

```
term  : holtype → type
Abs   : {A,B} (term A → term B) → term (A ⇒ B)         "λ 3"
Comb  : {A,B} term (A ⇒ B) → term A → term B           "3 ' 4"
equal : {A} term A ⇒ (A ⇒ bool)                        "2 = 3"
```

Note that the MMT notation language does not distinguish between LF and HOL arguments. For example, in the notation for application, 3 and 4 refer to the two HOL arguments; the two LF arguments A and B in positions 1 and 2 are defined to be implicit by not mentioning them. This is a novel feature that we get back to in Sect. 3.1.

The shallow polymorphism of HOL LIGHT corresponds exactly to the LF function space, which we can see in the declaration of equality: It takes one LF argument A for the HOL type and two HOL arguments of type A.

HOL LIGHT theorems are sequents $F_1, \ldots, F_n \vdash F$ using a type variable context $a_1, \ldots, a_n$. In LF, we use the judgments-as-types principle and represent the assumptions as an LF context. The above sequent corresponds to the type $\{a_1, \ldots, a_n\} \vdash F_1 \to \ldots \to \vdash F_n \to \vdash F$, and the valid proofs are exactly the terms of that type. The complete set of theorem construction rules is represented as:

```
thm      : term bool → type                              "⊢ 1"
REFL     : {A,X:term A} ⊢ X = X
TRANS    : {A,X,Y,Z:term A} ⊢ X = Y → ⊢ Y = Z → ⊢ X = Z
MP       : p,q ⊢ p = q → ⊢ p → ⊢ q
BETA     : {A,B,F:term A → term B,X:term A} ⊢ (λ F)'X = (F X)
MK_COMB  : {A,B, F,G:term A⇒B, X,Y:term A}
    ⊢ F = G → ⊢ X = Y → ⊢ F'X = G'Y
ABS      : {A,B, S,T:term A → term B}
    ({x: term A} ⊢ (S x) = (T x)) → ⊢ λ S = λ T
DEDUCT_ANTISYM_RULE : {p,q} (⊢ p → ⊢ q) → (⊢ q → ⊢ p) → ⊢ p = q
```

The above seven rules are the only ones needed for HOL LIGHT theorem construction steps. The three remaining ones (INST_TYPE, INST and ASSUME) are naturally represented by the contexts and substitutions of LF.

In [HKR12], we introduced the MMT feature of extension declarations. Moreover, in [HRK14], we introduced LFS, a variant of LF with arity polymorphism. If we apply extension declarations to LFS, we can also formalize the two extension principles of HOL LIGHT– definitions and type definitions:

```
extension definition =
    [n: nat] [A: holtypeⁿ → holtype] [a: {T: holtypeⁿ} term (A T)]
    c    : {T} term (A T)
    DEF  : {T} ⊢ (c T) = (a T)

extension new_basic_type_definition =
    [n:nat] [A: holtypeⁿ → holtype]
    [P: {T: holtypeⁿ} term (A T) ⇒ bool]
    [w: {T: holtypeⁿ} term (A T)]
    [nonempty: {T: holtypeⁿ} ⊢ (P T) ' (w T)]
    B        : holtypeⁿ → holtype
    abs      : {T} term (B T) ⇒ (A T)
    rep      : {T} term (A T) ⇒ (B T)
    rep_abs  : {T} {b: term (B T)} ⊢ (rep T) ' ((abs T) ' b) = b
    abs_rep  : {T} {a: term (A T)} ⊢
                    (P T) ' a = ((abs T) ' ((rep T) ' a) = a)
```

Here `nat` is the type of natural numbers. Arity polymorphism means that declarations may take a `nat` argument and use it to form different types. In our case, we use holtype^n to represent (essentially) n-tuples of `holtype`.

The `extension` keyword introduces declarations of named extension principles. Such a declaration λ-binds some arguments and then returns a list of declarations. The returned list is displayed using indentation.

The extension principle named `definition` takes an n-ary type constructor `A` and a polymorphic term `a` of that type. It returns a new constant `c` with an axiom `DEF` that makes `c` equal to `a`. The extension principle named `new_basic_type_definition` takes an n-ary type constructor `A` and a polymorphic unary predicate `P` on it. It returns a new n-ary type constructor `B` that is axiomatized to be isomorphic to the subtype of `A` defined by `P`. It also takes a polymorphic term `w` and a proof `nonempty` that `w` witnesses the non-emptiness of the new type.

2.2   Exporting the HOL Light Library

In order to export the HOL LIGHT library to MMT we need the information about the defined types, constants, theorems, and notations. The defined types and constants are stored by the kernel and the complete list can be accessed in any HOL LIGHT state together with the arity of types and the most general types of constants. The only mechanism for obtaining the defined theorems in HOL LIGHT is the `update_database` mechanism [HZ]. It accesses the internal OCAML data structures to extract name-value pairs for all defined values of the type `thm` in an arbitrary OCAML state.

Typically in order to export a library the above functionalities are invoked once [KK13], at the end of a development giving a complete list of values to export. A second processing of the developments is necessary to additionally discover the order in which the theorems are defined with the help of dependencies. This has been described in detail in [KU14] together with the splitting of theorems that are large conjunctions into separate named conjuncts.

Here we additionally preserve the separation of concepts into files. This means that we need to patch the HOL LIGHT theory loader to store a stack of names of loaded files and call the database functions at each file entry and exit. With this additional information, the second processing of the development can recover the division of concepts between files and properly order the types, constants and theorems in the files creating an ordered export list.

For each HOL LIGHT file in the export list, we produce an OMDoc [Koh06] file containing a single `theory`. This theory declares one `constant` for each exported theorem and several for each application of an extension principle. For every defined n-ary type constructor, an LF-constant of type $\text{holtype} \to \ldots \to \text{holtype} \to \text{holtype}$ is produced. For each HOL LIGHT constant of type `A`, an LF constant of type `term A` is produced. Similarly for each theorem asserting `F` (this includes axioms and the axioms generated by definitions and type definitions), a constant of type `thm F` is exported. In case of polymorphism, all free type variables `a` are universally quantified using {`a:holtype`}.

```
<constant name="PRE"><type>
  <om:OMOBJ xmlns:om="http://www.openmath.org/OpenMath"><om:OMA>
    <om:OMS module="LF" name="apply"></om:OMS>
    <om:OMS module="Kernel" name="term"></om:OMS>
    <om:OMA>
      <om:OMS module="LF" name="apply"></om:OMS>
      <om:OMS module="Kernel" name="fun"></om:OMS>
      <om:OMS module="nums" name="num"></om:OMS>
      <om:OMS module="nums" name="num"></om:OMS>
    </om:OMA>
  </om:OMA></om:OMOBJ>
</type></constant>
```

Fig. 1. OMDoc encoding of the HOL Light predecessor constant `PRE`

The writing of the particular types and terms is a straightforward transformation of the type and term structure of HOL into its corresponding MMT/LF tree, and we show only one example. Fig. 1 shows the export of the application of the `definition` extension principle to define the constant `PRE`.

For the defined types and constants, for which HOL Light provides notations, we additionally include the information about the fixity, precedence, and delimiter symbol. For the constants, which the HOL Light parser and printer treat in a special way (like `If` and `UNIV`), mixfix notations are exported. Details are described in Section 3.1.

Proofs in our encoding of the HOL Light logic are MMT definitions. However, we do not export the complete proof object and use the definition to store only dependency information. Then the definition is the application of a special constant (corresponding to an informal "by" operator) to all the theorems it depends on. This allows for mapping the HOL Light dependencies to MMT dependencies directly.

The results of some extension principles are not registered at the toplevel. Many definitions are internally introduced using the Hilbert operator ε in the definiens. Only the derived definitions — without ε — are registered at the toplevel. Complete dependency information is available for all theorems except for such derived definitions.

The complete OMDoc created for the core library of HOL Light repository version 182 consists of 22 files with the total of 2801 MMT constants. The constants correspond to 254 applications of extension principles and 2514 toplevel conjuncts. 95 notations are exported. The OMDoc encoding takes 75.2MB. On an IntelM 2.66 GHz CPU, it takes 6 minutes to prepare the export lists (this is mostly computation of dependencies) and 17 seconds to export the library. The exported library is hosted at `http://gl.mathhub.info/HOLLight/basic` as a part of the Open Archive of Formalizations.

3   A Library Browser for HOL Light

3.1   Notations

HOL LIGHT has no formal notation language that users could use to declare notations for new functions and type operators. But it does maintain three tables (for infix, unary prefix, and binding operators) that parser and printer use to convert between external and internal syntax. Our export represents these in terms of MMT's general notation language. Moreover, HOL LIGHT uses ad hoc notations for a few library symbols, most of which we can also map to MMT notation declarations.

Below we describe a few improvements we have made to MMT in order to better mimic and improve the notations of the HOL LIGHT library.

Nested Higher-Order Abstract Syntax. A major drawback of using a logical framework is the associated encoding overhead due to the higher-order abstract syntax (HOAS). This overhead does not necessarily cause performance issues, but it is much more work for theory and implementation, which is why current exports usually do not use it. This is particularly apparent when representing languages based on simple type theory like HOL LIGHT, which already use HOAS themselves.

For example, the export of the universal quantifier yields the following LF declaration:

```
! : {A} term ((A ⇒ bool) ⇒ bool)
```

Thus, we export the formula $!x.(p\,x)$ as ! Abs [x:term A] Comb p x, i.e., with two nested levels of binding (the LF-λ [] and the HOL-λ Abs), two nested levels of application (the LF-application whitespace and the HOL-application Comb), and two nested levels of type attributions (the LF-type attribution : and the HOL-type attribution term). Moreover, Abs and Comb take two LF-arguments for the involved HOL types, which we have not even shown here.

We now capture the complete absence of HOAS in human-facing syntax by ignoring HOAS in the MMT notation language. For example, we export the notation of forall as ! V2 . 3 where V2 represents the second argument (a bound variable), and 3 the third argument (the scope of the binding). (The unmentioned first argument is the type and remains implicit.) The fact that ! is a unary function for LF and a (different) unary function for HOL LIGHT is ignored and relegated to the type system.

We have added a generic component to MMT that uses these no-HOAS notations and systematically converts expressions between the three nesting levels of no, single, and double HOAS.

Unicode Notations. HOL LIGHT notations use keyboard-friendly multi-character delimiters such as /\ and ==> instead of ∧ and ⟹. We export all notations using the corresponding Unicode character, which is far superior for browsing.

However, to retain the ability to parse HOL LIGHT expressions with the MMT parser, we have added an optional lexing rule in MMT that converts multi- into single-character delimiters.

2-Dimensional Notations It is desirable to use MathML's 2-dimensional display (e.g., subscripts and fractions) for browsing. Therefore, we have extended MMT with a feature that permits declaring an optional second notation for every constant. If present, it is used for presentation only.

These notations amount to a text syntax for a small subset of MathML. Specifically, they are of the form

$$N ::= (arg \mid var \mid delim \mid implarg)^* \mid N \circ N \mid (N) \qquad \circ ::= \_ \mid \_\_ \mid \,\hat{}\, \mid \,\hat{}\,\hat{}\, \mid /$$

1-dimensional notations N consist of a sequence of *arg*ument positions (e.g., 3 above), *var*iable positions (e.g, V2 above), and *delim*iters (e.g., ! and . above). 2-dimensional notations may now additionally use subscript $\_$, superscript $\hat{}$, underscript $\_\_$ and overscript $\hat{}\,\hat{}$ as well as fraction $/$.

Moreover, we use *implarg* to explicitly position implicit arguments, which can be switched on or off interactively in the browser. For example, for the same-cardinality operator =_c that relates two sets of type A $\Rightarrow$ bool, we use the notation 2 = _ c ^ I1 3, which results in the display $s =_c^A t$ (with A being initially hidden).

Where possible, MMT constructs 2-dimensional notations automatically (e.g., HOL LIGHT's division a/b is rendered as a fraction). But in most cases, they have to be added manually because they are not a part of HOL LIGHT. Our export provides nine such 2-dimentional notations for division, power (including powersets) and cardinality compararison operations. An example rendering is shown in Fig. 2.

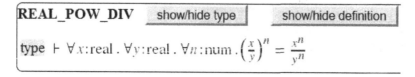

Fig. 2. Rendering Using 2-Dimensional Notations

3.2 Interactive HTML

MMT can produce static HTML pages, as well as serve these interactively. Fig. 3 shows an overview of the appearance of the HOL LIGHT library in the interactive web browser. A navigation bar on the left shows the files, and the main area shows the rendering of the respective declarations.

All formulas are presented by converting double-HOAS content MathML into no-HOAS presentation MathML using MMT's notation-based presentation engine. mrow elements are used to mark up the content structure and JavaScript

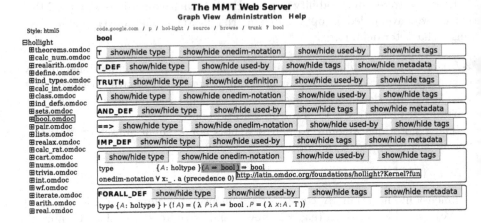

Fig. 3. The HOL LIGHT Theory Browser

to ensure that only full subterms can be selected. Every presentation element carries special attributes that identify which subterm it presents. This can be seen as a form of parallel markup between the presentation in the client and the content on the server.

For example, Fig. 4 uses this to infer the type of a subterm. Here, the parallel markup identifies the selected subterm in the double-HOAS representation so that the generic type inference of MMT can be used. After type inference, the result is rendered as no-HOAS presentation again.

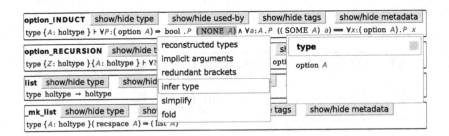

Fig. 4. Type Inference

Fig. 5 shows an example of the generic MMT parser in action. The exported notations are used to mimic HOL LIGHT's concrete input syntax and the result is rendered as presentation MathML. This is a very useful feature to integrate into a theory browser because it permits writing example expressions to better understand the browsed theory.

We also added a dependency-based navigation interface that is very convenient when working with a large library. Incoming dependencies (which theorems were used) are already explicit in our export and are shown as the definition of a

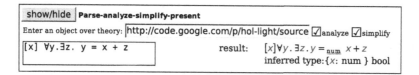

Fig. 5. Parser

theorem. Outgoing dependencies (where is this theorem used) are computed by MMT and added explicitly into the HTML. Both carry cross-references for navigation. This can be seen in Fig. 6.

| LENGTH_APPEND | show/hide type | show/hide definition | show/hide used-by | show/hide tags | show/hide me |

definition « using list_INDUCT , APPEND_conjunct0 , LENGTH_conjunct0 , ADD_CLAUSES_conjunct0 , APPEND_conjunct1 , LE
used by   APPEND_EQ_NIL

| MAP_APPEND | show/hide type | show/hide definition | show/hide used-by | show/hide tags | show/hide metada |

definition « using list_INDUCT , MAP_conjunct0 , APPEND_conjunct0 , MAP_conjunct1 , APPEND_conjunct1 , REFL_CLAUSE ,
used by   MAP_REVERSE

Fig. 6. Outgoing and Incoming Dependencies

4   Searching the HOL Light Library

MMT already includes a build tool for easily exporting content in different formats. The latter include building a term index that can be read by the Math-WebSearch system [KŞ06], which then constructs a substitution tree index of all terms in the library.

We have already used this for Mizar in [IKRU13], and we have now substantially expanded MMT with the ability to also query MathWebSearch. This is harder than it sounds because it requires constructing queries by parsing user input and infer all implicit arguments (or replace them with additional query variables). We have also added a corresponding frontend to the library browser.

Our new MMT search interface also allows queries based on other aspects than substitution, in particular we can use regular expression queries on identifiers. This is a seemingly minor feature that goes a long way in large libraries where users typically develop sophisticated naming schemes for theorems. For example the Coq-developments in [GAA+13] append letters to theorem names as a kind of tagging mechanism.

Fig. 7 shows an example query. We search for expressions x MOD p = y MOD p for arbitrary x,y,p, and we filter the results to be from a theory whose name contains arith.

It is straightforward to add other search criteria, such as filtering to theorems whose proof uses a certain theorem, or only returning results that match on toplevel.

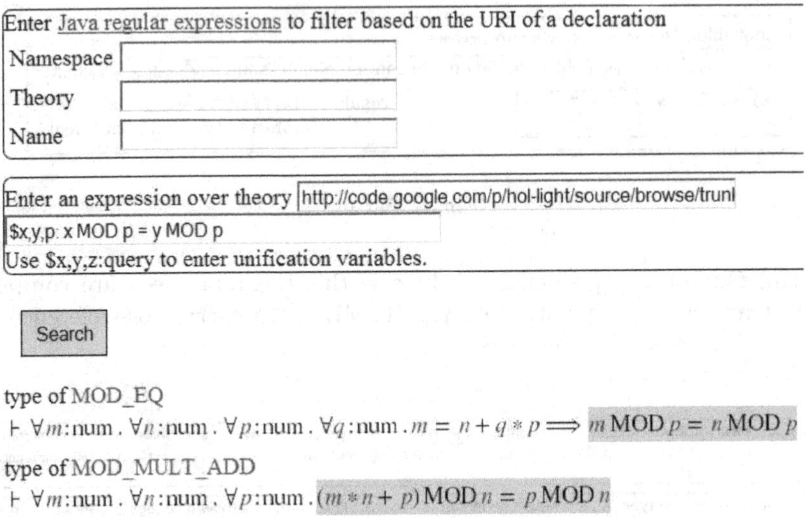

Enter Java regular expressions to filter based on the URI of a declaration

Namespace	
Theory	
Name	

Enter an expression over theory | http://code.google.com/p/hol-light/source/browse/trunl

$x,y,p: x MOD p = y MOD p
Use $x,y,z:query to enter unification variables.

Search

type of MOD_EQ
$\vdash \forall m:\text{num} . \forall n:\text{num} . \forall p:\text{num} . \forall q:\text{num} . m = n + q * p \implies m \,\text{MOD}\, p = n \,\text{MOD}\, p$

type of MOD_MULT_ADD
$\vdash \forall m:\text{num} . \forall n:\text{num} . \forall p:\text{num} . (m * n + p) \,\text{MOD}\, n = p \,\text{MOD}\, n$

Fig. 7. Search

5   Conclusion and Future Work

We have demonstrated that generic MKM services – in our case: browsing and search – can be utilized for libraries of proof assistants. However, this is contingent upon a good export of the library, which requires a substantial investment.

Each export requires two parts: A modification of the proof assistant that enables exporting the high-level structure needed by the MKM services; and a representation of the logic and library in an MMT-like interchange language. Our work constitutes the second step and was enabled by previous work providing the first step. Even though HOL LIGHT is one the simplest proof assistants in this respect, the latter goes all the way back to [OS06] and has only become available recently.

Therefore, we believe a major value of our work is in describing the data flow and the architecture necessary to connect deduction and knowledge management systems, and to kick off future work that applies the same approach to other systems.

An Open Archive of Formalizations. In a collaboration with Michael Kohlhase, the second author has initiated a project of collecting the libraries of proof assistants using MMT as the common representation language. After representing Mizar [IKRU13], the present work is the second library present in the OAF. We will continue this effort and develop it into a central hub for publishing and sharing libraries.

MathWiki. This serves as a next step towards a universal mathematical library that integrates the libraries of various proof assistants with each other and with the semi-formal proofs done by mathematicians initiated in [CK07].

The idea of wiki-style editing provided both for formal and informal text allows to identify overlaps and transport ideas. It can also be extended with human-annotated or machine-generated correspondences between concepts in formal libraries [GK14] in order to navigate across libraries. Semantic annotations, both human- and machine-generated, enhance the search capabilities, which can be further strengthened by learning-assisted automated reasoning techniques [KU14].

Library Refactoring. The axiomatic method encapsulates definitions and theorems in the smallest theory in which they make sense. This abstraction from the prerequisites has the advantage that results can be moved easily between theories and logics, thus maximizing reuse. This is a cornerstone of the mathematical method, and it has motivated formalized mathematics early on. For example, it is at the center of IMPS [FGT93] and motivated Reynolds' seminal work [Rey83].

Yet, HOL LIGHT and most other systems with large libraries use the definitional method that conservatively extends a single theory. We expect a rising interest in refactoring libraries according to the axiomatic method. For HOL LIGHT, we already see the beginnings in the systematic manual refactoring of the OpenTheory project [Hur09], which can inspire and evaluate future automated refactoring methods.

As a first step for example, we can introduce a new theory for every type definition and turn all theorems derived from the extension principle into axioms. Currently this is tricky for HOL LIGHT though because we can only tell if a theorem depends on *some* extension principle but not on which one. A clustering analysis of all theorems can provide further pointers to automated refactoring. Similarity analysis across libraries may also prove helpful to spot overlap, which is typically worthy of being refactored into a separate theory.

More generally, future work will develop logic-independent refactoring tools that can be applied generically to large definitional libraries. Exports like ours that represent libraries in universal formats while preserving the structure and semantics are the crucial prerequisite to apply such refactoring tools.

Library Integration The definitional method is also the reason why the integration of libraries across logics has so far been extremely difficult. Moving a definition d from one library to another is not desirable if (as is typical) the target library already uses a different definition d' for the same concept c. In those cases, we can abstract from the definitions and translate between libraries via partial theory morphisms.

Here the partiality of a morphism stems from dropping d, which permits translating c to its counterpart. This requires a dependency analysis because dropping d invalidates all results that depended on d. These must be dropped as well or reestablished in the target system.

Theoretically, this integration approach has been suggested in [RKS11]. Independently, its feasibility has been demonstrated in [KK13], which amounts to giving ad hoc partial morphisms that translate the HOL LIGHT library into the Isabelle/HOL library.

Future work can systematically write and verify these morphisms in MMT. This will yield verified integration functions even without reproving the translated theorems in the target system.

Acknowledgments. This work has been partially supported by the Austrian Science Fund (FWF): P26201.

References

AA12.   Adams, M., Aspinall, D.: Recording and refactoring HOL Light tactic proofs. In: Proceedings of the IJCAR Workshop on Automated Theory Exploration (2012), http://homepages.inf.ed.ac.uk/smaill/atxwing/atx2012_submission_9.pdf

BCC⁺04.   Buswell, S., Caprotti, O., Carlisle, D., Dewar, M., Gaetano, M., Kohlhase, M.: The Open Math Standard, Version 2.0. Technical report, The Open Math Society (2004), http://www.openmath.org/standard/om20

CK07.   Corbineau, P., Kaliszyk, C.: Cooperative repositories for formal proofs. In: Kauers, M., Kerber, M., Miner, R., Windsteiger, W. (eds.) MKM/Calculemus 2007. LNCS (LNAI), vol. 4573, pp. 221–234. Springer, Heidelberg (2007)

FGT93.   Farmer, W., Guttman, J., Thayer, F.: IMPS: An Interactive Mathematical Proof System. Journal of Automated Reasoning 11(2), 213–248 (1993)

GAA⁺13.   Gonthier, G., et al.: A Machine-Checked Proof of the Odd Order Theorem. In: Blazy, S., Paulin-Mohring, C., Pichardie, D. (eds.) ITP 2013. LNCS, vol. 7998, pp. 163–179. Springer, Heidelberg (2013)

GK14.   Gauthier, T., Kaliszyk, C.: Matching concepts across HOL libraries. In: Watt, S.M., Davenport, J.H., Sexton, A.P., Sojka, P., Urban, J. (eds.) CICM 2014. LNCS, vol. 8543, pp. 270–284. Springer, Heidelberg (2014)

Hal05.   Hales, T.: Introduction to the Flyspeck Project. In: Coquand, T., Lombardi, H., Roy, M. (eds.) Mathematics, Algorithms, Proofs. Internationales Begegnungs- und Forschungszentrum für Informatik (IBFI), Schloss Dagstuhl, Germany (2005)

Har96.   Harrison, J.: HOL Light: A Tutorial Introduction. In: Srivas, M., Camilleri, A. (eds.) FMCAD 1996. LNCS, vol. 1166, pp. 265–269. Springer, Heidelberg (1996)

HHP93.   Harper, R., Honsell, F., Plotkin, G.: A framework for defining logics. Journal of the Association for Computing Machinery 40(1), 143–184 (1993)

HKR12.   Horozal, F., Kohlhase, M., Rabe, F.: Extending MKM Formats at the Statement Level. In: Jeuring, J., Campbell, J.A., Carette, J., Dos Reis, G., Sojka, P., Wenzel, M., Sorge, V. (eds.) CICM 2012. LNCS (LNAI), vol. 7362, pp. 65–80. Springer, Heidelberg (2012)

HRK14.   Horozal, F., Rabe, F., Kohlhase, M.: Flexary Operators for Formalized Mathematics. In: Watt, S.M., Davenport, J.H., Sexton, A.P., Sojka, P., Urban, J. (eds.) CICM 2014. LNCS, vol. 8543, pp. 315–330. Springer, Heidelberg (2014)

Hur09.   Hurd, J.: OpenTheory: Package Management for Higher Order Logic The-
          ories. In: Reis, G.D., Théry, L. (eds.) Programming Languages for Mecha-
          nized Mathematics Systems, pp. 31–37. ACM (2009)
HZ.   Harrison, J., Zumkeller, R.: update_database module. Part of the HOL
          Light distribution
IKRU13.   Iancu, M., Kohlhase, M., Rabe, F., Urban, J.: The Mizar Mathematical
          Library in OMDoc: Translation and Applications. Journal of Automated
          Reasoning 50(2), 191–202 (2013)
KK13.   Kaliszyk, C., Krauss, A.: Scalable LCF-style proof translation. In: Blazy,
          S., Paulin-Mohring, C., Pichardie, D. (eds.) ITP 2013. LNCS, vol. 7998,
          pp. 51–66. Springer, Heidelberg (2013)
Koh06.   Kohlhase, M.: OMDoc – An Open Markup Format for Mathematical Doc-
          uments [version 1.2]. LNCS (LNAI), vol. 4180. Springer, Heidelberg (2006)
KŞ06.   Kohlhase, M., Sucan, I.: A Search Engine for Mathematical Formulae. In:
          Calmet, J., Ida, T., Wang, D. (eds.) AISC 2006. LNCS (LNAI), vol. 4120,
          pp. 241–253. Springer, Heidelberg (2006)
KU14.   Kaliszyk, C., Urban, J.: Learning-assisted automated reason-
          ing with Flyspeck. Journal of Automated Reasoning (2014),
          http://dx.doi.org/10.1007/s10817-014-9303-3
KW10.   Keller, C., Werner, B.: Importing HOL Light into Coq. In: Kaufmann, M.,
          Paulson, L.C. (eds.) ITP 2010. LNCS, vol. 6172, pp. 307–322. Springer,
          Heidelberg (2010)
NSM01.   Naumov, P., Stehr, M.-O., Meseguer, J.: The HOL/NuPRL proof trans-
          lator - A practical approach to formal interoperability. In: Boulton,
          R.J., Jackson, P.B. (eds.) TPHOLs 2001. LNCS, vol. 2152, pp. 329–345.
          Springer, Heidelberg (2001)
OAA13.   Obua, S., Adams, M., Aspinall, D.: Capturing hiproofs in HOL light. In:
          Carette, J., Aspinall, D., Lange, C., Sojka, P., Windsteiger, W. (eds.)
          CICM 2013. LNCS (LNAI), vol. 7961, pp. 184–199. Springer, Heidelberg
          (2013)
OS06.   Obua, S., Skalberg, S.: Importing HOL into Isabelle/HOL. In: Furbach,
          U., Shankar, N. (eds.) IJCAR 2006. LNCS (LNAI), vol. 4130, pp. 298–302.
          Springer, Heidelberg (2006)
Pau94.   Paulson, L.C.: Isabelle: A Generic Theorem Prover. LNCS, vol. 828.
          Springer, Heidelberg (1994)
Pit93.   Pitts, A.: The HOL logic. In: Gordon, M.J.C., Melham, T.F. (eds.) Intro-
          duction to HOL: A Theorem Proving Environment for Higher Order Logic.
          Cambridge University Press (1993)
Rab13.   Rabe, F.: The MMT API: A Generic MKM System. In: Carette, J., As-
          pinall, D., Lange, C., Sojka, P., Windsteiger, W. (eds.) CICM 2013. LNCS
          (LNAI), vol. 7961, pp. 339–343. Springer, Heidelberg (2013)
Rey83.   Reynolds, J.: Types, Abstraction, and Parametric Polymorphism. In: In-
          formation Processing, pp. 513–523. North-Holland, Amsterdam (1983)
RK13.   Rabe, F., Kohlhase, M.: A Scalable Module System. Information and Com-
          putation 230(1), 1–54 (2013)
RKS11.   Rabe, F., Kohlhase, M., Sacerdoti Coen, C.: A Foundational View on In-
          tegration Problems. In: Davenport, J.H., Farmer, W.M., Urban, J., Rabe,
          F. (eds.) Calculemus/MKM 2011. LNCS (LNAI), vol. 6824, pp. 107–122.
          Springer, Heidelberg (2011)

TGMW10.   Tankink, C., Geuvers, H., McKinna, J., Wiedijk, F.: Proviola: A tool for proof re-animation. In: Autexier, S., Calmet, J., Delahaye, D., Ion, P.D.F., Rideau, L., Rioboo, R., Sexton, A.P. (eds.) AISC/Calculemus/MKM 2010. LNCS (LNAI), vol. 6167, pp. 440–454. Springer, Heidelberg (2010)

TKUG13a. Tankink, C., Kaliszyk, C., Urban, J., Geuvers, H.: Communicating formal proofs: The case of Flyspeck. In: Blazy, S., Paulin-Mohring, C., Pichardie, D. (eds.) ITP 2013. LNCS, vol. 7998, pp. 451–456. Springer, Heidelberg (2013)

TKUG13b. Tankink, C., Kaliszyk, C., Urban, J., Geuvers, H.: Formal mathematics on display: A wiki for Flyspeck. In: Carette, J., Aspinall, D., Lange, C., Sojka, P., Windsteiger, W. (eds.) CICM 2013. LNCS (LNAI), vol. 7961, pp. 152–167. Springer, Heidelberg (2013)

Wie09.   Wiedijk, F.: Stateless HOL. In: Hirschowitz, T. (ed.) TYPES. EPTCS, vol. 53, pp. 47–61 (2009)

Automated Improving of Proof Legibility in the Mizar System*

Karol Pąk

Institute of Computer Science,
University of Bialystok, Poland
pakkarol@uwb.edu.pl

Abstract. Both easily readable and obscure proof scripts can be found in the bodies of formalisations around formal proof checking environments such as Mizar. The communities that use this system try to encourage writing legible texts by making available various solutions, e.g., by introduction of phrases and constructs that make formal deductions look closer to the informal ones. Still, many authors do not want to invest additional efforts in enhancing readability of their scripts and assume this can be handled automatically for them. Therefore, it is desirable to create a tool that can automatically improve legibility of proofs. It turns out that this goal is non-trivial since improving features of text that enhance legibility is in general NP-complete.

The successful application of SMT technology to solving computationally difficult problems suggests that available SMT solvers can give progress in legibility enhancement. In this paper we present the first experimental results obtained with automated legibility improving tools for the Mizar system that use Z3 solver in the backend.

Keywords: Operations on languages, Legibility of proofs, Proof assistants, SMT solvers.

1 Introduction

1.1 Motivations

Analysing examples of declarative natural deduction proof scripts, especially long and complicated ones, it can be observed that the proofs are often formulated in a chaotic way. During the analysis of formalised proofs, we can obviously find that some authors of proof scripts spend a lot of time on their readability, but there are other ones who tend to create deductions that are correct for computers, neglecting the legibility of their proof scripts. They believe that no one, with the exception of a proof checker, will want to analyse them. The experience of big proof development efforts [8] shows that adapting or modifying the existing proofs is unavoidable and requires reading of proof scripts [10]. Additionally, any

---
* The paper has been financed by the resources of the Polish National Science Center granted by decision n°DEC-2012/07/N/ST6/02147.

S.M. Watt et al. (Eds.): CICM 2014, LNAI 8543, pp. 373–387, 2014.

attempt to analyse the details of the proof scripts created in this way, according to the opinion of some proof writers, is extremely difficult or even impossible.

The problem of formal deductions illegibility, which at first sight does not seem to be difficult, led to the abandonment or collapse of many formalisation projects such as these carried by Bourbaki or Whitehead and Russell [28]. The formalisation has become feasible only after the availability of computers [29] increased, but even now it is still quite difficult. Therefore, it comes as no surprise that communities around formal proof checking environments such as Mizar [18] and Isabelle/Isar [27] attempt to resolve this problem and often implement various solutions aimed at raising the legibility of proof scripts. Adaptation of informal mathematical language constructs to formal ones is the most natural direction for research. Many of these implemented solutions [14,26] are also known from the programming language editor frameworks, where they proven their usefulness (B. A. Myers shows that syntax highlighting or hint system in programming languages can save up to 35% of time spent for search in the code that is supposed to be modified [13]). These efforts are also focused on the visualisation of proof scripts in the linked HTML form [25].

1.2   Proposed Approach

In this paper we focus on another, still underdeveloped approach that concentrates on modification of order between independent steps written in a proof script. Based on a result of a long experience with the Mizar system and development of Mizar Mathematical Library (MML) [19], an analysis of different opinions shared by users of MML as well as models of human cognitive perception of read material, we can conclude that there is a close relationship between the reasoning legibility and grouping of steps into linear, directly connected fragments. Naturally, legibility has different meanings for different people, but in general the results obtained in this research complies with the expectations. This research precisely demonstrates how important for human being "local deductions" are in the analysing process of mathematical proofs.

This can be summarised with the Behaghel's First Law that *elements that belong close together intellectually should be placed close together* [1]. This law is also recognised in modern scientific literature concerning human perception [15]. Note that every information used in justification of a step had to be derived before in the proof, but often we can manipulate the location of these premises in deduction. With Behaghel's law in mind, we assume that a step where at least part of required information is available in close neighborhood of the step is more intelligible than a step in which all used information is far away in the proof scripts. Obviously, close neighborhood can be defined in different ways, therefore we parametrise this notion with a fixed number n that counts the number of preceding steps that are considered to be our close neighborhood.

An important case emerges for $n = 1$. In this case the general method of referencing by means of labels can be replaced in the Mizar system by the **then** construction. Additionally, in case where every reference to a step can be replaced by the **then** construction, the label attached to this step in unnecessary and can be removed from proof scripts. In consequence, the number of labels in the proof script can be brought smaller. Note that the human short-term memory is the capacity for holding a small amount of information in mind in an active state, and its capacity is 7 ± 2 elements [3,17]. However, this capacity is prone to training [5], and in the case of research on Mizar it has been shown that it is in the range 5-10 [16]. Therefore, it comes as no surprise that we want to choose an ordering of steps that maximises the number of references in close neighbourhoods of reference targets, where we understand by a close neighbourhood proof steps that are supposed to be directly available from within the short-term memory.

In this paper we present the first experimental results obtained with auto-mated legibility improving tools for Mizar that aim at realisation of the above-described legibility criteria. Since improving legibility in most cases in NP-hard [22], we also present the results obtained with the application of a SMT-solver Z3 [4]. The impact of these criteria and the other ones has been already recognised by the scientific community of people who write proof scripts in Mizar [20,21] and in other systems [2,12,24].

In Section 2 we introduce the notion of an abstract model of proofs and its linearisation. In Section 3 we discuss selected methods to improve legibility and we show the complexity of two so far open cases, transforming to them known NP-complete problems. In Section 4 we discuss translations of legibility problems to the Z3-solver and application of its responses to proof scripts. Then in Section 5 we report the statistical results obtained for the MML database. Finally, Section 6 concludes the paper and discusses the future work.

2   Graph Representation of Proofs

Mizar [19] is a mathematical language and a proof checker for the language that are based on natural deduction created by S. Jakowski and F. B. Fitch [6,11]. In this paper, we do not concentrate on a full explanation of Mizar. We will just focus on a few easy-to-observe Mizar proof structure features presented in an example (Fig. 1), where the statement *a square matrix is invertible if and only if it has non-zero determinant* is proved. This example is contained in an article by Pąk and Trybulec [23].

Note that symbols 0_K, 1_K represent the additive identity element and the multiplicative identity element of a field K, respectively. Additionally, the Mizar system uses only ASCII characters, therefore operations such as $^{-1}$, $\cdot$, $(\cdot)^T$ are represented in Mizar as ", *, @ respectively, the labels MATRIX_6, MATRIX_7, MATRIX11 are identifiers of Mizar articles. An abstract model of the proof scripts (see Fig. 2) was considered in detail in [20], but for our purposes we detail only a sketch of its construction, focusing mainly on one-level deductions that ignore nested lemmas. Generally we call a DAG $P = \langle V, O \cup M \rangle$ *abstract proof graph* if O, M are disjoint families of arcs, called *ordered arcs* and *meta-edges* respectively, and O contains a distinguished set of arcs $\mathfrak{R}(P) \subseteq O$, the elements of which are called *references*. The vertices of P represent reasoning steps, ordered arcs represent all kinds of additional constraints that force one step to precede another

```
1:    theorem Th34:
          for n be Nat, M be Matrix of n,K st n >= 1 holds
              M is invertible iff Det M <> 0_K
          proof
2:            let n be Nat, M be Matrix of n,K such that
      A1:       n >= 1;
3:            thus M is invertible implies Det M <> 0_K
              proof
4:                assume M is invertible;
5:                then consider Minv be Matrix of n,K such that
      A2:             M is_reverse_of Minv by MATRIX_6:def 3;
6:    A3:         M · Minv = 1.(K,n) by A2,MATRIX_6:def 2;
7:                Det 1.(K,n) = 1_K by A1,MATRIX_7:16;
8:                then Det M · Det Minv = 1_K by A1,A3,MATRIX11:62;
9:                thus then Det M <> 0_K;
              end;
10:           assume
      A4:       Det M <> 0_K;
11:           then
      A5:       M · ( (Det M)^{-1} · (Matrix_of_Cofactor M)^T ) = 1.(K,n) by Th30;
12:               (Det M)^{-1} · (Matrix_of_Cofactor M)^T · M = 1.(K,n) by A4,Th33;
13:           then M is_reverse_of (Det M)^{-1} · (Matrix_of_Cofactor M)
                  by A5,MATRIX_6:def 2;
14:           thus then thesis by MATRIX_6:def 3;
          end;
```

Fig. 1. The example of proof script written in the Mizar style

one, and meta-edges represent dependence between a step that as a justification contains nested reasoning and each step of this reasoning. In Fig. 2 arrows $\twoheadrightarrow$ represent meta-edges. Elements of $\mathfrak{R}(P) \subseteq O$ correspond to solid arrows and represent the information flow between a step (the head of the arc, e.g., the vertex 8) that use in the justification premises formulated in a previously justified step (the tail of the arc, e.g., vertices 2, 6, 7). Other ordered arcs correspond to dashed arrows and represent, e.g., the dependence between a step that introduces a variable into the reasoning and a step that uses this variable in an expression (e.g., the arc $\langle 5, 6 \rangle$ that corresponds to the use of the variable Minv). Ordered arcs represent also the order of special kind reasoning steps in the Jaśkowski-style natural-deduction proofs (for more detail see [9]) such as steps that introduce quantifier or implication; or indicate a conclusion (e.g., arcs $\langle 2, 3 \rangle$, $\langle 4, 9 \rangle$). Clearly, every abstract proof graph does not contain labeled vertices and multiple arcs. Labels and multiple arcs visible in Fig. 2 have been used here only to aid the reader and simplify their identification.

It is easily seen that digraph $\langle V, M \rangle$ is a forest, i.e., a disjointed union of trees, in which every connected maximal tree is an arborescence (i.e., a rooted tree where all arcs are directed from leaves to the root). Additionally, using the notion of meta-edges we can define formal equivalent of one-level deductions.

Definition 1. *Let $P = \langle V, O \cup M \rangle$ be an abstract proof graph and D be a subgraph of P induced by a set of vertices. We call D one-level deduction, if D is induced by the set of all roots in the forest $\langle V, M \rangle$ or it is induced by $N^-_{\langle V,M \rangle}(v)$ for some vertex $v \in V$.*

Naturally, one-level deductions do not have meta-edge. Consequently, focusing only on such deductions results in a simplified model of proof graphs. Negative

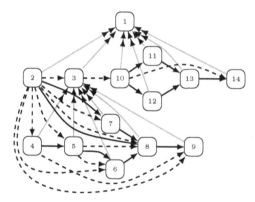

Fig. 2. The abstract proof graph illustrating the structure of reasoning presented in Fig. 1

consequence of this simplification is that hidden dependencies between steps in one-level deductions may occur. Therefore, we have to carefully add such dependencies to these digraphs. These dependencies was considered in detail in [22]. Here we remind only that hidden dependency occur between vertices v, u if

(i)   there exists a common one-level deduction that contains v and u,
(ii) the step that corresponds to u is justified by a nesting lemma,
(iii) a step s of this nesting lemma uses in the expression a variable that is introduced in a step $\overline{v}$ or the justification of s refers to the statement formulated in $\overline{v}$, where $\overline{v}$ corresponds to v.

Additionally in the case, where the justification of s refers to the statement formulated in $\overline{v}$ we call $\langle u, v \rangle$ *extended reference*.

Let $D = \langle V_D, O_D \rangle$ be a one-level deduction of P. To simplify we assume that the set of ordered arcs of D contains also every hidden dependency between vertices of V_D, and $\mathfrak{R}(D)$ contains only original references of P that connect vertices of V_D. The set of reference and extended reference arcs of G is denoted by $\overline{\mathfrak{R}}(D)$.

To study the general case, without the Mizar context, we assume only relation between distinguished sets of arcs in D that $\mathfrak{R}(D) \subseteq \overline{\mathfrak{R}}(D) \subseteq O_D$. Therefore, in the following considerations as a one-level deduction we take a DAG $\mathcal{D} = \langle \mathcal{V}, \mathcal{A} \rangle$, with two distinguished sets $\mathcal{A}_1 \subseteq \mathcal{A}_2 \subseteq \mathcal{A}$ that correspond to $\mathfrak{R}(\mathcal{D})$ and $\overline{\mathfrak{R}}(\mathcal{D})$, respectively. For simplicity, we assume also that $\mathcal{A}_1$-references can be replaced in the Mizar system by the **then** construction.

We identify here a modification of independent steps order to improve proof legibility with a modification of a topological sorting of $\mathcal{D}$. By *topological sorting*, called also *linearisation*, we mean a one-to-one function $\sigma : \mathcal{V} \to \{1, 2, \ldots |\mathcal{V}|\}$ such that $\sigma(u) < \sigma(v)$ for each arc $\langle u, v \rangle \in \mathcal{A}$. We denote by $TS(\mathcal{D})$ the set of all topological sortings. Let us consider $\sigma \in TS(\mathcal{D})$. We call a vertex v of $\mathcal{V}$ a $\textbf{then}^{\mathcal{A}_1}(\sigma)$-*step* if v corresponds to a vertex that is linked by $\mathcal{A}_1$ to the directly preceding step in the linearisation σ (e.g., steps 5, 8, 9 in Fig. 1) or more precisely:

$$v \in \mathbf{then}^{A_1}(\sigma) \iff \sigma(v) \neq 1 \wedge \langle \sigma^{-1}(\sigma(v)-1), v \rangle \in A_1. \tag{1}$$

We call a directed path $p = \langle p_0, p_1, \ldots p_n \rangle$ of $\mathcal{D}$ a σ^{A_1}–*linear reasoning* if p_k is a $\mathbf{then}^{A_1}(\sigma)$–step (i.e., $\sigma(p_k) = 1 + \sigma(p_{k-1})$ and $\langle p_{k-1}, p_k \rangle \in A_1$) for each $k = 1, 2, \ldots, n$ (p_0 does not have to be a $\mathbf{then}^{A_1}(\sigma)$–step). A σ^{A_1}–linear reasoning P is *maximal* if it is not a subsequence of any other σ^{A_1}–linear reasoning. In our considerations we also use a function that maps vertices of consecutive maximal σ^{A_1}–linear reasoning in linearisation σ to consecutive natural numbers. Let $\mathfrak{Then} : TS(\mathcal{D}) \times 2^{\mathcal{A}} \times \mathcal{V} \to \{1, 2, \ldots, |\mathcal{V}|\}$ be given by the following recursive definition:

$$\mathfrak{Then}_{\sigma}^{A_1}(v) = \begin{cases} 1 & \text{if } \sigma(v) = 1, \\ \mathfrak{Then}_{\sigma}^{A_1}(\sigma^{-1}(\sigma(v)-1)) & \text{if } v \in \mathbf{then}^{A_1}(\sigma), \\ \mathfrak{Then}_{\sigma}^{A_1}(\sigma^{-1}(\sigma(v)-1)) + 1 & \text{if } v \notin \mathbf{then}^{A_1}(\sigma), \end{cases} \tag{2}$$

for an arbitrary $v \in \mathcal{V}$, $A_1 \subseteq \mathcal{A}$, and $\sigma \in TS(\mathcal{D})$.

Now we define a formal equivalent of a label in proof graph formalism. We will say that a vertex v of $\mathcal{V}$ has to have a label or is a *labeled vertex* in linearisation σ if v is the tail of at most one $\mathcal{A}_2 \setminus \mathcal{A}_1$-arc (e.g., steps 2 in Fig. 1) or is the tail of at most one $\mathcal{A}_1$-arc that corresponds to a link which does not connect two steps located directly one after the other in the linearisation σ (e.g., step 6 in Fig. 1). We write $\mathbf{lab}^{A_1, A_2}(\sigma)$ for the set of all labeled vertices given by

$$v \in \mathbf{lab}^{A_1, A_2}(\sigma) \iff \underset{u \in \mathcal{V}}{\exists} \left(\langle v, u \rangle \in \mathcal{A}_2 \setminus \mathcal{A}_1 \vee (\langle v, u \rangle \in \mathcal{A}_1 \wedge \sigma(v) + 1 \neq \sigma(u)) \right) \tag{3}$$

where $v \in \mathcal{V}$. We define also a function $\mathfrak{Lab} : TS(\mathcal{D}) \times 2^{\mathcal{A}} \times 2^{\mathcal{A}} \times \mathcal{V} \to \{1, 2, \ldots, |\mathcal{V}|\}$ that associates the number of all labeled vertices u such that $\sigma(u) < \sigma(v)$, with every vertex $v \in \mathcal{V}$, defined as:

$$\mathfrak{Lab}_{\sigma}^{A_1, A_2}(v) = \begin{cases} 0 & \text{if } \sigma(v) = 1, \\ \mathfrak{Lab}_{\sigma}^{A_1, A_2}(\sigma^{-1}(\sigma(v)-1)) & \text{if } \sigma(v) \neq 1 \wedge v \notin \mathbf{lab}^{A_1, A_2}(\sigma), \\ \mathfrak{Lab}_{\sigma}^{A_1, A_2}(\sigma^{-1}(\sigma(v)-1)) + 1 & \text{if } \sigma(v) \neq 1 \wedge v \in \mathbf{lab}^{A_1, A_2}(\sigma). \end{cases} \tag{4}$$

At last we define a metric $\mathrm{d}_{\sigma} : \mathcal{V} \times \mathcal{V} \mapsto \mathbb{N}$ for a linearisation σ, which is called σ–distance and is determined by $\mathrm{d}_{\sigma}(v, u) = |\sigma(v) - \sigma(u)|$ for each $v, u \in \mathcal{V}$.

3 Methods of Improving Legibility

According to the approach presented in Section 1.2 we can formally define criteria of reasoning linearisation that quantify the legibility of obtained proof scripts. Since steps that refer to the preceding reasoning statements are perceived as more intelligible, we expect that the set $\mathbf{then}^{A_1}(\sigma)$ has the largest cardinality for the selected linearisation σ (1st MIL). Obviously, the length of σ–linear reasoning is also important for proof readers, therefore the average length of σ–linear reasoning should also have the maximal value. Note that optimisation of this determinant is realised by 1st MIL defined below, since such average length is equal to $\frac{|V_G|}{|V_G| - |\mathbf{then}^{A_1}(\sigma)|}$.

The 1st Method of Improving Legibility (1st MIL):
INSTANCE: A DAG $\mathcal{D} = \langle \mathcal{V}, \mathcal{A} \rangle$, a subset $\mathcal{A}_1$ of $\mathcal{A}$, a positive integer $K \leq |\mathcal{V}|$.
QUESTION: Does there exist a topological sorting σ of $\mathcal{D}$ for which $|\mathbf{then}^{\mathcal{A}_1}(\sigma)| \geq K$?

In addition we consider a parameterised version of the above-mentioned problem, since a close neighborhood of a step in proof scripts can be extended from the directly preceding step to last n preceding steps.

The 1st Method of Improving Legibility for n (1st MIL$_n$):
INSTANCE: A DAG $\mathcal{D} = \langle \mathcal{V}, \mathcal{A} \rangle$, a subset $\mathcal{A}_2$ of $\mathcal{A}$, a positive integer $K \leq |\mathcal{A}_2|$.
QUESTION: Does there exist a topological sorting σ of $\mathcal{D}$ for which

$$\mathbf{then}^{\mathcal{A}_2}_{\leq n}(\sigma) := \{\langle v, u \rangle \in \mathcal{A}_2 : d_\sigma(v, u) \leq n\}$$

has size at most K?

It is also desirable that fragments of reasoning that are maximal σ–linear subsequences should be pieces of reasoning with dense information flow.
Therefore, this flow has to be maximal in legible proof scripts or equivalently, the information flow between maximal σ–linear reasonings has to be minimal. This can be formulated as follows:

The 2st Method of Improving Legibility (2st MIL):
INSTANCE: A DAG $\mathcal{D} = \langle \mathcal{V}, \mathcal{A} \rangle$, subsets $\mathcal{A}_1 \subseteq \mathcal{A}_2 \subseteq \mathcal{A}$, a positive integer $K \leq |\mathcal{A}_2|$.
QUESTION: Does there exist a topological sorting σ of $\mathcal{D}$ for which

$$\{\langle v, u \rangle \in \mathcal{A}_2 : \mathbf{Then}^{\mathcal{A}_1}_\sigma(v) \neq \mathbf{Then}^{\mathcal{A}_1}_\sigma(u)\}$$

has size at most K?

In a similar way we obtain that the number of labeled vertices in the selected linearisation σ should be the smallest (3rd MIL), the same as sum of all σ–distances between vertices linked by $\mathcal{A}_2$-arcs (4th MIL).

The 3rd Method of Improving Legibility (3nd MIL):
INSTANCE: A DAG $\mathcal{D} = \langle \mathcal{V}, \mathcal{A} \rangle$, subsets $\mathcal{A}_1 \subseteq \mathcal{A}_2 \subseteq \mathcal{A}$, a positive integer $K \leq |\mathcal{V}|$.
QUESTION: Does there exist a topological sorting σ of $\mathcal{D}$ for which $|\mathbf{lab}^{\mathcal{A}_1, \mathcal{A}_2}(\sigma)| \leq K$?

The 4th Method of Improving Legibility (4th MIL):
INSTANCE: A DAG $\mathcal{D} = \langle \mathcal{V}, \mathcal{A} \rangle$, a subset $\mathcal{A}_2$ of $\mathcal{A}$, a positive integer $K \leq \binom{|\mathcal{V}|+1}{3}$.

QUESTION: Does there exist a topological sorting σ of $\mathcal{D}$ for which

$$\sum_{\langle u, v \rangle \in \mathcal{A}_2} d_\sigma(v, u) \leq K?$$

Referring to the short-term memory limitation of humans we formulated the last method that maximises the number of more intelligible references. We rely here on the assumption that a reference to a labeled vertex is more intelligible, if the number of statements that correspond to labeled vertices between linked by this reference vertices can be remembered by a reader.

The 5th Method of Improving Legibility for n (5th MIL$_n$):
INSTANCE: A DAG $\mathcal{D} = \langle \mathcal{V}, \mathcal{A} \rangle$, subsets $\mathcal{A}_1 \subseteq \mathcal{A}_2 \subseteq \mathcal{A}$, a positive integer $K \le |\mathcal{A}_2|$.
QUESTION: Does there exist a topological sorting σ of $\mathcal{D}$ for which

$$\mathfrak{R}^{\mathcal{A}_1,\mathcal{A}_2}(\sigma) := \{\langle v,u\rangle \in \mathcal{A}_2 : \mathfrak{Lab}_\sigma^{\mathcal{A}_1,\mathcal{A}_2}(u) - \mathfrak{Lab}_\sigma^{\mathcal{A}_1,\mathcal{A}_2}(v) \le n\}$$

has size at last K?

It is easy to see that the 4th MIL is NP-complete, since it generalises a known NP-complete problem Directed Optimal Linear Arrangement (see GT43 in [7] for $\mathcal{A}_2 = \mathcal{A}$). NP-completeness has been shown also for problems 1st, 2nd MIL, even for restricted instances $\mathcal{A}_1 = \mathcal{A}_2 = \mathcal{A}$ [22]. Additionally, for every instance of these problems in this case, there exists a Mizar proof script that contains a one-level deduction whose structure is that instance [22]. Therefore, realisation of 1st, 2nd, and 4th methods for one-level deductions potentially occurring in MML is NP-hard. In consequence, realisations of such methods for abstract proof graphs are also NP-hard, as one-level deductions are its substructures.

The problem 3rd MIL is also NP-complete in the general case, but for instances limited to ones that can actually occur in MML it is solvable in polynomial time [22]. Consequently, it is possible to effectively minimise the number of labels in Mizar proof scripts. These limits are a consequence of an additional syntax restriction of Mizar to use the **then** construct.

We show in Theorems 1, 2 below that problems 1st MIL$_n$ and 5th MIL$_n$ are also NP-complete. However, we do not focus on details of these proofs and we present only sketches.

Theorem 1. *The* 1st MIL *problem is reducible to the* 1st MIL$_n$ *problem for each natural number n.*

Proof. It is easily seen that the 1st MIL and the 1st MIL$_1$ are equivalent, if we take $\mathcal{A}_1 = \mathcal{A}_2$. Therefore, we can clearly assume that $n > 1$. Let $\mathcal{D} = \langle \mathcal{V}, \mathcal{A} \rangle$, $\mathcal{A}_1 \subseteq \mathcal{A}$, $K \le |\mathcal{V}|$ be an instance I of the 1st MIL. We can clearly construct in logspace a digraph $\mathcal{D}' = \langle \mathcal{V} \cup \mathcal{V}', \mathcal{A} \cup \mathcal{A}' \rangle$, a subset $\mathcal{A}'_2 = \mathcal{A}_1 \cup \mathcal{A}'$, and $K' = K + (n-1) \cdot |\mathcal{V}|$ as follows:

$$\mathcal{V}' := \{v_i : v \in \mathcal{V} \wedge 1 \le i < n\}, \quad \mathcal{A}' := \{\langle v_i, v\rangle : v \in \mathcal{V} \wedge 1 \le i < n\} \tag{5}$$

and consider it to be an instance I' of the 1st MIL$_n$. Let us take $\sigma \in TS(\mathcal{D})$ that is a solution of the 1st MIL problem for I and define a topological sorting $\sigma' \in TS(D')$ as $\sigma'(v) := n \cdot \sigma(v)$, $\sigma'(v_i) := i + (n-1) \cdot \sigma(v)$ for each $v \in \mathcal{V}$ and $1 \le i < n$. Obviously, $\mathcal{A}' \subseteq \mathbf{then}_{\le n}^{\mathcal{A}'_2}(\sigma')$ since a segment that contains all vertices of the form v_i for $i = 1, 2, \ldots, n-1$ directly precedes v for each $v \in \mathcal{V}$. Additionally, $d_\sigma(v,u) = 1$ if and only if $d_{\sigma'}(v,u) = n$, hence finally $|\mathbf{then}_{\le n}^{\mathcal{A}'_2}(\sigma')| \le K'$ and σ' is a solution of I'.

Now let $\sigma' \in TS(\mathcal{D}')$ be a solution of the 1st MIL$_n$ problem for I'. Note that the maximal value of $\sigma'(\mathcal{V}')$ is obtained for a vertex of $\mathcal{V}$. Denote it is as v. Let i be the index for which $\sigma'(v_i)$ has the smallest value among each $i = 1, 2, \ldots, n-1$.

We show that we can move all vertices of $\mathcal{V}'_v := \mathcal{V}' \setminus \{v, v_1, v_2, \ldots, v_{n-1}\}$ that are located between v_k and v, before v_k in σ' for each $k = 1, 2, \ldots, n-1$ so that the arrangement between vertices of $\mathcal{V}'_v$ is preserved and the size of $\mathbf{then}_{\leq n}^{\mathcal{A}'_2}(\sigma'_1)$ is not reduced, where σ'_1 is the new linearisation obtained in this way. Note that to compare size of $\mathbf{then}_{\leq n}^{\mathcal{A}'_2}(\sigma')$ and $\mathbf{then}_{\leq n}^{\mathcal{A}'_2}(\sigma'_1)$ we can clearly compare only $\mathcal{A}'_2$-arcs in-going to v included therein, which can be at most n in both sets. Suppose, contrary to our claim, that $\mathbf{then}_{\leq n}^{\mathcal{A}'_2}(\sigma')$ has more such arcs. But $\mathbf{then}_{\leq n}^{\mathcal{A}'_2}(\sigma'_1)$ contains $\langle v_j, v \rangle$ for each $j = 1, 2, \ldots, n-1$, hence $\mathbf{then}_{\leq n}^{\mathcal{A}'_2}(\sigma')$ contains exactly n such arcs.

Thus, $\mathbf{then}_{\leq n}^{\mathcal{A}'_2}(\sigma')$ contains the arcs $\langle u, v \rangle$ that have to be $\mathcal{A}'_2$-arcs, where u is the last vertex of $\mathcal{V}$ before v in σ. But then $d_{\sigma'_1}(u, v) = n$ and in consequence $\langle u, v \rangle \in \mathbf{then}_{\leq n}^{\mathcal{A}'_2}(\sigma'_1)$. This contradicts our assumption that $\mathbf{then}_{\leq n}^{\mathcal{A}'_2}(\sigma')$ has more arcs in-going to v than $\mathbf{then}_{\leq n}^{\mathcal{A}'_2}(\sigma'_1)$.

In a similar way, we can arrange vertices before the second, the third, the fourth, $\ldots$, $|\mathcal{V}|$-th up to the last vertex of $\mathcal{V}$ in $\sigma'_1, \sigma'_2, \ldots, \sigma'_{|\mathcal{V}|-1}$, creating a sequence $\sigma'_2, \sigma'_3, \ldots, \sigma'_{|\mathcal{V}|}$, respectively. Additionally, a topological sort $\sigma_{|\mathcal{V}|} \in TS(\mathcal{D})$ that preserves the order of vertices of $\mathcal{V}$ in $\sigma'_{|\mathcal{V}|}$ is a solution of I, since

$$K' \leq \mathbf{then}_{\leq n}^{\mathcal{A}'_2}(\sigma') \leq \mathbf{then}_{\leq n}^{\mathcal{A}'_2}(\sigma'_{|\mathcal{V}|}) = \mathbf{then}^{\mathcal{A}_1}(\sigma_{|\mathcal{V}|}) + (n-1) \cdot |\mathcal{V}|, \qquad (6)$$

and the proof is completed.

Theorem 2. *The* 1st MIL_n *problem is reducible to the* 5th MIL_n *problem.*

Proof. Let $\mathcal{D} = \langle \mathcal{V}, \mathcal{A} \rangle$, $\mathcal{A}_2 \subseteq \mathcal{A}$, $K \leq |\mathcal{A}_2|$ be an instance I of the 1st MIL. Let $\mathcal{D}' = \langle \mathcal{V} \cup \mathcal{V}', \mathcal{A} \cup \mathcal{A}' \cup \mathcal{A}'' \rangle$, $\mathcal{A}'_1 = \emptyset$, $\mathcal{A}'_2 = \mathcal{A}_2 \cup \mathcal{A}'$, $K' = K$ defined as follows:

$$\mathcal{V}' := \{v_1, v_2, \ldots, v_{n+1}\}, \qquad \mathcal{A}' := \{\langle u, v_{n+1} \rangle : u \in \mathcal{V}\}, \\ \mathcal{A}'' := \{\langle u, v_1 \rangle : u \in \mathcal{V}\} \cup \{\langle v_i, v_{i+1} \rangle : 1 \leq i \leq n\}, \qquad (7)$$

be an instance I' of the 5th MIL_n. Let us consider $\sigma \in TS(\mathcal{D}')$. Note that $\mathcal{A}' \cap \mathfrak{R}^{\mathcal{A}'_1, \mathcal{A}'_2}(\sigma) = \emptyset$. Indeed, for every $\mathcal{A}'$-arcs $\langle u, v_{n+1} \rangle$ we have $d_\sigma(u, v_{n+1}) > n$, since $\sigma(u) < \sigma(v_1) < \sigma(v_2) < \ldots < \sigma(v_{n+1})$. The main task of $\mathcal{A}'$-arcs is labeling every vertex of $\mathcal{V}$. Consequently, we obtain that $\mathfrak{Lab}_\sigma^{\mathcal{A}'_1, \mathcal{A}'_2}(w_1) - \mathfrak{Lab}_\sigma^{\mathcal{A}'_1, \mathcal{A}'_2}(w_2) = \sigma(w_1) - \sigma(w_2)$ for each $w_1, w_2 \in \mathcal{V}$. Note also that every vertex of $\mathcal{V}'$ has to be located after all vertices of $\mathcal{V}$ in σ and in unique order ($\sigma(v_i) = |\mathcal{V}| + i$ for each $i = 1, 2, \ldots, n+1$). Hence $\mathfrak{R}^{\mathcal{A}'_1, \mathcal{A}'_2}(\sigma) = \mathbf{then}_{\leq n}^{\mathcal{A}_2}(\sigma) = \mathbf{then}_{\leq n}^{\mathcal{A}_2}(\sigma_{|\mathcal{V}})$, where $\sigma_{|\mathcal{V}}$ is a topological sorting of $\mathcal{D}$ obtained by restricting σ to $\mathcal{V}$. Now the proof is immediate.

4   Automated Improving of Legibility as Support for Mizar Proof Authors

When we restrict our focus to methods based on a modification of the order of independent steps, the presented techniques agree in most cases with the needs of

Mizar users. However, it is controversial what the particular hierarchy of criteria should be applied, i.e., which criteria should be considered to be more important. Additionally, it can be observed in proof scripts of MML that application of a method to improve legibility can degrade the parameter optimised by another method. Therefore, we created a flexible application which can be used even by users with conflicting hierarchies of criteria for legibility. To use this application on a proof script, author needs simply to indicate a one-level deduction by typing a pragma `::$IL` $\langle strategy \rangle$ $\{\langle method \rangle\}^*$ at the beginning of this deduction (e.g., directly after `proof`, see Fig. 2) and run it, where:

$$\langle strategy \rangle ::= \text{Z3} : \langle time \rangle \mid \text{BF} : \langle number \rangle \mid \text{auto} : \langle number \rangle : \langle time \rangle ,$$
$$\langle method \rangle ::= \langle criterion \rangle : \langle condition \rangle \ [: \langle parameter \rangle] ,$$
$$\langle criterion \rangle ::= \text{Then} \mid \text{Flow} \mid \text{Lab} \mid \text{SumRef} \mid \text{LabRef} ,$$
$$\langle condition \rangle ::= \text{=} \mid \text{<} \mid \langle number \rangle,$$
$$\langle parameter \rangle ::= \langle number \rangle .$$

This application tries to find in three stages a more legible linearisation of a deduction decorated with this pragma. First, we create an abstract proof graph of this deduction. In the second stage, depending on the specified strategy, we describe the proof graph and the selected methods as a list of assertions and we check satisfiability by the Z3 solver for a given number of seconds (`Z3` : $\langle time \rangle$); use a brute-force attack with the limited number of checked linearisations (`BF` : $\langle number \rangle$); or check the number of linearisations and depending on the result we select the first (for greater than the limit) or the second (otherwise) strategy (`auto` : $\langle number \rangle$: $\langle time \rangle$). In the third stage, we adapt the obtained solution to the considered proof script, if we get a more legible linearisation. Criteria correspond to the 1st MIL_n, 2nd, 3rd, 4th MILs, 5th MIL_n problems respectively. Moreover, `Then` criterion with parameter 0 corresponds to the 1st MIL problem. The condition field set to `<` instructs the tool to search for a linearisation in which the optimised parameter given as the $\langle criterion \rangle$ tag is improved when compared with the initial situation. The condition field set to `=` instructs the tool not to make the parameter worse. Finally, condition field set to $\langle number \rangle$ defines the degree of the criterion in selected hierarchy.

The choice of the brute-force method is a consequence of the fact that 96,2% of one-level deductions in MML have at most one million possible linearisations that can be checked in this way "effectively". A vast majority of such deductions is located on deeply nested levels of proofs. Note that the other 3,8% of deductions cannot be omitted in the process of improving legibility, since this small group in MML is mainly located on shallow levels of proofs theorems (46% of this group is located on the first level, 25% on the second one, 14% on the third one). Additionally, when we try to analyse the main idea of reasoning we often concentrate the focus more on the first few levels than deeper ones (the deepest level of MML, 21, is used in the Mizar article JGRAPH_6).

Our translation of proof graph structure $\mathcal{D} = \langle \mathcal{V}, \mathcal{A} \rangle$, $\mathcal{A}_1 \subseteq \mathcal{A}_2 \subseteq \mathcal{A}$ for an optimisation method strongly depends on the method. Generally, we define the search linearisation as follows:

```
1:  (declare-const n Int) (assert (= n "|𝒱|"))
2:  (declare-fun S (Int) Int) (declare-fun Sinv (Int) Int)
3:  (assert (forall ((x Int))
        (=> (and (<= 1 x) (<= x n)) (and (<= 1 (S x)) (<= (S x) n)))))
4:  (assert (forall ((x Int) (y Int))
        (! (=> (and (and (<= 1 x) (<= x n)) (= (S x) (S y))) (= x y))
        :pattern ((S x) (S y)))))
5:  (assert (forall ((x Int))
        (! (=> (and (<= 1 x) (<= x n)) (= x (Sinv (S x)))) :pattern ((S x)))))
```

where $Sinv$ corresponds to S^{-1} and is introduced only for selected MILs problems. We also assert $(< (S\ \text{"x"})\ (S\ \text{"y"}))$ where $\langle x, y \rangle \in \mathcal{A}$ and every path of $\mathcal{D}$ directed form x to y has length at most 1.

The choice of the interpretation of MIL problems in Z3 assertions, generally requires to carry out initial research on the impact of the translation choice on the time of solution search. We often obtain significant reduction of the searching time when we replace an assertion with a quantifier (e.g., $\forall_{1 \leq x \leq n} \varphi(x)$) by a list of assertions that represent individual cases (e.g., $\varphi(1), \varphi(2), \ldots, \varphi(n)$). As an illustration, let us consider the 1st interpretation of the 1st MIL defined as follows:

```
1:  (declare-fun T (Int) Bool)
2:  (assert (= (T (Sinv 1)) false))
3:  (assert (forall ((x Int)) (= (T (Sinv x))
        (ite (and (and (< 1 x) (<= x n)) (= (select A1 (Sinv (- x 1)) (Sinv x)) true))
        true false))))
4:  (declare-fun SumT (Int) Int)
5:  (assert (= (SumT 0) 0))
6:  (assert (forall ((x Int)) (= (SumT x) (ite (and (and (< 0 x) (<= x n)) (= (T x) true))
        (+ 1 (SumT (- x 1))) (SumT (- x 1))))))
```

where A1 is the incidence matrix of the family of $\mathcal{A}_1$-arcs ((declare-const A1 (Array Int Int Bool))). An analysis of 285 one-level deductions contained in the proof script of [23] (e.g., Fig. 1) shows that we can speed up the search process (see Fig. 3) on average 37 times if we replace the assertion in the 6th row by a sequence of assentations like:

```
(assert (= (SumT "i+1")
       (ite (= (T (Sinv "i+1")) true ) (+ i (SumT "i")) (SumT "i"))))
```

for $i = 1, 2, \ldots, n$ (the 2nd interpretation). Note that we replace a quantifier in this assertion by a list of individual cases and we count the number of "*true*" values of T in the order determined by S^{-1}. This result can be improved further on average 2.57 times, if we remove an intermediate function T, using a sequence of assentations like:

```
(assert (= (SumT "i+1")
       (ite (= (select A1 (Sinv "i") (Sinv "i+1")) true) (+ 1 (SumT "i")) (SumT "i"))))
```

(the 3rd interpretation). Note that the number of unresolved cases that remain unresolved after 10 minutes was reduced from 31 to 10 and 7 respectively. Additionally, the average time to solve newly resolved cases were 38.2 and 16.9 seconds respectively.

Similar speed up in searches was obtained with removal of quantifiers from interpretations of the other methods. It comes as no surprise since functions defined in one interpretation are used in other ones. Note that the function SumT is used indirectly in the interpretation of the 2nd MIL problem, since there exists a relationship between the values of SumT and a relation determined by belonging of two vertices to the same maximal linear reasoning.

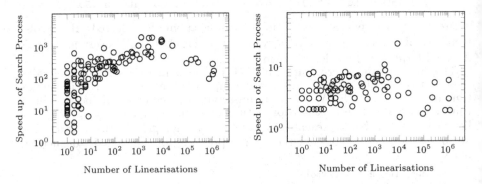

Fig. 3. The proportion between times of linearisation search with Z3 strategy realised through 1st and 3rd interpretation (on the left side) and through 2nd and 3rd interpretation (on the right side) on one-level deductions in proof scripts of [23]

As an illustration, let us consider one-level deductions contained in the abstract proof graph presented in Fig. 2, induced by vertices 2, 3, 10–14 and 4–9 respectively. Note that the largest number of **then**-steps is equal to 3 in both deductions and this this value is obtained in the proof script presented in Fig. 1. However, the Z3 solver verify the possibility of increasing the number of **then**-steps in the 1st interpretation of the 1st MIL for 6.10 and 0.43 seconds, in the 2nd interpretation for 0.08 and 0.06 seconds, and in the 3rd interpretation for 0.02 and 0.01 seconds respectively.

5   Statistical Results

The effectiveness of Z3 and BF strategies was studies on the MML database version 5.22.1191 including 208590 one-level deductions. We present here only statistical results that were obtained for two most popular criteria that correspond to the 1st and the 4th MIL problem. The results for other criteria have proved to be similar to those selected ones. This resemblance is mainly due to similar problem translation to Z3 solver. Running the optimisation program on the whole MML base with pragmas:

(1) ::$IL Z3:600 Then:<:0 ,   (2) ::$IL Z3:600 SumRef:< ,
(3) ::$IL BF:1000000 Then:<:0 ,   (4) ::$IL BF:1000000 SumRef:< ,

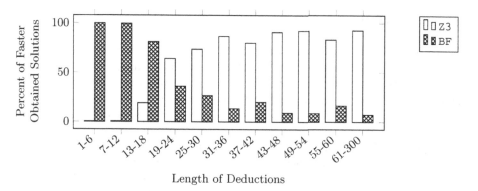

Fig. 4. The percent of one-level deductions, where the search time with Z3:600 and BF:10000000 strategies is compared depending on the deduction length, for criterion SumRef:<. Naturally, we limited results to cases, in which at least one strategy solved the problem.

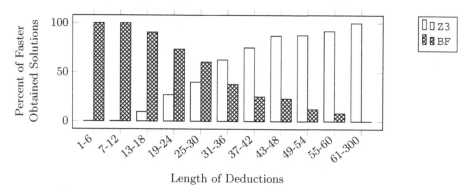

Fig. 5. The percent of one-level deductions, where the search time with Z3:600 and BF:10000000 strategies is compared depending on the deduction length, for criterion Then:<:0. Clearly, we limited results to cases, in which at least one strategy solved the problem.

takes 62.5h, 16.4h, 2.4h, and 16 min respectively, on a platform with 16 Intel Xeon E5520 Processors and 24 GB of RAM. Using both strategies only 1.92%, 0.03% of problems were left unsolved, for Then and SumRef criterion respectively. Clearly, BF strategy turns out to be significantly better than Z3, but this holds only whole MML base is taken as the subject of the test case. When we take into consideration only deductions that possess at least ten million possible linearisations, the effectiveness in both cases becomes comparable. Additionally, the effectiveness of the strategy Z3 is more highlighted when we analyse it in terms of the deductions length (see Fig. 4, 5). As we expected, application of SMT technology to long deductions, where the computationally hardest part of legibility improvement is concentrated, was fully justified. Additionally, this situation is evident even for deductions that have at least 25 steps.

6   Conclusions

In this paper we describe a next stage in the research on methods that improve proof legibility based on the modifying the order of independent deduction steps. The Mizar users need a "push-button" tool that automatically facilitates localisation of premises in deduction and highlights local subdeductions. However, creating such a tool is a non-trivial task, since the realisation of these expectations leads to NP-hard optimisation.

We presented initial results obtained with such a tool that uses the Z3 solver. This research showed that such tools can improve the legibility of deductions, even in the case they are long. Definitely, our result suggests the need of further work on the choice of interpretation of MIL problems in Z3 to speed up the search process

Note that we encounter computationally difficult problems, similarly as for MIL, when we try to improve the legibility of proof scripts based, e.g. on automated look-up of passages from long reasoning and extracting them as a lemma. Therefore, by successful application of the SMT-solver Z3 to solving MIL problems we we expect similar results to other methods of legibility enhancement.

References

1. Behaghel, O.: Beziehungen zwischen Umfang und Reihenfolge von Satzgliedern. Indogermanische Forschungen 25, 110–142 (1909)
2. Blanchette, J.C.: Redirecting Proofs by Contradiction. In: Third International Workshop on Proof Exchange for Theorem Proving, PxTP 2013. EPiC Series, vol. 14, pp. 11–26. EasyChair (2013)
3. Cowan, N.: The magical number 4 in short-term memory: A reconsideration of mental storage capacity. Behavioral and Brain Sciences 24(1), 87–114 (2001)
4. de Moura, L., Bjørner, N.: Z3: An efficient SMT solver. In: Ramakrishnan, C.R., Rehof, J. (eds.) TACAS 2008. LNCS, vol. 4963, pp. 337–340. Springer, Heidelberg (2008)
5. Ericsson, K.A.: Analysis of memory performance in terms of memory skill. Advances in the psychology of human intelligence, vol. 4. Lawrence Erlbaum Associates Inc., Hillsdale (1988)
6. Fitch, F.B.: Symbolic Logic: an Introduction. The Ronald Press Co. (1952)
7. Garey, M.R., Johnson, D.S.: Computers and Intractability: A Guide to the Theory of NP-Completeness. A Series of Books in the Mathematical Science. W. H. Freeman and Company, New York (1979)
8. Gonthier, G.: Formal Proof—The Four-Color Theorem. Notices of the AMS 55(11), 1382–1393 (2008)
9. Grabowski, A., Korniłowicz, A., Naumowicz, A.: Mizar in a Nutshell. Journal of Formalized Reasoning 3(2), 153–245 (2010)
10. Grabowski, A., Schwarzweller, C.: Improving Representation of Knowledge within the Mizar Library. Studies in Logic, Grammar and Rhetoric 18(31), 35–50 (2009)
11. Jaśkowski, S.: On the Rules of Supposition in Formal Logic. Studia Logica (1934), Warszawa Reprinted in Polish Logic, McCall, S., ed. Clarendon Press, Oxford (1967)

12. Kaliszyk, C., Urban, J.: PRocH: Proof Reconstruction for HOL Light. In: Bonacina, M.P. (ed.) CADE-24. LNCS (LNAI), vol. 7898, pp. 267–274. Springer, Heidelberg (2013)
13. Ko, A.J., Myers, B.A., Coblenz, M.J., Aung, H.H.: An Exploratory Study of How Developers Seek, Relate, and Collect Relevant Information during Software Maintenance Tasks. IEEE Transactions on Software Engineering 32(12), 971–988 (2006)
14. Korniłowicz, A.: Tentative Experiments with Ellipsis in Mizar. In: Jeuring, J., Campbell, J.A., Carette, J., Dos Reis, G., Sojka, P., Wenzel, M., Sorge, V. (eds.) CICM 2012. LNCS (LNAI), vol. 7362, pp. 453–457. Springer, Heidelberg (2012)
15. Levy, R.: Expectation-based syntactic comprehension. Cognition 106, 1126–1177 (2008)
16. Matuszewski, R.: On Automatic Translation of Texts from Mizar-QC language into English. Studies in Logic, Grammar and Rhetoric 4 (1984)
17. Miller, G.A.: The Magical Number Seven, Plus or Minus Two: Some Limits on Our Capacity for Processing Information. Psychological Review 63, 81–97 (1956)
18. Naumowicz, A., Korniłowicz, A.: A Brief Overview of MIZAR. In: Berghofer, S., Nipkow, T., Urban, C., Wenzel, M. (eds.) TPHOLs 2009. LNCS, vol. 5674, pp. 67–72. Springer, Heidelberg (2009)
19. Naumowicz, A., Korniłowicz, A.: A Brief Overview of MIZAR. In: Berghofer, S., Nipkow, T., Urban, C., Wenzel, M. (eds.) TPHOLs 2009. LNCS, vol. 5674, pp. 67–72. Springer, Heidelberg (2009)
20. Pąk, K.: The Algorithms for Improving and Reorganizing Natural Deduction Proofs. Studies in Logic, Grammar and Rhetoric 22(35), 95–112 (2010)
21. Pąk, K.: Methods of Lemma Extraction in Natural Deduction Proofs. Journal of Automated Reasoning 50(2), 217–228 (2013)
22. Pąk, K.: The Algorithms for Improving Legibility of Natural Deduction Proofs. PhD thesis, University of Warsaw (2013)
23. Pąk, K., Trybulec, A.: Laplace Expansion. Formalized Mathematics 15(3), 143–150 (2008)
24. Smolka, S.J., Blanchette, J.C.: Robust, Semi-Intelligible Isabelle Proofs from ATP Proofs. In: Third International Workshop on Proof Exchange for Theorem Proving, PxTP 2013. EPiC Series, vol. 14, pp. 117–132. EasyChair (2013)
25. Urban, J.: XML-izing Mizar: Making Semantic Processing and Presentation of MML Easy. In: Kohlhase, M. (ed.) MKM 2005. LNCS (LNAI), vol. 3863, pp. 346–360. Springer, Heidelberg (2006)
26. Urban, J.: MizarMode - An Integrated Proof Assistance Tool for the Mizar Way of Formalizing Mathematics. Journal of Applied Logic 4(4), 414–427 (2006)
27. Wenzel, M.: The Isabelle/Isar Reference Manual. University of Cambridge (2011)
28. Whitehead, A.N., Russell, B.: Principia Mathematica to *56. Cambridge Mathematical Library. Cambridge University Press (1910)
29. Zammit, V.: On the Readability of Machine Checkable Formal Proofs. PhD thesis, The University of Kent at Canterbury (March 1999)

A Vernacular for Coherent Logic*

Sana Stojanović[1], Julien Narboux[2], Marc Bezem[3], and Predrag Janičić[1]

[1] Faculty of Mathematics, University of Belgrade, Serbia
[2] Icube, UMR 7357 CNRS, University of Strasbourg, France
[3] Institute for Informatics, University of Bergen, Norway
sana@matf.bg.ac.rs, narboux@unistra.fr,
marc.bezem@ii.uib.no, janicic@matf.bg.ac.rs

Abstract. We propose a simple, yet expressive proof representation from which proofs for different proof assistants can easily be generated. The representation uses only a few inference rules and is based on a fragment of first-order logic called coherent logic. Coherent logic has been recognized by a number of researchers as a suitable logic for many everyday mathematical developments. The proposed proof representation is accompanied by a corresponding XML format and by a suite of XSL transformations for generating formal proofs for Isabelle/Isar and Coq, as well as proofs expressed in a natural language form (formatted in LaTeX or in HTML). Also, our automated theorem prover for coherent logic exports proofs in the proposed XML format. All tools are publicly available, along with a set of sample theorems.

1 Introduction

Mathematics can be done on two different levels. One level is rather informal, based on informal explanations, intuition, diagrams, etc., and typical for everyday mathematical practice. Another level is formal mathematics with proofs rigorously constructed by rules of inference from axioms. A large portion of mathematical logic and interactive theorem proving is aimed at linking these two levels. However, there is still a big gap: mathematicians still don't feel comfortable doing mathematics formally and proof assistants still don't provide enough support for dealing with large mathematical theories, automating technical problems, translating from one formalism to another, etc. We consider the following issue: there are several very mature and popular interactive theorem provers (including Isabelle, Coq, Mizar, HOL-light, see [29] for an overview), but they still cannot easily share the same mathematical knowledge. This is a significant problem, because there are increasing efforts in building repositories of formalized mathematics, but — still developed within specific proof assistants. Building a mechanism for translation between different proof assistants

---

* The first, second and the fourth author were partly supported by the Serbian-French Technology Co-Operation grant EGIDE/"Pavle Savić" 680-00-132/2012-09/12. The first and the fourth author are partly supported by the grant ON174021 of the Ministry of Science of Serbia.

S.M. Watt et al. (Eds.): CICM 2014, LNAI 8543, pp. 388–403, 2014.
© Springer International Publishing Switzerland 2014

is non-trivial because of many deep specifics of each proof assistant (there are some recent promising approaches for this task [13]). Instead of developing a translation mechanism, we propose a proof representation and a corresponding XML-based format. The proposed proof representation is light-weight and it does not aim at covering full power of everyday mathematical proofs or full power of first order logic. Still, it can cover a significant portion of many interesting mathematical theories. The underlying logic of our representation is coherent logic, a fragment of first-order logic. Proofs in this format can be generated in an easy way by dedicated, coherent logic provers, but in principle, also by standard theorem provers. The proofs can be translated to a range of proof assistant formats, enabling sharing the same developments.

We call our proof representation "coherent logic vernacular". *Vernacular* is the everyday, ordinary language (in contrast to the official, literary language) of the people of some country or region. A similar term, *mathematical vernacular* was used in 1980's by de Bruijn within his formalism proposed for trying to *put a substantial part of the mathematical vernacular into the formal system* [10]. Several authors later modified or extended de Bruijn's framework. Wiedijk follows de Bruijn's motivation [28], but he also notices:

> It turns out that in a significant number of systems ('proof assistants') one encounters languages that look almost the same. Apparently there is a canonical style of presenting mathematics that people discover independently: something like a natural mathematical vernacular. Because this language apparently is something that people arrive at independently, we might call it the mathematical vernacular.

We find that this language is actually closely related to a proof language of coherent logic, which is a basis of our proof representation presented in this paper.

Our proof representation is developed also with *readable proofs* in mind. Readable proofs (e.g., textbook-like proofs), are very important in mathematical practice. For mathematicians, the main goal is often, not only a trusted, but also a clear and intuitive proof. We believe that coherent logic is very well suited for automated theorem proving with a simple production of readable proofs.

2 Background

In this section, we give a brief overview of interactive theorem proving and proof assistants, of coherent logic, which is the logical basis for our proof representation, and of XML, which is the technical basis for our proof format.

2.1 Interactive Theorem Proving

Interactive theorem proving systems (or *proof assistants*) support the construction of formal proofs by a human, and verify each proof step with respect to the given underlying logic. The proofs can be written either in a *declarative* or in a *procedural* proof style. In the procedural proof style, the proof is described by a sequence of commands which modify the incomplete proof tree. In the declarative proof style the formal document includes the intermediate statements. Both styles are avaible in HOL-Light, Isabelle [27] and Coq proof assistants whereas only the declarative style is available in Mizar, see [30] for a recent discussion. The procedural proof style is more popular in the Coq community.

Formal proofs are typically much longer than "traditional proofs".[1] Progress in the field can be measured by proof scripts becoming shorter and yet contain enough information for the system to construct and verify the full (formal) proof. "Traditional proofs" can often hardly be called proofs, because of the many missing parts, informal arguments, etc. Using interactive theorem proving uncovered many flaws in many published mathematical proofs (including some seminal ones), published in books and journals.

2.2   Coherent Logic

Coherent logic (CL) was initially defined by Skolem and in recent years it gained new attention [3,11,4]. It consists of formulae of the following form:

$$A_1(x) \wedge \ldots \wedge A_n(x) \Rightarrow \exists y (B_1(x, y) \vee \ldots \vee B_m(x, y)) \tag{1}$$

which are implicitly universally quantified, and where $0 \leq n$, $0 \leq m$, x denotes a sequence of variables $x_1, x_2, \ldots, x_k$ $(0 \leq k)$, A_i (for $1 \leq i \leq n$) denotes an atomic formula (involving zero or more of the variables from x), y denotes a sequence of variables $y_1, y_2, \ldots, y_l$ $(0 \leq l)$, and B_j (for $1 \leq j \leq m$) denotes a conjunction of atomic formulae (involving zero or more of the variables from x and y). For simplicity, we assume that there are no function symbols with arity greater than zero (so, we only consider symbols of constants as ground terms).

The definition of CL does not involve negation. For a single atom A, $\neg A$ can be represented in the form $A \Rightarrow \bot$, where $\bot$ stands for the empty disjunction, but more general negation must be expressed carefully in coherent logic. In order to reason with negation in general, new predicate symbols are used to abbreviate subformulas. Furthermore, for every predicate symbol R (that appears in negated form), a new symbol $\overline{R}$ is introduced that stands for $\neg R$, and the following axioms are postulated (cf. [19]): $\forall x (R(x) \wedge \overline{R}(x) \Rightarrow \bot)$, $\forall x (R(x) \vee \overline{R}(x))$.

CL allows existential quantifications of the conclusion of a formula, so CL can be considered to be an extension of resolution logic. In contrast to the resolution-based proving, the conjecture being proved is kept unchanged and directly proved (refutation, Skolemization and transformation to clausal form are not used). Hence, proofs in CL are natural and intuitive and reasoning is

---

[1] The ratio between the length of formal proof *script* and the length of the informal proof is often called the *de Bruijn factor* [2]. It varies for different parts of mathematics and for different systems, and is currently often around 4. The de Bruijn factor can be below 1 if a lot of automation can be used. It can also be well over 10 when the informal proof is rather sketchy.

constructive. Readable proofs (in the style of forward reasoning and a variant of natural deduction) can easily be obtained [3].

A number of theories and theorems can be formulated directly and simply in CL. In CL, constructive provability is the same as classical provability. It can be proved that any first-order formula can be translated into a set of CL formulas (in a different signature) preserving satisfiability [19] (however, this translation does not always preserve constructive provability).

Coherent logic is semi-decidable and there are several implemented semi-decision procedures for it [3]. ArgoCLP [24] is a generic theorem prover for coherent logic, based on a simple proof procedure with forward chaining and with iterative deepening. ArgoCLP can read problems given in TPTP form[2] [25] and can export proofs in the XML format that we describe in this paper. These proofs are then translated into target languages, for instance, the Isar language or natural language thanks to appropriate XSLT style-sheets.

2.3   XML

Extensible Markup Language (XML)[3] is a simple, flexible text format, inspired by SGML (ISO 8879), for data structuring using tags and for interchanging information between different computing systems. XML is primarily a "metalanguage"— a language for describing other customized markup languages. So, it is not a fixed format like the markup language HTML—in XML the tags indicate the semantic structure of the document, rather than only its layout. XML is a project of the World Wide Web Consortium (W3C) and is a public format. Almost all browsers that are currently in use support XML natively.

There are several schema languages for formaly specifying the structure and content of XML documents of one class. Some of the main schema languages are DTD (*Data Type Definition*), XML Schema, Relax, etc. [17]. Specifications in the form of schema languages enable automatic verification ("validation") of whether a specific document meets the given syntactical restrictions.

Extensible style-sheet language transformation (XSLT)[4] is a document processing language that is used to transform the input XML documents to output files. An XSLT style-sheet declares a set of rules (templates) for an XSLT processor to use when interpreting the contents of an input XML document. These rules tell to the XSLT processor how that data should be presented: as an XML document, as an HTML document, as plain text, or in some other form.

3   Proof Representation

The proposed proof representation is very usable and expressive, yet very simple. It uses only a few inference rules, a variant of the rules given in [4]. Given a set of coherent axioms AX and a coherent conjecture $A_1(x) \wedge \ldots \wedge A_n(x) \Rightarrow$

---

[2] http://www.cs.miami.edu/~tptp/
[3] http://www.w3.org/XML/
[4] http://www.w3.org/Style/XSL/

$\exists y(B_1(x,y) \vee \ldots \vee B_m(x,y))$, the goal is to prove, using the rules given below, the following (where a denote a vector of new symbols of constants):

$$AX, A_1(a) \wedge \ldots \wedge A_n(a) \vdash \exists y(B_1(a,y) \vee \ldots \vee B_m(a,y))$$

The rules are applied in a forward manner, so they can be read from bottom to top. In the rules below we assume:

- $ax \in AX$ is a formula of the form (1) (page 390);
- a, b, c denote vectors of constants (possibly of length zero);
- in the rule mp, b are fresh constants;
- x and y denote vectors of variables (possibly of length zero);
- $A_i(x)$ $(B_i(x,y))$ have no free variables other than from x (and y);
- $A_i(a)$ are ground atomic formulae;
- $B_i(a,b)$ and $B_i(c)$ are ground conjunctions of atomic formulae;
- $\underline{\Phi}$ denotes the list of conjuncts in Φ.

$$\frac{\Gamma, ax, A_1(a) \wedge \ldots \wedge A_n(a), B_1(a,b) \vee \ldots \vee B_m(a,b) \vdash P}{\Gamma, ax, A_1(a) \wedge \ldots \wedge A_n(a) \vdash P} \quad mp \ (modus \ ponens)$$

$$\frac{\Gamma, B_1(c) \vdash P \quad \ldots \quad \Gamma, B_n(c) \vdash P}{\Gamma, B_1(c) \vee \ldots \vee B_n(c) \vdash P} \quad cs \ (case \ split)$$

$$\frac{}{\Gamma, B_i(a,b) \vdash \exists y(B_1(a,y) \vee \ldots \vee B_m(a,y))} \quad as \ (assumption)$$

$$\frac{}{\Gamma, \bot \vdash P} \quad efq \ (ex \ falso \ quodlibet)$$

None of these rules change the goal P, which helps generating readable proofs as the goal can be kept implicit. Note that the rule mp actually combines universal instantiation, conjunction introduction, modus ponens, and elimination of (zero or more) existential quantifiers. This seems a reasonable granularity for an inference step, albeit probably the maximum for keeping proofs readable. Compared to [20] which defines the notion of obvious inference rule by putting constraints on an automated prover, our position is: the obvious inferences are the ones defined by the inference rules above. Compared to the rules given in [4], we choose to separate the *case split* rule (disjunction elimination) and the *ex falso quodlibet* rule from the single combined rule in [4], in order to improve readability. Case distinction (split) is an important way of structuring proofs that deserves to be made explicit. Also, *ex falso quodlibet* could be seen as a *case split* with zero cases, but this would be less readable.

Any coherent logic proof can be represented in the following simple way (mp is used zero or more time, cs involves at least two other *proof* objects):

$$proof ::= mp^* \ (cs(proof^{\geq 2}) \mid as \mid efq)$$

4   XML Suite for CL Vernacular

The proof representation described in Section 3 is used as a basis for our XML-based proof format. It is developed as an interchange format for automated and interactive theorem provers. Proofs (for Coq and Isabelle/Isar) that are produced from our XML documents are fairly readable. The XML documents themselves can be read by a human, but much better alternative is using translation to human readable proofs in natural language (formatted in LaTeX, for instance). The proof representation is described by a DTD `Vernacular.dtd`. As an illustration, we show some fragments:

```
...
<!--******** Theory **************-->
<!ELEMENT theory (theory_name, signature, axiom*) >
<!ELEMENT theory_name (#PCDATA)>
<!ELEMENT signature (type*, relation_symbol*, constant*) >
<!ELEMENT relation_symbol (type*)>
<!ATTLIST relation_symbol name CDATA #REQUIRED>
<!ELEMENT type (#PCDATA)>
<!ELEMENT axiom (cl_formula)>
<!ATTLIST axiom name CDATA #REQUIRED>
...
```

The above fragment describes the notion of theory. (Definitions, formalized as pairs of coherent formulae, are used as axioms.) A file describing a theory could be shared among several files with theorems and proofs.

```
...
<!--******** Theorem *************-->
<!ELEMENT theorem (theorem_name, cl_formula, proof+)>
<!ELEMENT theorem_name (#PCDATA)>
<!ELEMENT conjecture (name, cl_formula)>

<!--******** Proof *************-->
<!ELEMENT proof (proof_step*, proof_closing, proof_name?)>
<!ELEMENT proof_name EMPTY>
<!ATTLIST proof_name name CDATA #REQUIRED>

<!--******** Proof steps *************-->
<!ELEMENT proof_step (indentation,modus_ponens)>
<!ELEMENT proof_closing (indentation, (case_split|efq|from),
        (goal_reached_contradiction|goal_reached_thesis))>
...
```

The above fragment describes the notion of a theorem and a proof. As said in Section 3, a proof consists of a sequence of applications of the rule modus ponens and closes with one of the remaining proof rules (*case split*, *as*, or *efq*). Within the last three, there is the additional information on whether the proof closes by $\perp$ (by detecting a contradiction) or by detecting one of the disjuncts from the goal. This information is generated by the prover and can be used for better readability of the proof but also for some potential proof transformations. Within each proof step there is also the information on indentation. This information, useful for better layout, tells the level of subproofs and as such can be, in principle, computed from the XML representation. Still, for convenience and simplicity of the XSLT style-sheets, it is stored within the XML representation.

We implemented XSL transformations from XML format to Isabelle/Isar (`VernacularISAR.xls`), Coq (`VernacularCoqTactics.xls`), and to a natural language (English) in LaTeX form and in HTML form (`VernacularTex.xls` and `VernacularHTML.xls`).

The translation from XML to the Isar language is straightforward and each of our proof steps is trivially translated into Isar constructs.[5] Naturally, we use native negation of Isar (and Coq) instead of defined negation in coherent logic.

The translation to Coq has been written in the same spirit as the Isar output despite the fact proofs using tactics are more popular in Coq than declarative proofs. We refer to the assumptions by their statement instead of their name (for example: by `cases on (A = B \/ A <> B)`). Moreover, when we can, we avoid to refer to the assumptions at all. We did not use the declarative proof mode of Coq because of efficiency issues. We use our own tactics to implement the inference rules of CL to improve readability. Internally, we use an Ltac tactic to get the name of an assumption. The forward reasoning proof steps consist of applications of the `assert` tactic of Coq. Equality is translated into Leibniz equality.

The translation to LaTeX and HTML includes an additional XSLT style-sheet that optionally defines specific layout for specific relation symbols (so, for instance, $(A, B) \cong (C, D)$ can be the layout for `cong(A,B,C,D)`).

The developed XSLT style-sheets are rather simple and short — each is only around 500 lines long. This shows that transformations for other target languages (other theorem provers, like Mizar and HOL light, LaTeX with other natural languages, MathML, OMDoc or TPTP) can easily be constructed, thus enabling wide access to a single source of mathematical contents.

Our automated theorem prover for coherent logic ArgoCLP exports proofs in the form of the XML files that conforms to this DTD. ArgoCLP reads an input theory and the conjecture given in the TPTP form (assuming the coherent form of all formulae and that there are no function symbols or arity greater than 0). ArgoCLP has built-in support for equality (during the search process, it uses an efficient union-find structure) and the use of equality axioms is implicit in generated proofs. The generated XML documents are simple and consist of three parts: `frontpage` (providing, for instance, the author of the theorem, the prover used for generating the proof, the date), `theory` (providing the signature and the axioms) and, organized in chapters, a list of conjectures or theorems with their proofs. This way, some contents (`frontpage` and `theory`) can be shared by a number of XML documents. On the other hand, this also enables simple

---

[5] The system Isabelle has available a proof method `coherent` based on a internal theorem prover for coherent logic. Our Isar proofs do not use this proof method.

construction of bigger collections of theorems. The following is one example of
an XML document generated by ArgoCLP:

```
<?xml version="1.0" encoding="UTF-8"?>
<!DOCTYPE main SYSTEM "Vernacular.dtd">
<?xml-stylesheet href="VernacularISAR.xsl" type="text/xsl"?>

<main>
<xi:include href="frontpage.xml" parse="xml"
    xmlns:xi="http://www.w3.org/2003/XInclude"/>
<xi:include href="theory_thm_4_19.xml" parse="xml"
    xmlns:xi="http://www.w3.org/2003/XInclude"/>

<chapter name="th_4_19">
<xi:include href="proof_thm_4_19.xml" parse="xml"
    xmlns:xi="http://www.w3.org/2003/XInclude"/>
</chapter>
</main>
```

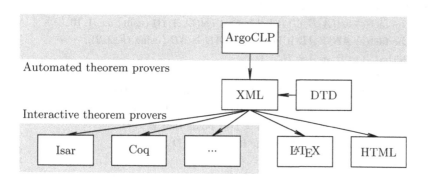

Fig. 1. Architecture of the presented framework

The overall architecture of the framework is shown in Figure 1.[6]

5   Examples

Our XML suite for coherent logic vernacular is used for a number of proofs
generated by our prover ArgoCLP. In this section we discuss proofs of theorems from the book *Metamathematische Methoden in der Geometrie*, by Wolfram Schwabhäuser, Wanda Szmielew, and Alfred Tarski [21], one of the twenty-
century mathematical classics. The theory is described in terms of first-order
logic, it uses only one sort of primitive objects — points, has only two primitive

---

[6] The whole of our XML suite, along with a collection of theorems is available online
from http://argo.matf.bg.ac.rs/downloads/software/clvernacular.zip.

predicates (*cong* or arity 4 and *bet* of arity 3, intuitively for congruence and betweenness) and only eleven axioms. The majority of theorems from this book are in coherent logic or can be trivially transformed to belong to coherent logic. After needed transformations, the number of theorems in our development (238) is somewhat larger than in the book [23].

Here we list a proof of one theorem (4.19) from Tarski's book. The theorem was proved by ArgoCLP (using the list of relevant axioms and theorems produced by a resolution theorem prover), the proof was exported in the XML format, and then transformed to a proof in natural language by appropriate XSL transformation (($A, B) \cong (C, D)$ is an infix notation for $cong(A, B, C, D)$ and it denotes that the pairs of points (A, B) and (C, D) are congruent, $bet(A, B, C)$ denotes that the point B is between the points A and C, $col(A, B, C)$ denotes that the points A, B and C are collinear).

Theorem 1 (th_4_19). *Assuming that $bet(A, B, C)$ and $AB \cong AD$ and $CB \cong CD$ it holds that $B = D$.*

  Proof:

1. It holds that $bet(B, A, A)$ (using th_3_1).
2. From the fact(s) $bet(A, B, C)$ it holds that $col(C, A, B)$ (using ax_4_10_3).
3. From the fact(s) $AB \cong AD$ it holds that $AD \cong AB$ (using th_2_2).
4. It holds that $A = B$ or $A \neq B$.
   5. Assume that: $A = B$.
      6. From the fact(s) $AD \cong AB$ and $A = B$ it holds that $AD \cong AA$.
      7. From the fact(s) $AD \cong AA$ it holds that $A = D$ (using ax_3).
      8. From the fact(s) $A = B$ and $A = D$ it holds that $B = D$.
      9. The conclusion follows from the fact(s) $B = D$.
   10. Assume that: $A \neq B$.
       11. It holds that $A = C$ or $A \neq C$.
          12. Assume that: $A = C$.
             13. From the fact(s) $bet(A, B, C)$ and $A = C$ it holds that $bet(A, B, A)$.
             14. From the fact(s) $bet(A, B, A)$ and $bet(B, A, A)$ it holds that $A = B$ (using th_3_4).
             15. From the fact(s) $A \neq B$ and $A = B$ we get contradiction.
          16. Assume that: $A \neq C$.
             17. From the fact(s) $A \neq C$ it holds that $C \neq A$.
             18. From the fact(s) $C \neq A$ and $col(C, A, B)$ and $CB \cong CD$ and $AB \cong AD$ it holds that $B = D$ (using th_4_18).
             19. The conclusion follows from the fact(s) $B = D$.
       20. The conclusion follows in all cases.
21. The conclusion follows in all cases.
    QED

Below is the same proof in Isabelle/Isar form:

```
lemma th_4_19 :  assumes  "bet A B C" and "cong A B A D" and
"cong C B C D"  shows "(B = D)"

proof -

have "bet B A A"  by (rule th_3_1)
from 'bet A B C' have "col C A B" by (rule ax_4_10_3)
from 'cong A B A D' have "cong A D A B" by (rule th_2_2)
have "A = B ∨ A ~= B" by (subst disj_commute, rule excluded_middle)
  show ?thesis
  proof(cases "A = B")
    case True
      from 'cong A D A B' and 'A = B' have "cong A D A A" by simp
      from 'cong A D A A' have "A = D" by (rule ax_3)
      from 'A = B' and 'A = D' have "B = D" by simp
      from 'B = D' show ?thesis by assumption
    next
      case False
      have "A = C ∨ A ~= C" by (subst disj_commute, rule
excluded_middle)
        show ?thesis
        proof(cases "A = C")
          case True
            from 'bet A B C' and 'A = C' have "bet A B A" by simp
            from 'bet A B A' and 'bet B A A' have "A = B" by (rule
th_3_4)
            from 'A ~= B' and 'A = B' have "False" by (rule notE)
            from this show ?thesis by (rule FalseE)
          next
            case False
            from 'A ~= C' have "C ~= A" by (rule not_sym)
            from 'C ~= A' and 'col C A B' and 'cong C B C D' and
'cong A B A D' have "B = D" by (rule th_4_18)
            from 'B = D' show ?thesis by assumption
          qed
  qed
qed
end
```

Below is the same proof in Coq form:

Theorem th_4_19 : ∀ (*A:point*) (*B:point*) (*C:point*) (*D:point*), (*bet A B C* ∧ *cong A B A D* ∧ *cong C B C D*) → *B* = *D*.
Proof.
 intros.
 assert (*bet B A A*) by *applying* (*th_3_1 B A*) .
 assert (*col C A B*) by *applying* (*ax_4_10_3 A B C*) .
 assert (*cong A D A B*) by *applying* (*th_2_2 A B A D*) .
 assert (*A* = *B* ∨ *A* ≠ *B*) by *applying* (*ax_g1 A B*) .
 by *cases on* (*A* = *B* ∨ *A* ≠ *B*).
 - {
   assert (*cong A D A A*) by (*substitution*).
   assert (*A* = *D*) by *applying* (*ax_3 A D A*) .
   assert (*B* = *D*) by (*substitution*).
   conclude.
   }
 - {
   assert (*A* = *C* ∨ *A* ≠ *C*) by *applying* (*ax_g1 A C*) .
   by *cases on* (*A* = *C* ∨ *A* ≠ *C*).
   - {
     assert (*bet A B A*) by (*substitution*).
     assert (*A* = *B*) by *applying* (*th_3_4 A B A*) .
     assert (**False**) by (*substitution*).
     contradict.
     }
   - {
     assert (*C* ≠ *A*) by (*substitution*).
     assert (*B* = *D*) by *applying* (*th_4_18 C A B D*) .
     conclude.
     }
   }
Qed.

From the set of individual theorems (238), the prover ArgoCLP completely automatically proved 85 (36%) of these theorems and generated proofs in the XML format. We created a single XML document that contains all proved theorems and other theorems tagged as conjectures. The whole document matches the original book by Schwabhäuser, Szmielew, and Tarski and can be explored in the LaTeX (or PDF) form, HTML or as Isabelle or Coq development.[7]

---

[7] Translating the XML document with 85 proofs by to Isabelle, Coq, HTML, LaTeX (and then to PDF) takes altogether around 20s on a PC with AMD Opteron 6168. The resulting Isabelle document is verified in 30s, and the Coq document in 6s.

6   Related Work

In [28], Wiedijk proposes a mathematical vernacular that is in a sense the common denominator of the proof languages of Hyperproof, Mizar and Isabelle/Isar. We agree with his conclusion in the last sentence of the quotation in the introduction, but we think that the three proof languages were *not* discovered independently. Natural deduction has been introduced by the Polish logicians Łukasiewicz and Jaśkowski in the late 1920's, in reaction on the formalisms of Frege, Russell and Hilbert. The term *natural deduction* seems to have been used first by Gentzen, in German:

> *Ich wollte zunächst einmal einen Formalismus aufstellen, der dem wirklichen Schließen möglichst nahe kommt. So ergab sich ein "Kalkül des natürlichen Schließens".* (First of all I wanted to set up a formalism that comes as close as possible to actual reasoning. Thus arose a "calculus of natural deduction".)—Gentzen, Untersuchungen über das logische Schließen (Mathematische Zeitschrift 39, pp.176–210, 1935)

The qualifier *natural* was of course particularly well-chosen to express that the earlier formalisms were unnatural! As this was indeed the case, natural deduction quickly became the predominant logical system, helped by the seminal work by Gentzen on cut-elimination. (Ironically, this technical work in proof theory is best carried out with proofs represented in *sequent calculus*, using natural deduction on the meta-level.)

It should thus not come as a surprise that the vernacular we propose also is based on natural deduction. One difference with Wiedijk's vernacular is that ours is based on coherent logic instead of full first-order logic. This choice is motivated in Section 2.2 (easier semi-decision procedure and more readable proofs). Another difference is that Wiedijk allows proofs to be incomplete, whereas we stress complete proof objects. This difference is strongly related to the fact that Wiedijk's vernacular is in the first place an input formalism for proof construction, whereas our vernacular is an output formalism for proof presentation and export of proofs to different proof assistants. As far as we know, the mathematical vernacular proposed by Wiedijk's has not been implemented on its own, although Hyperproof, Mizar and Isabelle/Isar are developed using the same ideas.

A number of authors independently point to this or similar fragments of first-order logic as suitable for expressing significant portions of standard mathematics (or specifically geometry), for instance, Avigad et.al. [1] and Givant and Tarski et.al. [26,21] in the context of a new axiomatic foundations of geometry. A recent paper by Ganesalingam and Gowers [12] is also related to our work. Their goal is comparable to ours: full automation combined with human-style output. They propose inference rules which are very similar to our coherent logic based proof system. For example, their rule *splitDisjunctiveHypothesis* corresponds to the rule *case split*, *deleteDoneDisjunct* corresponds to *as*, *removeTarget* corresponds to *as* (with length of **y** greater than 0), *forwardsReasoning* corresponds to the rule *mp*. Yet, some rules they proposed are not part of our set of rules. The logic they use is full first-order, with a plan to include second-order features (this would also be perfectly possible for coherent logic, which is the first-order fragment of *geometric* logic, which is in turn a fragment of higher-order logic, see [8]). Upon closer inspection, the paper by Ganesalingam and Gowers seems to stay within the coherent fragment, and proofs by contraposition and contradiction are delegated to future work. We find some support for our approach in the observation by Ganesalingam and Gowers that it will be hard to avoid that such reasoning patterns are applied in "inappropriate contexts". On the other hand, the primary domain of application of their approach is metric space theory so far, with the ambition to attack problems in other domains as well. It would be very interesting to test the two approaches on the same problem sets. One difference is that [12] insists on proofs being faithful to the thought processes, whereas we would be happy if the prover finds a short and elegant proof even after a not-so-elegant proof search. Another difference is that we are interested in portability of proofs to other systems. To our knowledge, the prover described in [12] is not publicly available.

Compared to OMDoc [16], our proof format is much more specific (as we specify the inference rules we use) and has less features. It can be seen as a specific set of `methods` elements of the `derive` element of OMDoc.

An alternative to using coherent logic provers would be using one of the more powerful automated theorem provers and exploiting existing and ongoing work on proof reconstruction and refactoring (see, for example, [22,5,14]). This is certainly a viable option. However, reconstructing a proof from the log of a highly optimized prover is difficult. One problematic step is deskolemization, that is, proof reconstruction from a proof of the skolemized version of the problem. (The most efficient provers are based on resolution logic, and clausification including skolemizing is the first step in the solution procedure.) What can be said about this approach in its current stage is that more theorems can be proved, but their proofs can still be prohibitively complicated (or use additional axioms). It has been, however, proved beneficial to use powerful automated theorem provers as preprocessors, to provide hints for ArgoCLP.

The literature contains many results about exchanging proofs between proof assistant using deep or shallow embeddings [18,15]. Boessplug, Cerbonneaux and Hermant propose to use the $\lambda\Pi$-calculus as a universal proof language which can express proof without losing their computational properties [9]. To our knowledge, these works do not focus on the readability of proofs.

7   Conclusions and Further Work

Over the last years a lot of effort has been invested in combining the power of automated and interactive theorem proving: interactive theorem provers are now equipped with trusted support for SAT solving, SMT solving, resolution method, etc [7,6]. These combinations open new frontiers for applications of theorem proving in software and hardware verification, but also in formalization of mathematics and for helping mathematicians in everyday practice. Exporting proofs in formats such as the presented one opens new possibilities for exporting readable mathematical knowledge from automated theorem provers to interactive theorem provers. In the presented approach, the task of generating object-level proofs for proof assistants or proofs expressed in natural language is removed from theorem provers (where it would be hard-coded) and, thanks to the interchange XML format, delegated to simple XSLT style-sheets, which are very flexible and additional XSLT style-sheets (for additional target formats) can be developed without changing the prover. Also, different automated theorem provers can benefit from this suite, as they don't have to deal with specifics of proof assistants.

The presented proof representation is not intended to serve as "the mathematical vernacular". However, it can cover a significant portion of many interesting mathematical theories while it is very simple.

Often, communication between an interactive theorem prover and an external automated theorem prover is supported by a verified, trusted interface which enables direct calling to the prover. On the other hand, our work yields a common format which can be generated by different automated theorem provers and from which proofs for different interactive theorem provers can be generated. The advantage of our approach relies on the fact that the proof which is exported is not just a certificate, it is meant to be human readable.

The current version of the presented XML suite does not support function symbols of arity greated than 0. For the future work, we are planning to add that support to the proof format and to our ArgoCLP prover.

In the current version, for simplicity, the generated Isar and Coq proofs use tactics stronger than necessary. We will try to completely move to basic proofs steps while keeping simplicity of proofs. Beside planning to further improve existing XSLT style-sheets, we are also planning to implement support for additional target languages such as OMDoc.

Acknowledgements. We are grateful to Filip Marić for his feedback and advices on earlier phases of this work.

References

1. Avigad, J., Dean, E., Mumma, J.: A Formal System for Euclid's Elements. The Review of Symbolic Logic (2009)
2. Barendregt, H., Wiedijk, F.: The Challenge of Computer Mathematics. Philosophical Transactions of the Royal Society 363(1835), 2351–2375 (2005)
3. Bezem, M., Coquand, T.: Automating Coherent Logic. In: Sutcliffe, G., Voronkov, A. (eds.) LPAR 2005. LNCS (LNAI), vol. 3835, pp. 246–260. Springer, Heidelberg (2005)
4. Bezem, M., Hendriks, D.: On the Mechanization of the Proof of Hessenberg's Theorem in Coherent Logic. Journal of Automated Reasoning 40(1) (2008)
5. Blanchette, J.C.: Redirecting Proofs by Contradiction. In: Blanchette, J.C., Urban, J. (eds.) Third International Workshop on Proof Exchange for Theorem Proving, PxTP 2013, Lake Placid, NY, USA, June 9-10. EPiC Series, vol. 14, pp. 11–26. EasyChair (2013)

6. Blanchette, J.C., Böhme, S., Paulson, L.C.: Extending Sledgehammer with SMT Solvers. Journal of Automated Reasoning 51(1), 109–128 (2013)
7. Blanchette, J.C., Bulwahn, L., Nipkow, T.: Automatic Proof and Disproof in Isabelle/HOL. In: Tinelli, C., Sofronie-Stokkermans, V. (eds.) FroCoS 2011. LNCS (LNAI), vol. 6989, pp. 12–27. Springer, Heidelberg (2011)
8. Blass, A.: Topoi and Computation. Bulletin of the EATCS 36, 57–65 (1998)
9. Boespflug, M., Carbonneaux, Q., Hermant, O.: The $\lambda\Pi$-calculus Modulo as a Universal Proof Language. In: Second Workshop on Proof Exchange for Theorem Proving (PxTP). CEUR Workshop Proceedings, vol. 878, pp. 28–43. CEUR-WS.org (2012)
10. de Bruijn, N.G.: The Mathematical Vernacular, a Language for Mathematics with Typed Sets. In: Dybjer, et al. (eds.) Proceedings of the Workshop on Programming Languages (1987)
11. Fisher, J., Bezem, M.: Skolem Machines and Geometric Logic. In: Jones, C.B., Liu, Z., Woodcock, J. (eds.) ICTAC 2007. LNCS, vol. 4711, pp. 201–215. Springer, Heidelberg (2007)
12. Ganesalingam, M., Gowers, W.T.: A fully automatic problem solver with human-style output. CoRR, abs/1309.4501 (2013)
13. Kaliszyk, C., Krauss, A.: Scalable LCF-Style Proof Translation. In: Blazy, S., Paulin-Mohring, C., Pichardie, D. (eds.) ITP 2013. LNCS, vol. 7998, pp. 51–66. Springer, Heidelberg (2013)
14. Kaliszyk, C., Urban, J.: PRocH: Proof Reconstruction for HOL Light. In: Bonacina, M.P. (ed.) CADE-24. LNCS (LNAI), vol. 7898, pp. 267–274. Springer, Heidelberg (2013)
15. Keller, C., Werner, B.: Importing HOL Light into Coq. In: Kaufmann, M., Paulson, L.C. (eds.) ITP 2010. LNCS, vol. 6172, pp. 307–322. Springer, Heidelberg (2010)
16. Kohlhase, M.: An OMDoc primer. In: Kohlhase, M. (ed.) OMDoc – An Open Markup Format for Mathematical Documents [version 1.2]. LNCS (LNAI), vol. 4180, pp. 33–34. Springer, Heidelberg (2006)
17. Lee, D., Chu, W.W.: Comparative analysis of six xml schema languages. SIGMOD Record 29(3), 76–87 (2000)
18. Obua, S., Skalberg, S.: Importing HOL into Isabelle/HOL. In: Furbach, U., Shankar, N. (eds.) IJCAR 2006. LNCS (LNAI), vol. 4130, pp. 298–302. Springer, Heidelberg (2006)
19. Polonsky, A.: Proofs, Types and Lambda Calculus. PhD thesis, University of Bergen (2011)
20. Rudnicki, P.: Obvious inferences. Journal of Automated Reasoning 3(4), 383–393 (1987)
21. Schwabhäuser, W., Szmielew, W., Tarski, A.: Metamathematische Methoden in der Geometrie. Springer, Berlin (1983)
22. Smolka, S.J., Blanchette, J.C.: Robust, Semi-Intelligible Isabelle Proofs from ATP Proofs. In: Blanchette, J.C., Urban, J. (eds.) Third International Workshop on Proof Exchange for Theorem Proving, PxTP 2013. EPiC Series, vol. 14, pp. 117–132. EasyChair (2013)
23. Stojanović, S., Narboux, J., Janičić, P.: Synergy Between Interactive and Automated Theorem Proving in Formalization of Mathematical Knowledge: A Case Study of Tarski's Geometry (submitted for publication, 2014)
24. Stojanović, S., Pavlović, V., Janičić, P.: A Coherent Logic Based Geometry Theorem Prover Capable of Producing Formal and Readable Proofs. In: Schreck, P., Narboux, J., Richter-Gebert, J. (eds.) ADG 2010. LNCS (LNAI), vol. 6877, pp. 201–220. Springer, Heidelberg (2011)

25. Sutcliffe, G.: The TPTP Problem Library and Associated Infrastructure: The FOF and CNF Parts, v3.5.0. Journal of Automated Reasoning 43(4), 337–362 (2009)
26. Tarski, A., Givant, S.: Tarski's system of geometry. The Bulletin of Symbolic Logic 5(2) (June 1999)
27. Wenzel, M.: Isar - A Generic Interpretative Approach to Readable Formal Proof Documents. In: Bertot, Y., Dowek, G., Hirschowitz, A., Paulin, C., Théry, L. (eds.) TPHOLs 1999. LNCS, vol. 1690, pp. 167–183. Springer, Heidelberg (1999)
28. Wiedijk, F.: Mathematical Vernacular. Unpublished note (2000), http://www.cs.ru.nl/~freek/notes/mv.pdf
29. Wiedijk, F. (ed.): The Seventeen Provers of the World. LNCS (LNAI), vol. 3600. Springer, Heidelberg (2006)
30. Wiedijk, F.: A Synthesis of the Procedural and Declarative Styles of Interactive Theorem Proving. Logical Methods in Computer Science 8(1) (2012)

An Approach to Math-Similarity Search

Qun Zhang and Abdou Youssef

Department of Computer Science
The George Washington University
Washington DC, 20052, USA

Abstract. The unique structural syntax and the variety of semantic equivalences of mathematic expressions make it a challenge for a keyword-based text search engine to effectively meet the users' search needs. Many existing math search solutions focus on exact search where the notational matching determines the relevance rank, while the structural similarity and mathematical semantics are often missed out or not addressed adequately. One important research question is how to effectively and efficiently find math expressions that are similar to a user's query, and how to do relevance ranking of hits by similarity. This paper focuses on (1) conceptualizing similarity between mathematical expressions, (2) defining metrics to measure math similarity, (3) utilizing those metrics for math similarity search, and (4) evaluating performance to validate advantage of the proposed math similarity search. Our results show that the performance of math-similarity search is superior to that of keyword-based math search.

Keywords: math search, similarity search, similarity metric, similarity ranking, Strict Content MathML.

1   Introduction

More and more math knowledge has become available on the Web, and search is a gate to such vast treasure of digital mathematics content [11]. Even though Information Retrieval technology has reached maturity, math retrieval is still in its nascent stages, and many challenges remain. Those challenges are due in part to the significant difference of math knowledge from other textual documents. A math expression is often written in a symbolic language with several levels of abstraction, and often contains rich structural information. Additionally, notational ambiguities, and syntactical and semantic equivalences, make math knowledge harder to search. Furthermore, similarity search in math needs to capture not only the taxonomically similar operation or function names but also the hierarchically similar structures (we will use the term "function" to encompass both "function" and "operator"). For example, $x^2 + y^2 + z^2$ is expected by the user to match $a^2 + b^2 + c^2$ due to the structural similarity of the two expressions. The great inference on the structural aspect and semantic aspects of math expressions calls for a search engine that is capable of detecting and measuring similarity between mathematical constructs.

S.M. Watt et al. (Eds.): CICM 2014, LNAI 8543, pp. 404–418, 2014.

Most "first-generation" math search systems are full text-search-based math search systems that treat math objects as linear strings. However, this approach often misses out the structural information of the math expressions, and makes it nearly impossible to find a semantically similar math expression. On the other hand, there are XML-based math search solutions that identify the common sub-paths between the query expression and the candidate expressions. However, XML-based search methods often limit search to exact matches without systematically measuring the structural similarity or the semantic similarity between the query expression and the candidate expression. Similarity search enables users to find additional knowledge, discover latent relationships to different fields, and compensate for false recognition [13].

In this paper, we will lay out certain fundamental facts about math-similarity search to find, for a given user query math expression, the math expressions that are structurally and semantically similar to the query. The specific goals of this paper are:

1. Conceptualize math similarity in a way that makes it possible to measure and utilize similarity in math search;
2. Develop and study math similarity metrics to measure the similarity between two math expressions;
3. Develop algorithms for computing math-similarity metrics;
4. Leverage the NIST Digital Library of Mathematical Functions [1] to build "ground truth" of math queries and corresponding matching expressions with human experts' knowledge input;
5. Implement a ranking comparison metric to benchmark the results of a math search against the "ground truth".

The rest of the paper starts with a brief summary of the related work in Section 2. It then elaborates our research work in Section 3, and draws conclusions in Section 4.

2 Background

Existing math search engines can be categorized as text-based and structure-based. Text-based math search engines extend full-text search to achieve math awareness by transforming math expressions into either equivalent linear text tokens or expanded bags of text tokens. Miller, Youssef, et al. [6], [14], [15] developed the first generation of an equation-based math search system as part of the DLMF project at NIST. They developed an innovative TexSN (i.e. Textualization, Serialization/Scoping, and Normalization) process to convert math to text, and built a math search engine on top of existing text search technology. However the conversion process loses considerable structural, and captures little semantics. Additionally, its relevance ranking leaves room for improvement. Because it is one of the few deployed math search engines that are available for us, we leverage it for performance evaluation.

Some other text-based math search engines include Mathdex [7], EgoMath [8], and MIaS [11], ActiveMath [5]. They all took advantage of the mature and optimized text search engines that are already available. But like the DLMF they are forced to transform math expressions into the form that the text search engine can effectively process, leading to the destruction of much of the native structures of the expressions, and thus preventing truly structural or similarity search from taking place. Structure-based math search systems, on the other hand, use a radically different approach based on emerging XML-based technologies and markup languages. Those math search systems analyze the structure inherent in the content representations, and statistically identify the math expressions that have the most common sub-structures with the query expressions.

Kohlhase et al. [4] implemented MathWebSearch which leverages the semantic information that resides in the structured math equation written in MathML or OpenMath. With the adoption of the unique substitution tree indexing technique, it provides the full support of alpha-equivalence matching and sub-equation matching. However, MathWebSearch does not provide relevance ranking or similarity search.

Other structure-based math search engines include DFS & BFS Index of MathML DOM [2], Waterloo Math Retrieval System [3]. They often leverage the metadata to extract semantic annotations. But most of them either simply rank the candidate hits by basic statistical methods such as count of the occurrences of the matching sub-structures, or not pay enough attention to the matching function to calculate the similarity score between the math expressions [13].

The paramount challenge of math search is to identify relevant results by finding expressions that are similar to a query expression while allowing for difference in variable names, order, and structure. However, the lack of a definition for similarity between math expressions, and the inadequacy of exact-match searching, makes the problem of math search even harder [3]. To the best of our knowledge, there are very few efforts in math similarity search for MathML encoded expressions; Yokoi and Aizawa [13]'s work is by far the only significant one. They introduced a similarity measure that is based on the "Subpath Set" of Content MathML syntactic trees. A "Subpath Set" is defined as "the paths from the root to the leaves and all the sub-paths of those paths". Trees whose "Subpath Sets" overlap with each other are considered to be similar. The significance of their approach is that, rather than the notational similarity of tokens that the conventional math search engines evaluate, they focused on the structural similarity of MathML expressions, which we do as well. But they miss the semantic aspect in the similarity measure. Due to the numerous variations of Content MathML expressions to express one math expression, without sufficient normalization it is hard for the search engine to find semantically equivalent expressions which only differ syntactically from the query expression. Additionally, little performance evaluation was done in the aspect of ranking.

In the latest W3C release of MathML, MathML 3, a subset of Content MathML is defined: Strict Content MathML. This uses a minimal, but sufficient, set of elements to represent the meaning of a mathematical expression in a uniform and unambiguous structure [12].

Strict Content MathML requires only 10 XML Elements to be understood by MathML 3 processors, namely: *m:apply, m:bind, m:bvar, m:csymbol, m:ci, m:cn, m:cs, m:share, m:semantics, m:cerror,* and *m:cbytes.* This provides a great economy for implementation. On the other hand, MathML 3 assigns semantics to content markup by defining a mapping from arbitrary Content MathML to Strict Content MathML, and W3C even laid out a nine-step algorithm [12] to transform an arbitrary Content MathML expression into a Strict Content MathML counterpart. We limit our work to math expressions that can be encoded with Strict Content MathML. Given all these special characteristics of Strict Content MathML, it is chosen for the MathML search implementation in our research.

3   Our Approach to Math-Similarity Search

To the best of our knowledge, there is no solution available to address the similarity measurement of the Strict Content MathML expressions. This motivated us to start the research effort by addressing similarity and taking the structure-based approach to implementing semantics-sensitive math-aware similarity search with native math language MathML as query input.

3.1   Research Problem

Our research problem is defined as follows: Given a math expression that is encoded in Strict Content MathML, identify a list of structurally and semantically similar math expressions from a library of Strict Content MathML encoded math expressions, and sort the list by similarity according to some similarity measure. Specifically, the tasks of our research include: identify conceptual factors to math similarity, deduce math similarity metric, implement the math similarity metric, evaluate and refine the math similarity metric.

3.2   Math Similarity Factors

Influenced by the Multidimensional Relevance Metric proposed by [15], we came up with the vector model based multidimensional similarity metric which takes all the factors into consideration during similarity measurement. The following five factors are identified and evaluated:

1. **Taxonomic Distance of Functions** Taxonomy defines the hierarchical groups, i.e. taxa, to be referenced for grouping individual items. Taxonomic Distance is a measure of taxonomic similarity between two mathematical terms. In a taxonomy, it is intuitive to assign more similarity to two terms belonging to the same category than to terms belonging to different categories. In our search, terms that belong to the same Content Dictionary (CD)

are attributed a higher similarity value than terms that belong to different CDs.

For future consideration, even within the same Content Dictionary, some finer-granularity hierarchy could be superimposed to further differentiate the functions for the more precise similarity measurement.

2. **Data Type Hierarchical Level** The node of a MathML expression is of a data type, such as a constant number, a variable, a function (e.g. multiplication, log, etc.), or a function of function (e.g. integral, diff, etc.). Different data types contribute different levels of significance to the math expression. To illustrate, here is an example, Query Q: $a + 2$, one of the expression E_1: $a + 3$, and another expression E_2: $\log_a 2$. Expression E_1 "matches" Q at the function level, while E_2 "matches" Q at the variable and constant level. Intuitively, similarity at the function level is more important than at variable or constant level. Thus E_1 is more similar to Q than E_2 is. By reference to the Common LISP types design, we organize these different data types into a partially ordered hierarchy of types defined by the subset relationship [10]. That is, variables and constants are at the lowest level, function is at the higher level, and function of function is at the highest level. The premise is that the higher the data type is in the hierarchy, the higher the significance of that element is to the whole expression. Note that there are more data type levels in data type which can be considered in future work, but in this work we limit ourselves to two levels: function level, and argument level.

3. **Match-Depth** Naturally each MathML expression is expressed in an XML tree structure. The nodes at the higher level of the MathML expression tree decide how the expression starts, and largely determine the nature of the whole expression. Further down the tree, the nodes depict the characteristics of the expression in more detail and more locality. We claim that the similarity at the higher level matters more than at the lower level. In other words, the more deeply nested the query is in an expression, the less similarity there is between the query and that expression. An example is given in Fig. 1. Tree-wise, Q "matches" E_1 at a higher level in the tree than it does

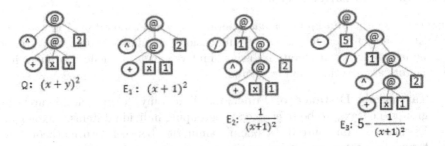

Fig. 1. Illustration of Math Similarity Factor: Depth

to E_2, and Q "matches" E_2 at a higher level in the tree than it does to E_3. Intuitively, E_1 is more similar to Q than E_2 is, and E_2 is more similar to Q than E_3 is. This illustrates that high-level matches correspond to stronger similarity than lower-level matches.

To incorporate the match-depth element into our similarity metrics, we propose to represent match-depth as a similarity-decaying multiplicative factor. It is a decaying factor because the bigger the depth, the smaller the multiplicative factor should be in order to cause the similarity to be smaller. One can utilize different models for this decay factor, such as exponential decay, linear decay, quadratic decay, or constant decay. The different models produce different degrees of penalty for depth difference. As for which model to choose for math similarity search, it depends on the type of application of the math search. For those knowledge discovery oriented math search applications, the structural similarity of math expressions is more important, thus the exponential decay model can be a good choice.

4. **Query Coverage** In actual use, how much of the query expression Q is "covered" in the returned expression E is very important. The following example gives an intuitive illustration: There is a query Q: $(x + y)^2$, an expression E_1: $(x + y)^2 + (x - y)^2$, and another expression E_2: $x + y$. Q is intuitively more similar to E_1 than to E_2. Generally, the higher the query coverage is, the higher the significance is.

5. **Formula vs. Expression** If an expression has at the root level a relational operator (e.g., $=$, $\geq$), it is treated as a "formula"; otherwise, a "non-formula". Typically in math content, formulas are more significant and more informative than non-formula expressions, and therefore more weight should be given to the former than to the latter.

Note that, strictly speaking, this is not really a similarity factor, instead it is a relevance ranking factor. But it is incorporated into our similarity measure, because our similarity measure is our relevance ranking formula.

This concludes all the factors that are considered for similarity measure. Next a similarity metric is defined to take those five factors into account for math similarity measure.

3.3   Math Similarity Metric

We take parse trees as the primary model representing math expressions, and focus especially on Strict Content MathML parse trees. The notion of similarity between two math expressions will be defined in terms of their corresponding parse trees T_1 and T_2, and the similarity measure between them, denoted $sim(T_1, T_2)$, will be defined and computed recursively based on the height of the Strict Content MathML parse tree as explained next.

1. **For two trees T_1 and T_2 of same height 0** In this case, both trees T_1 and T_2 are singleton leaves, the similarity $sim(T_1, T_2)$ is defined as:

(a) If T_1 and T_2 are constants
   i. $sim(T_1, T_2) = 1$, if $T_1 = T_2$.
   ii. $sim(T_1, T_2) = \delta$, if $T_1 \neq T_2$, where $0 \leq \delta < 1$.
   δ is one of the parameters that are optimized experimentally.

(b) If T_1 and T_2 are variables
   i. $sim(T_1, T_2) = 1$, if $T_1 = T_2$.
   ii. $sim(T_1, T_2) = \epsilon$, if $T_1 \neq T_2$, where $0 \leq \epsilon \leq 1$.
   Because the choice of symbol used for a variable name is immaterial in most cases, ϵ is simply set to 1 as the initial value in our implementation prior to the optimization process. Our research focuses on the context-free evaluation; otherwise, similarity of two variables can depend on not only value, but location and role, which can be an interesting topic for future work.

(c) If T_1 and T_2 are functions, the taxonomic distance is leveraged to measure the similarity between the two functions.
   i. $sim(T_1, T_2) = 1$, if T_1 and T_2 are the same function.
   ii. $sim(T_1, T_2) = \mu$, if T_1 and T_2 are are functions of same category in the taxonomy, where $0 < \mu < 1$. μ is one of the parameters that are optimized experimentally.
   iii. $sim(T_1, T_2) = 0$, if T_1 and T_2 are functions that belong to different categories in the taxonomy.

(d) If T_1 and T_2 belong to different data types
   i. $sim(T_1, T_2) = \theta$, if one tree is a constant and the other is a variable, where $0 \leq \theta < 1$.
   ii. $sim(T_1, T_2) = 0$, if one tree is a function and the other is a constant or variable.

2. **For two trees T_1 and T_2 of same height $h \geq 1$** In this case, the trees T_1 and T_2 are composed of function apply operator @ as root, a left-most child node representing function, followed by a list of argument nodes which are sub-trees, as illustrated in Fig. 2. Naturally the similarity between T_1 and T_2 is affected by the similarity between the two function node f_1 and f_2, and by the similarity between the two lists of argument nodes. p is the number of argument nodes in T_1, while q is the number of argument nodes in T_2. We treat T_1 as the query expression, T_2 as an expression in the database. Because functions are more important than arguments, the similarity between T_1 and T_2 is defined as a weighted sum:

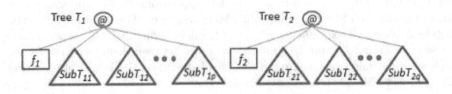

Fig. 2. Illustration of two trees T_1 and T_2 of same height $h \geq 1$

$sim(T_1, T_2) = \alpha \cdot sim(f_1, f_2) + \beta \cdot sim(\{SubT_{11}, SubT_{12}, \ldots, SubT_{1p}\},$
$\{SubT_{21}, SubT_{22}, \ldots, SubT_{2q}\}),$
where α and β are weighting factors that capture the significance of the similarity contribution from each child node of the tree. Weighting factor $\alpha = \frac{\omega}{p+\omega}$, and $\beta = \frac{1}{p+\omega}$, where ω is boost value for the leftmost child being a function data type as opposed to argument. We take $\omega > 1$. Using p instead of q takes the query coverage factor into account.

The similarity between the two lists of argument nodes,
$sim(\{SubT_{11}, SubT_{12}, \ldots, SubT_{1p}\}, \{SubT_{21}, SubT_{22}, \ldots, SubT_{2q}\}),$
is a compound value,
$0 \leq sim(\{SubT_{11}, SubT_{12}, \ldots, SubT_{1p}\}, \{SubT_{21}, SubT_{22}, \ldots, SubT_{2q}\}) \leq p.$
The measure of the similarity between the two lists of argument nodes depends on the commutative nature of the functions.

(a) If f_1 and f_2 are non-commutative functions, the order of the arguments is observed. The similarity between the two lists is the sum of the similarities between the corresponding available pairs of argument nodes with one from each tree:
$sim(\{SubT_{11}, SubT_{12}, \ldots, SubT_{1p}\}, \{SubT_{21}, SubT_{22}, \ldots, SubT_{2q}\}) =$
$\sum_{i=1}^{min(p,q)} sim(SubT_{1i}, SubT_{2i})$

(b) If f_1 and f_2 are commutative functions, an argument node in T_1 can be paired with any argument node in T_2. To find the best pairing between the 2 lists of argument nodes, the permutations of the argument nodes are taken into consideration, which can be very costly to compute. In this research, we apply the Greedy Approximation algorithm as described in Fig. 3 to find a solution that is close to the optimum similarity value.

```
Greedy_Similarity (T₁, T₂)
{
    sim(SubT₁ , SubT₂) = 0;
    P = { i | 1 ≤ i ≤ p };
    Q = { j | 1 ≤ j ≤ q };
    while ( P ≠ Ø and Q ≠ Ø ) {
        sim(SubT₁ , SubT₂) = sim(SubT₁ , SubT₂) + max {sim(SubT₁ᵢ , SubT₂ⱼ} | i ∈ P, j ∈ Q};
        P = P - { i };
        Q = Q - { j };
    }
    sim (T₁ , T₂) = (ω·sim(f₁ ,f₂) + sim(SubT₁ , SubT₂) ) / (p + ω);
    return sim (T₁ , T₂);
}
```

Fig. 3. Greedy Algorithm to find Similarity of Two Trees with Commutative Functions

In this case, the similarity between the two lists of arguments is defined as:

$sim(\{SubT_{11}, SubT_{12}, \ldots, SubT_{1p}\}, \{SubT_{21}, SubT_{22}, \ldots, SubT_{2q}\})$

$= \max\{(\sum_{i=1}^{p}(sim(SubT_{1i}, SubT_{2t(q,p,i)})))\}$, where $t(q,p,i)$ is the i-th element of a p-permutation of q.

$\approx \sum_{i=1}^{min(p,q)} \max (\ \{sim(SubT_{1i}, SubT_{2\varphi(i)})\}$, by applying greedy approximation, $\varphi(i) = 1, 2, \ldots, q$ and $\varphi(i) \notin \{\varphi(1), \varphi(2), \ldots, \varphi(i-1)\}$.

It is noted that other approximation algorithms can be used to replace the proposed Greedy algorithm for more optimum approximation and/or less computational complexity. This is not to be addressed in this research, but deferred for future research.

(c) If f_1 is commutative function, and f_2 is non-commutative function, or vice versa, we argue that this case should be the same as the above case with both f_1 and f_2 are commutative functions. Because for example, if we have query tree $Q : 5 - 2$, expression $E_1 : 5 + 2$, and expression $E_2 : 2 + 5$, then we should have $sim\ (Q, E_1) = sim\ (Q, E_2)$ which we will not have if we do not "permute" the sub-trees of the tree with commutative function. Thus, in this case, the similarity between the two lists of arguments is defined the same as the case with both functions being commutative. The example in Fig.4 is given for illustration.

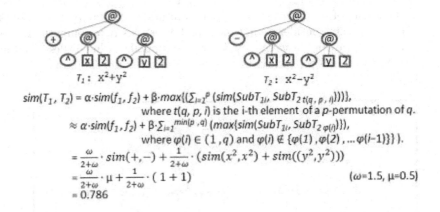

$T_1 : x^2 + y^2$ $T_2 : x^2 - y^2$

$sim(T_1, T_2) = \alpha \cdot sim(f_1, f_2) + \beta \cdot max\{(\sum_{i=1}^{p}(sim(SubT_{1i}, SubT_{2\,t(q,\,p,\,i)})))\}$,

where $t(q, p, i)$ is the i-th element of a p-permutation of q.

$\approx \alpha \cdot sim(f_1, f_2) + \beta \cdot \sum_{i=1}^{min(p,\,q)} (max\{sim(SubT_{1i}, SubT_{2\,\varphi(i)})\})$,

where $\varphi(i) \in (1, q)$ and $\varphi(i) \notin \{\varphi(1), \varphi(2), \ldots \varphi(i-1)\}$).

$= \dfrac{\omega}{2+\omega} \cdot sim(+, -) + \dfrac{1}{2+\omega} \cdot (sim(x^2, x^2) + sim((y^2, y^2)))$

$= \dfrac{\omega}{2+\omega} \cdot \mu + \dfrac{1}{2+\omega} \cdot (1 + 1)$ $(\omega{=}1.5, \mu{=}0.5)$

≈ 0.786

Fig. 4. T_1 with commutative function, and T_2 with non-commutative function

3. **For two trees T_1 and T_2 of different heights** If one of the two trees, say T_1, is of $height(T_1) = h$, and the other tree T_2 is of $height(T_2) \geq h+1$, then the match between T_1 and T_2 can be at the highest level of T_2, or nested in the T_2, and the best match of these two possibilities is taken. In other words, to measure the similarity between T_1 and T_2, not only the similarity between T_1 and T_2 at their root level is evaluated, but also the similarity between entire tree T_1 and each single sub-tree of T_2, that is $sim(T_1, SubT_{2j})$, in this

case because the match is nested, the match-Depth Penalty is applied. Then we choose whichever the larger value as the final similarity measure. Thus, in this case, the similarity between the two trees T_1 and T_2 is defined as shown in Fig 5.

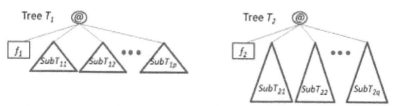

$sim(T_1, T_2) = max\{(\alpha \cdot sim(f_1, f_2) + \beta \cdot sim(\{SubT_{11}, SubT_{12} \ldots SubT_{1p}\}, \{SubT_{21}, SubT_{22} \ldots SubT_{2q}\})\},$
$\quad dp \cdot max\{ sim(T_1, SubT_{21}), sim(T_1, SubT_{22}), \ldots, sim(T_1, SubT_{2q}) \}\},$
where dp is the Depth Penalty factor.

Fig. 5. Similarity Metric for two trees T_1 and T_2 of different heights

We recursively keep comparing the first tree T_1 with the sub-trees of the second tree T_2 , till the two trees under evaluation are of the same height, in which case the similarity metric is already defined.

3.4   Performance Evaluation

To our best knowledge, there is no standard benchmark Strict Content MathML encoded documents set together with a set of standard sample queries that can be used to evaluate MathML search engine's performance. This makes it a challenge to quantitatively compare the performance of versions math similarity metrics as well as various math search engines.

1. **Evaluation Methodology** As the DLMF math digital library and search engine are among the few available and easily accessible, this research lever-ages the DLMF as the source for mathematical expressions repository, and we compare our similarity search approach to the DLMF search system. To our knowledge there is no Strict Content MathML encoding of the DLMF; therefore, a significant subset of the DLMF is hand-crafted into Strict Con-tent MathML encoding in this research. The methodology of how we build the dataset and evaluate the performance of the proposed similarity metrics is depicted below.

   On the one hand, the queries with varying degrees of mathematical com-plexity and length were selected. Table 1 lists the test queries that we used. For each query in the test set, we identify the expected relevant expressions from DLMF source repository, and further rank them manually by a group of human experts, which are then named as "ground truth". On the other hand, each query expression is compared with the expressions in the DLMF

Table 1. The Test Queries

Query ID	Query Expression	Query ID	Query Expression
Q1	e^z	Q2	$\tan(z)$
Q3	$\int_a^b f(x)dx$	Q4	$f(z_0) = \frac{1}{2\pi i}\int_c \frac{f(z)}{z-z_0}dz$
Q5	$\ln(1+z) = z - \frac{z^2}{2} - \frac{z^3}{2} - \cdots$	Q6	$\frac{d}{dx}\int_a^x f(t)dt = f(x)$
Q7	$\int_c f(z)\,dz = 0$	Q8	$f(z) = c_0 + c_1 z + c_2 z^2 + \cdots$
Q9	$\lim \frac{\sin(z)}{z}$	Q10	$A\cosh(az) + B\sinh(az)$
Q11	$\sin^2 x + \cos^2 x = 1$	Q12	$\sin(x+y) = \sin x \cos y + \cos x \sin y$
Q13	$\int \sin(x)\,dx = -\cos(x)$	Q14	$\cosh(x) \le (\frac{\sinh(x)}{x})^3$
Q15	$\sinh(z) = \frac{e^z - e^{-z}}{2}$	Q16	$\delta(x-a)$
Q17	$a^2 + b^2$	Q18	$\frac{a(1-x^n)}{1-x} = a + ax + ax^2 + \cdots + ax^{n-1}$
Q19	$\sum a_j b_j \le (\sum a_j^p)^{1/p}(\sum b_j^q)^{1/q}$	Q20	$F(x) = \frac{1}{\sqrt{2\pi}}\int f(t)e^{ixt}dt$
Q21	$\Gamma(z)$	Q22	$\det[a_{ij}]$
Q23	$a \cdot b = \sum a_j b_j$	Q24	$\sqrt{a^2 + b^2}$
Q25	$e^{i\pi} + 1 = 0$	Q26	$e^x < \frac{1}{1-x}$
Q27	$\int \frac{dx}{1+e^x}$	Q28	$(f * g)(t)$
Q29	$\lim_{x\to\infty} x^a e^{-x}$	Q30	$\frac{d}{dx}\arctan(\sin(x^2))$
Q31	$\sin^2 x + \cos^2 x$	Q32	$\sin(x+y)$
Q33	$\int \sin(x)\,dx$	Q34	$a + ax + ax^2 + \cdots + ax^{n-1}$
Q35	$\sum a_j b_j \le$	Q36	$\int_c \frac{f(z)}{z-z_0}dz$
Q37	$\sin(u) + \sin(v)$	Q38	$\cos(u) + \cos(v)$
Q39	$\frac{d}{dz}\text{arccot}(z)$	Q40	$\lim_{n\to\infty}\left(1+\frac{1}{n}\right)^n$

repository, and a similarity value is computed with the proposed similarity metric. Afterwards, this list of expressions is ordered by the similarity measurement.

Up to this point, for any given query, there are three hit lists: one from "ground truth", one from DLMF site returned by DLMF search, and another ranked by the proposed similarity metric. In order to quantitatively evaluate the performance, this research proposes to compare the three lists of results to figure out the correlation between the proposed Math-Similarity Search (MSS) result list and the "ground truth" list, and the correlation between the DLMF search result list and the "ground truth" list. Our comparison is done with respect to recall and relevance ranking.

To evaluate the quality of the relevance ranking, the two classical rank correlation coefficient metrics, namely, Kendall's tau (τ) and Spearman's rho (ρ) are used. In statistics, τ is used to measure the extent of agreement between two lists of measurements, while ρ is the standard correlation coefficient of statistical dependence between two variables. In general, the magnitude of τ is less than the value of ρ. τ focuses more on the relative order of the hits (which came before which), whereas ρ focuses more on absolute order (where each hit ranked). Both metrics are implemented in this research to complement each other in the ranking comparison.

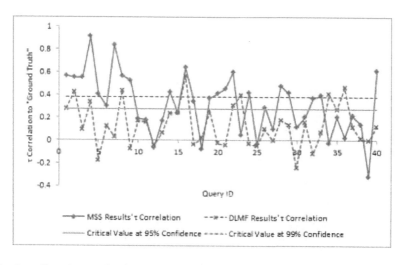

Fig. 6. τ Correlation Analysis of MSS Results vs. DLMF Results over 40 Queries

2. Performance Evaluation Results The performance evaluation of the proposed search shows that both the recall and the ranking based on our proposed similarity metric align better with the "ground truth" than that of DLMF search. Figure 6 and Fig. 7 indicate that the search results of most of the 40 queries in our evaluation that are returned by the proposed MSS search have better correlation to "ground truth" than those of DLMF, with respect to τ metric and ρ metric. That validates the advantage of the proposed MSS over the DLMF search with respect to relevance ranking.

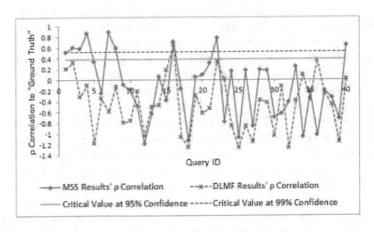

Fig. 7. ρ Correlation Analysis of MSS Results vs. DLMF Results over 40 Queries

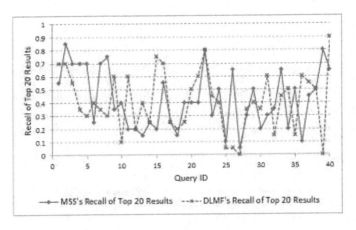

Fig. 8. Recall of MSS Top 20 Results vs. DLMF's Top 20 Results over 40 Queries

Figure 8 and Fig. 9 indicate that with respect to the recall of the top 20 results, the MSS does not differ significantly from the DLMF search. However, with respect to the recall of the top 10 results, the MSS search shows better performance than the DLMF search does.

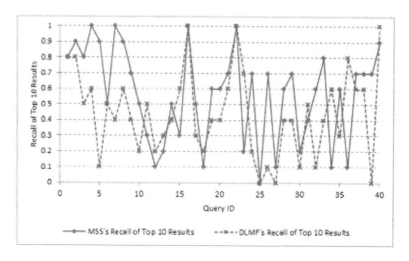

Fig. 9. Recall of MSS Top 10 Results vs. DLMF's Top 10 Results over 40 Queries

4   Conclusions and Future Work

In order to effectively and efficiently find math expressions that are similar to a user's query, this paper conceptualizes math similarity between mathematical expressions with more weight to structural similarity and mathematical semantics than the mere notational matching that many existing math search solutions focus on. Further, this paper proposes a semantic-sensitive math-similarity metric to measure the math similarity. With the availability of Strict Content MathML which represents math in disambiguated uniform structure, an algorithm is developed to compute the math similarity between any two Strict Content MathML encoded math expressions. Additionally, a "ground truth" of math queries and corresponding matching expressions is constructed by leveraging DLMF, and is used as a benchmark for performance evaluation. Comparing with the existing non-similarity based math search techniques, primarily the DLMF math search, the proposed math-similarity search does show the performance advantage with respect to both recall and relevance ranking.

However, many parameters of the proposed similarity metric are yet to be optimized, including taxonomic distance values (e.g. μ, θ) between functions, function nodes type booster value ω, depth penalty decay model and its parameters, query coverage factor, etc. We plan to address them in the near future. Other future directions include: (1) Address normalization in the context of Strict Content MathML. (2) Cover in the similarity search the remaining elements of Strict Content MathML that are not covered in this research, such as "*m:bind*" and "*m:share*". (3) Leverage the sample queries and benchmark dataset that are to be produced from the NTCIR-11 [9] ongoing math task, for more thorough and more objective performance evaluation.

References

1. The Digital Library of Mathematical Functions (DLMF), the National Institute of Standards and Technology (NIST), http://dlmf.nist.gov/
2. Hashimoto, H., Hijikata, Y., Nishida, S.: Incorporating Breadth First Search for Indexing MathML Objects. In: IEEE International Conference on Systems, Man and Cybernetics, SMC 2008 (2008)
3. Kamali, S., Tompa, F.: A New Mathematics Retrieval System. In: Proceedings of the 19th ACM International Conference on Information and Knowledge Management, CIKM 2010. ACM, New York (2010)
4. Kohlhase, M., Sucan, I.: A Search Engine for Mathematical Formulae. In: Calmet, J., Ida, T., Wang, D. (eds.) AISC 2006. LNCS (LNAI), vol. 4120, pp. 241–253. Springer, Heidelberg (2006)
5. Libbrecht, P., Melis, E.: Methods to Access and Retrieve Mathematical Content in ACTIVEMATH. In: Iglesias, A., Takayama, N. (eds.) ICMS 2006. LNCS (LNAI), vol. 4151, pp. 331–342. Springer, Heidelberg (2006)
6. Miller, B., Youssef, A.: Technical Aspects of the Digital Library of Mathematical Functions. Annals of Mathematics and Artificial Intelligence 38(1-3), 121–136 (2003)
7. Miner, R., Munavalli, R.: An Approach to Mathematical Search Through Query Formulation and Data Normalization. In: Kauers, M., Kerber, M., Miner, R., Windsteiger, W. (eds.) MKM/Calculemus 2007. LNCS (LNAI), vol. 4573, pp. 342–355. Springer, Heidelberg (2007)
8. Miutka, J., Galambo, L.: Mathematical Extension of Full Text Search Engine Indexer. In: Proceedings of Information and Communication Technologies: From Theory to Applications, ICTTA 2008, IEEE Catalog number CFP08577, Syria, pp. 207–208 (2008)
9. The 11th National Institute of Informatics Testbeds and Community for Information access Research Workshop (2013-2014), http://ntcir-math.nii.ac.jp/
10. Reddy, A.: Features of Common Lisp (2008),
    http://random-state.net/features-of-common-lisp.html
11. Sojika, P., Liška, M.: The Art of Mathematics Retrieval. In: Proceedings of the ACM Conference on Document Engineering, DocEng 2011, Mountain View, CA, pp. 57–60 (2011)
12. Mathematical Markup Language (MathML) Version 3.0 (3rd edn.), World Wide Web Consortium, http://www.w3.org/TR/MathML3/
13. Yokoi, K., Aizawa, A.: An Approach to Similarity Search for Mathematical Expressions using MathML. In: Towards a Digital Mathematics Library, Grand Bend, Ontanrio, Canada, pp. 27–35. Masaryk University Press, Brno (2009)
14. Youssef, A.: Information Search and Retrieval of Mathematical Contents: Issues and Methods. In: The ISCA 14th Int'l Conf. on Intelligent and Adaptive Systems and Software Engineering (IASSE 2005), Toronto, Canada, July 20-22 (2005)
15. Youssef, A.S.: Methods of Relevance Ranking and Hit-content Generation in Math Search. In: Kauers, M., Kerber, M., Miner, R., Windsteiger, W. (eds.) MKM/Calculemus 2007. LNCS (LNAI), vol. 4573, pp. 393–406. Springer, Heidelberg (2007)

Digital Repository of Mathematical Formulae[*]

Howard S. Cohl[1], Marjorie A. McClain[1], Bonita V. Saunders[1],
Moritz Schubotz[2], and Janelle C. Williams[3]

[1] Applied and Computational Mathematics Division, National Institute
of Standards and Technology (NIST), Gaithersburg, Maryland, USA
{howard.cohl,marjorie.mcclain,bonita.saunders}@nist.gov
[2] Database Systems and Information Management Group,
Technische Universität, Berlin, Germany
schubotz@tu-berlin.de
[3] Department of Mathematics and Computer Science, Virginia State University,
Petersburg, Virginia, USA
janelle.williams35@gmail.com

Abstract. The purpose of the NIST Digital Repository of Mathematical Formulae (DRMF) is to create a digital compendium of mathematical formulae for orthogonal polynomials and special functions (OPSF) and of associated mathematical data. The DRMF addresses needs of working mathematicians, physicists and engineers: providing a platform for publication and interaction with OPSF formulae on the web. Using MediaWiki extensions and other existing technology (such as software and macro collections developed for the NIST Digital Library of Mathematical Functions), the DRMF acts as an interactive web domain for OPSF formulae. Whereas Wikipedia and other web authoring tools manifest notions or descriptions as first class objects, the DRMF does that with mathematical formulae. See http://gw32.iu.xsede.org/index.php/Main_Page.

1 Introduction

Compendia of mathematical formulae have a long and rich history. Many scientists have developed such repositories as books and these have been extremely useful to scientists, mathematicians and engineers over the last several centuries (see [2, 3, 5, 6, 9, 13, 15] for instance). While there may be some overlap of formulae in different compendia, one often needs to be familiar with many different compendia to find a specific desired formula. Online compendia of mathematical formulae exist, such as the NIST Digital Library of Mathematical Functions (DLMF), subsets of Wikipedia, and the Wolfram Functions Site. We hope to take advantage of the best aspects of these online efforts while also incorporating powerful new features that a community-arm of scientists should find beneficial. Our strategy is to start with validated and trustworthy special function data from the NIST DLMF, while adding Web 2.0 capabilities which will

---

[*] Official contribution of the National Institute of Standards and Technology; not subject to copyright in the United States.

S.M. Watt et al. (Eds.): CICM 2014, LNAI 8543, pp. 419–422, 2014.

encourage community members to discuss mathematical data associated with formulae. These discussions will include internally hyperlinked proofs as well as mathematical connections between formulae in the repository.

The online repository will be designed for a mathematically literate audience and should (1) facilitate interaction among a community of mathematicians and scientists interested in formulae data related to orthogonal polynomials and special functions (OPSF); (2) be expandable, allowing the input of new formulae; (3) be accessible as a standalone resource; (4) have a user friendly, consistent, and hyperlinkable viewpoint and authoring perspective; and (5) contain easily searchable mathematics and take advantage of modern MathML tools for easy to read, scalably rendered mathematics. It is the desire of our group to build a tool that brings the above features together in a public website for mathematicians, scientists and engineers. We refer to this web tool as the Digital Repository of Mathematical Formulae (DRMF).

Our project was motivated by the goal of creating an interactive online compendium of mathematical formulae. This need was addressed in SIAM Activity Group OPSF-Net discussions, such as Dmitry Karp (OPSF-Net 18.4, Topic #5). In that OPSF-Net edition, there were two related posts (OPSF-Net 18.4, Topics #6,#7) with a follow-up post in OPSF-Net 18.6, Topic #3.

2   Implementation

In our project, we have taken advantage of the free and open source MediaWiki wiki software as well as tools developed within the DLMF project [14], such as LaTeXML and the DLMF LaTeX macros. DLMF macros (and extensions as necessary) tie specific character sequences to unique mathematical objects such as special functions, orthogonal polynomials, or to other mathematical symbols associated with these. The DLMF macros are hence used to define OPSF within DRMF and through LaTeXML, their corresponding rendered mathematical symbols. Furthermore, the use of DLMF macros as linked to their definitions within the DLMF, allows for easy access to precise OPSF definitions for the symbols used within the LaTeX source for OPSF formulae. The committed use of DLMF macros guarantees a mathematical and structural consistency throughout the DRMF. As a web tool, the DRMF provides (1) formula interactivity, (2) formula home pages, (3) centralized bibliography, and (4) mathematical search. The DRMF shares the core DLMF component, LaTeXML, which (through the MediaWiki Math extension) processes Wikitext math markup written in LaTeX to produce XML and HTML. For formula interactivity and menus linked to formulae, we have utilized the JOBAD interactivity framework and are investigating the use of MathJax [4]. We have also incorporated the MediaWiki: Math and MathSearch [16] extensions. Within the DRMF, we will develop technology for users to interact with formulae using a clipboard, which allows for easy copy/paste of formula source representations (to include LaTeX with DLMF macros; presentation or content MathML; as well as input formats for computer algebra systems such as Mathematica, Maple, and Sage).

The DRMF treats formulae as first class objects, describing them in formula home pages that currently contain: (1) a rendered description of the formula itself (required); (2) bibliographic citation (required); (3) open section for proofs (required); (4) list of symbols used and links to their definitions corresponding to the DLMF macros (required); (5) open section for notes relevant to the formula (e.g., formula name, if the formula is a generalization or specialization of some other formula, growth or decay conditions, links to errata pages, etc.); (6) open section for external links; (7) substitutions with definitions required to understand the formula; and (8) constraints the formula must obey. For each formula home page there is a corresponding talk page, and we are incorporating a strategy for handling the insertion of formula errata. A major resource in our ability to implement effective and precise OPSF search will be the use of the DLMF macros in building the LaTeX source for OPSF formulae and related mathematical data.

Next, we present an overview of the seed resources, which we plan to incorporate within DRMF. We have been given permission and are seeding the DRMF with data from the NIST DLMF [14]. We have also been given permission to and are seeding LaTeX formulae data from [11] (KLS). We will also incorporate Tom Koornwinder's companion of recent arXiv published additions to KLS [12]. We have also been given permission to incorporate seed formula data from [5, 6] (BMP). Efforts to upload BMP data, as well as any book data without existing LaTeX source, will prove extremely difficult, since this effort will rely on the use of mathematical optical character recognition (OCR) software such as InftyReader to produce LaTeX source for these formulae. Mathematical OCR is still in its nascence and this effort is currently under consideration for feasibility of use. We are in communication with other authors and publishers to gain access and permission for other proven sources of mathematical OPSF formulae such as [1, 7, 9, 10] and we are are excited about the prospect of seeding proof data by Victor Moll and collaborators (see for instance [8]). For LaTeX source where DLMF macros are not present (such as KLS), we are developing tools which automate DLMF macro replacements. Seeding and generating symbol lists are accomplished by converting LaTeX source into Wikitext, in an automated fashion. We use `Pywikibot` to automate the uploading of Wikitext pages to our demo site.

Acknowledgements. [1] We are deeply indebted to Deyan Ginev for sharing with us his expansive vision and especially for his support in the development of our proof of concept. Without his guidance and coding, our present demonstration would not be possible. We would also like to thank Bruce Miller at NIST for his invaluable contributions regarding LaTeXML. We would also like to express our deep gratitude to the KWARC group at Jacobs University, Bremen,

---

[1] The mention of specific products, trademarks, or brand names is for purposes of identification only. Such mention is not to be interpreted in any way as an endorsement or certification of such products or brands by the National Institute of Standards and Technology, nor does it imply that the products so identified are necessarily the best available for the purpose. All trademarks mentioned herein belong to their respective owners.

Germany, and especially to its group leader, Michael Kohlhase, for his advice and for access to his group's mathweb server for our initial DRMF development. We would also like to thank Dan Lozier, Tom Koornwinder, Dmitry Karp, Dan Zwillinger, Victor Moll, and Hans Volkmer for offering their advice and for valuable discussions.

References

[1] Andrews, G.E., Askey, R., Roy, R.: Special Functions. Encyclopedia of Mathematics and its Applications, vol. 71. Cambridge University Press, Cambridge (1999)

[2] Brychkov, Y.A.: Handbook of Special Functions: Derivatives, Integrals, Series and Other Formulas. Chapman & Hall/CRC Press, Boca Raton (2008)

[3] Byrd, P.F., Friedman, M.D.: Handbook of Elliptic Integrals for Engineers and Physicists. Die Grundlehren der Mathematischen Wissenschaften in Einzeldarstellungen mit Besonderer Berücksichtigung der Anwendungsgebiete, Bd LXVII. Springer, Berlin (1954)

[4] Cervone, D.: Mathjax: A Platform for Mathematics on the Web. Notices of the American Mathematical Society 59(2), 312–316 (2012)

[5] Erdélyi, A., Magnus, W., Oberhettinger, F., Tricomi, F.G.: Higher Transcendental Functions, vols. 1-3 (1981)

[6] Erdélyi, A., Magnus, W., Oberhettinger, F., Tricomi, F.G.: Tables of Integral Transforms, vols. 1-2 (1954)

[7] Gasper, G., Rahman, M.: Basic Hypergeometric Series, 2nd edn. Encyclopedia of Mathematics and its Applications, vol. 96. Cambridge University Press, Cambridge (2004), With a foreword by Richard Askey

[8] Glasser, L., Kohl, K.T., Koutschan, C., Moll, V.H., Straub, A.: The integrals in Gradshteyn and Ryzhik. Part 22: Bessel-K functions. Scientia. Series A. Mathematical Sciences. New Series 22, 129–151 (2012)

[9] Gradshteyn, I.S., Ryzhik, I.M.: Table of Integrals, Series, and Products, 7th edn. Elsevier/Academic Press, Amsterdam (2007)

[10] Ismail, M.E.H.: Classical and Quantum Orthogonal Polynomials in one Variable. Encyclopedia of Mathematics and its Applications, vol. 98. Cambridge University Press, Cambridge (2005), With two chapters by Walter Van Assche, With a foreword by Richard A. Askey

[11] Koekoek, R., Lesky, P.A., Swarttouw, R.F.: Hypergeometric Orthogonal Polynomials and their q-analogues. Springer Monographs in Mathematics. Springer, Berlin (2010), With a foreword by Tom H. Koornwinder

[12] Koornwinder, T.H.: Additions to the formula lists in "Hypergeometric orthogonal polynomials and their q-analogues", by Koekoek, Lesky and Swarttouw. arXiv:1401.0815 (2014)

[13] Magnus, W., Oberhettinger, F., Soni, R.P.: Formulas and Theorems for the Special Functions of Mathematical Physics, 3rd enlarged edn. Die Grundlehren der mathematischen Wissenschaften, Band 52. Springer-Verlag New York, Inc., New York (1966)

[14] Olver, F.W.J., Lozier, D.W., Boisvert, R.F., Clark, C.W. (eds.): NIST Handbook of Mathematical Functions. Cambridge University Press, Cambridge (2010), Companion to http://dlmf.nist.gov

[15] Prudnikov, A.P., Brychkov, Y.A., Marichev, O.I.: Integrals and Series, vols. 1-5. Gordon & Breach Science Publishers, New York (1986)

[16] Schubotz, M.: Making Math Searchable in Wikipedia. In: Conferences on Intelligent Computer Mathematics, abs/1304.5475 (2013)

NNexus Reloaded

Deyan Ginev[1] and Joseph Corneli[2]

[1] Computer Science, Jacobs University Bremen, Germany
[2] Knowledge Media Institute, The Open University, UK

Abstract. Interlinking knowledge is one of the cornerstones of online collaboration. While wiki systems typically rely on links supplied by authors, in the early 2000s the mathematics encyclopedia at Planet-Math.org introduced a feature that provides automatic linking for previously defined concepts. The NNexus software suite was developed to support the necessary subtasks of concept indexing, concept discovery and link-annotation. In this paper, we describe our recent reimplementation and revisioning of the NNexus system.

1 NNexus 1.0 – Introduction

PlanetMath.org is a mathematics digital library, built "the commons-based peer production way" [Kro03]. Like Wikipedia, which launched the same year, Planet-Math has been created by volunteer contributors from around the world. However, unlike Wikipedia, PlanetMath focuses solely on mathematics. Since its launch, it has used custom software both to support the display of mathematical expressions, and to facilitate the integration of new user-contributed content. One of the features designed to assist in content integration was an *autolinking service*. This service allowed authors to write without concerning themselves with wiki-style links to technical concepts that had already been added to the corpus. Instead, these links would be added automatically – and links would be recalculated and adjusted automatically as the encyclopedia grew, using a sophisticated caching and expiry system. The system provided an example of named entity recognition [NS07], where the entities to identify in submitted text are article titles, the names of terms defined in the articles, and any known synonyms. The process of adding links to named entities in text has come to be known as "wikification" [Rat+11].

In 2006, NNexus 1.0 began the process of decoupling autolinking from PlanetMath, and provided integration with other corpora (Wikipedia, Mathworld) on a demonstration basis [GKX06], an effort that has matured with the current release.

2 NNexus 2.0 – Reload, Refresh, Refactor

The primary goal of our rebuild was to decouple fully from the old Noosphere system on PlanetMath.org. A strong contributing motivation was that Noosphere was in the process of being deprecated on PlanetMath and replaced by

S.M. Watt et al. (Eds.): CICM 2014, LNAI 8543, pp. 423–426, 2014.

the new Planetary system [CD12]. The new NNexus works with Planetary, but also functions in a stand-alone fashion, and is published as a software library on the Comprehensive Perl Archive Network (CPAN)[1]. It has been refactored to operate either as a web service[2], or programmatically via an API. NNexus accepts arbitrary HTML input and performs concept discovery against its concept index, followed by a serialization of the mined data, either as stand-off metadata or by in-place embedding.

Concept indexing is performed by NNexus' built-in web crawler. It is based on a plugin architecture. Every indexed web resource requires its own indexer class, which contains the custom rules for detecting the concept definitions in the page. For example, PlanetMath's key terms are found in RDFa metadata that has been deposited in the encyclopedia pages, whereas the Digital Library of Mathematical Functions lists its defined concepts in its index as bold-anchored elements.

The current NNexus release ships with a database that integrates the concepts from seven web resources for mathematical concepts. These include the three best-known web resources for mathematics – Wolfram's MathWorld; PlanetMath.org; and Wikipedia – as well as Springer's Encyclopedia of Mathematics; the Digital Library of Mathematical Functions (DLMF); the nLab (which focuses on category theory); and the recently created MathHub.info.

At the time of writing, the NNexus index contains just under 50,000 unique concepts in its index. With the introduction of client-side tools for embedding NNexus [Gin13], we can also report successful auto-linking in third-party platforms such as arXiv.org and Zentralblatt MATH.

3   Concept Discovery

The NNexus implementations to date have only scratched the surface of the knowledge discovery problem. NNexus performs longest-token matching, aided by classic preprocessing techniques (stopword lists, morphological normalization) to discover all possible concept candidates. Concepts are considered discovered if there is an exact match between the linguistically normalized input document and the identically normalized concept index. When a concept A is a substring of a concept B, since they both match at the same starting point of the input, preference is given to the longer string B. To demonstrate, take A to be "fundamental groupoid" and B to be "fundamental groupoid functor".

This simplistic approach leads to false positive hits, for instance, in words that have multiple part-of-speech uses or words that have both technical and everyday meanings. Accordingly, each of the following examples becomes a candidate for linking, even though the words in the right-hand column are not being used in a technical sense.

---

[1] Run **cpan NNexus** to install the software locally.

[2] A demonstration instance is available at http://nnexus.mathweb.org.
Sending an HTML snippet as the body of a POST request will return the link-annotated snippet back, embedded in a thin JSON wrapper.

*"Let G be a **group**"*   *"**group** the numbers in rows"*
*"**chain** in a graph"*   *"**chain** made of steel"*
*"**permanent** of a matrix"* *"using a **permanent** marker"*

This particular phenomenon is less observable as the length of the concept grows, both because longer words are less frequently overloaded and because multi-word concepts rarely have non-technical meanings. Inversely, the problem is particularly challenging in concepts with short single-word names.

Another challenge is disambiguating between overloaded concept names, used differently in different scientific areas. To address that, NNexus does not immediately return all of the named entities it discovers. Instead, it first uses a clustering algorithm, based on a distance metric between the classes of the MSC [Ame09] categorization scheme, which determines a kernel of closely related concepts. We have observed our distance metric is effective in separating concepts from typically disjoint subfields of science, but less successful in making fine-grained distinctions in subfields that tend to have a lot of mutual connections. For example, "entanglement" is a concept both in quantum mechanics and graph theory. NNexus is able to tell which meaning is intended by contextually clustering with the rest of the discovered concepts. However, generic concepts such as "equivalence" tend to be redefined in closely related subfields, and NNexus cannot tell these apart.

In addition to these technical limitations, NNexus is limited by the quality of the metadata provided by its indexed sources. For example, as pointed out by one of the reviewers of this paper, PlanetMath currently has no article on "classical logic", and links to this term are currently being directed to PlanetMath's article on quantum logic. This looked like a rather strange error until we realized that the quantum logic article includes a definition of the term "classical logic", in contravention of the "one main concept per article" norm.

4   NNexus 3.0 Revolution – An Outlook

Auto-linking continues to be a useful tool around PlanetMath. For instance, Planetary added support for contributing problems and problem sets, and technical terms in problems are linked to definitions drawn from the encyclopedia. We plan to add a PlanetMath feature where, given a contributed piece of text, a small "course packet" of preliminaries would be built on the fly, created out of auto-linked encyclopedia articles. Thanks to the metadata in the links provided by NNexus, we will be able to consider both "incoming" and "outgoing" links – this means we can discover applications of a concept as well as simpler concepts.

Some other efforts that we plan to explore include autolinking in math blogs, such as the blogs indexed on http://mathblogging.org/. NNexus could build a "term cloud" of technical terms from across the math blogosphere, providing a useful access method that parallels the familiar tag cloud.

The main challenge ahead is to solve the problem of reliable concept discovery. The immediate goal is to achieve reliable disambiguation of overloaded concept words (such as "set" or "group"), possibly by employing the help of a part-of-speech tagger. A complementary idea is to improve precision by augmenting longest-token matching with weights derived by statistical term-likelihood analysis. As statistical term-likelihood methods do not depend on an a priori fixed lexicon, they could also be used to detect concepts that are not yet included in the index. That would allow us to enable another desirable feature – the automatic creation of dangling links (similar to Wikipedia's "red links").

Deeper scrutiny of mathematical formulas and terms will allow us to link occurrences of math constants in MathML, both globally, for symbols like the reduced Planck constant $\hbar$, and locally, following the annotation of the corresponding natural language term, such as "Assume a cyclic group $\mathbb{Z}_{mn}$ ...".

References

[Ame09]   American Mathematical Society. Mathematics Subject Classification MSC 2010 (2009), http://www.ams.org/mathscinet/msc/

[CD12]   Corneli, J., Dumitru, M.A.: PlanetMath/Planetary. In: Davenport, J., et al. (eds.) Joint Proceedings of the 24th OpenMath Workshop, the 7th Workshop on Mathematical User Interfaces (MathUI), and the Work in Progress Section of the Conference on Intelligent Computer Mathematics, Bremen, Germany, July 9-13. CEUR Workshop Proceedings, vol. 921, pp. 66–72. Aachen (2012), http://ceur-ws.org/Vol-921/wip-02.pdf

[Gin13]   Ginev, D.: NNexus Glasses: a drop-in showcase for wikification. In: Lange, C., et al. (eds.) Joint Proceedings of the MathUI, OpenMath, PLMMS and ThEdu Workshops and Work in Progress at the Conference on Intelligent Computer Mathematics 2013, Bath, UK, July 8-12. CEUR Workshop Proceedings, vol. 1010. Aachen (2013), http://ceur-ws.org/Vol-1010/paper-13.pdf

[GKX06]   Gardner, J., Krowne, A., Xiong, L.: NNexus: Towards an automatic linker for a massively-distributed collaborative corpus. In: International Conference on Collaborative Computing: Networking, Applications and Worksharing, CollaborateCom 2006, pp. 1–3. IEEE (2006) ISBN: 1424404290

[Kro03]   Krowne, A.: Building a digital library the commons-based peer production way. D-Lib Magazine 9(10) (2003)

[NS07]   Nadeau, D., Sekine, S.: A survey of named entity recognition and classification. Lingvisticae Investigationes 30(1), 3–26 (2007)

[Rat+11]   Ratinov, L., et al.: Local and Global Algorithms for Disambiguation to Wikipedia. In: The 49th Annual Meeting of the Association for Computational Linguistics: Human Language Technologies, ACL 2011 (2011), http://cogcomp.cs.illinois.edu/papers/RRDA11.pdf

E-books and Graphics with LaTeXml*

Deyan Ginev[1], Bruce R. Miller[2], and Silviu Oprea[3]

[1] Computer Science, Jacobs University Bremen, Germany
[2] National Institute of Standards and Technology, Gaithersburg, MD, USA
[3] Department of Computer Science, University of Oxford, Oxford, UK

Abstract. Marked by the highlights of native generation of EPUB E-books and Tikz support for creating SVG images, we present an annual report of LaTeXML development in 2013. LaTeXML provides a reimplementation of the TeX parser, geared towards preserving macro semantics; it supports an array of output formats, notably HTML5, EPUB, XHTML and its own LaTeX-near XML. Other highlights include enhancing performance when used inside high-throughput build-systems, via incorporating a native ZIP archive workflow, as well as a simplified installation procedure that now allows to deploy LaTeXML as a cloud service. To this end, we also introduce an official plugin-based scheme for publishing new features that go beyond the core scope of LaTeXML, such as web services or unconventional post-processors. The software suite has now migrated to GitHub and we welcome forks and patches from the wider FLOSS community.

1 Introduction

Another busy year of LaTeXML[1] development has gone by; while we've not completely accomplished all the tasks we'd hoped for (c.f. [1]), we've finished others including some we hadn't originally planned. While it was originally developed for NIST's Digital Library of Mathematical Functions[2], where it continues to serve, we continue to find additional applications. One, carried out this year, was the natural extension of the system to generate EPUB documents; the first converter, to our knowledge, natively generating EPUB from TeX. Including MathML, along with Daisy[3] support of audio rendering of math, EPUB is a major step forward for accessibility. Two planned milestones were also completed, namely: supporting the Tikz, a large, elaborate graphics package in which one draws complex diagrams, plots and other 2D and 3D graphics using TeX markup; as well as completing a community-facing project reorganization. Together, these features are hoped to extend the reach of MKM technologies.

Before we delve into details, a little background about LaTeXML may be in order. Two main approaches are currently used to generate HTML from TeX. The first, exemplified by tex4ht, uses the actual TeX engine to process the source

---

* The rights of this work are transferred to the extent transferable according to title 17 U.S.C. 105.

[1] See http://dlmf.nist.gov/LaTeXML/
[2] See http://dlmf.nist.gov
[3] See http://www.daisy.org/

while redefining certain commands to drop \special data into the normal dvi
output. An alternative dvips then deciphers that augmented dvi to infer and
construct the appropriate HTML. In the second approach, used by LaTeXML, the
program emulates TeX for the most part but interprets some macros specially,
producing XML directly.

The first approach has the advantage of (usually) allowing the processing of arbitrary TeX and LaTeX packages, although the resulting HTML may not reflect the
intended structure nor semantics. The challenges are in the TeX programming
necessary to insert the \specials, generating valid, indeed even well-formed,
HTML, and in recovering sufficient semantic structure from the dvi.

The second approach gives more direct control of the generated output. It is
easier to extend to new XML structures, and being fundamentally XML aware, it
produces valid XML. LaTeXML uses an intermediate XML format preserving the semantic structure. A feature of LaTeXML 'bindings' (LaTeXML's re-implementation
of LaTeX packages) is control sequences defined to be "Constructors", directly
constructing the XML representation of their content. The challenge lies in emulating TeX sufficiently well to process complex packages, or alternatively, to
develop LaTeXML-specific bindings for them.

In either approach, LaTeX packages that define macros with semantic intent
must be dealt with individually or else the semantics will be lost.

2   Reorganization

We have reorganized both our code development and our code base. In the first
sense, we have moved our repository to GitHub[4] where you can more conveniently browse our code, or obtain the latest version. We have also ported our
Trac tickets to GitHub's Issues, preserving all bug and feature requests.

Along with the move to GitHub came opportunities to share code and development calling for clearer coding standards. We committed to code quality and
formatting by adopting perltidy and perlcritic policies, which were adapted
to the polyglot of TeX, Perl, XML, XSLT, automatically enforced by git.

In the second sense, we have reorganized the code itself to more clearly separate the modules related to the separate phases of processing. At the same
time, we enable "conversion as an API", offering a connection and code sharing
between those phases when more complex processing is called for, such as carrying a single TeX source file through the full processing to HTML, or even EPUB
(see §3). In particular, it provides better support for daemonized processing,
foundational to batch conversions and web service deployments.

This reorganization positions us to develop a plugin architecture allowing
modular extensions covering both new LaTeX styles and bindings, but also enhanced postprocessing for more sophisticated applications such as sTeX. We have
already refactored three flavors of LaTeXML web servers, an alternative grammar
for math parsing, as well as an extension for converting TeX formulas into queries
for the MathWebSearch search engine, all hosted on GitHub as separate repositories. The true power of the new contribution model is revealed when combined

---

[4] See https://github.com/brucemiller/LaTeXML

with Perl's CPAN distribution and dependency management system, which will allow for single command installation of any LaTeXML-based project and its full dependency tree.

3   E-books

The newest version of EPUB, version 3, is primarily a packaging of HTML pages representing chapters or sections into a structured ZIP archive. The big step forward for the scientific community is that it now calls for the use of MathML to represent mathematics. Since LaTeXML is already generating HTML, with embedded MathML, and allows that output to be split into multiple pages as specified by the user, it seemed an obvious and natural extension to generate EPUB documents. Moreover, the web-service spin-off projects had already called for and drafted the compression of the resulting directory of generated content into a ZIP archive. Thus, with appropriate rearrangement of the pieces, and the addition of a Manifest of the correct structure, we have all the basic components needed to generate EPUB documents. We have generated a number of EPUB documents and successfully validated them against the official idpf validator[5].

We subsequently considered to also add support for Amazon's proprietary mobi E-book format. However, at the time of writing the mobi ecosystem is transitioning to the new Amazon Kindle Format 8 (AKF8), which aims to more fully align with EPUB 3.0. Finally, the lack of an open ecosystem around the format prevented us from repeating the quick and painless design process for the EPUB output, so we did not venture further.

4   Graphics

Given the challenges of developing LaTeXML bindings for complex LaTeX packages, we were skeptical when Michael Kohlhase initially posed the challenge: Was LaTeXML's engine good enough to implement the TikZ package and generate SVG? The package is so large and complex, not to mention its development so fast-moving, that creating LaTeXML-specific bindings for all its many commands is impractical. However, TikZ is designed to pass all processed graphics through a relatively small driver layer, and in fact already has a tex4ht driver for producing SVG! Providing we can faithfully emulate all the TeX processing that leads to that driver layer, we may have a chance; presumably any semantics implied by TikZ markup isn't so critical, but the expected SVG obviously is.

The main tasks, then, were to implement LaTeXML bindings for just that driver, covering universal graphics primitives such as points, lines and angles; then improve LaTeXML's engine to cope with the sophisticated TeX macro usage in the higher layers of pgf and TikZ.

Ultimately, we succeeded beyond our expectations. Although the results are not perfect, LaTeXML now successfully processes 3/4 of the first page[6] of TikZ

---

[5] See http://validator.idpf.org/
[6] See http://www.texample.net/tikz/examples/all/

examples on the TeXample.net website, generating valid HTML5, with text and MathML combined. In contrast, `tex4ht` succeeds on slightly more than half the examples, often producing invalid markup, and doesn't support MathML embedded in the SVG. It must be admitted, however, that LaTeXML is *very* slow at processing `TikZ` markup! Converting the 'signpost' example from TeXample.net required almost 2 minutes, whereas `tex4ht` needed only 2 seconds (albeit with incomplete math). `pdflatex` converts it to pdf in less than half a second.

In the process, we have further improved the fidelity of the TeX emulation, introduced a (currently very rudimentary) mechanism for estimating the size of displayed objects and exercised the integration of both MathML and SVG into HTML. Additionally, LaTeXML now has its own TeX profiler, which offers binding developers per-macro feedback on exclusive runtimes, helping to identify core conversion bottlenecks. These improvements are beneficial even outside the graphics milestone and contribute to an overall better LaTeXML ecosystem.

Areas needing further work are `TikZ`' matrix structure which currently clashes with LaTeXML's handling of alignments; inaccuracies of LaTeXML's sizing of objects; and, of course, examples involving other exotic packages not yet known to LaTeXML. We plan to test against the entire suite of examples at TeXample.net to discover other weaknesses and further improve the module.

Beyond `TikZ`, we are hoping to leverage this experience and apply it to supporting the `xy` package, another popular and powerful system. It seems to have a less well-defined driver layer and we are in the early stages of discovering the smallest set of macros that could serve that function. Nevertheless, we have had some preliminary, proof-of-concept, success. We already have minimal support for the `pstricks` package, but with its Postscript oriented design, it is more time consuming to develop further bindings.

5   Outlook

The initial success with `TikZ` processing is quite gratifying, but it needs refinement, and we look forward to testing on a larger scale. We also intend to extend our reach to the `xy` packages. Other E-book formats such as AKF8 should be possible with specializations of manifest generation and other fine tuning. Surprisingly, generating Word and OpenOffice formats shares many features with E-books; of course finding the documentation and writing the XSLT transformations from LaTeXML's native XML to Word's will be challenging.

Our move to GitHub, the code reorganization and the plugin contribution model should make it easier for users to use and adapt the system, as well as to contribute back patches and improvements that will help our development.

Reference

1. Ginev, D., Miller, B.R.: LaTeXml 2012 - A year of LaTeXml. In: Carette, J., Aspinall, D., Lange, C., Sojka, P., Windsteiger, W. (eds.) CICM 2013. LNCS (LNAI), vol. 7961, pp. 335–338. Springer, Heidelberg (2013)

System Description: MathHub.info

Mihnea Iancu, Constantin Jucovschi, Michael Kohlhase, and Tom Wiesing

Computer Science, Jacobs University Bremen, Germany
http://kwarc.info

Abstract. We present the MathHub.info system, a development environment for active mathematical documents and an archive for flexiformal
mathematics. It offers a rich interface for reading, writing, and interacting with mathematical documents and knowledge. The core of the
MathHub.info system is an archive for flexiformal mathematical documents and libraries in the OMDoc/MMT format. Content can be au-
thored or archived in the source format of the respective system, is
versioned in GIT repositories, and transformed into OMDoc/MMT for
machine-support and further into HTML5 for reading and interaction.

1 Introduction

As the field of Mathematical Knowledge Management (MKM) and Digital Mathematical Libraries (DML) mature, we need to shift attention from experimenting
with semantic services on small and practice data sets to supporting management of large corpora of mathematical knowledge and documents. In the past,
MKM and DML have latched onto existing corpora/libraries ranging from digi-
tized mathematical articles (e.g. EuDML [EUD]) over semi-structured representations generated from LaTeX sources [Sta+10] to fully formal theorem prover
libraries, e.g. the Mizar Mathematical Library [MizLib]. But so far management
support for and semantic services deployed on such libraries have been essentially insular, existing cross-library methods remain experiments, and have not
been implemented into usable systems.

This insularity also applies to the work in the KWARC group. For instance, we
have i) designed a cross-library representation language: OMDoc (Open
Mathematical Documents), [Koh06], ii) developed a meta-logical framework
MMT (A Module System for Mathematical Theories) [RK13; Cod+11]) that allows to represent the logical languages underlying the theorem prover libraries,
and relate them to each other by logic morphisms, iii) built libraries of formalized mathematics in OMDoc/MMT either by manually creating content (e.g.
LATIN [Cod+11]) or by implementing transformations from existing theorem
prover libraries (e.g. Mizar Mathematical Library [Ian+13]), and iv) have used
OMDoc as the basis for active mathematical documents in the Planetary system [Koh12]. But we have not integrated all of these or made them available to
mathematicians in one comprehensive environment.

To change this situation – and to provide a realistic base for our own cross-
library research and development efforts – we have started work to realize a universal archiving solution for formal and informal mathematical corpora/libraries.

S.M. Watt et al. (Eds.): CICM 2014, LNAI 8543, pp. 431–434, 2014.

We present the MathHub.info system and its design goals in this paper. Math-Hub.info must satisfy two conflicting goals: On the one hand, it must be so generic that it is open to all logics and implementations; on the other hand, it must be aware of the semantics of the formalized content so that it can offer meaning-ful services. These services must be independent of both the formal system and the implementation used to produce the library and offer a uniform high-level interface for both users and machines to access the combined library.

We claim that MathHub.info will resolve two major bottlenecks in the current state of the art. It will provide a permanent archiving solution that not all systems and user communities can afford to maintain separately. And it will establish a standardized and open library format that serves as a catalyst for comparison and thus evolution of systems.

Concretely, we see three ways the formal methods and mathematical knowl-edge management communities can benefit from MathHub.info: *i*) users can view formerly disparate developments in a common, neutral framework and compare them, *ii*) system developers can import libraries from other logical systems to extend the reach of formalizations and avoid duplicate development *iii*) the ex-istence of a library management system (and importable content) can lower the entry hurdle for developing new logic-based systems. In the next section we present the current system architecture and realization, and Section 3 concludes the paper.

2   System Architecture and Realization

MathHub.info is realized as an instance of the Planetary System [Koh12], which we have substantially extended in the course of the work reported here.

The system architecture has three main components: *i*) a versioned *data store* holding the source documents *ii*) a *semantic service provider* that imports the source documents and provides services for them *iii*) and a *frontend* that makes the sources and the semantic services available to users. Specifically, we use the GitLab repository manager [GL] as the data store, the MMT API as the semantic service provider and Drupal as the frontend.

Figure 1 shows the detailed architecture.

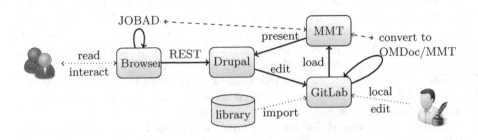

Fig. 1. The MathHub.info Architecture

In this setup, Drupal serves as a container management system[1] that supplies uniform theming, user management, discussion forums, etc. GitLab on the other hand, provides versioned storage of the content documents, and organizes them into repositories owned by users and groups. The advantage of this setup is that we can combine two methods for accessing the contents of MathHub.info: *i*) an online, web-based editing/interaction workflow for the casual user, in the spirit of the Planetary system and *ii*) (new) an offline editing/authoring workflow based on a GIT working copy. The latter is important for power authors and for bulk editing jobs. A user can fork or pull the relevant repositories from GitLab, edit them and submit them back to MathHub either via a pull request to the repository masters or a direct commit/push. As the content is usually highly networked and distributed across multiple GIT repositories, we have developed a command line tool `lmh` (local MathHub) that manages working copies across repository borders.

In the web-based system, semantic services (notation-based, presentation, definition lookup, relational navigation, dependency management, etc.) are provided by MMT and are made available to the user, primarily by dedicated JOBAD [GLR09] modules. Note that even though the active document functionalities and semantic editing support in MathHub.info are based on OMDoc/MMT representation of the content, the authors interact with the content in the source format. Both of these representations are versioned in GitLab and are converted into OMDoc/MMT by custom transformers. `lmh` also supports running these transformers locally and previewing HTML5 renderings of the generated OMDoc/MMT.

In order to to deal with flexiformal mathematical content in OMDoc, we have also extended the MMT API, which was previously restricted to fully formal content. In the extended MMT API, each MMT service works whenever it is theoretically applicable (e.g. type checking when there exists type information, change management when there is dependency information, etc.).

3 Conclusion

MathHub.info is deployed at `http://mathhub.info` and has reached a state, where it can be used for initial experiments and resources, but has not been scaled much beyond 10 000 documents and a couple of dozens or users and repositories yet.

Specifically, we are currently hosting a test set of formal and informal mathematical content to develop and evaluate system functionality; concretely: *i*) the SMGloM termbase with ca. 1500 small STEX files containing definitions of mathematical terminology and notation definitions. *ii*) ca. 6500 files with STEX-encoded teaching materials (slides, course notes, problems, and solutions) in Computer Science, *iii*) the LATIN logic atlas with ca. 1000 meta-theories and logic morphisms, *iv*) the Mizar Mathematical Library of ca. 1000 articles with ca. 50.000

---

[1] Drupal and similar systems self-describe as content management systems, but they actually only manage the documents without changing their internal structure.

theorems, definitions, and proofs, and *v*) a part of the HOL Light Library with 22 theories and over 2800 declarations. Already now, it is unique in its class in that it gives a unified interface to multiple theorem prover libraries together with linguistic and educational resources. Now that the ground work has been laid, we anticipate the rapid integration of new semantic services, editing support and new content.

Acknowledgements. This work has been partially funded by DFG under Grant KO 2428/13-1. The authors acknowledge that the MathHub system builds on a long series of experiments in system integration in the KWARC group and that the design and implementation would not have been possible without substantial discussions in the group.

References

[Cod+11] Codescu, M., Horozal, F., Kohlhase, M., Mossakowski, T., Rabe, F.: Project Abstract: Logic Atlas and Integrator (LATIN). In: Davenport, J.H., Farmer, W.M., Urban, J., Rabe, F. (eds.) Calculemus/MKM 2011. LNCS (LNAI), vol. 6824, pp. 289–291. Springer, Heidelberg (2011)

[EUD] EuDML – The European Digital Mathematics Library, http://eudml.eu (visited on August 02, 2011)

[GL] GitLab, http://gitlab.org (visited on February 24, 2014)

[GLR09] Giceva, J., Lange, C., Rabe, F.: Integrating Web Services into Active Mathematical Documents. In: Carette, J., Dixon, L., Coen, C.S., Watt, S.M. (eds.) Calculemus/MKM 2009. LNCS (LNAI), vol. 5625, pp. 279–293. Springer, Heidelberg (2009)

[Ian+13] Iancu, M., et al.: The Mizar Mathematical Library in OMDoc: Translation and Applications. Journal of Automated Reasoning 50(2), 191–202 (2013), doi:10.1007/s10817-012-9271-4

[Koh06] Kohlhase, M.: OMDoc – An Open Markup Format for Mathematical Documents [version 1.2]. LNCS (LNAI), vol. 4180. Springer, Heidelberg (2006)

[Koh12] Kohlhase, M.: The PLANETARY Project: Towards eMath3.0. In: Jeuring, J., Campbell, J.A., Carette, J., Dos Reis, G., Sojka, P., Wenzel, M., Sorge, V. (eds.) CICM 2012. LNCS (LNAI), vol. 7362, pp. 448–452. Springer, Heidelberg (2012), arXiv:1206.5048[cs.DL]

[MizLib] Mizar Mathematical Library, http://www.mizar.org/library (visited on September 27, 2012)

[RK13] Rabe, F., Kohlhase, M.: A Scalable Module System. Information & Computation (230), 1–54 (2013)

[Sta+10] Stamerjohanns, H., et al.: Transforming large collections of scientific publications to XML. Mathematics in Computer Science 3(3), 299–307 (2010), Autexier, S., Sojka, P., Suzuki, M. (eds.)

Developing Corpus-Based Translation Methods between Informal and Formal Mathematics: Project Description

Cezary Kaliszyk[1], Josef Urban[2,*], Jiří Vyskočil[3,**], and Herman Geuvers[2]

[1] University of Innsbruck, Austria
[2] Radboud University Nijmegen, The Netherlands
[3] Czech Technical University, Czech Republic

Abstract. The goal of this project[1] is to (i) accumulate annotated informal/formal mathematical corpora suitable for training semi-automated translation between informal and formal mathematics by statistical machine-translation methods, (ii) to develop such methods oriented at the formalization task, and in particular (iii) to combine such methods with learning-assisted automated reasoning that will serve as a strong semantic component. We describe these ideas, the initial set of corpora, and some initial experiments done over them.

1 Introduction and Motivation Ideas

Formal mathematics and automated reasoning are in some sense at the top of the complexity ladder of today's precise ("neat")AI corpora and techniques. Many of us believe that practically all mathematical theorems can be precisely formulated and that their proofs can be written and verified formally, and that this carries over to a lot of the knowledge accumulated by other exact sciences. Given this unmatched expressivity and coverage, automated reasoning over formal mathematics then amounts (or aspires) to being the generic problem-solving technique for arbitrary problems that are expressed in a sufficiently "neat" (formal) language and non-contradictory setting.

The last ten years have brought significant progress in formalization of mathematics and in automated reasoning methods for such formalized corpora. Some graduate textbooks have been formalized, and we have produced general reasoning methods that can often automatically find previous relevant knowledge and prove many smaller steps and lemmas in such textbooks without the necessity to manually provide any further hints or guidance.

However, even routine formalization is today still quite laborious, and the uptake of formalization among mathematicians (and other scientists) is very limited. There is a lot of cognitive processing involved in formalization that is uncommon to majority of today's mathematicians: formalization is a nontrivial

---

* Funded by NWO grant *Knowledge-based Automated Reasoning*.
** Supported by the Grant Agency of Czech Republic Project GACR P103/12/1994.
[1] http://mws.cs.ru.nl/~mptp/inf2formal

S.M. Watt et al. (Eds.): CICM 2014, LNAI 8543, pp. 435–439, 2014.

skill to learn, and it takes time. As a result, more than 100 years after Turing's birth, most of mathematical (and scientific) knowledge is still largely inaccessible to deep semantic computer processing.

We believe that this state of affairs can be today helped by automatically *learning* how to formalize ("semanticize") informal texts, based on the knowledge available in existing large formal corpora. There are several reasons for this belief:

1. Statistical machine learning (data-driven algorithm design) has been responsible for a number of recent AI breakthroughs, such as web search, query answering (IBM Watson), machine translation (Google Translate), image recognition, autonomous car driving, etc. As soon as there are enough data to learn from, data-driven algorithms can automatically learn complicated sets of rules that would be often hard to program and maintain manually.
2. With the recent progress of formalization, reasonably large corpora are emerging that can be (perhaps after additional annotation) used for experiments with machine learning of formalization. The growth of such corpora is only a matter of time, and automated formalization might gradually "bootstrap" this process, making it faster and faster.
3. Statistical machine learning methods have already turned out to be very useful in deductive AI domains such as automated reasoning in large theories (ARLT), thus disproving conjectures that its inherent undecidability makes mathematics into a special field where data-driven techniques cannot apply.
4. Analogously, strong semantic ARLT methods are likely to be useful in the formalization field also for complementing the statistical methods that learn formalization. This could lead to hybrid understanding/thinking AI methods that self-improve on large annotated corpora by cycling between (i) statistical prediction of the text disambiguation based on learning from existing annotations and knowledge, and (ii) improving such knowledge by confirming or rejecting the predictions by the semantic ARLT methods.

The last point (4) is quite unique to the domain of (informal/formal) mathematics, and a good independent reason to start with this AI research. There is hardly any other domain where natural language processing (NLP) could be related to such a firm and expressive semantics as mathematics has, which is additionally to a reasonable degree already checkable with existing ITP and ARLT systems. It is not unimaginable that if we gradually manage to learn how mathematicians (ab)use the normal imprecise vocabulary to convey ideas in the semantically well-grounded mathematical world, such semantic grounding of the natural mathematical language (or at least its underlying mechanisms) will then be also helpful for better semantic treatment of arbitrary natural language texts.

2   Approach

The project is in the phase of preparing and analysing suitable corpora, extracting interesting datasets from them on which learning methods can be tried, collecting basic statistics about the corpora. and testing initial learning approaches on them. Initially we consider the following corpora:

1. **The various HOL Light developments:** in particular Flyspeck and Multi-variate, for which we have a strong ARLT online service available [2], and which is also in the case of Flyspeck and Multivariate aligned (by Hales) with the informal Flyspeck book. This is the main corpus we have so far worked on. We have already written programs that collect the links between the informal and formal Flyspeck parts (theorems and definitions), and used such annotations for example for the joint informal/formal HTML presentation of Flyspeck [5]. Currently there are about 250-400 theorems mapped (using the guid tag defined by Hales), however we still need to improve our searching mechanism to find all the mapped informal/formal pairs in various parts of the library. In addition to the aligned theorems, Hales has also aligned over 200 concepts, which can be used as the ground level (dictionary) for the statistical translation algorithms. It is likely that further annotation of the texts will be useful, possibly also with some refactoring of the informal and formal parts so that they better correspond to each other. Most of the extraction/alignment chain is now automated so we can update our data after such transformations of the source texts. We export the aligned theorems in several formats: parsed LATEXvia LATEXML (using libxml for querying), the original HOL text, bracketed HOL text suitable for parsing into external tools, internal (parsed and type-annotated) representation of the HOL theorems in a Lisp-like notation and in a XML notation, and also representation of each theorem in the (Prolog-parsable) THF TPTP format, containing type declarations of all constants recursively used by the theorems.

2. **The Mizar/MML library:** and in particular its mapping to the book Compendium of Continuous Lattices [1] (CCL) and a smaller mapping to Engelking's General Topology provided by Bancerek.[2] This is a potential large source of informal/formal pairs, however the MML has been developing quickly, and updating the mapping might be necessary to align the books with the current MML for which we have a strong online ARLT service [6,3]. We have also obtained the corresponding LATEX sources of the CCL book from Cambridge University Press, however we have not yet clarified the possible publication of the data extracted.

3. **The ProofWiki and PlanetMath informal corpora:** We have the XML and LATEX dumps of these wikis and have used them for initial experiments with disambiguation of informal texts in the student project *Mathifier*,[3] motivated by the NLP work on Wikipedia disambiguation [4]. One relatively surprising preliminary result of this project is quite good performance (75%) of the naive disambiguation algorithm using just the most frequent mathematical meaning without any additional context information. Another initial exploration was done on ProofWiki, whose relatively strict proof style is quite close to the Jaskowski-style natural deduction used in Mizar. We have measured this by mapping all math expressions and references in the ProofWiki sentences to just one generic expression/reference, and counted the frequency of various proof sentences. The

---

[2] http://fm.uwb.edu.pl/mmlquery/fillin.php?filledfilename=t.mqt&argument=number+1

[3] http://mws.cs.ru.nl/~urban/Mathifier/

results[4] again show great homogeneity of the corpus, where most of the proof discourse can be superficially mapped to Mizar natural deduction quite economically. Apart from defining and experimenting with such proof-level translation patterns, the main work on these corpora will be their mapping (possibly automated by using frequency analysis) to the Mizar and HOL Light corpora, in particular general topology that is developed a lot in ProofWiki and MML.

2.1 Methods, Tools and Planned Experiments

There is a lot of relevant NLP research in (i) machine translation (algorithms that directly translate between two languages) (ii) word-sense disambiguation (algorithms that determine the exact meaning of (sets of) words in sentences), and (iii) part-of-speech tagging and phrasal and dependency parsing . The most successful statistical methods (e.g., n-gram-based) require much larger corpora of aligned data than we currently have, however some smarter algorithms such as chart-parsing (the CYK) algorithm with probabilistic grammars (PCFGs) should be usable already on the current scale of our data, perhaps complemented by leaner memory-based approaches such as k-nearest neighbor in the MBT toolkit.[5] Currently, we have started experimenting with the Stanford parser,[6] the Moses toolkit,[7] and our own Prolog/Perl implementation of the (lexicalized) CYK algorithm on a subset of 500 formal (bracketed) Flyspeck expressions about trigonometric functions. Such initial experiments concern relaxing of the precise disambiguated formal texts by adding more ambiguity. For example whenever a casting functor (such as Cx or &) has to be used in the formal text, we can remove it, and measure the success of the probabilistic parsing getting the right formal meaning. Once such experiments produce good results, the next step in this direction is learning the alignment of the informal/formal text/trees using for example the tree-based learning in the Moses toolkit. The work with established tools such as the Stanford parser and Moses will likely be complemented by our custom implementations that take advantage of the domain knowledge. For example we can add immediate pruning of potential parse trees in the CYK algorithm (or any chart parser) by using the HOL Light (Hindley-Milner) type system or the Mizar (soft, dependent) type system at each step of the algorithm.

References

1. Bancerek, G., Rudnicki, P.: A Compendium of Continuous Lattices in MIZAR. J. Autom. Reasoning 29(3-4), 189–224 (2002)
2. Kaliszyk, C., Urban, J.: HOL(y)Hammer: Online ATP service for HOL Light. CoRR, abs/1309.4962 (2013), Accepted in Mathematics in Computer Science
3. Kaliszyk, C., Urban, J.: MizAR 40 for Mizar 40. CoRR, abs/1310.2805 (2013)

---

[4] http://mizar.cs.ualberta.ca/~mptp/fpk1/opaqcounts1.txt

[5] http://ilk.uvt.nl/mbt/

[6] http://nlp.stanford.edu/software/lex-parser.shtml

[7] http://www.statmt.org/moses/

4. Ratinov, L., Roth, D., Downey, D., Anderson, M.: Local and global algorithms for disambiguation to Wikipedia. In: ACL (2011)
5. Tankink, C., Kaliszyk, C., Urban, J., Geuvers, H.: Formal mathematics on display: A wiki for Flyspeck. In: Carette, J., Aspinall, D., Lange, C., Sojka, P., Windsteiger, W. (eds.) CICM 2013. LNCS (LNAI), vol. 7961, pp. 152–167. Springer, Heidelberg (2013)
6. Urban, J., Rudnicki, P., Sutcliffe, G.: ATP and presentation service for Mizar formalizations. J. Autom. Reasoning 50, 229–241 (2013)

System Description: A Semantics-Aware LaTeX-to-Office Converter

Lukas Kohlhase and Michael Kohlhase

Mathematics/Computer Science
Jacobs University Bremen

Abstract. We present a LaTeX-to-Office conversion plugin for LaTeXML that can bridge the divide between publication practices in the theoretical disciplines (LaTeX) and the applied ones (predominantly Office). The advantage of this plugin over other converters is that LaTeXML conserves enough of the document- and formula structure, that the transformed structures can be edited and processed further.

1 Problem and State of the Art

Technical documents from the STEM fields (Science, Technology, Engineering, and Mathematics) augment the text with structured objects – images, mathematical/chemical formulae, diagrams, and tables – that carry essential parts of the information. There are two camps with different techniques for authoring documents. The more theoretical disciplines (Mathematics, Physics, and Computer Science) prefer LaTeX, while the more applied ones (e.g. Life Sciences, Chemistry, Engineering) use Office Suites almost exclusively. Transforming between these two document formatting approaches is non-trivial: The TeX/LaTeX paradigm relies on in-document macros to "program" documents, empowering authors to automate document aspects and leading to community-supplied domain-specific extensions via LaTeX packages. Office suites rely on document styles that adapt visual parameters of the underlying document markup either document-wide or for individual elements.

This incompatibility of document preparation approaches causes friction in cross-paradigm collaboration as each camp deems their approach vastly superior and the other's insufferable. In this paper, we will discuss the transformation from TeX/LaTeX to Office documents. The converse direction would also be useful, but uses different methods.

copy from PDF	paste (libreoffice)
$h_{\mu_\varphi}(f) + \int_X \varphi d\mu_\varphi = \sup_{\mathcal{M}(f,X)} \{h_\mu(f) + \int_X \varphi d\mu\},$	$h_{\mu_\varphi}(f)+\boxed{\cdot}\ \varphi d\mu_\varphi = \sup\{h_\mu(f)+\boxed{\cdot}\ \varphi d\mu\},$

Fig. 1. Copy & Paste into Word Processors

There are several methods to transform papers from LaTeX to an Office word processor. The first method is to just generate a PDF file and then open this file

S.M. Watt et al. (Eds.): CICM 2014, LNAI 8543, pp. 440–443, 2014.
© Springer International Publishing Switzerland 2014

in Word/LibreOffice or copy/paste a fragment. This achieves the goal of looking like the desired PDF document, just in Office. There are two problems with this route: *i*) mathematical formulae are not preserved (see Figure 1) *ii*) even if the result looks OK the results have lost their links (e.g. for citations/references or label/ref), or become difficult to edit, because they do not conform to the styling system of the word processor. The fundamental problem is that this process converts only the appearance of the document and loses all meaning that was encoded in the markup macros that were expanded during PDF generation. This is especially blatant when looking at the math in a document, which is either treated as text or images and cannot be edited/processed further. The same holds true for references, they are essentially treated as parts of text with a linked number in front of them, complicating adding new references substantially.

The other way of transforming LaTeX to Word, by transforming the LaTeX source file directly, avoids these problems. latex2rtf [L2R] is a widely used system that uses a custom parser to convert a non-trivial fragment of LaTeX to the RTF format understood by most office systems. The system works well, but coverage is limited by the LaTeX parser and the aging RTF format. TeX4ht [T4H], which uses the TeX parser itself and seeds the output with custom directives that are parsed to create HTML has a post-processor that generates ODF. Its coverage of LaTeX is unlimited, but the intermediate format HTML somewhat limits the range of document fragments that can be generated.

Here we present a similar approach, only that we extend the backend of the LaTeXML system [LTX] to generate WML – the file format of MS Word – and ODT – that of Libre- and OpenOffice. Like latex2rtf, the LaTeXML system directly parses LaTeX source files. The main difference to TeX4ht is that LaTeXML generates an XML representation that is structurally near to the LaTeX sources and thus preserves the author-supplied semantics for further processing. Coverage for TeX primitives is complete, semantics-preserving LaTeXML bindings are available for most commonly used LaTeX packages.

2   The Office Formats

WML and ODT follow the same architectural paradigm: they are both zip-packaged directories of XML files that contain document content, metadata, and styling. We will use WML in our presentation here and point out differences in ODT as we go along.

The main content of a WML document – text, document structure, placement of images, tables etc. – is represented by special content markup elements in an XML file document.xml. All elements contain styling information, usually by referencing a style element in the file style.xml, which can be modified by adding local settings in children of the properties child. The other important kind of file are the .rels files, which are again XML. These files contain relationship elements, which detail the relations between elements in document.xml and external resources (e.g. for hyperlinks) or resources in the WML package (e.g. the image data files). The WML package additionally contains miscellaneous XML files; e.g. `settings.xml`, which is used to make the state of the word processor applications persistent and `fonttable.xml`, which contains extra information about fonts.

Of special interest is the representation of mathematical formulae. WML uses a proprietary XML format for presentation markup together with a variant of TEX markup that is used for user input. The expression of the left is the –slightly abridged – representation

```
<omml:oMath>
  <r><t>1.5</t></r>
  <sSup>
    <e><r><t>10</t></r></e>
    <sup><r><t>7</t></r></sup>
  </sSup>
  1.5\times 10^{7}
</omml:oMath>
```

of 1.5×10^7. The ODT format treats formulae as external objects; every single one has a subdirectory in the package which contains a presentation MathML file (for external communication), a user input file in the venerable StarOffice format, and an image of the formula (for display in the word processor).

3   Transformation

To create the WML/ODT files we first transform the `.tex` file to an intermediate XML-based LTXML format using LaTeXML. Then we use an XSLT stylesheet to generate `document.xml`. For LTXML elements that do not have a direct counterpart in WML we adapt existing WML elements. For instance, a LaTeX quote environment is represented by a WML p ("paragraph") element with a special style "quote" we added to `styles.xml`. This allows the user to later semantically work with the document, e.g. by changing all quotes to red. For WML formulae, we use a stylesheet supplied by Microsoft to transform the MathML generated by LaTeXML

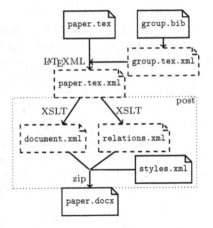

Fig. 2. The Transformation Process

to the WML math format, for ODT formulae we make use of MathML and image generation in LaTeXML. The file `document.xml.rels` is generated by XSLT from `.tex.xml` and is placed into the directory structure the by the LaTeXML post-processor together with other supporting files such as images and some static files that are independent of the input document. The main file of interest here is `styles.xml`, which contains the style information of the document. This had to be adapted manually recreate the visual appearance of the PDF files generated by LaTeX. Finally the LaTeXML post-processor zips documents into the final WML/ODT file.

The user does not see all these transformation, generation, and packaging steps: given a LaTeX paper, all she has to do is type

```
latexmlc paper.tex --destination=paper.docx
```

A transformation to ODT can be specified by choosing the destination `paper.odt`.

4   Conclusion

We have presented a LaTeXML plugin that transforms LaTeX papers into Office documents in a one-line system call. The result of converting the formula from Figure 1 to

$$h_{\mu_\varphi}(f) + \int_X \varphi d\mu_\varphi = \sup_{M(f,X)} \{h_\mu(f) + \int_X \varphi d\mu\},$$

Fig. 3. Converted Formula in MS Word

MS Word is on the right. With the recent web front-end of LaTeXML, it will be simple to extend this to a web service. The LaTeXML Word Processing plugin is public domain and is available from GitHub at [L2O]. The conversion makes crucial use of the fact that LaTeXML preserves more of the document and formula semantics than other systems that process LaTeX documents, this ensures that the core process in the transformation – the translation of LaTeXML XML to Office XML (WML or ODF) has enough information to generate the respective target document structures. The biggest limitations of the current transformation are that *i)* we cannot currently generate the text-based input format (StarMath or the WML TeX variant) and *ii)* citations and references are only partially converted into the "semantic" formats. This makes it difficult to edit formulae/references in the respective word processors after transformation. For ODF formulae, we want to make use of the TeXMaths plugin for Libreoffice, which uses LaTeX instead of StarMath for user input of formulae – but hides it in the comment area of the images which makes handling more difficult.

In the future we want to develop an "office package" for LaTeXand a corresponding LaTeXML binding, which allows the direct markup of higher-level structures – e.g. document metadata in LaTeX documents, so that it can be transferred to the office documents. Similarly, we want to extend the transformation to carry over even more semantics from the sTeX format into semantically extended office formats like CPoint or CWord; this would finally give us a way to cleanly interface the currently LaTeX-based document methods in the KWARC group to applied STEM disciplines.

References

[L2O]   GitHub repository, https://github.com/KWARC/LaTeXML-Plugin-Doc
[L2R]   LaTeX to RTF converter, http://sourceforge.net/projects/latex2rtf/ (visited on August 01, 2010)
[LTX]   Miller, B.: LaTeXML: A LaTeX to XML Converter, http://dlmf.nist.gov/LaTeXML/ (visited on December 03, 2013)
[T4H]   TeX4ht: LaTeX and TeX for Hypertext, http://www.tug.org/applications/tex4ht/mn.html (visited on August 01, 2010)

Math Indexer and Searcher Web Interface

Towards Fulfillment of Mathematicians' Information Needs

Martin Líška, Petr Sojka, and Michal Růžička

Masaryk University, Faculty of Informatics, Botanická 68a, Brno, Czech Republic
{martin.liski,mruzicka}@mail.muni.cz, sojka@fi.muni.cz
https://mir.fi.muni.cz/

Abstract. We are designing and developing a web user interface for digital mathematics libraries called WebMIaS. It allows queries to be expressed by mathematicians through a faceted search interface. Users can combine standard textual autocompleted keywords with keywords in the form of mathematical formulae in LaTeX or MathML formats. Formulae are shown rendered by the web browser on-the-fly for users' feedback. We describe WebMIaS design principles and our experiences deploying in the European Digital Mathematics Library (EuDML). We further describe the issues addressed by formulae canonicalization and by extending the MIaS indexing engine with Content MathML support.

Keywords: search interface, math-aware search, digital mathematical library, formulae canonicalization, WebMIaS, MIaS, EuDML, MathML.

1 The Need for a Math-Aware Search Interface

Scalable search facilities now have the status of killer application on the web and are in high demand among the users of digital mathematics libraries (DML). There are some papers in DMLs which contain more formulae than words. With this in mind, we are designing and implementing the math-aware search engine, Math Indexer and Search (MIaS) [8] supporting a presentation form of mathematics, since the vast majority of scholarly literature in math has only been available in optically recognized presentation formats.

MIaS has been developed primarily for use in EuDML [1]. Since there is no established math-aware user interface, we were faced with the task of designing and implementing one. To gain acceptance across the wider community of potential DML users, the main design goal was ease of use. Having the entry barrier as low as possible is important for attracting new users.

The only available formulae search which does not have format of sources of documents under control was LaTeXsearch.com interface by Springer. It allows only one LaTeX formula as a query. As the same formula can be written in many ways in TeX, string hashing is used to match the query with formulae in documents written in LaTeX. While most mathematicians are used to writing a query in LaTeX, there are problems with this approach as formulae similarity cannot be defined as a metric on LaTeX formulae strings. Other qualities of formulae, such as their structure should be taken into

S.M. Watt et al. (Eds.): CICM 2014, LNAI 8543, pp. 444–448, 2014.

account, as well as textual phrases denoting the content sought. Furthermore, allowing users to type longer LaTeX formulae with immediate visual feedback simplifies the use.

For EuDML, we have added on-the-fly rendering of math, as autodetected in LaTeX and MathML formats. We have added facets for searching in different document fields [6]. Most importantly, we have had the privilege of mining EuDML search logs for user search scenarios which has shown how users have striven to find the information they require. For example, an interesting observation was that Content MathML has started to appear in the math search box. New LaTeXML converter [4] allows the development of new corpora of math texts with math representation in both Presentation and Content MathML, an example of which is the database available for NTCIR-10 Math Task [5] (100,000 arXiv documents). The most challenging problems have included the normalization of math notations coming from different sources, typically a typed or copy-pasted query, and heterogeneous document formats. Development of a robust math canonicalizer emerged as a must for the success of the new math search paradigm to be supported by the DML search user interface.

2   User Interface for Math Information Retrieval

Users are accustomed to forming search strategies with minimal effort using words as queries for documents represented as bags of words. For EuDML we have designed an advanced search form   `http://eudml.org/search`
to allow faceted searches with one facet designed for inputting math formulae. On
`http://mir.fi.muni.cz/webmias`
we maintain a link to the latest version of the development version of WebMIaS to discuss possible DML users' search migration paths and strategies, and to get feedback from the user community. The WebMIaS search interface in Figure 1 observes several design principles and qualities:

formulae in TeX. Mathematicians know and use compact LaTeX math notation. Autodetection of MathML is also in place. To convert LaTeX queries into MIaS-supported MathML, we switched the converter from Tralics to LaTeXML, which is able to convert the user input into mixed Presentation-Content MathML.
on-the-fly formulae rendering. Formulae rendering allows quick feedback when writing the query—users know what they want when they see it. Robust live rendering of copy-pasted MathML is provided means of MathJax. Users are also warned when writing an invalid TeX query.
pop-up help. Pop-up windows inform users about the interface.
domain-specific auto-completion. Frequent collocations and terms from the DML domain are suggested for text queries.
facets. Adding facets allows natural filtering (by language, author,...) of search results to achieve high precision.
snippets with query coloring. Snippets are shown in hit lists. Matched words and formulae are colored in the snippets for a quicker first look evaluation of the results.
scoring and debugging. Scoring of computed relevance to a query is shown for every hit. In the development interface, one can deduce document score computation.

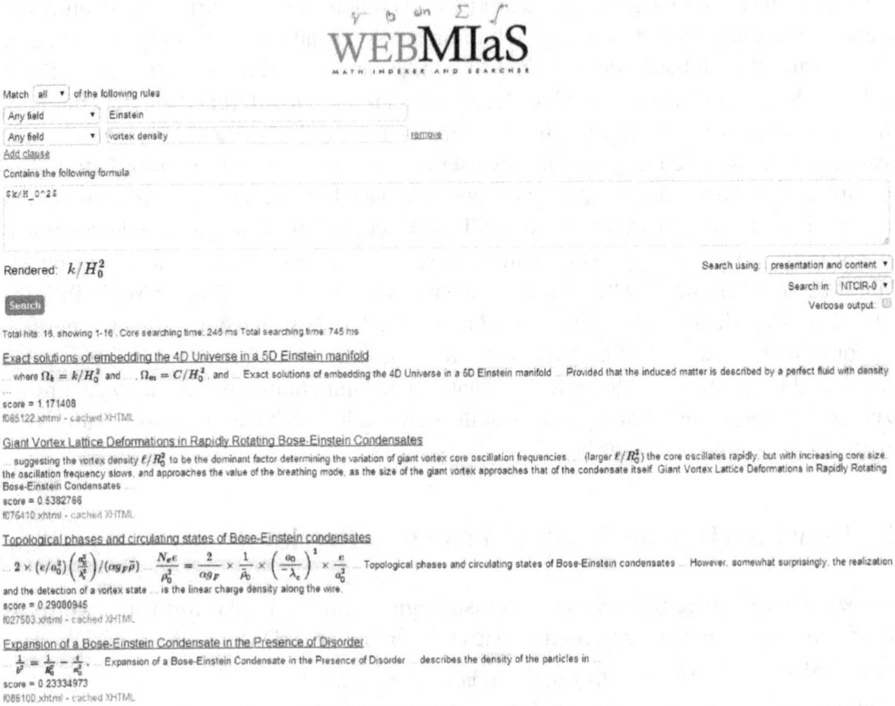

Fig. 1. WebMIaS user interface

Mining the EuDML and WebMIaS query search logs reveals quite different, often contradictory user demands. While some users prefer exact searches of visually remembered formulae, others demand semantic specification of terms representing them. For example, we got a request to constrain a search to $E = mc^2$, where m represents mass. The first request could be fulfilled by an exact Presentation MathML retrieval. However, the latter needed semantic tagging which is usually absent in full-text XMLs and may only be approximated by indexing disambiguated Content MathML.

For testing the search behaviour we indexed 100,000 math papers from the NTCIR-10 Math task [7]. To formulate queries containing math, it emerged that new strategies will have to be developed. It can be expected that it will take some time before math search users learn them. To find the balance between word and math search and between exact, proximity and subtree search, several MIaS indexing parameters have to be set. These parameters differ from collection to collection. Currently set parameters are being evaluated for the current collection.

As MathML created by different content generators such as InftyReader, Tralics, or LATEXML differ significantly. To prevent the rapid growth of the math index containing mathematical formulae [8], their *canonical representations* with the same meaning need to be chosen and indexed. As the results produced by available normalization tools for MathML were not reliable enough, we concluded that it was necessary to develop our own normalization tool that would become part of both indexing and searching in the (Web)MIaS system [2].

To help us to see the changes made by the normalization tool we are generating HTML reports of its inputs and results. Samples of these reports are available for a DML-CZ paper. The LaTeX source code of the paper was transformed to XHTML+ MathML by several tools. Examples of normalized MathML can be seen for Tralics and LaTeXML[1].

When querying using math formulae [3] one has to decide on the *formulae similarity metrics* to allow not only exact formulae matches. Subformulae, formulae expressed in different notation or even similar formulae with different variables need to be considered as hits with lower scores. These metrics are used for weighting similar formulae in documents in the same way that *term frequency* is used for weighting standard word hits.

The link to the WebMIaS interface provided above also shows our development tools that allow us to debug MIaS indexing and querying. With the verbose output enabled, the computation process of document scores can be inspected to see how words and (sub)formulae affect the ranking. The option to take only Presentation MathML, only Content MathML or both into account is also available. Even though Content MathML may give better, semantically related results most of the time, there are cases where visual fidelity is sought and Presentation MathML is preferred.

In addition to the web user interface, it is also possible to use WebMIaS remotely via web services: its general usage is described in the OpenSearch standard. Details of API and WebMIaS OpenSearch description document can be found on the WebMIaS home page above. The web services are particularly useful for remote automated searching, as for example, for the purposes of system evaluation or for a programmable search.

3   Conclusions and Future Work

We have described the WebMIaS math-aware user interface designed to allow searching DMLs with keyword and formulae faceted searching. It has been applied in the EuDML project to the users' satisfaction. Further deployment of the WebMIaS user interface is in preparation for further DMLs as DML-CZ or arXiv.

Acknowledgements. This work has been financed in part by the EU through its Competitiveness and Innovation Programme (Information and Communication Technologies Policy Support Programme, "Open access to scientific information", Grant Agreement No. 250503).

References

1. Borbinha, J., Bouche, T., Nowiński, A., Sojka, P.: Project euDML – A first year demonstration. In: Davenport, J.H., Farmer, W.M., Urban, J., Rabe, F. (eds.) Calculemus/MKM 2011. LNCS (LNAI), vol. 6824, pp. 281–284. Springer, Heidelberg (2011)

---

[1] https://mir.fi.muni.cz/mathml-normalization/samples/

2. Formánek, D., Líška, M., Růžička, M., Sojka, P.: Normalization of digital mathematics library content. In: Davenport, J., et al. (eds.) 24th OpenMath Workshop, 7th MathUI Workshop, and Intelligent Computer Mathematics Work in Progress, Aachen. CEUR Workshop Proceedings, vol. 921, pp. 91–103 (2012), http://ceur-ws.org/Vol-921/wip-05.pdf
3. Kamali, S.: Querying Large Collections of Semistructured Data. Ph.D. thesis (advisor: Frank Tompa), University of Waterloo (2013)
4. LaTeXML Project, A.: LaTeX to XML Converter (2014), http://dlmf.nist.gov/LaTeXML/
5. Líška, M., Sojka, P., Růžička, M.: Similarity Search for Mathematics: Masaryk University team at the NTCIR-10 Math Task. In: Kando, N., et al. (eds.) NTCIR-10 online Proceedings (2013), http://research.nii.ac.jp/ntcir/workshop/OnlineProceedings10/pdf/NTCIR/MATH/06-NTCIR10-MATH-LiskaM.pdf
6. Líška, M., Sojka, P., Růžička, M.: Mravec, P.: Web Interface and Collection for Mathematical Retrieval. In: Sojka, P., Bouche, T. (eds.) Proceedings of DML 2011, Masaryk University, Bertinoro, Italy, pp. 77–84 (2011), http://hdl.handle.net/10338.dmlcz/702604
7. NTCIR Project: NTCIR Pilot Task: Math Task (2012), http://ntcir-math.nii.ac.jp/
8. Sojka, P., Líška, M.: The Art of Mathematics Retrieval. In: Proc. of the ACM Conference on Document Engineering, DocEng 2011, pp. 57–60. ACM, Mountain View (September 2011), http://doi.acm.org/10.1145/2034691.2034703

SAT-Enhanced Mizar Proof Checking

Adam Naumowicz

University of Białystok, Institute of Informatics,
Sosnowa 64, 15-887 Białystok, Poland
adamn@mizar.org
http://alioth.uwb.edu.pl/~adamn

Abstract. In this paper we present an experimental extension of the Mizar system employing an external SAT solver to strengthen the notion of obviousness of the Mizar proof checker. The presented extension is based on a version of MiniSAT, called Logic2CNF. The SAT-enhanced Mizar checker is programmed to automatically spawn a new Logc2CNF process whenever it needs to justify any goal that involves equalities based on Boolean operations.

Keywords: Mizar, SAT, Boolean operations.

1 Introduction

Mizar [4,14] is a proof checker accompanied by a large library of formal proofs based on set theory [16]. Although its user input language is being developed to resemble standard mathematics as much as possible, the amount of details a user must currently provide to make the system check the text's full logical correctness is still too big. The strength of the proof checking system is the most important factor responsible for maintaining the de Bruijn factor on a low level [10]. Therefore all sorts of automations that strengthen the checker and at the same time make the formalized proofs more compact are most welcome by the Mizar users' community. Numerous new techniques are being currently developed to make the Mizar checker stronger (cf. [3,8,9]). There is also active research on combining Mizar with external automated theorem provers [7,17]. Still, the most widely-used and simplest method is based on Mizar "requirements" [11,12], which provide a way to implement specific procedures that make the checker handle certain simple mathematical objects, frequently used in typical Mizar texts. This includes special treatment of Boolean operations on sets, complex arithmetic and the like. Since the Mizar library is built on top of set theory axioms, the usage of various set-based constructs is ubiquitous in the library. Apart from "articles" devoted to sets *per se*, there are many more abstract ones that heavily use sets for constructing some models or examples (e.g. in geometry, lattice theory or graph theory). Therefore the automation of processing sets is beneficial for most of the current library (enabling to reduce its size) but most importantly for future developments. Hard-coding specific checking of Boolean operations, known as "requirements BOOLE", was implemented quite

S.M. Watt et al. (Eds.): CICM 2014, LNAI 8543, pp. 449–452, 2014.

450   A. Naumowicz

early in the history of MIZAR development and it remained partial and not very efficient [13]. In this paper we present the extension which exploits the natural correspondence between propositional formulae and Boolean operations on sets in order to eliminate the need of referencing definitions of these operations and all sorts of lemmas based on them in MIZAR proofs.

2   The Underlying SAT System

Various optimization differences between numerous SAT tools available today are quite irrelevant for solving the simple task mentioned above, so any of the popular SAT solvers should suit our purposes equaly well. Therefore, the basic criteria for choosing the underlying system for implementing this extension are ease of use and a simple interface. One of the most natural candidates was the MiniSAT[1] system developed by Niklas Eén and Niklas Sörensson, which is a minimalistic SAT solver that supports a standard DIMACS input notation. The system is successfully used in a number of other projects, because it is relatively easy to modify, well-documented, highly efficient and designed for integration as a backend to other tools. It is also released under an open source license which allows it to be coupled with MIZAR without any legal issues [1].

For ease of implementing the interface needed to interact with the MIZAR proof checker in a way very similar to the interface previously implemented for Gröbner bases computation [12], we decided to use a MiniSAT variant developed by Edd Barrett, called Logic2CNF[2]. Logic2CNF uses a small input language for logic input from file/stdin. It then converts it to CNF, solves it using built-in MiniSAT and reports the results as assignments of input literal names. Logic2CNF is written in portable C/C++ and it supports Linux, OpenBSD, Solaris, and OSX. We also succeeded in compiling a Windows version using the Cygwin environment to support all main platforms for which MIZAR is precompiled, so there is no obstacle with the extension should it become part of the standard MIZAR distribution and be used to perform revision of the MIZAR library using the standard methodology developed by the MIZAR Library Committee [5,6].

The simple interface we implemented uses the Logic2CNF input language to construct a formula in which each propositional variable represents equality classes made up of all available terms. If a term represents a Boolean operation, the input stream is appended by a corresponding logical formula (e.g. set intersection yields a conjunction in the input and so on). Then, every instance of a negated equality in a given inference (MIZAR's checker is a disprover, so this means testing if some two sets are equal) is checked whether it logically entails the conjunction of all previously stored formulae. This is naturally obtained by negating the entailment and computing its satisfiability with a call to

---

[1] MiniSAT is available for download at http://minisat.se/.
[2] Logic2CNF is available for download at
http://projects.cs.kent.ac.uk/projects/logic2cnf/trac/wiki/WikiStart.

Logic2CNF. From the user's point of view, the proposed extension works automatically and there's no need to directly call the external tool. The Mizar verifier (as well as other MIZAR tools that use its checker) just spawns a new Logc2CNF process whenever it is needed to justify any goal that involves Boolean operations.

Using the standard MIZAR tools such as `relprem` (for eliminating unnecessary references) and `trivdemo` (reducing simple proofs to straightforward justifications), `checklab` (detecting unused labels) and `inacc` (removing unused text fragments), the encyclopedic article `XBOOLE_1` containing many facts on Boolean operations of sets can be significantly reduced. In particular, 32 out of 117 theorems become obvious for the checker[3] and simply by removing the unnecessary proofs the article's size can be reduced by 35%. As for the whole library, `relprem` reported in total over 1600 unnecessary references in 22% of all the library articles.

As mentioned in Section 1, Logic2CNF is primarily developed for Unix-like platforms. The compilation for Linux or BSD-based systems is fairly straightforward, one can also find some pre-compiled versions for selected platforms on the developers' web-site.

The prototype of the extended MIZAR system equipped with the interface to Logic2CNF and pre-compiled for main supported software platforms can be downloaded from the author's web-site: `http://mizar.uwb.edu.pl/~softadm/boolesat`.

Please note that all the MIZAR verification tools assume that the executable named `logic2cnf` (or `logic2cnf.exe` for Windows) is available in the search PATH. Otherwise, an error will occur. For the Windows platform, users can also find there the necessary libraries of the Cygwin environment.

3   Conclusion and Future Plans

The demonstrated application of a SAT solver proved quite useful in strengthening the MIZAR checker's processing of Boolean operations on sets. However, the extension briefly described here can be further developed, e.g. to also enable automation in processing of inclusions and equalities by means of solving the corresponding implications and equivalences. As demonstrated by the integration of SAT and SMT solvers in the context of other proof assistants such as e.g. Coq [2], Isabelle or HOL [18], there is a big potential in integrating solvers not only for strengthening the checker, but also on other levels of proof checking. This can include the reimplementation of the PRECHECKER module responsible for calculating propositional relations between atomic formulae used in inference steps. In the context of the huge MIZAR library, another possible direction is to apply SAT-based techniques in the process of refactoring the library for better legibility and organization (cf. [15]).

---

[3] The article contains also theorems which are not completely related to Boolean operations, and therefore they are not obvious even for the SAT-enhanced checker.

References

1. Alama, J., Kohlhase, M., Mamane, L., Naumowicz, A., Rudnicki, P., Urban, J.: Licensing the Mizar Mathematical Library. In: Davenport, J.H., Farmer, W.M., Urban, J., Rabe, F. (eds.) MKM 2011 and Calculemus 2011. LNCS, vol. 6824, pp. 149–163. Springer, Heidelberg (2011)
2. Armand, M., Faure, G., Grégoire, B., Keller, C., Théry, L., Werner, B.: A modular integration of SAT/SMT solvers to Coq through proof witnesses. In: Jouannaud, J.-P., Shao, Z. (eds.) CPP 2011. LNCS, vol. 7086, pp. 135–150. Springer, Heidelberg (2011)
3. Caminati, M.B., Rosolini, G.: Custom automations in Mizar. Journal of Automated Reasoning 50(2), 147–160 (2013)
4. Grabowski, A., Korniłowicz, A., Naumowicz, A.: Mizar in a nutshell. Journal of Formalized Reasoning 3(2), 153–245 (2010)
5. Grabowski, A., Schwarzweller, C.: Revisions as an essential tool to maintain mathematical repositories. In: Kauers, M., Kerber, M., Miner, R., Windsteiger, W. (eds.) MKM/CALCULEMUS 2007. LNCS (LNAI), vol. 4573, pp. 235–249. Springer, Heidelberg (2007)
6. Grabowski, A., Schwarzweller, C.: Towards automatically categorizing mathematical knowledge. In: Proceedings of Federated Conference on Computer Science and Information Systems – FedCSIS 2012, Wroclaw, Poland, September 9–12, pp. 63–68 (2012)
7. Kaliszyk, C., Urban, J.: Mizar 40 for Mizar 40. CoRR, abs/1310.2805 (2013)
8. Korniłowicz, A.: Tentative experiments with ellipsis in Mizar. In: Jeuring, J., Campbell, J.A., Carette, J., Dos Reis, G., Sojka, P., Wenzel, M., Sorge, V. (eds.) CICM 2012. LNCS (LNAI), vol. 7362, pp. 453–457. Springer, Heidelberg (2012)
9. Korniłowicz, A.: On rewriting rules in Mizar. Journal of Automated Reasoning 50(2), 203–210 (2013)
10. Naumowicz, A.: An example of formalizing recent mathematical results in Mizar. Journal of Applied Logic 4(4), 396–413 (2006)
11. Naumowicz, A.: Evaluating prospective built-in elements of computer algebra in Mizar. Studies in Logic, Grammar and Rhetoric 10(23), 191–200 (2007)
12. Naumowicz, A.: Interfacing external CA systems for Gröbner bases computation in Mizar proof checking. International Journal of Computer Mathematics 87(1), 1–11 (2010)
13. Naumowicz, A., Byliński, C.: Improving Mizar texts with properties and requirements. In: Asperti, A., Bancerek, G., Trybulec, A. (eds.) MKM 2004. LNCS, vol. 3119, pp. 290–301. Springer, Heidelberg (2004)
14. Naumowicz, A., Korniłowicz, A.: A brief overview of MIZAR. In: Berghofer, S., Nipkow, T., Urban, C., Wenzel, M. (eds.) TPHOLs 2009. LNCS, vol. 5674, pp. 67–72. Springer, Heidelberg (2009)
15. Pąk, K.: Methods of lemma extraction in natural deduction proofs. Journal of Automated Reasoning 50(2), 217–228 (2013)
16. Trybulec, A., Korniłowicz, A., Naumowicz, A., Kuperberg, K.: Formal mathematics for mathematicians. Journal of Automated Reasoning 50(2), 119–121 (2013)
17. Urban, J., Rudnicki, P., Sutcliffe, G.: ATP and presentation service for Mizar formalizations. J. Autom. Reasoning 50(2), 229–241 (2013)
18. Weber, T.: Integrating a SAT solver with an LCF-style theorem prover. In: Proceedings of the Third International Workshop on Pragmatical Aspects of Decision Procedures in Automated Reasoning, PDPAR (2005)

A Framework for Formal Reasoning about Geometrical Optics

Umair Siddique and Sofiène Tahar

Department of Electrical and Computer Engineering,
Concordia University, Montreal, Quebec, Canada
{muh_sidd,tahar}@ece.concordia.ca
http://hvg.ece.concordia.ca/projects/optics/

Abstract. Recently, optics technology has emerged as a promising solution by resolving critical bottlenecks in conventional electronic systems. Its application domain spans over diverse fields ranging from laser surgeries to space telescopes. In this paper, we describe an ongoing project which aims at building a theorem proving based framework for the formal reasoning about geometrical optics, an essential theory required in the design and analysis of optical systems. Mainly, we present the motivation of our work, a road-map to achieve our goals, current status of the project and future milestones.

1 Motivation and Background

Generally, optical systems are composed of different components (e.g., mirrors and lenses) which process light to achieve desired functionalities such as light amplification, ultrashort pulse generation and astronomical imaging. In order to model and analyze the behavior of such systems, light can be characterized at three levels of abstraction, i.e., ray, electromagnetic and quantum [4]. Geometrical optics (also known as ray optics) describes light as a collection of straight lines which linearly propagates through optical systems. On the other hand, electromagnetic and quantum optics characterize light as a coupled vector field and a stream of photons, respectively. The analysis of engineering optical systems (e.g., refractometry of cancer cells and optical networks) using geometrical optics is an integral part of their design life-cycle. Traditional optical system analysis techniques like paper-and-pencil based proofs and numerical algorithms have some known limitations of human-error proneness and incompleteness, respectively, which impeded their usage in the designing of critical optical systems which may result in the loss of human lives (e.g., laser surgeries) or heavy financial loss (e.g., Hubble Telescope failure [1]). We therefore propose theorem proving based formal methods for the accurate and scalable analysis of optical systems.

In this paper, we present details of an ongoing project[1] to develop a formal reasoning support for the analysis of geometrical optics. We use the HOL

---

[1] http://hvg.ece.concordia.ca/projects/optics/rayoptics.htm

S.M. Watt et al. (Eds.): CICM 2014, LNAI 8543, pp. 453–456, 2014.

Light theorem prover to formalize the underlying theories of geometrical optics. The main reasons of our choice are the existence of rich multivariate analysis libraries as well as the active projects like Flyspeck [3]. This project is part of larger program on the formal analysis of different forms of optics (i.e., ray, wave, electromagnetic and quantum) [2].

2   Formal Analysis Framework

The proposed framework, given in Figure 1, outlines the main idea to formally model and prove that the optical systems model satisfies the system specification. The whole framework can be decomposed into four major parts which are

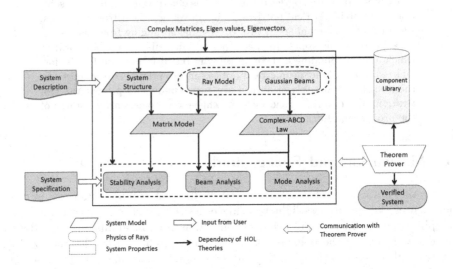

Fig. 1. Formal Analysis Framework

depicted by different shapes and colors as shown in Figure 1. First, the formalization of some complex linear algebra concepts such as complex matrices and eigenvalues; Second, the formalization of optical system structure; Third, modeling of rays and Gaussian beams and last, the formalization of the properties of optical systems such as stability, mode and output beam analysis. The two inputs to the framework are the description of the optical system and specification, i.e., the spatial organization of various components and their parameters (e.g., radius of curvature of mirrors and distance between the components etc.). The first step in conducting formal analysis is to construct a formal model of the given system in higher-order logic. In order to facilitate this step, we require a formalization of optical system structures which consist of definitions of optical interfaces (e.g., plane or spherical) and optical components (e.g., lenses and mirrors). The second step is to formalize the physical concepts of ray and

Gaussian beams. Building on these fundamentals, the next step is to derive the matrix model of the optical system which is basically a multiplication of the matrix models of individual optical components. This step also includes the formalization of the complex ABCD-Law of geometrical optics which describes the input-output relation of the given ray and Gaussian beams parameters.

Furthermore, in order to facilitate the modeling of system properties and reasoning about their satisfaction in the given system model, we provide their formal definitions and most frequently used theorems. These properties are *stability* which ensures the confinement of rays within the system, *beam analysis* which provides the basis to derive the suitable parameters of Gaussian beams for a given system structure and *mode analysis* which is necessary to evaluate the field distributions inside the optical system. Finally, we develop a library of frequently used optical components such as thin lenses, thick lenses and mirrors. Such a library greatly facilitates the formalization of new optical systems which are composed of these components as shown in Figure 1. The output of the proposed framework is the formal proof that certifies that the system implementation meets its specification. The verified systems will then also be available in the library for future use either independently or as part of a larger optical system. In practice, optical components are two dimensional and it is compulsory

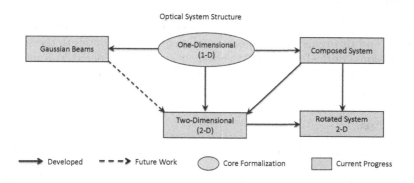

Fig. 2. Formalization Flow of Optical Systems Structure

to consider the rotational effects of individual components. Another important aspect is to consider the fact that optical systems are composed of small subsystems which are configured in a particular way to achieve desired functionalities. In our framework, we consider all of the above mentioned requirements in a systematic way as shown in Figure 2.

3 Current Status and Future Milestones

So far, we have developed a core formalization of geometrical optics [7] as shown in Figure 2. We aslo developed a library of frequently used optical components

U. Siddique and S. Tahar

(such as thin lens, thick lens and dielectric plate) [7] and the formalization of optical resonators [6,7]. We showed the effectiveness of developed theories by the formal analysis of practical optical resonators like Fabry Pérot resonator with fiber-rod lens and Z-shaped resonators [6,7]. Moreover, we developed a generalized procedure for the formal stability analysis of optical resonators usable by physicists and optical engineers (details can be found in [5]). Recently, we have formalized two-dimensional and composed optical systems along with the formalization of Gaussian beams.

Finally, we outline the major tasks to achieve the remaining milestones until the end of this research project. According to our assessment, it would require additional 1.5 years (total project duration of 3.5 years) by an expert user of HOL Light with a sufficient background of geometrical optics. This time-line includes the extensions and revisions of existing formalization along with the dissemination of developed results. Following is the list of main tasks:

- Formalization of eigenray stability for periodic optical systems [8].
- Formalization of Gaussian beams in 2-D as shown in Figure 2.
- Formalization of misaligned optical systems.
- Extension of Gaussian beam formalization to handle laser mode-locking.

4 Conclusion

The main contribution of this project is a comprehensive framework of formal definitions and theorems ranging from one-dimensional ray optics to the rotated two-dimensional optical systems. Moreover, it can be considered as a one step towards an ultimate goal of applying formalized reasoning in new domains such as biology, physics and mechanical engineering.

References

1. The Hubble Space Telescope Optical System Failure Report. Technical report, NASA (1990)
2. Afshar, S.K., Siddique, U., Mahmoud, M.Y., Aravantinos, V., Seddiki, O., Hasan, O., Tahar, S.: Formal Analysis of Optical Systems. Mathematics in Computer Science 8(1) (2014), http://arxiv.org/abs/1403.3039
3. Hales, T.C.: Introduction to the Flyspeck Project. In: Mathematics, Algorithms, Proofs. Dagstuhl Seminar Proceedings, vol. 05021 (2005)
4. Saleh, B.E.A., Teich, M.C.: Fundamentals of Photonics (2007)
5. Siddique, U., Aravantinos, V., Tahar, S.: A New Approach for the Verification of Optical Systems. In: Optical System Alignment, Tolerancing, and Verification VII. SPIE, vol. 8844, pp. 88440G-1–88440G-14 (2013)
6. Siddique, U., Aravantinos, V., Tahar, S.: Formal Stability Analysis of Optical Resonators. In: Brat, G., Rungta, N., Venet, A. (eds.) NFM 2013. LNCS, vol. 7871, pp. 368–382. Springer, Heidelberg (2013)
7. Siddique, U., Aravantinos, V., Tahar, S.: On the Formal Analysis of Geometrical Optics in HOL. In: Ida, T., Fleuriot, J. (eds.) ADG 2012. LNCS, vol. 7993, pp. 161–180. Springer, Heidelberg (2013)
8. Siegman, A.E.: Lasers, 1st edn. University Science Books (1986)

Author Index